Lecture Notes in Physics

Volume 916

The Lecture Notes in Physics

The series Lecture Notes in Physics (LNP), founded in 1969, reports new developments in physics research and teaching-quickly and informally, but with a high quality and the explicit aim to summarize and communicate current knowledge in an accessible way. Books published in this series are conceived as bridging material between advanced graduate textbooks and the forefront of research and to serve three purposes:

- to be a compact and modern up-to-date source of reference on a well-defined topic
- to serve as an accessible introduction to the field to postgraduate students and nonspecialist researchers from related areas
- to be a source of advanced teaching material for specialized seminars, courses and schools

Both monographs and multi-author volumes will be considered for publication. Edited volumes should, however, consist of a very limited number of contributions only. Proceedings will not be considered for LNP.

Volumes published in LNP are disseminated both in print and in electronic formats, the electronic archive being available at springerlink.com. The series content is indexed, abstracted and referenced by many abstracting and information services, bibliographic networks, subscription agencies, library networks, and consortia.

Proposals should be sent to a member of the Editorial Board, or directly to the managing editor at Springer:

Christian Caron
Springer Heidelberg
Physics Editorial Department I
Tiergartenstrasse 17
69121 Heidelberg/Germany
christian.caron@springer.com

More information about this series at http://www.springer.com/series/5304

Yutaka Yoshida • Guido Langouche
Editors

Defects and Impurities in Silicon Materials

An Introduction to Atomic-Level Silicon Engineering

Editors
Yutaka Yoshida
Shizuoka Institute of Science
and Technology
Fukuroi, Japan

Guido Langouche
Nuclear solid state physics
Katholieke Universiteit Leuven
(KU Leuven)
Leuven, Belgium

ISSN 0075-8450 ISSN 1616-6361 (electronic)
Lecture Notes in Physics
ISBN 978-4-431-55799-9 ISBN 978-4-431-55800-2 (eBook)
DOI 10.1007/978-4-431-55800-2

Library of Congress Control Number: 2016930107

Springer Tokyo Heidelberg New York Dordrecht London

Printed on acid-free paper

Springer Japan KK is part of Springer Science+Business Media (www.springer.com)

This book is dedicated to the memory of Prof. George Rozgonyi.

Preface

This book provides the basic physics behind crystal growth, modeling, and evaluation techniques used in silicon materials science and especially emphasizes the importance of the fascinating atomistic insights into defects and impurities as well as their dynamic behavior in silicon materials. During the last 20 years these insights have become directly accessible by newly developed experimental methods, first-principles calculations, modeling, and other computational techniques. Central to the success of semiconductor technology is the ability to control the electrical properties of silicon material. The role of intentionally added dopants to produce the optimum concentration of holes and electrons for a given electronic application is well established. However, the presence of unintentional impurities and structural defects in the lattice can have a dramatic effect on the electronic properties of silicon and are therefore a topic of attention throughout this book.

Accordingly, the book addresses young researchers, scientists, and engineers in the related industry. The main purpose is to offer to readers: (1) an in-depth coverage of the basic physics of defects in silicon materials, (2) an introduction to atomistic modeling as well as characterization techniques related to defects and impurities in silicon materials, and (3) an overview of the wide range of research topics in this field.

In Chap. 1, Hartmut Bracht discusses diffusion of self- and dopant atoms in silicon. The diffusion studies comprise experiments on self- and dopant diffusion performed (both) separately and simultaneously. The mathematical treatment of diffusion–reaction mechanisms is introduced in order to explain what information regarding atomic mechanisms and properties of point defects can be deduced from diffusion experiments performed under various experimental conditions. Computer simulations of atomic transport in silicon are compared with experimental self- and dopant profiles, and the mechanisms that determine the diffusion of self- and dopant atoms in silicon are summarized. Finally, yet unsolved questions concerning the properties of point defects in silicon and the diffusion behavior in three-dimensional confined silicon structures are addressed.

In Chap. 2, José Coutinho provides the grounds for defect modeling in silicon materials using density functional methods. His chapter starts with a review of the theoretical framework and tools, including the relevant methods to treat the exchange-correlation interactions. It then describes how to step up from total energies, electron densities, and Kohn–Sham states to the actual defect calculations. Particular emphasis is given to the calculation of spectroscopic observables such as electrical levels, local vibrational modes, spin densities, migration barriers, and defect response to uniaxial stress. The defect survey starts with a description of elemental intrinsic centers. Understanding their fundamental properties will be crucial to unravel the features of many substitutional and interstitial impurities involving dopants, transition metals, carbon, oxygen, or hydrogen. In the last section the latest developments in modeling defects in silicon nanostructures are discussed. While holding great promises regarding the fascinating optical emission/absorption properties, nano-silicon presents many challenges, particularly with regard to defect control, doping, and electrical transport.

Chapter 3 is devoted to reviewing techniques that characterize and quantify the properties of defects and impurities in silicon materials and devices in terms of their effect on free carriers and their recombination and generation behavior. Anthony R. Peaker and Vladimir P. Markevich explore, in particular, the application of deep-level transient spectroscopy and its many variants to electronic grade silicon during critical process steps. The effect of ion implantation damage and process contamination is considered in relation to specific devices for signal processing and power control. In the case of solar grade silicon, problems associated with different types of silicon material are considered including the use of inexpensive impure silicon feedstock. The complexities of the interaction of unwanted impurities with extended defects in cast multi-crystalline silicon and the problems in evaluating the defects concerned is discussed. The physics of carrier recombination at these defects is presented and related to methods to map minority carrier lifetime in cast silicon. The status and limitations of the techniques for qualification of material for solar cells is discussed.

In Chap. 4, Jan Vanhellemont, Kozo Nakamura, Eiji Kamiyama, and Koji Sueoka are concerned with single-crystal growth of Si and Ge. The so-called Voronkov criterion defines a critical value Γ_{crit} of the ratio $\Gamma = v/G$ of the pulling rate v over the thermal gradient G at the melt–solid interface of a growing crystal. For $\Gamma > \Gamma_{crit}$, the crystal is vacancy-rich and can contain large vacancy clusters that are detrimental for gate-oxide performance and for thin-film epitaxial growth. For $\Gamma < \Gamma_{crit}$, the crystal is self-interstitial-rich and in the worst case will contain dislocation clusters. For Γ Γ_{crit}, the crystal is free of grown-in intrinsic point defect clusters and optimal for device processing. The important impact of thermal stress σ_{th} at the melt–solid interface and crystal doping on Γ_{crit} is clarified. As σ_{th} increases with increasing crystal diameter, controlling G and v become a real challenge for the development of future 450-mm diameter, defect-free Si crystals.

Chapter 5, written by Bing Gao and Koichi Kakimoto, discusses large-scale numerical modeling for silicon single-crystal growth for large-scale integrated systems and solar cells, where the goal is to reduce the number of light elements and dislocations from the crystals. For an accurate prediction of carbon and oxygen impurities in crystalline silicon material for solar cells, a global simulation of coupled oxygen and carbon transport in a unidirectional solidification furnace is implemented. Both the gas flow and the silicon melt flow are considered in this chapter. The effects of flow rate and pressure on the impurities are examined. To effectively control dislocations in crystalline silicon, the relationship between the locations of activated dislocations and the cooling flux direction is studied numerically from the perspective of activation of slip systems. A model of dislocation propagation in a crystal during crystal growth and cooling processes is developed. The first part describes how to calculate stress and strain in the crystal. Then, the mechanism of dislocation propagation is explained by using the Alexander–Haasen–Sumino model. This part explains a non-linear system of dislocation propagation, which is important to predict the dislocation density quantitatively. Finally, preferential slip systems or dislocation propagation are discussed.

In Chap. 6, Gudrun Kissinger reports on nucleation and growth of oxygen precipitates from the point of view of classical nucleation theory and on possibilities and limits of their detection. The initial states of oxygen precipitation as suggested by ab initio calculation are described. Results on the impact of intrinsic point defects, doping, and co-doping are presented. A second focus of this chapter is directed towards the impact of grown-in oxygen precipitate nuclei in silicon wafers on oxygen precipitation, and creation of high-quality defect denuded zones during device processing. Conventional and modern methods of thermal processing and their impact on oxygen precipitation are discussed. Methods to determine the getter efficiency of oxygen precipitates and their results are described.

In Chap. 7, Takashi Sekiguchi and Jun Chen describe electron-beam-induced current (EBIC) and cathode-luminescence (CL) techniques, which have been used for the electrical/optical characterization of extended defects in Si. For both techniques, they use scanning electron microscopes (SEM) for electron beam irradiation. The electric current induced at the internal circuit and light emission are used for imaging of EBIC and CL, respectively. They classify the dislocations and grain boundaries (GBs): clean dislocations are not per se electrically active, but do become active after metallic decoration. Large-angle (LA) GBs also behave like dislocations. The coherency of GB and the degree of contamination are the major factors that determine the electrical activities of LA-GBs. Small-angle (SA) GBs have certain carrier recombination activities at room temperature. Owing to dislocation bundles, SA-GBs emit D-lines and can be distinguished in the D-line imaging in the CL mode. The SA-GBs are classified by these D-lines according to the character and misorientation angle. EBIC/CLs have been extensively used to characterize multi-crystalline Si for photovoltaic applications.

Chapter 8 by Guido Langouche and Yutaka Yoshida explains that the hyperfine interaction between an atomic nucleus and its surrounding charge and electromagnetic-field distribution is extremely sensitive to the atomic and electronic configuration of this atom. In the field of defects and impurities in semiconductors, the study of their hyperfine interaction can therefore contribute substantially to their identification and characterization. The introduction of radioactive isotopes as impurity atoms allows probing the hyperfine interaction at extremely low quantities of such impurities. Several dedicated nuclear methods such as Mössbauer spectroscopy, perturbed angular correlations, and low-temperature nuclear orientation allow measuring the hyperfine interaction at the nuclear site of the impurity atom by analyzing the radiation emitted by these probe nuclei. The emission channeling technique, on the other hand, allows studying the precise lattice-site location of the probe atoms from the channeling behavior of the particles emitted by these probe nuclei. The chapter focuses on Mössbauer spectroscopy, which has been applied to study ^{57}Fe solute atoms in Si wafers and solar cells by developing new techniques.

Finally, in Chap. 9, Werner Bergholz focuses on defect engineering in silicon materials employing insights gained from the preceding, more fundamental chapters. First, the fundamentals are explained, e.g., the analogy between the chemistry of ions in water as a "substrate" and the "chemistry" of point defects in silicon, the most perfect, purest solid material available. The role of point defects in the nucleation of extended defects and their interaction with extended defects is then explained, along with what kind of "levers" are available to control the defect scenarios. Secondly, the application of these fundamental defect engineering principles are explained for various technology areas. The focus is on microelectronics and photovoltaics to complement other contributions that address the important area of crystal growth. Topics in microelectronics include the impact of oxidation, ion implantation, etching, thermal processes/gettering, and potential-induced degradation (PID) in photovoltaics.

This book was proposed as a follow-up project of the 7th Forum on the Science and Technology of Silicon Materials (Silicon Forum) held from October 19 to 22, 2014, in Hamamatsu, Japan. Prof. George Rozgonyi was one of the chapter authors of this book project, but he regrettably passed away on November 24, 2014. He had worked in our research field for over 50 years and made many important contributions to the understanding of defects in silicon. Prof. A. R. Peaker of the University of Manchester, UK, therefore suggested dedicating this book to him, and all the authors agreed with such a valuable suggestion. Subsequently, Prof. Fumio Shimura, who was an executive adviser of the silicon forum, and worked at North Carolina State University (NCSU) in Raleigh with George for a long time, agreed to provide "Personal Reminiscences About George Rozgonyi".

Shizuoka Institute of Science and Technology, Shizuoka, Japan Yutaka Yoshida
University of Leuven (KU Leuven), Leuven, Belgium Guido Langouche
July 30, 2015

Personal Reminiscences About George Rozgonyi

Fumio Shimura

It was November 25, 2014, when I received the following e-mail from Norrish, George's wife:

> "Good night, sweet prince, and flights of angels sing thee to thy rest."
> GEORGE ARTHUR ROZGONYI
> April 24, 1937–November 24, 2014
> "When he shall die, cut him out in little stars; and he shall make the face of heaven so fine that all the world will be in love with night . . . ."

George Arthur Rozgonyi was born in Brooklyn, N.Y. He graduated from St. Peter's Preparatory School in Jersey City, N.J. He went on to receive his B.S. and M.A. degrees from the University of Notre Dame. In 1963, he obtained his Ph.D. in Aero-Space Sciences from the University of Arizona. He joined AT&T Bell Telephone Laboratories in Murray Hill, N.J., where he had a distinguished career of over 19 years.

In 1982, Prof. Rozgonyi transitioned to academia and joined the Materials Science and Engineering faculty at North Carolina State University (NCSU) in Raleigh. He became an active researcher and teacher and mentored many graduate students and post-doctoral fellows. He also organized numerous scientific conferences, workshops, and consortia.

In his research Prof. Rozgonyi focused on the important field of defect engineering of silicon for integrated circuit (IC) and photovoltaic (PV) applications, impurity control via intrinsic and extrinsic gettering processes, and dynamics of point defect/extended defect. The Rozgonyi group developed a set of defect diagnostic tools, which are available for crystal growth and device processing. Over the years, the lab built by Prof. Rozgonyi exposed hundreds of graduate, undergraduate, and post-doctoral students to real-world lab experiences that enhanced their basic physics, chemistry, and materials science knowledge base. The research results were presented at numerous invited lectures at international conferences and were documented in more than 400 scientific papers. Prof. Rozgonyi was responsible for advances that have played an important role in the development of integrated circuits, solar cells, and light-emitting diodes, all of which are in everyday use today.

Through his work, Prof. Rozgonyi was well known in the scientific community and was often invited to join various research teams. He spent sabbaticals at the Max Planck Institute in Stuttgart, Germany; at C.N.E.T. in Grenoble, France; and at universities in South Africa. He served on the Executive Committee, Electronics Division, of the Electrochemical Society and won numerous awards during his career, including the Electrochemical Society Award for "... advancing our fundamental understanding of defects in semiconductor device processing." He was also selected as a Fellow of the Electrochemical Society and IBM Faculty Scholar.

One of Prof. Rozgonyi's great strengths was his ability to organize events and bring together researchers from around the world to work on important topics. He was the founding director of two National Science Foundation Industry/University Cooperative Research Centers: the Silicon Wafer Engineering and Defect Studies (SiWEDS) center and the Silicon Solar Consortium (SiSoC), which supported collaborative research between eight universities, two national labs (NREL and Sandia), and 27 industrial sponsors.

Prof. Rozgonyi is survived by his wife, Norrish Munson Rozgonyi, and his three daughters and their families. Through the course of his life, he touched and positively influenced many people. His family can be very proud of what he achieved in his time – he created an important body of research work. He is remembered well and will be missed.

As an Executive Adviser for The Forum on the Science and Technology of Silicon Materials 2014, October 19–22, Hamamatsu, Japan, I asked George to deliver an invited talk on "Defects in silicon," and he willingly accepted our invitation; however, very regretfully, I received the following e-mail from George on June 11, 2014:

> Fumio:
>
> I am with the whole family at the beach house. In wiped-out recovery stage from second of six monthly chemo-therapy treatments which finished last Friday. Difficult to predict October, but I would say right now I couldn't do it comfortably and if you need an answer now it would have to be that getting another speaker is the only choice.
>
> George

The last time I saw George was at The 6th International Symposium on Advanced Science and Technology of Silicon Materials, November 19–23, 2012, Hawaii. Unfortunately, his physical condition made me realize I would have to say my final good-bye to him in the near future. I indeed grieve over his death, and I am so sad to realize the truth.

I first became aware of George Rozgonyi through his scientific paper in 1976, when I was working on micro defects and intrinsic gettering in silicon at the NEC Central Research Laboratories, Japan. His paper, titled "The identification, annihilation, and suppression of nucleation sites responsible for silicon epitaxial stacking faults" (J. Electrochem. Soc. 1976), co-authored with Deysher and Pearce from Bell Laboratories, was indeed the pioneering work in the field of defect engineering for silicon crystals. Life is interesting indeed, because 10 years later I also transitioned to academia. I joined the Materials Science and Engineering faculty

Professor Rozgonyi at The 6th International Symposium on Advanced Science and Technology of Silicon Materials, November 19–23, 2012, Hawaii (Courtesy of Prof. A. Ogura, Meiji University)

at NCSU, due to the strong push and support of Prof. Rozgonyi, who had joined NCSU 5 years before. I worked closely there with Prof. Rozgonyi as a research professor for about 7 years, until the fall of 1993, when I returned to Japan.

In closing my personal reminiscences about George Rozgonyi, I would like to express my sincere thanks to him. He gave me a chance to have a wonderful time in North Carolina; indeed, I do not hesitate to say that the time spent in North Carolina was the best in my life. Now, simply, I would like to thank him very much for everything he gave me. George, you are immortal in my mind.

Finally, I would like to thank Dr. Zbigniew Radzimski, who prepared the precise personal history of George Rozgonyi.

Shizuoka Institute of Science and Technology, Japan Fumio Shimura

Contents

Chapter 1
Diffusion and Point Defects in Silicon Materials

Hartmut Bracht

Abstract This chapter aims to provide a basic understanding on the complex diffusion behavior of self-, dopant-, and selected metal atoms in silicon (Si). The complexity of diffusion in Si becomes evident in the shape of self- and foreign-atom diffusion profiles that evolves under specific experimental conditions. Diffusion studies attempt to determine from the diffusion behavior not only the mechanisms of atomic transport but also the type of the point defects involved. This information is of pivotal interest to control the diffusion and activation of dopants during the fabrication of Si-based devices and, from a more fundamental scientific point of view, for comparison to the predictions of theoretical calculations on the properties of point defects in Si. In general, diffusion research relies both on experimental methods to accurately determine diffusion profiles established under well-defined conditions. The analysis of diffusion profiles that can be based on either analytical or numerical solutions of the considered diffusion-reaction equations provides first information about possible diffusion mechanisms. To identify the mechanisms of diffusion, studies under different experimental conditions have to be performed. This chapter on diffusion in Si starts with an introduction on the significance of diffusion research in semiconductors to determine the properties of atomic defects. Diffusion in solids is treated from a phenomenological and atomistic point of view. Experiments designed to investigate the diffusion of self- and foreign atoms are presented and typical self- and foreign-atom profiles obtained after diffusion annealing under specific conditions are illustrated. The mathematical treatment of diffusion-reaction mechanisms is introduced to understand the shape of diffusion profiles and the meaning of the diffusion coefficient deduced from experiments. Modeling of self-, dopant-, and metal-atom diffusion is described that aims at a consistent interpretation of atomic transport processes in Si based on unified properties of the native point defects involved. Finally, till unsolved questions on the properties of point defects in bulk Si and on the diffusion behavior in three-dimensional confined Si structures are addressed.

H. Bracht (✉)
Institute of Materials Physics, University of Münster, Wilhelm-Klemm-Str. 10, 48149 Münster, Germany
e-mail: bracht@uni-muenster.de

Y. Yoshida, G. Langouche (eds.), *Defects and Impurities in Silicon Materials*,
Lecture Notes in Physics 916, DOI 10.1007/978-4-431-55800-2_1

Keywords Self-diffusion • Dopant diffusion • Metal diffusion • Point defects in silicon • Reaction mechanism • Isotope structures

1.1 Introduction

Over the last six decades our daily life has been revolutionized by the invention of Si-based electronic devices. The key for this development was the preparation and growth of high purity single crystals as well as the ability to control the impurity level, i.e., the defects incorporated in Si on the atomic scale. Till these days the defect density determines the integrity of Si devices for e.g. electronic and photovoltaic applications [1]. Further improvement of Si-based electronic devices requires an in-depth understanding on the properties of atomic defects and their interactions. The properties of atomic defects concern their type, structure, and charge states as well as their thermodynamic properties, i.e., their free enthalpy of formation and migration. Although atomistic calculations based on density functional theory (DFT) [2–15], tight binding molecular dynamics (TBMD) [16–19] and molecular dynamics (MD) simulations [20–23] were and are still increasingly used to predict the stability, mobility, and electronic properties of defects, the relevance of the theoretical results must finally be verified experimentally, since it remains unclear, how far theoretical data, mostly deduced for zero Kelvin, are also applicable for higher temperatures.

Theoretical calculations of the structure and formation energy of point defects in solids are most valuable for comparison with spectroscopic results gained e.g. from electron paramagnetic resonance (EPR) studies [24–26], optical methods such as infrared (IR) and photoluminescence spectroscopy [27], deep level transient spectroscopy (DLTS) [28], and perturbed angular correlation (PAC) experiments [29]. These methods provide results for temperatures which range from cryogenic to room temperature. More general applicable spectroscopic methods for studying point defects at temperatures relevant for device processing are hardly available. An exception is the positron annihilation spectroscopy (PAS) [30–32] and the Mössbauer spectroscopy (MS) [33]. Whereas the former method is highly capable for the investigation of vacancy-like defects in condensed matter, the latter method is mainly applicable to investigate the preferred incorporation of iron (Fe). Unfortunately PAS fails in the case of Si because the concentration of vacancies in thermal equilibrium is below the detection limit of the method [34]. On the other hand MS can provide valuable information about the occupancy of Fe on interstitial and substitutional lattice sites [35] but this method is less suited for studying diffusion phenomena in general as MS is practically restricted to the radioactive isotope ^{57}Fe.

Another capable method for studying point defect properties at elevated temperatures is diffusion in solids. This is highlighted by the present contribution that explains the origin of characteristic diffusion profiles in Si and describes what kind of information can be deduced from the diffusion of self- and foreign-atoms.

First a phenomenological and microscopic description of diffusion in solids is given in Sects. 1.3 and 1.4. Then the diffusion mechanisms mostly relevant for elemental semiconductors are presented. Direct and indirect diffusion mechanisms are introduced that highlight the significance of native point defects in atomic transport processes. The mathematical description of diffusion in solids with emphasis on Si is presented considering diffusion both under electronically intrinsic and extrinsic doping conditions. Examples on self-, dopant and metal diffusion experiments are given. The analyses of the experimental diffusion profiles provide valuable information about the properties of the point defects involved in the diffusion process. These examples also demonstrate the methods that are widely used to analyze diffusion in semiconductors. Special emphasis is paid to the interrelation between self- and foreign-atom diffusion that becomes directly evident in the fast diffusion of some mainly substitutionally dissolved metals and the simultaneous diffusion of self- and dopant atoms in isotopically modulated Si structures. Overall, diffusion studies pursue the goal to identify the mechanisms of diffusion and to determine the properties of the point defects involved. This understanding on the diffusion and interaction of point defects in semiconductors is of fundamental significance to control the diffusion and the electrical activation of dopants in the fabrication of electronic devices. Accordingly, point defects in semiconductors can be considered as the "salt of the soup". They not only determine the atomic transport but also the electronic, optical, and mechanical properties of semiconductors.

1.2 Defects in Semiconductors

Defects in elemental semiconductors such as Si and germanium (Ge) significantly affect their electrical properties. The same holds for binary group III–V (GaAs, GaSb, GaN, etc.), group II-VI (ZnO, ZnSe, CdTe, etc.) and ternary and quaternary compound semiconductors. Point defects such as impurities, which are mainly dissolved on lattice sites and introduce shallow acceptor or donor centers, make the material highly conductive and therewith suitable for electronic applications. In addition, defects which give rise to deep level centers affect the electrical properties of semiconductors. Such defects, like Au in Si, act as recombination centers for electrons and holes and are commonly used to reduce the lifetime of minority electrons and holes in high frequency devices. On the other hand high concentrations of deep level centers are undesirable because the effective doping concentration of the material would be strongly reduced by these compensating centers.

During the fabrication of electronic devices appropriate processing steps have to be performed to minimize the concentration of unintentionally introduced defects. In Si device technology, the concentration of detrimental transition metal contaminants (Fe, Cu, etc.) are reduced by gettering treatments. Transition metal precipitates and also D-defects in silicon wafers, the latter are considered to be vacancy clusters [36, 37], are responsible for the degradation of the gate oxide integrity of MOS structures [38–40]. This illustrates that not only the incorporation of point defects but also the formation of extended defects such as dislocations, stacking faults and agglomerates of foreign atoms or native point defects must be controlled during wafer processing and, in particular, already during Si crystal growth.

Charged native point defects in Si are generally not considered to alter the electrical properties of the material significantly. Their concentration is believed to be sufficiently low as this is supported by direct measurements of the native defect concentration with positron annihilation experiments [34] and results deduced from diffusion studies (see e.g. [41]). However, compared to elemental semiconductors the concentration of native point defects in compound semiconductors can be several orders of magnitude higher because the formation of these defects depends on the partial pressure of the components over the compound [42–44].

The controlled incorporation of point defects in semiconductors is one of the main tasks in the production of electronic devices. Homogeneous doping is generally achieved by adding a controlled amount of the dopant element to the melt or to the gas phase of epitaxial layer deposition systems. However, the fabrication of electronic devices like diodes, transistors, or complex integrated circuits requires spatially inhomogeneous dopant distributions. Such distributions are formed by the deposition of dopants on or implantation beneath the surface followed by a high temperature diffusion step. In order to tune the semiconductor devices to the desired functionality the diffusion induced dopant distribution must be predictable and as accurate as possible. This requires a detailed knowledge of the atomic mechanisms of dopant diffusion that comprises information on the type of native point defects involved in dopant diffusion, their charge states and formation and migration enthalpies and entropies.

The diffusion mechanisms considered in this chapter are mainly representative for self- and foreign-atom diffusion in Si. It is noteworthy that the mechanisms are also applicable to diffusion in other elemental semiconductors and even to diffusion in compound semiconductors since the atomic transport in semiconductor compounds is often restricted to the sublattice of one of the constituents.

Before the atomic transport of self- and foreign-atoms in Si is treated in Sect. 1.5, first a phenomenological and atomistic view on diffusion is given in the following.

1.3 Phenomenological Treatment of Diffusion

Diffusion describes a process where an initial inhomogeneous distribution of matter in a media becomes homogenously distributed. Accordingly diffusion is reflected by the transport of matter. This transport acts in the direction to remove existing concentration gradients [45]. In many diffusion processes the concentration C_{A} of a foreign-atom A is low compared to the number density C_o of the matrix atoms, i.e., $C_{\mathrm{A}} \leq 10^{-2}$at%. In this nearly ideal dilution Fick's first law

$$\mathbf{j}_{\mathrm{A}}(\mathbf{r}, t) = -D_{\mathrm{A}} \nabla C_{\mathrm{A}}(\mathbf{r}, t) \tag{1.1}$$

describes the transport of a particle A in an existing concentration gradient (see e.g. [46]). $\mathbf{j}_A$ denotes the diffusion flux in units of $[m^{-2}s^{-1}]$, D_A the diffusion coefficient in units of $[m^2 s^{-1}]$, and C_A the number density, i.e. concentration, in units of $[m^{-3}]$. $\nabla = (\frac{\partial}{\partial x}, \frac{\partial}{\partial y}, \frac{\partial}{\partial z})$ is the nabla operator in the unit $[m^{-1}]$. The minus sign in Eq. (1.1) considers that the particle transport is opposite to the direction of the concentration gradient. In general, D_A is a second order tensor because the particle flux $\mathbf{j}_A$ and concentration gradient ∇C_A can be directed differently. In crystals with cubic symmetry, such as in Si and Ge, the diffusion is isotrop and accordingly the diffusion coefficient is a scalar quantity.

In the case the number of particles is conserved during diffusion of A, i.e., no loss of particles occurs due to e.g. aggregation, segregation or other interactions with defects in the lattice, the combination of the continuity equation

$$\frac{\partial C_A(\mathbf{r}, t)}{\partial t} = -\nabla \cdot \mathbf{j}_A(\mathbf{r}, t) \tag{1.2}$$

with Fick's first law (1.1) yields Fick's second law

$$\frac{\partial C_A(\mathbf{r}, t)}{\partial t} = \nabla \cdot (D_A \nabla C_A(\mathbf{r}, t)). \tag{1.3}$$

For a concentration- and location-independent diffusion coefficient D_A can be placed before the nabla operator and we obtain

$$\frac{\partial C_A(\mathbf{r}, t)}{\partial t} = D_A \, \Delta C_A(\mathbf{r}, t) \tag{1.4}$$

where $\Delta = \frac{\partial^2}{\partial x^2} + \frac{\partial^2}{\partial y^2} + \frac{\partial^2}{\partial z^2}$ represents the Laplace operator in units of $[m^{-2}]$. Considering only diffusion in x-direction that is often realized by the diffusion of an element A into a solid from an infinite source on top of the surface or in experiments that consider the diffusion induced intermixing of layered structures, Fick's second law reads

$$\frac{\partial C_A(x, t)}{\partial t} = D_A \frac{\partial^2 C_A(x, t)}{\partial x^2}. \tag{1.5}$$

The solution of this second order linear partial differential equation is given by

$$C_A(x, t) = C_A^{\circ} \operatorname{erfc} \frac{x}{2\sqrt{D_A t}} \tag{1.6}$$

in the case of diffusion into a semi-infinite solid with constant surface concentration $C_A(x = 0, t) = C_A^{\circ}$. erfc in Eq. (1.6) represents the complementary error function (erfc $= 1 -$ erf). Solution (1.6) is typical for concentration-independent diffusion processes. Other solutions of Eq. (1.5) for different initial and boundary conditions are given in textbooks [46, 47] or can be calculated with the method of separation of variables as well as by means of Laplace and Fourier transforms.

Concentration profiles described by Eq. (1.6) are expected for the diffusion of particles A into a solid under constant surface concentrations when the type of the particle does not change during diffusion. This, strictly speaking, only holds for mainly interstitial dissolved foreign-atoms such as hydrogen (H) and copper (Cu) in Si. These interstitial foreign atoms diffuse via interstitial lattice sites and remain interstitial atoms during their diffusive jump from one interstice to another. The interstitial mechanism of diffusion is considered in Sect. 1.5.1. In the case the type of the particle changes in the diffusion process, due to interaction of A with other defects, Fick's law of diffusion are no longer valid because the number of particles is not conserved. Accordingly, the corresponding diffusion equations have to consider defect reactions. This increases significantly the mathematical complexity to calculate the solution of the differential equations, that is, in most cases the system of underlying diffusion equations can only be solved numerically. Typical examples of elements A that exhibit a complex diffusion behavior are n- and p-type dopant atoms in Si such as P and B, respectively, as well as amphoteric foreign-atoms such as Au and Zn in Si that occupy both interstitial and substitutional lattice sites. Diffusion processes that involve defect reactions are treated in Sect. 1.5.2. Analytical solutions of the differential equation system are derived for conditions that can be realized experimentally.

In the following Sect. 1.4 diffusion in solids is treated from an atomistic point of view. Subsequently, the mechanisms of diffusion are divided in direct and indirect mechanisms and treated in Sect. 1.5 with special emphasis on the most relevant mechanisms for Si.

1.4 Atomistic Description of Diffusion

The fundamental process of diffusion of atomic components is a jump between two adjacent lattice sites. Since the site exchange is associated with a deformation of the environment, the particle has to overcome a saddle point between the potential minimum of the initial and final position. The magnitude of the barrier is given by the difference in the free enthalpy between minimum and saddle point. Accordingly, the particle has to overcome an activation enthalpy to perform a successful diffusion jump. From this treatment of diffusive jumps an exponential temperature dependence of the diffusion coefficient D_A is deduced that is described by an Arrhenius equation

$$D_A = D_A^\circ \exp\left(-\frac{Q_A}{k_B T}\right) \tag{1.7}$$

where k_B and T are the Boltzmann constant and absolute temperature, respectively. D_A° is the pre-exponential factor, that is a product of the jump distant squared, the jump frequency, the correlation factor, an entropy term and a geometry factor. The specific expressions of D_A° and the activation enthalpy Q depend on the underlying diffusion mechanisms that are described in the next Sect. 1.5. In logarithmic representation Eq. (1.7) describes a linear relation between $\log D_A$ and $\frac{1}{T}$. Deviations from the Arrhenius equation can be due to a simultaneous occurrence of different contributions to diffusion or due to a temperature dependent activation enthalpy $Q(T)$.

1.5 Diffusion Mechanisms

Diffusion in solids can, in general, be described by means of direct and indirect diffusion mechanisms. Characteristic of the direct diffusion of atoms is that no native point defects are involved to assist the migration of the atom. On the other hand, native point defects are required for the indirect diffusion of atoms. In the following the diffusion of an atom A via direct and indirect diffusion mechanisms and the corresponding diffusion coefficients are considered.

1.5.1 Direct Diffusion Mechanisms

The interstitial diffusion is schematically illustrated in Fig. 1.1. This mechanism represents the direct diffusion of an interstitial foreign-atom A_i and describes the jump of interstitially dissolved foreign-atoms to the neighboring interstitial position as indicated by the arrows in Fig. 1.1.

In diluted systems the concentration of interstitial foreign-atoms is small compared to available interstitial sites. Then the interstitial diffusion is a purely statistical process. For cubic crystals the following diffusion coefficient D_{A_i} of the

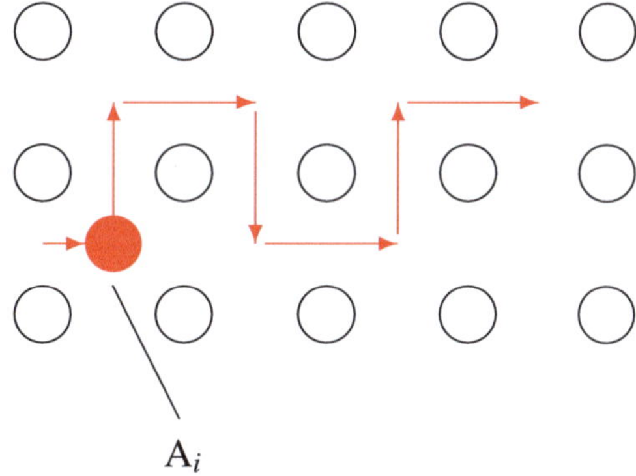

Fig. 1.1 Interstitial mechanism of the diffusion of interstitially dissolved foreign-atoms A_i

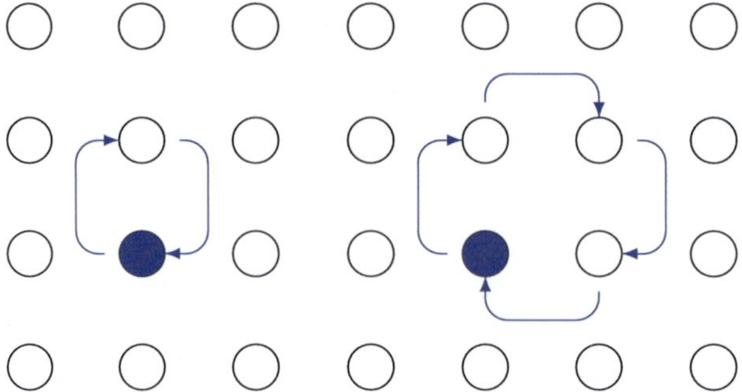

Fig. 1.2 Direct diffusion of substitutional atoms (self- or foreign-atoms) via a direct exchange

interstitial foreign-atom is derived [46, 48]

$$D_{A_i} = D^{\circ}_{A_i} \exp\left(-\frac{H^{M}_{A_i}}{k_B T}\right) \tag{1.8}$$

with the pre-exponential factor

$$D^{\circ}_{A_i} = g_{A_i} a_o^2 \nu_o \exp\left(\frac{S^{M}_{A_i}}{k_B}\right). \tag{1.9}$$

The geometry factor g_{A_i} includes the crystal structure and details about the atomistic diffusion process. In the case of direct diffusion via the interstice of a crystal with diamond structure the geometry factor equals $1/8$. a_o and ν_o are the lattice constant (Si: 5.431×10^{-10} m) and attempt frequency. The latter quantity is of the order of the Debye frequency ($\approx 10^{13}\,s^{-1}$). The activation enthalpy of direct interstitial diffusion in Eq. (1.8) equals the migration enthalpy $H^{M}_{A_i}$ of the interstitial foreign-atom. This quantity is of the order of 1 eV or even less. $S^{M}_{A_i}$ in Eq. (1.9) represents the corresponding migration entropy. Typical examples for interstitial diffusors in Si are hydrogen (H) and copper (Cu) (see [49, 50] and references therein).

The exchange of two atoms on substitutional sites or the exchange of atoms along a ring describe the direct diffusion of substitutional atoms. This is illustrated in Fig. 1.2. So far no experimental evidence has been found for this direct mechanism of diffusion in Si indicating that the diffusion of A_s via indirect mechanisms is energetically more favorable.

1.5.2 *Indirect Diffusion Mechanisms*

Native point defects such as vacancies (V) and self-interstitials (I) are always present even in single crystalline, high purity dislocation-free Si wafers. This is a consequence of the Gibbs free energy of the crystal under thermal equilibrium which is minimized when native defects are formed (see e.g. [46]). Diffusion mechanisms that involve native point defects as diffusion vehicle are called indirect mechanisms. Both the diffusion of self-atoms, i.e. self-diffusion, and the diffusion of substitutional foreign-atoms are mediated by native point defects. In the following the indirect mechanisms of self- and foreign-atom diffusion are introduced that are most relevant for Si.

1.5.2.1 Self-Diffusion

Vacancy Mechanism

The diffusion of self-atoms requires native point defects as diffusion vehicle. The mechanism involving vacancies is the so called vacancy mechanism that is schematically shown in Fig. 1.3.

In order that self-diffusion via vacancies can proceed, a vacancy must exist next to a tagged matrix atom. The probability to find a vacancy at a next nearest neighbor site is given by the concentration C_V^{eq} of V in thermal equilibrium normalized by the number density C_o of the matrix atoms (Si: $C_o = 5 \times 10^{22}\,\text{cm}^{-3}$). For thermal equilibrium conditions the vacancy concentration at a specific temperature is given by

$$C_V^{eq} = C_o \exp\left(-\frac{G_V^F}{k_B T}\right) \tag{1.10}$$

with the Gibbs free energy of vacancy formation G_V^F. This energy is interrelated via $G_V^F = H_V^F - T S_V^F$ with the formation enthalpy H_V^F and entropy S_V^F of V. Considering the jump of the vacancy in the case of tracer self-diffusion experiments the site exchange of the vacancy proceeds with equal probability to any next nearest

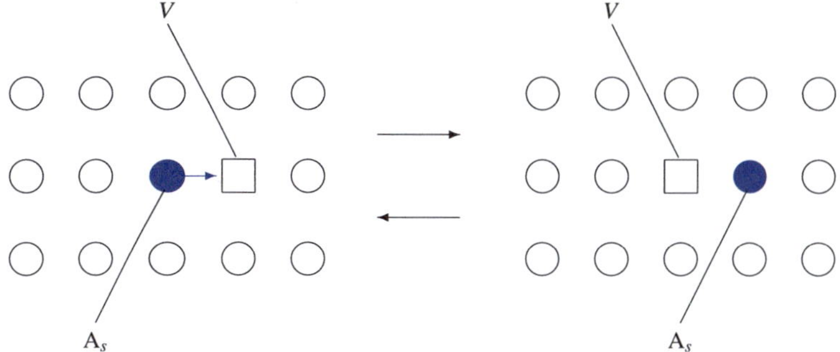

Fig. 1.3 Diffusion of self- and foreign-atoms via the vacancy mechanism. Radioactive or enriched stable isotopes can be used as suitable marker atoms (•)

neighbor site. In contrast, the jump of the tracer atom occurs with higher probability into the vacancy. The so called correlation factor f_V accounts for the preferred backward jump of the tracer atom. For solids with diamond structure the correlation factor for self-diffusion via V is exactly $f_V = 0.5$ [51].

The activation enthalpy $H_V^{SD} = H_V^F + H_V^M$ of self-diffusion via vacancies comprises both the enthalpy of formation H_V^F and migration H_V^M of the vacancy. In crystals with cubic symmetry the temperature dependence of the tracer self-diffusion coefficient is given by the following Arrhenius expression (see e.g. [46, 48])

$$D_V^{SD} = \frac{1}{C_o} f_V C_V^{eq} D_V = D_{Vo}^{SD} \exp\left(-\frac{H_V^{SD}}{k_B T}\right) \tag{1.11}$$

with the pre-exponential factor

$$D_{Vo}^{SD} = f_V g_V a_o^2 \nu_o \exp\left(\frac{S_V^{SD}}{k_B}\right). \tag{1.12}$$

The pre-exponential factor depends on the activation entropy $S_V^{SD} = S_V^F + S_V^M$ of self-diffusion via V that reflects the sum of the entropy of vacancy formation S_V^F and migration S_V^M. The other parameters in Eq. (1.11) have the same meaning as in Eq. (1.9). The geometry factor equals 1/8 for the case of self-diffusion via V in a diamond structure. The vacancy mechanism controls self-diffusion in Ge [52] and in many crystalline metals [53, 54].

Interstitialcy Mechanism

An other indirect mechanism of self-diffusion is the so called interstitialcy mechanism. This mechanism is schematically illustrated in Fig. 1.4 and relevant in particular for self-diffusion in Si [49].

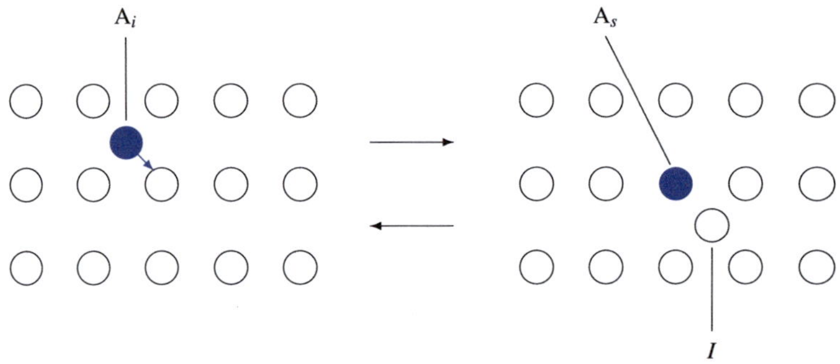

Fig. 1.4 The interstitialcy mechanism of self- and foreign-atom diffusion

In case of the interstitialcy mechanism a self-interstitial (I) mediates the diffusion of the matrix atoms, i.e., a self-interstitial pushes a neighboring host atom into the next interstitial position. Thereby the former self-interstitial becomes substitutional and forms a new self-interstitial. Subsequently the diffusion process is repeated with the newly formed I. In analogy to self-diffusion via V, self-diffusion via I is described by an Arrhenius expression

$$D_I^{SD} = \frac{1}{C_o} f_I\, C_I^{\text{eq}}\, D_I = D_{Io}^{SD} \exp\left(-\frac{H_I^{SD}}{k_B\, T}\right) \tag{1.13}$$

with the pre-exponential factor

$$D_{Io}^{SD} = f_I\, g_I\, a_o^2\, \nu_o \exp\left(\frac{S_I^{SD}}{k_B}\right) \tag{1.14}$$

and

$$H_I^{SD} = H_I^F + H_I^M \quad , \quad S_I^{SD} = S_I^F + S_I^M \tag{1.15}$$

as well as with the concentration of I in thermal equilibrium

$$C_I^{\text{eq}} = C_o \exp\left(-\frac{G_I^F}{k_B\, T}\right) \tag{1.16}$$

where C_o represents the number of interstitial lattice sites. The parameters f_I, g_I, ν_o and the thermodynamic quantities H_I^F, H_I^M, S_I^F, S_I^M, and G_I^F have a corresponding meaning as the quantities in Eqs. (1.10), (1.11), and (1.12).

The diffusion of the tracer atom via the interstitialcy mechanism is also correlated. Starting with a tracer atom in interstitial position the first diffusional jump of the tracer atom occurs with equal probability in all possible directions. After the site exchange the tracer atom occupies a substitutional lattice site with a self-interstitial in close vicinity. Accordingly, a backward jump of the tracer atom is highly probable [48].

The correlation factor for self-diffusion via the interstitialcy mechanism for a crystalline solid with diamond structure was calculated by Compaan and Haven [55] to $f_I = 0.7273$. This factor corresponds to self-interstitials occupying tetrahedral interstitial lattice sites of the diamond structure. The geometry factor follows to $g_I = 1/4$ in this model considering next nearest neighbor jumps. More recent theoretical calculations of f_I consider that self-interstitials in the diamond lattice can exist in various configurations. In addition to tetrahedral interstitials also self-atoms on hexagonal interstices as well as split-interstitials, where two self-atoms share one lattice site, are predicted by theory. Accordingly, self-diffusion can proceed via different self-interstitial defects with their corresponding correlation factor. Following recent theoretical calculations the split interstitial is the most stable defect among the various possible self-interstitial configurations [5, 11, 20, 23]. For this configuration a correlation factor of $f_I = 0.59$ [56] was deduced. In the case the dominant configuration of I changes with temperature, f_I becomes temperature dependent [57]. The correlation factor f_I for interstitial-mediated self-diffusion in Si was calculated to change from 0.64 to 0.80 for temperatures between 1000 and 1100 °C [57]. This represents a change of f_I within 20 % in the considered temperature range that is hardly resolved due to the limited accuracy of self-diffusion experiments.

Whereas self-diffusion in closely packed metals and germanium is fully described by the vacancy mechanism, both the vacancy and interstitialcy mechanism contribute to self-diffusion in Si [58–67]. Till today the individual contributions of V and I to Si self-diffusion are a matter of debate [65–67].

1.5.2.2 Foreign-Atom Diffusion

The vacancy and interstitialcy mechanisms illustrated in Figs. 1.3 and 1.4 also describe the indirect diffusion of mainly substitutionally dissolved foreign-atoms. Isolated V and I approach substitutional impurities and form next-nearest AV and AI defect pairs due to Coulomb attraction and/or minimizing local strain. Theses processes are described by the following reactions

$$A_s + V \rightleftharpoons AV \tag{1.17}$$

$$A_s + I \rightleftharpoons AI. \tag{1.18}$$

It is emphasized that for a long-range migration of A_s via the vacancy mechanism, the AV pair must partially dissociate, and the vacancy has to diffuse to at least a third-nearest neighbor site in the diamond lattice and return to the substitutional impurity along a different path to complete the diffusion step. In the case that the interaction potential between the vacancy and the dopant extends beyond the third-nearest neighbor site, complete dissociation of the AV complex becomes less probable, and A and V diffuse as a pair [68]. This vacancy-assisted diffusion mechanism is also called E-center mechanism [69]. Dopant diffusion via the interstitialcy mechanism only occurs if the AI pair does not dissociate.

The activation enthalpy Q_A of foreign-atom diffusion via reactions (1.17) and (1.18) reflects the sum of the migration H^M_{AX} and formation enthalpy H^F_{AX} of the dopant-defect AX pair ($X \in \{V, I\}$) that mediates dopant diffusion, i.e., $Q_A = H^F_{AX} + H^M_{AX}$. Depending on the binding energy H^B_{AX} between the dopant and the native defect X, the formation enthalpy H^F_{AX} can be higher or lower than the formation enthalpy H^F_X of the isolated defect. For $H^F_{AX} > H^F_X$ ($H^F_{AX} < H^F_X$) a repulsive (attractive) interaction between A and X exists. The binding energy H^B_{AX} is the difference in energy to form an isolated defect X and a dopant-defect pair AX, i.e., $H^B_{AX} = H^F_X - H^F_{AX}$. This shows that Q_A comprises H^M_{AX}, H^F_X and H^B_{AX} and, accordingly, the temperature dependence of D_{A} reads

$$D_{\mathrm{A}} = D_{\mathrm{A}}^{\circ} \exp\left(-\frac{H_{\mathrm{AX}}^{M} + H_{X}^{F} - H_{\mathrm{AX}}^{B}}{k_B T}\right) \tag{1.19}$$

with the pre-exponential factor

$$D_{\mathrm{A}}^{\circ} = f\, g\, a_o^2\, \nu_o \exp\left(\frac{S_{\mathrm{AX}}^{M} + S_{X}^{F} - S_{\mathrm{AX}}^{B}}{k_B}\right). \tag{1.20}$$

The correlation factor f is now a complicated quantity which considers temperature-dependent site exchange frequencies under the impact of nearby foreign-atoms [70].

According to Eq. (1.19) the activation enthalpy Q_{A} of dopant diffusion via the vacancy (interstitialcy) mechanism is $Q_{\mathrm{A}} = H_{\mathrm{AV}}^{M} + H_{V}^{F} - H_{\mathrm{AV}}^{B}$ ($Q_{\mathrm{A}} = H_{\mathrm{A}I}^{M} + H_{I}^{F} - H_{\mathrm{A}I}^{B}$). Following Hu's analysis of dopant diffusion via vacancies, the activation enthalpy can be further decomposed as [71]

$$Q_{\mathrm{A}} = \underbrace{H_{V}^{\mathrm{F}} - \Delta H_1}_{H_{\mathrm{AV}}^{\mathrm{F}}} + \underbrace{H_{V}^{\mathrm{M}} + \Delta H_1 - \Delta H_3}_{H_{\mathrm{AV}}^{\mathrm{M}}}, \tag{1.21}$$

where H_{V}^{F} and H_{V}^{M} are the formation and migration enthalpy of vacancies far away from the dopant. The potential energy landscape of the vacancy as assumed by Hu [71] and Dunham and Wu [72] is illustrated in Fig. 1.5. ΔH_1, ΔH_2, and ΔH_3 denote the difference in potential energy of a vacancy far away and on first, second, and third nearest neighbor site from the dopant, respectively, with $\Delta H_1 = H_{\mathrm{AV}}^{B}$.

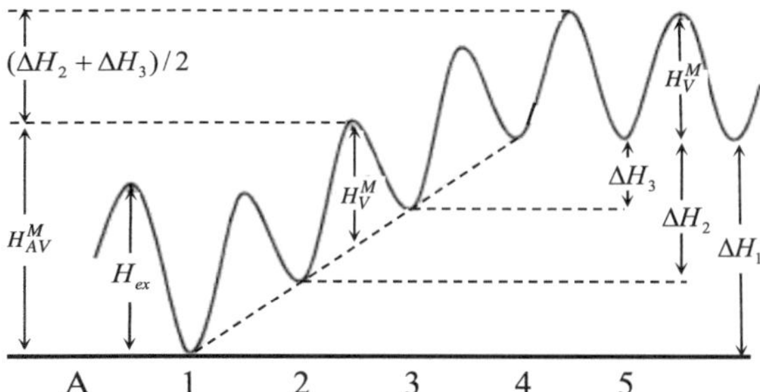

Fig. 1.5 Potential energy landscape of a vacancy close and far away from a dopant A [75]. $H_{\mathrm{AV}}^{\mathrm{M}}$ and H_{V}^{M} are the migration enthalpy of the dopant-vacancy pair and the isolated vacancy, respectively. ΔH_1, ΔH_2, and ΔH_3 denote the difference in potential energy of a vacancy far away and on first, second, and third nearest neighbor site from the dopant, respectively. H_{ex} is the enthalpy barrier for site exchange between dopant and vacancy

The more recent analysis of vacancy-mediated diffusion in a diamond structure performed by Dunham and Wu [72] lead to the following modification of Hu's approach

$$\begin{aligned} Q_{\mathrm{A}} &= H_{V}^{\mathrm{F}} + H_{V}^{\mathrm{M}} - \frac{\Delta H_2 + \Delta H_3}{2} \\ &= H_{V}^{\mathrm{F}} + H_{V}^{\mathrm{M}} - \overline{\Delta H_{23}}. \end{aligned} \tag{1.22}$$

Equation (1.22) accounts for a reduction in the migration barrier of the vacancy associated with changes in the binding energy between the sites. According to Dunham and Wu the difference between the activation enthalpy of self-diffusion $H_V^{SD} = H_V^F + H_V^M$ and dopant diffusion Q_{A} is the average of the vacancy binding energy $\overline{\Delta H_{23}}$ at the second and third nearest neighbor site from the dopant with respect to a vacancy far away (see Eq. (1.22)).

The binding energy between the dopant and the vacancy significantly affects the diffusion behavior of dopants, i.e., the higher the binding energy the lower is the activation enthalpy of dopant diffusion compared to self-diffusion. This becomes e.g. clearly evident in the diffusion behavior of n-type dopants such as phosphorus (P), arsenic (As) and antimony (Sb) in Ge [73]. The diffusion activation enthalpy of P, As, and Sb decreases from P to Sb although the size of the dopants increases from P to Sb. This peculiar diffusion behavior reflects in accordance with theoretical calculations [74] a binding energy between the n-type dopant and the vacancy in Ge that increases with increasing atomic size. A prerequisite for diffusion via the vacancy mechanism is that the AV pair must at least partially dissociate. In the diamond lattice the vacancy has to move to at least a third nearest neighbor site before it returns to the dopant to complete one diffusion step. Otherwise, as the V moves away from the donor atom the AV pair dissociates. In that case the donor atom is effectively immobilized and an interaction with another vacancy is required to migrate further [74].

The description of dopant diffusion demonstrates that the interaction of the dopant with the native defect strongly affects the diffusion activation enthalpy. An attractive (repulsive) interaction will result in an activation enthalpy that is lower (higher) compared to the activation enthalpy of self-diffusion. This also demonstrates that a comparison of dopant diffusion to self-diffusion is mandatory to understand the dopant diffusion in more detail.

In addition to the vacancy and interstitialcy mechanisms (1.17) and (1.18) the reactions

$$\mathrm{A}I + V \rightleftharpoons \mathrm{A}_s \tag{1.23}$$

$$\mathrm{A}V + I \rightleftharpoons \mathrm{A}_s \tag{1.24}$$

are considered for modeling dopant diffusion in Si [64, 76]. These reactions describe a dopant-defect pair assisted recombination of V and I that provide additional ways of recombination beside direct recombination of V and I via the Frenkel-pair reaction

$$V + I \rightleftharpoons \phi. \tag{1.25}$$

In Si some foreign-atoms such as gold (Au), platinum (Pt), and zinc (Zn) are incorporated both on interstitial and substitutional sites (see [49, 50] and references therein). The diffusion of these hybrid elements is a combination of direct and indirect diffusion. The diffusion of the atom in the interstitial configuration significantly exceeds the diffusion of the atom in the substitutional configuration, that is, the indirect diffusion of A_s can be neglected and the binding energy between foreign-atom and native defect is circumstantial. In the following diffusion models are introduced that describe the diffusion of hybrid elements.

Interstitial-Substitutional Diffusion Mechanisms

The treatment of interstitial-substitutional exchange mechanisms is reasonable for elements that are dissolved both on interstitial and substitutional sites. The transformation from the interstitial configuration A_i to the substitutional configuration A_s can only take place by means of a native point defect. The defect involved can be the vacancy as well as the self-interstitial. This is expressed on the one hand by the dissociative mechanism, that describes the A_i-A_s transformation via V, and on the other hand via the kick-out mechanism, that involves a self-interstitial for the A_i-A_s transformation

$$A_i + V \rightleftharpoons A_s \tag{1.26}$$

$$A_i \rightleftharpoons A_s + I. \tag{1.27}$$

These transformation reactions are schematically illustrated in Figs. 1.6 and 1.7. The figures reflect the diffusion behavior of hybrid elements which are mainly dissolved on substitutional sites but are highly mobile as interstitial defects A_i.

The dissociative mechanism was first proposed by Frank and Turnbull [77] to explain the diffusion behavior of copper (Cu) in Ge. Today this mechanism forms the basis for modeling the diffusion of various transition metals such as iron (Fe), cobalt (Co), and nickel (Ni) in Ge as well as to explain the diffusion behavior of Cu, silver (Ag) and gold (Au) in Ge. In particular, Cu diffusion experiments are well suited to extract information not only about the uncorrelated Ge self-diffusion coefficient [78–80] but also about the concentration and diffusivity of vacancies in Ge [81].

Gösele et al. [82] introduced the kick-out diffusion model (see Fig. 1.7) to explain the diffusion behavior of Au in Si. Based on reaction (1.27) Au diffusion profiles are consistently described and confirm the impact of self-interstitials I on Au diffusion in Si [83–86].

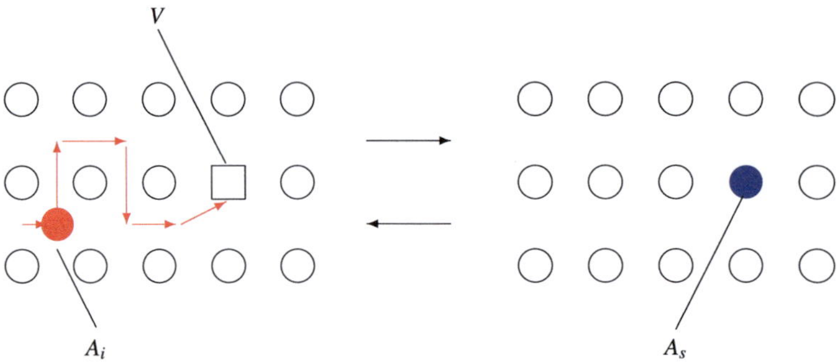

Fig. 1.6 Dissociative mechanism of the diffusion of foreign atoms that occupy both interstitial (*red filled circle*) and substitutional (*blue filled circle*) lattice sites. The interstitial-substitutional exchange is mediated by means of a vacancy V

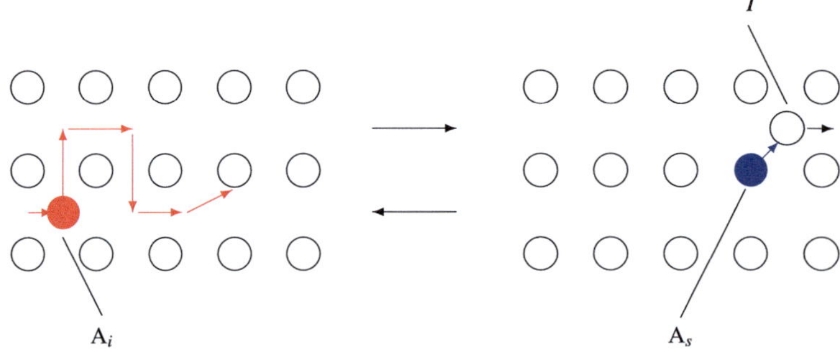

Fig. 1.7 Kick-out mechanism of the diffusion of foreign atoms that occupy both interstitial (*red filled circle*) and substitutional (*blue filled circle*) lattice sites. The interstitial-substitutional exchange is mediated by means of a self-interstitial I

In the following section the mathematical treatment of diffusion is introduced. Special emphasis is given to the diffusion of hybrid and dopant elements. The maximum solubility of typical hybrid elements such as Au, Pt, and Zn is of the order of $10^{17}\,\mathrm{cm}^{-3}$ at about 1573 K and decreases to about $10^{15}\,\mathrm{cm}^{-3}$ at 1173 K (see [50] and references therein). The incorporation of these hybrid elements introduces electronic defect states within the band gap of Si [50]. Due to their low solubility the free carrier concentration established at the particular diffusion temperature is not significantly affected. Accordingly, the diffusion of the hybrid elements can be considered to occur under spatial homogenous doping conditions, i.e, constant position of the Fermi level E_f. In the case when hybrid diffusion in undoped Si is considered, the diffusion occurs under electronically intrinsic conditions and the Fermi level is at its intrinsic position, i.e., at midgap position. In contrast, the maximum solubility of n- and p-type dopants in Si such as P, As, and B reaches a few atomic percent. The electron and hole concentration introduced by these dopants exceeds the intrinsic carrier concentration even at 1573 K. Accordingly, the position of the Fermi level changes along the dopant diffusion profile that is established when dopant diffusion proceeds from an infinite source into undoped Si. The dopant profile gives rise to an internal electric field that affects the diffusion of charged mobile defects. In the following the mathematical treatment of diffusion under both intrinsic and extrinsic diffusion conditions is considered.

1.6 Mathematical Description of Diffusion

The first paragraph of this section, i.e. Sect. 1.6.1, treats the mathematical description of the diffusion of hybrid elements in Si. The diffusion of these elements is fully described on the basis of interstitial-substitutional exchange mechanisms, i.e., the dissociative and kick-out mechanisms illustrated in Figs. 1.6 and 1.7 and formulated by the reactions (1.26) and (1.27). Possible charge states of the point defects involved are not considered because the position of the Fermi level is not affected during the diffusion of hybrid elements in Si.

The second paragraph, i.e. Sect. 1.6.2, treats the diffusion of n- and p-type dopant atoms in Si whose incorporation on mainly substitutional lattice site cause extrinsic doping conditions even at high temperatures. To demonstrate the impact of extrinsic doping on diffusion and the formation of point defects the simultaneous diffusion of self- and dopant atoms in Si is treated that can be realized experimentally by dopant diffusion in isotopically controlled Si multilayer structures. The diffusion mechanisms relevant for dopant diffusion are the vacancy and interstitialcy mechanisms illustrated in Figs. 1.3 and 1.4 and formulated by the reactions (1.17) and (1.18), respectively. In addition dopant-defect pair assisted recombination reactions (1.23) and (1.24) and possible charge states of the point defects involved in these reactions are taken into account to accurately describe and predict dopant diffusion in Si.

1.6.1 Diffusion of Hybrid Atoms

The full reactions representing the dissociative and kick-out mechanisms must consider beside the point defects indicated in reactions (1.26) and (1.27) also unoccupied interstitial sites Z and substitutional sites O occupied by the host atom. In case of the dissociative model a vacancy V is annihilated via the transformation of an interstitial foreign-atom to a substitutional foreign-atom and concurrently an unoccupied interstitial position is formed. On the other hand, in case of the kick-out reaction a substitutional lattice atom O is transformed to a self-interstitial I. Thus reactions (1.26) and (1.27) are fully described by

$$A_i + V \underset{k_{-1}}{\overset{k_{+1}}{\rightleftharpoons}} A_s + Z \tag{1.28}$$

$$A_i + O \underset{k_{-2}}{\overset{k_{+2}}{\rightleftharpoons}} A_s + I. \tag{1.29}$$

k_{+1}, k_{-1} and k_{+2}, k_{-2} represent the forward and backward reaction rates of the particular reaction in units of $[m^3 s^{-1}]$. The law of mass action for the reactions (1.28) and (1.29) combines the reaction rates with the equilibrium concentrations C_X^{eq} of the reaction partners. This yields

$$K_1 = \frac{k_{+1}}{k_{-1}} = \frac{C_{\mathrm{A}_s}^{\mathrm{eq}}\, C_o}{C_{\mathrm{A}_i}^{\mathrm{eq}}\, C_V^{\mathrm{eq}}} \tag{1.30}$$

$$K_2 = \frac{k_{+2}}{k_{-2}} = \frac{C_{\mathrm{A}_s}^{\mathrm{eq}}\, C_I^{\mathrm{eq}}}{C_{\mathrm{A}_i}^{\mathrm{eq}}\, C_o} \tag{1.31}$$

with the equilibrium constants K_1 and K_2 for the dissociative and kick-out reactions (1.28) and (1.29), respectively. $C_o = 5 \times 10^{22}\,\mathrm{cm}^{-3}$ is the concentration or atom density of Si atoms on substitutional lattice site that also equals the concentration of interstitial sites in Si. The concentration of interstitial and substitutional foreign-atoms is generally much lower than C_o. Accordingly, the concentration of matrix atoms and interstices remains nearly constant. In dynamical equilibrium the temporal change of the educts equals that of the products, i.e.,

$$k_{+1} C_{\mathrm{A}_i} C_V = k_{-1} C_{\mathrm{A}_s} C_o \tag{1.32}$$

$$k_{+2} C_{\mathrm{A}_i} C_o = k_{-2} C_{\mathrm{A}_s} C_I. \tag{1.33}$$

The formation of substitutional dissolved foreign-atoms per time unit via reactions (1.28) and (1.29) thus reads

$$\frac{\partial C_{\mathrm{A}_s}}{\partial t} = k_{+1} C_{\mathrm{A}_i} C_V - k_{-1} C_o C_{\mathrm{A}_s} + k_{+2} C_{\mathrm{A}_i} C_o - k_{-2} C_{\mathrm{A}_s} C_I. \tag{1.34}$$

The concentrations $C_X(\mathbf{r}, t)$ of the reactants $X \in \{\mathrm{A}_s, \mathrm{A}_i, V, I\}$ change not only by the exchange reactions but also by diffusion (see Eq. (1.3)). The following equations represent the full differential equation system that describes the diffusion of foreign-atoms in Si via the interstitial-substitutional exchange mechanisms (1.28) and (1.29)

$$\frac{\partial C_{A_s}}{\partial t} = \nabla \cdot (D_{A_s} \nabla C_{A_s}(\mathbf{r},t)) + k_{+1} C_{A_i} C_V - k_{-1} C_o C_{A_s} + k_{+2} C_{A_i} C_o - k_{-2} C_{A_s} C_I \quad (1.35)$$

$$\frac{\partial C_{A_i}}{\partial t} = \nabla \cdot (D_{A_i} \nabla C_{A_i}(\mathbf{r},t)) - k_{+1} C_{A_i} C_V + k_{-1} C_o C_{A_s} - k_{+2} C_{A_i} C_o + k_{-2} C_{A_s} C_I \quad (1.36)$$

$$\frac{\partial C_V}{\partial t} = \nabla \cdot (D_V \nabla C_V(\mathbf{r},t)) - k_{+1} C_{A_i} C_V + k_{-1} C_o C_{A_s} + k_V (C_V^{eq} - C_V) \quad (1.37)$$

$$\frac{\partial C_I}{\partial t} = \nabla \cdot (D_I \nabla C_I(\mathbf{r},t)) + k_{+2} C_{A_i} C_o - k_{-2} C_{A_s} C_I + k_I (C_I^{eq} - C_I). \quad (1.38)$$

These differential equations and Eqs. (1.30) and (1.31) represent the full equation system to describe the diffusion of hybrid elements in Si. One rate constant of each reaction can be replaced by means of the corresponding mass action equation (see Eqs. (1.30) and (1.31)). The terms $k_V(C_V^{eq} - C_V)$ and $k_I(C_I^{eq} - C_I)$ consider the formation and annihilation of V and I at internal sources and sinks of the crystal [82]. Stacking faults [87, 88], dislocations [86, 88], swirls [89] and defect clusters can act as sources and sinks of native defects. The quantities k_V and k_I denote the effectiveness of sources and sinks and strongly depend on the defect structure of the crystal. Considering a diffusion controlled supply or annihilation of native point defects the quantities k_V and k_I are described by [82]

$$k_V = \gamma_V \, \rho_V \, D_V \quad (1.39)$$

$$k_I = \gamma_I \, \rho_I \, D_I \quad (1.40)$$

where $\rho_{V,I}$ represents the density of the source and sink of the native defects in units of [m^{-2}]. The constants $\gamma_{V,I}$ characterize the effectiveness of e.g. dislocations and are of the order of 1 [90].

The coupled, nonlinear partial differential equations (1.35), (1.36), (1.37), and (1.38) describe the full mathematical treatment of foreign-atom diffusion via interstitial-substitutional exchange mechanisms (1.28) and (1.29) in three dimensions. Analytical solutions of this system of equations can be derived for specific conditions, i.e., for sufficient long diffusion times and appropriate boundary and initial conditions. Solutions for more general conditions can be calculated numerically. In the following analytical solutions of the differential equation system are derived for diffusion in one dimension.

1.6.1.1 Reduced Differential Equation System

In order to discuss analytical solutions of the differential equations (1.35), (1.36), (1.37), and (1.38) that describe the diffusion of hybrid elements such as Au [83–86, 91] and Zn [41, 91–94] in Si under special experimental conditions the equation system is reduced to one dimension, i.e., ($\nabla \rightarrow \frac{\partial}{\partial x}$ and $C(\mathbf{r}, t) \rightarrow C(x, t)$). This reduction is justified when the diffusion of foreign-atoms into Si crystal proceeds from a large surface covered with the foreign-atom of interest or from the gas phase. Moreover, the diffusion coefficients D_{A_s}, D_{A_i}, D_V, and D_I are moved in front of the differentiation. This approach requires concentration- and depth-independent diffusion coefficients. This is justified in the case of hybrid elements whose solubility is small in Si compared to the intrinsic carrier concentration at typical diffusion temperatures. Accordingly, the free carrier concentration introduced by the diffusion and incorporation of the hybrid elements in Si does not significantly affect the position of the Fermi level, that is, the Fermi level stays at its position under electronically intrinsic conditions. The other case when the free carrier concentration introduced by the foreign-atoms exceeds the intrinsic carrier concentration is treated in Sect. 1.6.2. Under the considered intrinsic conditions the equilibrium concentrations of the point defects involved in reactions (1.28) and (1.29) represent their concentrations for electronic intrinsic conditions. The equilibrium constants K_1 and K_2 are concentration independent and their temperature dependence is described by an Arrhenius equation [68].

The direct diffusion D_{A_s} of the substitutional foreign-atom A_s via direct exchange with a lattice atom or the ring mechanism (see Fig. 1.2) can be neglected compared to the indirect diffusion of A via the interstitial-substitutional exchange, i.e., $D_{\mathrm{A}_s} \approx 0$ effectively holds. Introducing normalized concentrations $\widetilde{C}_{\mathrm{X}} = C_{\mathrm{X}}/C_{\mathrm{X}}^{\mathrm{eq}}$ with $X \in \{\mathrm{A}_s, \mathrm{A}_i, V, I\}$, reduced diffusivities $D_{\mathrm{X}}^* = C_{\mathrm{X}}^{\mathrm{eq}} D_{\mathrm{X}} / C_{\mathrm{A}_s^{\mathrm{m}}}^{\mathrm{eq}}$, and considering Eqs. (1.30) and (1.31), the following differential equation system is obtained

$$\frac{\partial \widetilde{C}_{\mathrm{A}_s}}{\partial t} = \left(\frac{\partial \widetilde{C}_{\mathrm{A}_s}}{\partial t}\right)_V + \left(\frac{\partial \widetilde{C}_{\mathrm{A}_s}}{\partial t}\right)_I \tag{1.41}$$

$$\frac{C_{\mathrm{A}_i}^{\mathrm{eq}}}{C_{\mathrm{A}_s}^{\mathrm{eq}}} \frac{\partial \widetilde{C}_i}{\partial t} = D_{\mathrm{A}_i}^* \frac{\partial^2 \widetilde{C}_{\mathrm{A}_i}}{\partial x^2} - \left(\frac{\partial \widetilde{C}_{\mathrm{A}_s}}{\partial t}\right)_V - \left(\frac{\partial \widetilde{C}_{\mathrm{A}_s}}{\partial t}\right)_I \tag{1.42}$$

$$\frac{C_V^{\mathrm{eq}}}{C_{\mathrm{A}_s}^{\mathrm{eq}}} \frac{\partial \widetilde{C}_V}{\partial t} = D_V^* \frac{\partial^2 \widetilde{C}_V}{\partial x^2} - \left(\frac{\partial \widetilde{C}_{\mathrm{A}_s}}{\partial t}\right)_V + k_V(1 - \widetilde{C}_V) \tag{1.43}$$

$$\frac{C_I^{\mathrm{eq}}}{C_{\mathrm{A}_s}^{\mathrm{eq}}} \frac{\partial \widetilde{C}_I}{\partial t} = D_I^* \frac{\partial^2 \widetilde{C}_I}{\partial x^2} + \left(\frac{\partial \widetilde{C}_{\mathrm{A}_s}}{\partial t}\right)_I + k_I(1 - \widetilde{C}_I) \tag{1.44}$$

with

$$\left(\frac{\partial \widetilde{C}_{A_s}}{\partial t}\right)_V = k_{-1}C_o(\widetilde{C}_{A_i}\widetilde{C}_V - \widetilde{C}_{A_s}) \tag{1.45}$$

$$\left(\frac{\partial \widetilde{C}_{A_s}}{\partial t}\right)_I = k_{-2}C_I^{eq}(\widetilde{C}_{A_i} - \widetilde{C}_{A_s}\widetilde{C}_I). \tag{1.46}$$

Equations (1.41), (1.42), (1.43), (1.44), (1.45), and (1.46) form the basis for the simulation of the diffusion of Au, Pt, and Zn in Si. In the following section, solutions of the above differential equations are discussed that represent special cases of practical significance. These examples also demonstrate the benefit of the reduced equation system with the model parameters $D^*_{A_i}$, D^*_V, and D^*_I. These parameters mainly affect the diffusion behavior of hybrid atoms after sufficient long diffusion times. Finally, in Sect. 1.6.1.5 numerical solutions are presented to demonstrate the significance of the model parameters for the time evolution of the concentration profiles.

1.6.1.2 Dominance of the Dissociative Mechanism

The reaction rate constants k_{+1}, k_{-1}, k_{+2}, and k_{-2} of the interstitial-substitutional exchange reactions (1.28) and (1.29) determine the process that mediates the formation of A_s (see Eqs. (1.41), (1.45), and (1.46)). Considering the relation

$$k_{-1}C_o \gg k_{-2}C_I^{\mathrm{eq}} \tag{1.47}$$

and that V and I are in thermal equilibrium at the onset of diffusion, i.e., $\widetilde{C}_V(x,0) = \widetilde{C}_I(x,0) = 1$, the formation of A_s is mainly controlled by the dissociative mechanism (1.28). In this case the differential equation system (1.41), (1.42), (1.43), (1.44), (1.45), and (1.46) simplifies to equations for $\widetilde{C}_{A_s}$, $\widetilde{C}_{A_i}$, and $\widetilde{C}_V$ with in total six model parameters. A further reduction is achieved when diffusion in dislocation-free, i.e. $k_V = 0$, and highly dislocated Si is considered. These cases are treated in the following two paragraphs.

Diffusion in Dislocation-Free Crystals

The third term on the right hand side of Eq. (1.43), which expresses the contribution of internal sources and sinks to the formation of V, can be neglected for diffusion in a defect-free crystal ($k_V = 0$). In addition we assume that the transport capacity of interstitial foreign-atoms exceeds the transport capacity of V, i.e., $C^{\mathrm{eq}}_{A_i}D_{A_i} \gg C^{\mathrm{eq}}_V D_V$. These conditions lead to concentrations of A_i that approach thermal equilibrium

($\widetilde{C}_{\mathrm{A}_i} \approx 1$) for sufficient long diffusion times and Eqs. (1.41), (1.42), (1.43), and (1.45) can be reduced to the differential equation

$$\frac{C_V^{\mathrm{eq}}}{C_{\mathrm{A}_s}^{\mathrm{eq}}} \frac{\partial \widetilde{C}_V}{\partial t} = D_V^* \frac{\partial^2 \widetilde{C}_V}{\partial x^2} - \left(\frac{\partial \widetilde{C}_{\mathrm{A}_s}}{\partial t}\right)_V . \tag{1.48}$$

At sufficient long diffusion times, i.e. for $t \gg (k_{-1}C_o)^{-1}$, reaction (1.28) is in local equilibrium. According to the law of mass action, local equilibrium is expressed by

$$\frac{C_{\mathrm{A}_s}}{C_{\mathrm{A}_i}\, C_V} = \frac{C_{\mathrm{A}_s}^{\mathrm{eq}}}{C_{\mathrm{A}_i}^{\mathrm{eq}}\, C_V^{\mathrm{eq}}} \tag{1.49}$$

or

$$\frac{\widetilde{C}_{\mathrm{A}_s}}{\widetilde{C}_{\mathrm{A}_i}\, \widetilde{C}_V} = 1 \tag{1.50}$$

Assuming Eq. (1.50) and $\widetilde{C}_{\mathrm{A}_i} \approx 1$, $\widetilde{C}_V$ in Eq. (1.48) can be replaced by $\widetilde{C}_{\mathrm{A}_s}$. We obtain the single differential equation

$$\frac{\partial \widetilde{C}_{\mathrm{A}_s}}{\partial t} = D_V^{\mathrm{eff}} \frac{\partial^2 \widetilde{C}_{\mathrm{A}_s}}{\partial x^2} \tag{1.51}$$

with the effective diffusion coefficient

$$D_V^{\mathrm{eff}} = \frac{C_V^{\mathrm{eq}} D_V}{C_{\mathrm{A}_s}^{\mathrm{eq}} + C_V^{\mathrm{eq}}} \approx D_V^* \tag{1.52}$$

for $C_{\mathrm{A}_s}^{\mathrm{eq}} \gg C_V^{\mathrm{eq}}$. This relation holds for hybrid elements such as Au, Pt, and Zn in Si [41, 85, 86, 93–99] and Cu in Ge [78, 79, 81]. Moreover, these elements preferentially occupy the substitutional lattice site in the respective semiconductor, i.e., $C_{\mathrm{A}_s}^{\mathrm{eq}} \gg C_{\mathrm{A}_i}^{\mathrm{eq}}$ [50]. This ensures to measure mainly the distribution of A_s by means of both chemical and electrical profiling methods.

$D_V^{\mathrm{eff}} \approx D_V^* = C_V^{\mathrm{eq}} D_V / C_{\mathrm{A}_s}^{\mathrm{eq}}$ implies that the established diffusion profile of A_s is controlled by the diffusion of V or, strictly speaking, by the transport capacity $C_V^{\mathrm{eq}} D_V$. This diffusion mode is called native-defect controlled diffusion of A_s. In the case a constant surface concentration is established during annealing, the solution of Eq. (1.51) is given by a complementary error function (1.6) with $C_{\mathrm{A}}(x,t) = C_{\mathrm{A}_s}(x,t)$, $C_{\mathrm{A}}^{\circ} = C_{\mathrm{A}_s}^{\mathrm{eq}}$, and $D_{\mathrm{A}} = D_V^{\mathrm{eff}} \approx D_V^*$. Note, although Eq. (1.51) describes the diffusion of A_s, the apparent diffusion coefficient D_V^{eff} is interrelated with the properties of V and almost equals D_V^* that is one of the model parameters of the full differential equation system (1.41), (1.42), (1.43), (1.44), (1.45), and (1.46). Multiplying and dividing the parameter D_V^* with the boundary concentration $C_{\mathrm{A}_s}^{\mathrm{eq}}$ and the atom density C_o, respectively, yields the uncorrelated self-diffusion coefficient $C_V^{\mathrm{eq}} D_V / C_o$ (see Eq. (1.11)). Studies on Cu diffusion in dislocation-free Ge fully reveal this correlation between Cu diffusion and the V-mediated Ge self-diffusion coefficient and thus confirm the dissociative mechanism for the diffusion of Cu in Ge [78, 79, 81].

In the following the dissociative diffusion of hybrid elements in highly dislocated crystals is described. Based on the full system of differential equations again one differential equation for A_s with an effective diffusivity can be deduced.

Diffusion in Highly Dislocated Crystals

We consider a crystal with a high density of dislocations or other extended defects that can serve as efficient sources of V. Vacancies consumed by the transformation of A_i to A_s via reaction (1.28) can now be supplied by these internal sources much faster than via diffusion from the surface. As a consequence, the thermal equilibrium concentration of V is maintained, i.e. $C_V(x,t) = C_V^{eq}$ or $\widetilde{C}_V \approx 1$. Local equilibrium of reaction (1.28) is established after sufficient long times ($t \gg 1/k_{-1}C_o$), i.e. $\widetilde{C}_{A_s} \approx \widetilde{C}_{A_i}$ according to Eq. (1.50). Following these approximations, Eqs. (1.41), (1.42), (1.43), and (1.45) can be replaced by a single differential equation

$$\frac{\partial \widetilde{C}_{A_s}}{\partial t} = D_{A_i}^{eff} \frac{\partial^2 \widetilde{C}_{A_s}}{\partial x^2} \tag{1.53}$$

with the effective diffusion coefficient

$$D_{A_i}^{eff} = \frac{C_{A_i}^{eq} D_{A_i}}{C_{A_s}^{eq} + C_{A_i}^{eq}}. \tag{1.54}$$

With the relation $C_{A_s}^{eq} \gg C_{A_i}^{eq}$, that holds for mainly substitutional dissolved hybrid elements such as Cu in Ge and Au, Zn, and Pt in Si [50], the effective diffusion coefficient $D_{A_i}^{eff}$ equals to a good approximation $D_{A_i}^{*} = C_{A_i}^{eq} D_{A_i} / C_{A_s}^{eq}$. This quantity represents again one model parameter of the full differential equation system (1.41), (1.42), (1.43), (1.44), (1.45), and (1.46). For constant surface concentrations $C_{A_s}^{eq}$, the solution of Eq. (1.53) is given by the complementary error function [47]

$$\widetilde{C}_{A_s}(x,t) = \operatorname{erfc} \frac{x}{2\sqrt{D_{A_i}^{*} t}}. \tag{1.55}$$

Multiplying the parameter $D_{A_i}^{*}$ with the boundary concentration $C_{A_s}^{eq}$ yields the transport capacity $C_{A_i}^{eq} D_{A_i}$. This diffusion mode is named the foreign-atom controlled diffusion of A_s since the formation of A_s via reaction (1.28) is controlled by the supply of A_i.

The investigation of Cu diffusion in highly dislocated Ge yields data for $C_{A_i}^{eq} D_{A_i}$ that exceed $C_V^{eq} D_V$ and thus the Ge self-diffusion coefficient (see [79, 81] and references therein). This result is consistent with the diffusion of Cu in dislocation-free Ge that only provides information about the Ge self-diffusion when the relation $C^{eq}_{A_i} D_{A_i} \gg C^{eq}_{V} D_{V}$ is fulfilled (see Sect. 1.6.1.2) [78, 81].

In the following we consider that the kick-out mechanism (1.29) dominates the A_i-A_s exchange.

1.6.1.3 Dominance of the Kick-Out Mechanism

Again we consider that the concentrations of V and I are in thermal equilibrium at the onset of diffusion. The kick-out mechanism controls the formation of A_s in the case when the relation

$$k_{-2} C^{eq}_{I} \gg k_{-1} C_{o} \tag{1.56}$$

instead of Eq. (1.47) holds. Within this approximation the differential equation system (1.41), (1.42), (1.43), (1.44), (1.45), and (1.46) can be reduced to equations for $\widetilde{C}_{A_s}$, $\widetilde{C}_{A_i}$, and $\widetilde{C}_{I}$ with in total six model parameters. A further reduction is achieved for diffusion in dislocation-free, i.e. $k_I = 0$, and highly dislocated Si.

Diffusion in Dislocation-Free Crystals

Assuming a dislocation-free crystal, i.e. a crystal without internal sources of native point defects, self-interstitials can neither be formed nor annihilated at internal sources and sinks, respectively. Accordingly, the contribution on the right hand side of Eq. (1.44) that considers possible formation/annihilation of I at internal sources/sinks can be neglected ($k_I = 0$). A high transport capacity $C^{eq}_{A_i} D_{A_i}$ described by $C^{eq}_{A_i} D_{A_i} \gg C^{eq}_{I} D_{I}$ leads to $\widetilde{C}_{A_i} \approx 1$ for sufficient long times. With the condition for local equilibrium of reaction (1.29)

$$\frac{\widetilde{C}_{A_s} \widetilde{C}_{I}}{\widetilde{C}_{A_i}} = 1 \tag{1.57}$$

the differential equation

$$\frac{\partial \widetilde{C}_{A_s}}{\partial t} = \frac{\widetilde{C}_{A_s}^{\ 2}}{\widetilde{C}^{2}_{A_s} + C^{eq}_{I}/C^{eq}_{A_s}} \frac{\partial}{\partial x} \frac{1}{\widetilde{C}^{2}_{A_s}} D^{*}_{I} \frac{\partial \widetilde{C}_{A_s}}{\partial x} \tag{1.58}$$

with the reduced diffusion coefficient

$$D^{*}_{I} = \frac{C^{eq}_{I} D_{I}}{C^{eq}_{A_s}} \tag{1.59}$$

is obtained. Gösele et al. [82] demonstrate that the relation

$$\widetilde{C}^2_{A_s} \gg C_I^{eq}/C_{A_s}^{eq} \tag{1.60}$$

is fulfilled for the diffusion of Au in thin Si samples after sufficient long diffusion times. Accordingly, Eq. (1.58) can be replaced by

$$\frac{\partial \widetilde{C}_{A_s}}{\partial t} = \frac{\partial}{\partial x} D_I^{eff} \frac{\partial \widetilde{C}_{A_s}}{\partial x} \tag{1.61}$$

with

$$D_I^{eff} = \frac{1}{\widetilde{C}^2_{A_s}} D_I^*. \tag{1.62}$$

Equations (1.61) and (1.62) describe the native-defect controlled diffusion of A_s via the kick-out mechanism. The concentration dependence of D_I^{eff} leads to characteristic diffusion profiles that strongly differ from the native-defect controlled profiles of the dissociative model (see Eqs. (1.51) and (1.52)).

Based on Eq. (1.61) Gösele [82] deduced the relation

$$\widetilde{C}_{A_s}(\frac{d}{2}, t) = \left(4\pi D_I^* \frac{t}{d^2}\right)^{\frac{1}{2}} \tag{1.63}$$

for the concentration of A_s in the center of the crystal with thickness d and the reduced diffusion coefficient D_I^*. Equation (1.63) shows that the uncorrelated, I-mediated Si self-diffusion coefficient $C_I^{eq} D_I / C_o$ can be determined from the time evolution of the concentration of A_s in the center of the crystals after sufficient long diffusion times ($\widetilde{C}_{A_i} \approx 1$). Data of $C_I^{eq} D_I / C_o$ obtained from Au, Pt, and Zn diffusion studies (see e.g. [41, 83, 93, 100]) are in accord with the Si self-diffusion coefficients determined from direct tracer diffusion studies. This confirms the kick-out mechanism for the diffusion of these elements in Si and the contribution of I to Si self-diffusion [58].

Diffusion in Highly Dislocated Crystals

In crystals with a high density of dislocations or other efficient sources/sinks for self-interstitials, the thermal equilibrium concentration of I can be maintained during foreign-atom diffusion even when self-interstitials are consumed or formed via the kick-out reaction. Thus $\widetilde{C}_I(x,t) \approx 1$ holds and the differential equation system (1.41), (1.42), (1.43), (1.44), (1.45), and (1.46) can be reduced to

$$\frac{C_{A_i}^{eq}}{C_{A_s}^{eq}} \frac{\partial \widetilde{C}_i}{\partial t} = D_{A_i}^* \frac{\partial^2 \widetilde{C}_{A_i}}{\partial x^2} - \frac{\partial \widetilde{C}_{A_s}}{\partial t}. \tag{1.64}$$

For sufficient long times $((t \gg 1/k_{-2}C_I^{\mathrm{eq}}))$ the kick-out reaction (1.29) is in local equilibrium, i.e. $\widetilde{C}_{\mathrm{A}_s} = \widetilde{C}_{\mathrm{A}_i}$ (see Eq. (1.57)). Then the differential equation

$$\frac{\partial \widetilde{C}_{\mathrm{A}_s}}{\partial t} = D_{\mathrm{A}_i}^{\mathrm{eff}} \frac{\partial^2 \widetilde{C}_{\mathrm{A}_s}}{\partial x^2} \tag{1.65}$$

with the effective diffusion coefficient

$$D_{\mathrm{A}_i}^{\mathrm{eff}} = \frac{C_{\mathrm{A}_i}^{\mathrm{eq}} D_{\mathrm{A}_i}}{C_{\mathrm{A}_s}^{\mathrm{eq}} + C_{\mathrm{A}_i}^{\mathrm{eq}}} \tag{1.66}$$

is obtained. The diffusion coefficient $D_{\mathrm{A}_i}^{\mathrm{eff}}$ equals $D_{\mathrm{A}_i}^*$ for $C_{\mathrm{A}_s}^{\mathrm{eq}} \gg C_{\mathrm{A}_i}^{\mathrm{eq}}$. This, in particular, holds for both Au [86] and Zn [41] in Si. Concentration profiles of these elements measured after diffusion in highly dislocated Si fully confirm the impact of the defect density on the diffusion of these hybrid elements [41, 86]. The ratio $D_{\mathrm{A}_i}^*/D_I^* = C_{\mathrm{A}_i}^{\mathrm{eq}} D_{\mathrm{A}_i}/C_I^{\mathrm{eq}} D_I$ clearly exceeds unity and thus confirms the native-defect controlled diffusion of these elements in dislocation-free Si that requires the validity of $D_{\mathrm{A}_i}^* \gg D_I^*$. Note, diffusion in highly dislocated crystals via the kick-out and dissociative mechanisms provide the same effective diffusion coefficient $D_{\mathrm{A}_i}^{\mathrm{eff}}$ (see Eqs. (1.54) and (1.66)). This demonstrates that the type of native point defect involved in the interstitial-substitutional exchange can not be identified by means of diffusion studies with defect-rich materials because the concentration of V and I close to thermal equilibrium ($\widetilde{C}_{V,I} \approx 1$).

1.6.1.4 Occurrence of Both A_i-A_s Exchange Mechanisms

Both V and I exist in Si under thermal equilibrium conditions. Accordingly, beside the dissociative also the kick-out mechanisms can contribute to the native-defect controlled diffusion of hybrid foreign-atoms. The simultaneous occurrence of both A_i-A_s exchange mechanisms is considered in the following for local equilibrium and non-equilibrium conditions.

Diffusion Under Local Equilibrium Conditions

Assuming local equilibrium of reactions (1.28) and (1.29) described by Eqs. (1.50) and (1.57), respectively, and additionally $\widetilde{C}_{\mathrm{A}_i} \approx 1$, the following differential equation can be derived from the full differential equation system (1.41), (1.42), (1.43), (1.44), (1.45), and (1.46) for a defect-free crystal (without one-, two-, or three-dimensional defects),

$$\left(1 + \frac{C_V^{\mathrm{eq}}}{C_{\mathrm{A}_s}^{\mathrm{eq}}} + \frac{C_I^{\mathrm{eq}}}{C_{\mathrm{A}_s}^{\mathrm{eq}} \left(\widetilde{C}_{\mathrm{A}_s}\right)^2}\right) \frac{\partial \widetilde{C}_{\mathrm{A}_s}}{\partial t} = \frac{\partial}{\partial x} \left(\frac{1}{\left(\widetilde{C}_{\mathrm{A}_s}\right)^2} D_I^* + D_V^*\right) \frac{\partial \widetilde{C}_{\mathrm{A}_s}}{\partial x}. \tag{1.67}$$

With the relations $C_{A_s}^{eq} \gg C_V^{eq}$ and $\left(\widetilde{C}_{A_s}\right)^2 \gg C_I^{eq}/C_{A_s}^{eq}$, which hold for hybrid elements in Si under indiffusion conditions [82, 84], Eq. (1.67) reads

$$\frac{\partial \widetilde{C}_{A_s}}{\partial t} = \frac{\partial}{\partial x}\left(\frac{1}{\left(\widetilde{C}_{A_s}\right)^2} D_I^* + D_V^*\right)\frac{\partial \widetilde{C}_{A_s}}{\partial x} = \frac{\partial}{\partial x} D_{I,V}^{\text{eff}} \frac{\partial \widetilde{C}_{A_s}}{\partial x} \tag{1.68}$$

with the effective diffusion coefficient

$$D_{I,V}^{\text{eff}} = \frac{1}{\left(\widetilde{C}_{A_s}\right)^2} D_I^* + D_V^*. \tag{1.69}$$

The concentration dependence of $D_{I,V}^{\text{eff}}$ reveals that under in-diffusion conditions with $\widetilde{C}_{A_s}(x,t) < 1$ the concentration profile of substitutional foreign-atoms is mainly characterized by diffusion via the kick-out mechanism even for $D_V^* \approx D_I^*$. In typical diffusion experiments where the hybrid-atom penetrates from the surface into the Si bulk the resulting diffusion profile is mainly controlled by the kick-out reaction (see e.g. [41]). Thus D_I^* can be determined from native-defect controlled in-diffusion profile whereas D_V^* is hardly extracted. On the other hand, the diffusion parameter D_V^* controls $D_{I,V}^{\text{eff}}$ in the case when $\widetilde{C}_{A_s} > 1$ holds (see Eq. (1.69)). This condition is realized in out-diffusion experiments where the concentration of the substitutional foreign-atom in the bulk exceeds its concentration at the surface (see e.g. [101, 102]).

The diffusion of hybrid atoms under local equilibrium conditions is characterized by the validity of Eqs. (1.50) and (1.57). This implies that

$$\widetilde{C}_I \widetilde{C}_V = 1. \tag{1.70}$$

is fulfilled. Equation (1.70) represents local equilibrium of the Frenkel-pair reaction (1.25). This interrelation between interstitial-substitutional exchange mechanisms and the Frenkel-pair reaction shows that local equilibrium between V and I can be established via the exchange reactions (1.28) and (1.29) even when direct recombination of V and I is a more moderate process due to an energy or entropy barrier [103]. Moreover, it is noted that foreign-atom profiles representing local equilibrium conditions can be described equally by any combination of two of the three reactions (1.25), (1.28), and (1.29). This actually holds for numerous Au and Pt diffusion profiles in Si. Thus these profiles do not provide definite information about the preferred way of interstitial-substitutional exchange.

The full differential equation system given by Eqs. (1.41), (1.42), (1.43), (1.44), (1.45), and (1.46) has been treated in the limit of either the dissociative or the kick-out mechanism for diffusion times that ensures local equilibrium conditions. Under these conditions the differential equations can be reduced to a single differential equation of the form of Eq. (1.5) with an effective diffusion coefficient that approximately equals the reduced diffusion coefficient $D_X^* = C_X^{eq} D_X / C_{A_s^m}^{eq}$ with $X \in \{A_i, V, I\}$. This demonstrates that diffusion profiles of hybrid atoms are mainly sensitive to D_X^*. The other model parameters involved in the full mathematical formulation of diffusion via interstitial-substitutional exchange play only a minor role. However, these parameters become significant in concentration profiles established after short and intermediate diffusion times (see Sect. 1.6.1.5).

Diffusion Under Non-equilibrium Conditions

Diffusion studies under experimental conditions deviating from local equilibrium of the point defects involved in defect reactions can, in principle, differentiate between the microscopic mechanisms of atomic transport (see e.g. [104–110]). Such studies provide valuable information about the model parameters entering the differential equations that describe the diffusion of dopant elements in Si via indirect diffusion mechanisms. In general, the solution of differential equations is only defined under specific initial and boundary conditions. These conditions must be in accord with the conditions realized experimentally. This is mandatory for direct comparison of experimental and calculated diffusion profiles that aims to determine the model parameters of the underlying diffusion equations. In the following section the impact of the model parameters on the temporal evolution of defect concentration profiles are described. It is shown that diffusion studies covering different time regimes are very advantageous to determine the mechanisms of diffusion and thus the underlying model parameters.

1.6.1.5 Numerical Simulation of Foreign-Atom Diffusion via Interstitial-Substitutional Exchange

The differential equations (1.41), (1.42), (1.43), (1.44), (1.45), and (1.46) representing diffusion in Si via interstitial-substitutional exchange are solved numerically to demonstrate the impact of the model parameters on the time evolution of the diffusion of hybrid atoms. The solutions reveal various diffusion regions that are differently sensitive to the settings of the model parameters.

Simulation of Kick-Out Diffusion

The simulation of foreign-atom diffusion via interstitial-substitutional is restricted for the sake of clarity to the kick-out mechanism. This limitation is justified because the in-diffusion of hybrid elements such as Au [82, 84, 86], Pt [95, 100], and Zn [41, 93] in Si is mainly controlled by this reaction (see Eqs. (1.68) and (1.69)). The mathematical formulation of the kick-out mechanism is given by (see Eqs. (1.41), (1.42), (1.43), (1.44), (1.45), and (1.46) for comparison)

$$\frac{\partial \widetilde{C}_{\mathrm{A}_s}}{\partial t} = k_{-2} C_I^{\mathrm{eq}} \left(\widetilde{C}_{\mathrm{A}_i} - \widetilde{C}_{\mathrm{A}_s} \widetilde{C}_I \right) \tag{1.71}$$

$$\frac{C_{\mathrm{A}_i}^{\mathrm{eq}}}{C_{\mathrm{A}_s}^{\mathrm{eq}}} \frac{\partial \widetilde{C}_{\mathrm{A}_i}}{\partial t} = D_{\mathrm{A}_i}^* \frac{\partial^2 \widetilde{C}_{\mathrm{A}_i}}{\partial x^2} - \frac{\partial \widetilde{C}_{\mathrm{A}_s}}{\partial t} \tag{1.72}$$

$$\frac{C_I^{\mathrm{eq}}}{C_{\mathrm{A}_s}^{\mathrm{eq}}} \frac{\partial \widetilde{C}_I}{\partial t} = D_I^* \frac{\partial^2 \widetilde{C}_I}{\partial x^2} + \frac{\partial \widetilde{C}_{\mathrm{A}_s}}{\partial t} \tag{1.73}$$

and Eq. (1.31). The five model parameters $D_{\mathrm{A}_i}^*$, D_I^*, $C_{\mathrm{A}_i}^{\mathrm{eq}}/C_{\mathrm{A}_s}^{\mathrm{eq}}$, $C_I^{\mathrm{eq}}/C_{\mathrm{A}_s}^{\mathrm{eq}}$ and $k_{-2}C_I^{\mathrm{eq}}$ fully determine the solution of the differential equations under given initial- and boundary conditions. The boundary conditions

$$\widetilde{C}_{\mathrm{A}_i}(0,t) = \widetilde{C}_{\mathrm{A}_i}(d,t) = 1 \tag{1.74}$$

$$\widetilde{C}_I(0,t) = \widetilde{C}_I(d,t) = 1 \tag{1.75}$$

are considered for the numerical calculation of $\widetilde{C}_X$ with $X \in \{\mathrm{A}_s, \mathrm{A}_i, I\}$ in a finite crystal of thickness d. Equation (1.74) describes an infinite source of interstitial foreign-atoms at the surface. Moreover, Eq. (1.75) reflects an equilibrium concentration of I maintained at the surface. These equations represent conditions for in-diffusion of foreign-atoms that are deposited on top of or rapidly supplied via the gas phase to the sample surface. The boundary condition for A_s is given by the solution of Eq. (1.71) for $x = 0$ and $x = d$ taking into account Eqs. (1.74) and (1.75) to

$$\widetilde{C}_{\mathrm{A}_s}(0,t) = \widetilde{C}_{\mathrm{A}_s}(d,t) = 1 - \exp(-k_{-2}C_I^{\mathrm{eq}}t). \tag{1.76}$$

This solution indicates that the parameter $k_{-2}C_I^{\mathrm{eq}}$ controls the temporal increase of the A_s concentration at the surface.

The initial concentration of the diffusing element is generally very low prior to diffusion, i.e., below the detection limit of the method subsequently applied to measure the diffusion profile. Accordingly, the initial concentrations of interstitial and substitutional foreign-atoms within the semiconductor are lower than their equilibrium values $C_{\mathrm{A}_i}^{\mathrm{eq}}$ and $C_{\mathrm{A}_s}^{\mathrm{eq}}$ established at the surface in the course of diffusion. The following initial conditions are considered for the normalized concentrations of A_i and A_s

$$\widetilde{C}_{\mathrm{A}_s}(x,0) = \widetilde{C}_{\mathrm{A}_i}(x,0) = 10^{-10}. \tag{1.77}$$

The concentration of self-interstitials at the onset of diffusion, i.e., at time $t = 0$, is set to the equilibrium concentration C_I^{eq}

$$\widetilde{C}_I(x, 0) = 1. \tag{1.78}$$

These conditions described by Eqs. (1.77) and (1.78) fulfill local equilibrium of the kick-out reaction (see Eq. (1.57)) for $x > 0$ and $t = 0$.

The boundary and initial conditions formulated by Eqs. (1.74) (1.75), (1.76), (1.77), and (1.78) can be realized experimentally by Zn diffusion in Si [41, 111]. Equilibration of Si samples prior to Zn in-diffusion is achieved by separating the Zn-source from the Si samples and releasing Zn vapor after the equilibration process [111]. Corresponding diffusion profiles of substitutional Zn_s measured by means of the spreading resistance technique are presented in Sect. 1.7.1.

To demonstrate the impact of the model parameters on the diffusion of hybrid atoms via the kick-out mechanism, numerical solutions of Eqs. (1.71), (1.72), and (1.73) are presented in the following. We start from a set of model parameters that accurately describe the diffusion behavior of Zn in Si at 1481 K ($D^*_{A_i} = 4 \cdot 10^{-07}\,cm^2s^{-1}$, $D^*_I = 8 \cdot 10^{-08}\,cm^2s^{-1}$, $C^{eq}_{A_i}/C^{eq}_{A_s} = 0.03$, $C^{eq}_I/C^{eq}_{A_s} = 0.001$, $k_{-2}C^{eq}_I = 0.1\,s^{-1}$) [41]. Figures 1.8, 1.9, and 1.10 illustrate additional simulations with model parameters deviating from these reference values.

The significance and impact of the model parameter $k_{-2}C^{eq}_I$ on the diffusion of hybrid elements via the kick-out mechanism is illustrated in Fig. 1.8. Increasing $k_{-2}C^{eq}_I$ by one order of magnitude, i.e., from $0.1\,s^{-1}$ to $1.0\,s^{-1}$, mainly affects the normalized concentration $\widetilde{C}_{A_s} = C_{A_s}(x, t)/C^{eq}_{A_s}$ for diffusion times $t \leq 240\ s$. For longer times no impact of $k_{-2}C^{eq}_I$ on the diffusion profiles is evident. The parameter

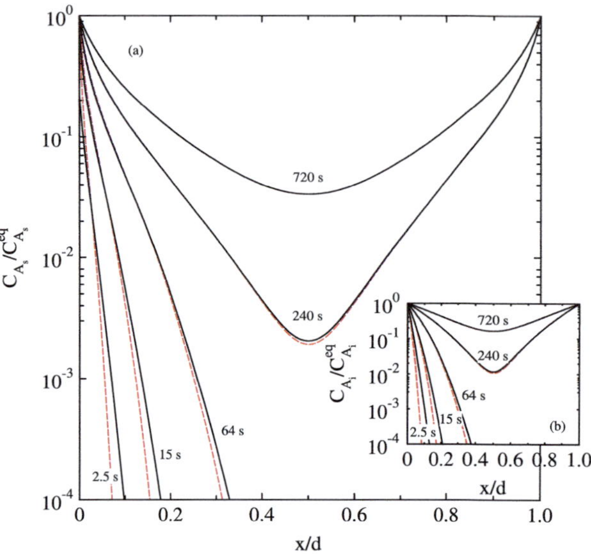

Fig. 1.8 Temporal evolution of the diffusion profiles of substitutional A_s (**a**) and interstitial A_i (**b**) foreign-atoms normalized by their respective thermal equilibrium concentration $C^{eq}_{A_s}$ and $C^{eq}_{A_i}$. The profiles illustrate the impact of the model parameter $k_{-2}C^{eq}_I$ (——: $k_{-2}C^{eq}_I = 0.1\ s^{-1}$) (- - - -: $k_{-2}C^{eq}_I = 1.0\ s^{-1}$)

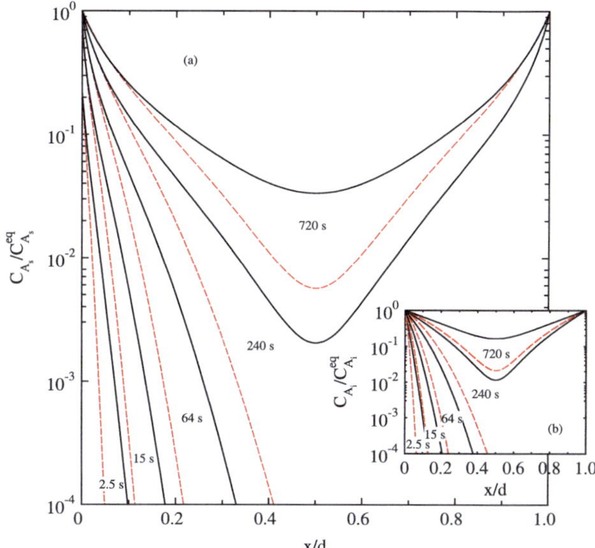

Fig. 1.9 Temporal evolution of the diffusion profiles of substitutional A_s (**a**) and interstitial A_i (**b**) foreign-atoms normalized by their respective thermal equilibrium concentration $C^{eq}_{A_s}$ and $C^{eq}_{A_i}$. The profiles illustrate the impact of the model parameter $C^{eq}_{A_i}/C^{eq}_{A_s}$. (—— : $C^{eq}_{A_i}/C^{eq}_{A_s} = 0.03$) (- - - - : $C^{eq}_{A_i}/C^{eq}_{A_s} = 0.3$)

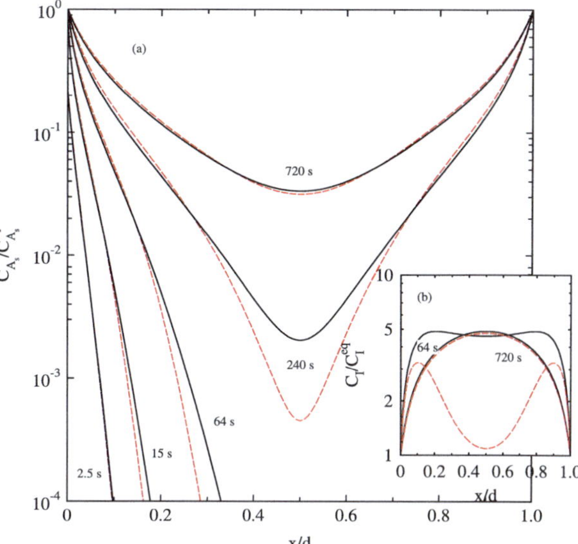

Fig. 1.10 Temporal evolution of the diffusion profiles of the substitutional A_s foreign-atom (**a**) and the self-interstitial I (**b**) normalized by their respective thermal equilibrium concentration $C^{eq}_{A_s}$ and C^{eq}_I. The profiles illustrate the impact of the model parameter $C^{eq}_I/C^{eq}_{A_s}$ (—— : $C^{eq}_I/C^{eq}_{A_s} = 0.001$) (- - - - : $C^{eq}_I/C^{eq}_{A_s} = 0.01$)

$k_{-2}C^{eq}_I$ reminds in Eq. (1.76) of the decay constant in the decay law of radioactive isotopes. In this analogy, $1/k_{-2}C^{eq}_I$ represents the time constant that describes the mean life-time for the transformation of interstitially to substitutionally dissolved foreign-atoms. In case the time constant is lowered, that is, $k_{-2}C^{eq}_I$ is increased, the mean time foreign-atoms stay on interstitial position decreases and accordingly, the penetration depth of both A_i and A_s is smaller due to the coupling between A_i and

A_s via the kick-out reaction. This is the main impact of $k_{-2}C_I^{eq}$ on the diffusion profiles of $\widetilde{C}_{A_s}$ and $\widetilde{C}_{A_i}$ illustrated in Fig. 1.8.

Equation (1.76) describes that higher values of $k_{-2}C_I^{eq}$ lead to increased A_s concentrations at the surface until the equilibrium concentration $C_{A_s}^{eq}$ is established. The interstitial-substitutional exchange via the kick-out reaction creates self-interstitials. Considering native-defect controlled diffusion of hybrid elements in dislocation-free crystals that prevails when $D_{A_i}^* > D_I^*$ holds, a supersaturation of I is obtained, i.e. $\widetilde{C}_I > 1$. This supersaturation hampers the formation of additional A_s via the exchange mechanism. Further formation of A_s is governed by the transport of I to the sample surface that is limited by the parameter D_I^*. Close to the surface, the concentration of I is already reduced due to out-diffusion. However, in the bulk the I concentration till increases due to interstitial-substitutional exchange. This describes that the native defect-controlled mode of diffusion is reached earlier at the surface than in the bulk. Therefore diffusion profiles for longer times still reveal a dependence of $k_{-2}C_I^{eq}$ at the diffusion front and/or at the center of the sample whereas the surface region is already insensitive to the parameter $k_{-2}C_I^{eq}$ (see Fig. 1.8: t = 64 s and 240 s). Finally, when the supersaturation of I has reached its maximum, the formation of A_s is fully controlled by out-diffusion of I to the surface. At this stage of diffusion the entire A_s profile is insensitive to $k_{-2}C_I^{eq}$ as supported by the $\widetilde{C}_{A_s}$ profile in Fig. 1.8 for 720 s.

The impact of the model parameter $C_{A_i}^{eq}/C_{A_s}^{eq}$ on the time evolution of A_s and A_i profiles is shown by Fig. 1.9. At constant $D_{A_i}^*$, an increase of $C_{A_i}^{eq}/C_{A_s}^{eq}$ is associated with a decrease of D_{A_i}. With increasing value of $C_{A_i}^{eq}/C_{A_s}^{eq}$ the penetration depth of A_i decreases and, due to the coupling between A_i and A_s via the kick-out reaction, also the penetration depth of A_s. The distribution of A_s is within a close surface region almost independent of this parameter. Here the removal of supersaturated self-interstitials by diffusion to the surface determines the profile shape. On the other hand, the supply of A_i and removal of I controls the formation of A_s at the transition region to the diffusion front. The diffusion front itself is affected both by diffusion and reaction processes. Finally, the diffusion of I fully controls further formation of A_s after sufficient long diffusion times that lead to $\widetilde{C}_{A_i}(x,t) \approx 1$. At this stage the A_s profiles are insensitive to the model parameter $C_{A_i}^{eq}/C_{A_s}^{eq}$ and fully described by the solution of Eq. (1.61). Since this case was already discussed in Sect. 1.6.1.3, it is not illustrated in Fig. 1.9.

The impact of $C_I^{eq}/C_{A_s}^{eq}$ on the diffusion of A_s for different times is illustrated by Fig. 1.10. The corresponding distributions of I for 64 s and 720 s are shown for comparison (see Fig. 1.10b). At the beginning of diffusion (after 2.5 s) the parameter does not affect the distribution of A_s. At this stage, the velocity of the formation of the reaction products is high and A_s and I are formed very shortly. With progressing formation of I via the kick-out reaction the impact of the backward reaction increases and the formation of the reaction products decreases, i.e., the formation of I per time unit decreases. At this stage of diffusion, which is characterized by times of 15 s to 240 s, the interstitial-substitutional exchange reaction controls the diffusion profiles and causes an increase in the I supersaturation. Since $C_I^{eq}/C_{A_s}^{eq}$

is a measure of the supersaturation established via A_i-A_s exchange, the diffusion profiles are sensitive to this parameter. With increasing supersaturation the transport of I to the surface increases. The drop of the I concentration to thermal equilibrium ($\widetilde{C}_I = 1$) at $x = 0$, which is illustrated in the insert of Fig. 1.10, reflects the property of the surface as sink for I. Finally, after 720 s, the I supersaturation decreases. Now the formation of A_s is mainly controlled by the transport of I to the surface that is mainly mediated by D_I^*. Overall, the parameter $C_I^{eq}/C_{A_s}^{eq}$ affects the temporal evolution of A_s profiles for times varying between very short and effectively long times.

The Dirichlet boundary condition (1.74) for interstitial dissolved foreign atoms can be established experimentally by a constant vapor pressure p_A of the foreign atom (see e.g. sulfur: [112, 113]; zinc: [41, 93]). The vapor pressure determines the thermodynamic activity of the foreign atom A at the surface. Accordingly, the concentrations $C_{A_i}^{eq}$ and $C_{A_s}^{eq}$ correspond to equilibrium states that depend on p_A. In contrast, C_I^{eq} and the diffusion coefficients D_I and D_{A_i} describe properties of I and A_i that are independent of p_A. The parameters D_I^* and $C_I^{eq}/C_{A_s}^{eq}$ depend via $C_{A_s}^{eq}$ on p_A whereas $D_{A_i}^*$, $C_{A_i}^{eq}/C_{A_s}^{eq}$ and $k_{-2}C_I^{eq}$ are independent of p_A. This becomes evident by Eq. (1.31). The mass action constant K_2 is, like any other reaction constant, independent of p_A.

The impact of p_A on hybrid diffusion can be demonstrated by numerical simulations with varying values of D_I^* and $C_I^{eq}/C_{A_s}^{eq}$. Calculated $\widetilde{C}_{A_s}$ profiles for different parameters are illustrated in Fig. 1.11. The corresponding profiles of A_i and I are displayed in Figs. 1.12 and 1.13, respectively. Simulations for $\alpha \equiv D_{A_i}^*/D_I^* = 25$ reflect Zn diffusion profiles in dislocation-free and highly dislocated Si at 1388 K for 240 s [41]. The profiles belonging to $\alpha \in \{5, 1, 0.2, 0.04\}$ simulate the impact of a decreasing vapor pressure on diffusion in a sample with a thickness of 4 millimeter. The significance of $D_{A_i}^*/D_I^*$ on the shape of the A_s profiles is demonstrated by Fig. 1.11. A convex shape within the surface area is obtained for $\alpha > 1$. This reflects the native-defect controlled diffusion mode that holds for $C_{A_i}^{eq}D_{A_i} \gg C_I^{eq}D_I$. Within

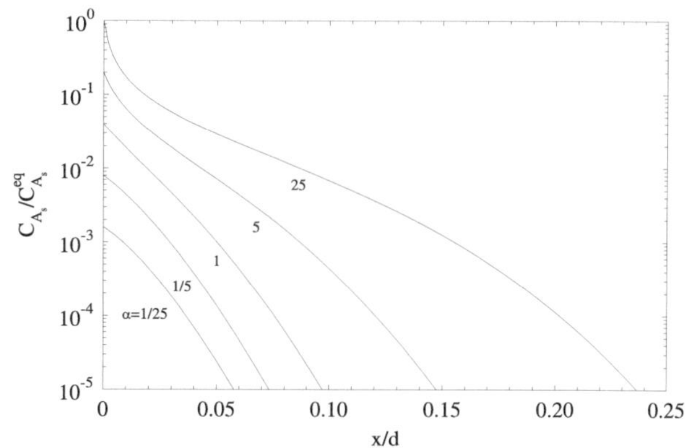

Fig. 1.11 Impact of the ratio $\alpha \equiv D_{A_i}^*/D_I^*$ on the normalized concentration profile of A_s

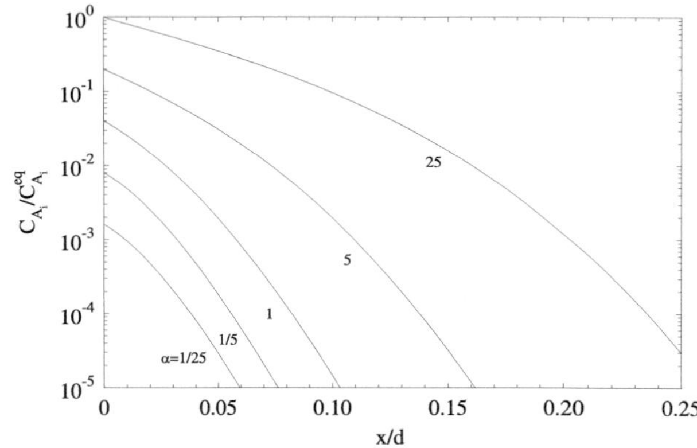

Fig. 1.12 Normalized concentration profiles of interstitial foreign-atoms A_i for different values of $\alpha \equiv D^*_{A_i}/D^*_I$

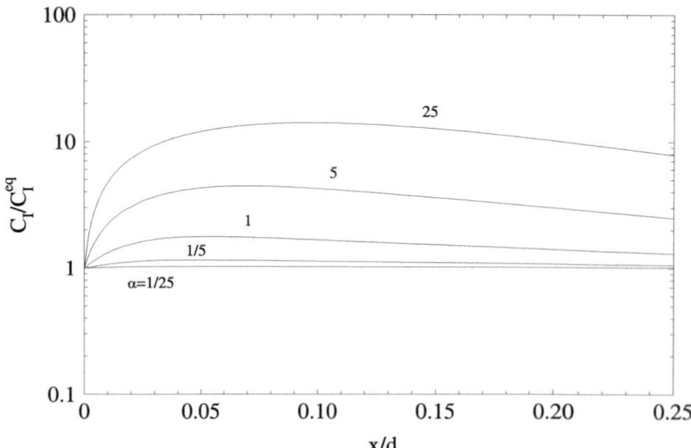

Fig. 1.13 Normalized concentration profile of self-interstitials I for different values of $\alpha \equiv D^*_{A_i}/D^*_I$

the approximation $\widetilde{C}_{A_i} \approx 1$, the relation

$$\widetilde{C}_{A_s} = \frac{1}{\widetilde{C}_I} \tag{1.79}$$

follows from Eq. (1.57). This is confirmed by the near-surface characteristic of the $\widetilde{C}_{A_s}$ and $\widetilde{C}_I$ profiles shown in Figs. 1.11 and 1.13 for $\alpha > 1$. Convex profiles reflect the significance of I for diffusion of the hybrid elements. In contrast, $\widetilde{C}_{A_s}$ profiles are concave for $D^*_{A_i} < D^*_I$ and are characterized by a negligible supersaturation of I, i.e. $\widetilde{C}_I \approx 1$. In this approximation Eq. (1.57) predicts $\widetilde{C}_{A_s} = \widetilde{C}_{A_i}$. The profiles shown in Figs. 1.11 and 1.12 for $\alpha < 1$ confirm this correlation.

Overall, Figs. 1.8, 1.9, 1.10, and 1.11 reveal a clear sensitivity of hybrid diffusion to all kick-out model parameters. This is summarized in Table 1.1 with $1/(k_{-2}C_I^{eq})$ as characteristic time for different diffusion stages. At the beginning of diffusion ($t < 1/(k_{-2}C_I^{eq})$) the A_s profiles not only depend on $D^*_{A_i}$ and D^*_I but are also sensitive to the parameters $k_{-2}C_I^{eq}$ and $C_{A_i}^{eq}/C_{A_s}^{eq}$. At intermediate times ($t \geq 1/(k_{-2}C_I^{eq})$) both $C_{A_i}^{eq}/C_{A_s}^{eq}$ and $C_I^{eq}/C_{A_s}^{eq}$ characterize the $\widetilde{C}_{A_s}$ profile. The near-surface curvature of $\widetilde{C}_{A_s}$ is a measure of the ratio $D^*_{A_i}/D^*_I$ that in particular affects the profiles for

Table 1.1 Dependence of $\widetilde{C}_{A_s}$ profiles in dislocation-free (⋆) and dislocated (∗) crystals on the model parameters of the kick-out mechanism

Diffusion stage	$D^*_{A_i}$	D^*_I	$\frac{C^{eq}_{A_i}}{C^{eq}_{A_s}}$	$\frac{C^{eq}_I}{C^{eq}_{A_s}}$	$k_{-2}C^{eq}_I$
$t < \frac{1}{k_{-2}C^{eq}_I}$	⋆ (∗)	⋆	⋆	–	⋆ (∗)
$t \geq \frac{1}{k_{-2}C^{eq}_I}$	⋆ (∗)	⋆	⋆	⋆	⋆ (∗)
$t \gg \frac{1}{k_{-2}C^{eq}_I}$	⋆ (∗)	⋆	⋆	–	–
$\widetilde{C}_{A_i} \approx 1$	–	⋆	–	–	–

$t \gg 1/(k_{-2}C^{eq}_I)$. Finally, according to Eq. (1.63), D^*_I can be obtained from the time evolution of the A_s concentration established in the center of the crystal after sufficient long diffusion times ($\widetilde{C}_{A_i} \approx 1$). The simulations of hybrid diffusion in dislocation-free crystals demonstrate that full information about the model parameters of the underlying diffusion-reaction mechanism are accessible from the time evolution of hybrid diffusion provided that effective short up to long diffusion times are realized experimentally.

Compared to hybrid diffusion in dislocation-free crystals, simultaneously performed experiments with highly dislocated material provide independent information about the parameters $D^*_{A_i}$ and $k_{-2}C^{eq}_I$. $\widetilde{C}_{A_s}$ profiles established in dislocated crystals for long times ($t \gg 1/(k_{-2}C^{eq}_I)$) are fully described by means of a complementary error function with $D^*_{A_i}$ (see Eq. (1.55)). The concave $\widetilde{C}_{A_s}$ curves are not controlled by I. These profiles are also very advantageous to determine the parameter $k_{-2}C^{eq}_I$ from the increase of the A_s concentration at the surface (see Eq. (1.76)), because the A_s boundary concentration is more accurately deduced from concave profiles in dislocated crystals than from convex profiles in dislocation-free material. Accordingly, $\widetilde{C}_{A_s}$ profiles in dislocated crystals can provide additional information about $k_{-2}C^{eq}_I$ and $D^*_{A_i}$. In the case hybrid diffusion in both dislocation-free and highly dislocated materials is considered, experimental $\widetilde{C}_{A_s}$ profiles in dislocation-free crystal mainly serve to determine D^*_I, $C^{eq}_{A_i}/C^{eq}_{A_s}$, and $C^{eq}_I/C^{eq}_{A_s}$. This is indicated in Table 1.1 where $k_{-2}C^{eq}_I$ and $D^*_{A_i}$ are also accessible by diffusion experiments in dislocated material. Experimental studies on this kind of hybrid diffusion in dislocation-free and highly dislocated Si are presented in Sect. 1.7.1 (see also [41]).

The numerical simulations presented in this section show that comprehensive diffusion experiments under equilibrium and non-equilibrium conditions are required to determine all model parameters of a diffusion-reaction system.

In the following the atomic mechanisms that mediate the diffusion of dopant elements are treated. Since the doping level of common p- and n-type dopants such as B, P, and As exceeds the intrinsic carrier concentration of Si and Ge, the impact of electronically extrinsic conditions on dopant diffusion must be considered. Again the classification into native-defect and foreign-atom controlled diffusion of A_s is used to describe the effective dopant diffusion coefficient.

1.6.2 *Diffusion of Dopant Atoms*

We consider that the solubility of substitutionally dissolved dopants A_s at diffusion temperature exceeds the intrinsic carrier concentration n_{in}. As a consequence, the Fermi level E_f deviates from its intrinsic position E_f^{in} and the material is denoted electronically extrinsic. Moving the Fermi level from its intrinsic position to a position under extrinsic conditions, the formation of charged point defects X is affected and therewith their equilibrium concentrations C_X^{eq} and transport capacities $C_X^{eq} D_X$ [114]. Moreover, the dopant profile creates an electric field that affects the diffusion of charged mobile defects. Accordingly, the charge states of defects must be added to the reaction mechanisms. Since dopant diffusion is mediated by indirect diffusion mechanisms such as the vacancy-, interstitialcy-, kick-out-, and dissociative mechanisms, only these mechanisms are considered exemplary in the following and analytical solutions of the underlying diffusion equations are derived for specific experimental conditions.

1.6.2.1 Reaction Mechanisms with Charge States

Indirect diffusion mechanisms that are relevant for dopant diffusion in Si are the vacancy-, interstitialcy-, dissociative-, and kick-out mechanisms that, respectively, are described by the following reactions

$$AV^{j} + (j - m - k)e^{-} \rightleftharpoons A_s^{m} + V^{k} \tag{1.80}$$

$$AI^{v} + (v - m - u)e^{-} \rightleftharpoons A_s^{m} + I^{u} \tag{1.81}$$

$$A_i^{q} + V^{k} + (q + k - m)e^{-} \rightleftharpoons A_s^{m} \tag{1.82}$$

$$A_i^{q} + (q - m - u)e^{-} \rightleftharpoons A_s^{m} + I^{u} \tag{1.83}$$

with the charge states $j, k, m, q, u, v \in \{0, \pm 1, \pm 2, \ldots\}$ and the electrons e^{-}.

In addition other reactions between point defects and defect pairs can be considered. Beside the Frenkel-pair reaction,

$$V^{k} + I^{u} + (k + u)e^{-} \rightleftharpoons \phi, \tag{1.84}$$

which describes the annihilation (formation) of charged V^k and I^u into (from) self-atoms on regular lattice sites, the reactions

$$AI^{v} + V^{k} + (v + k - m)e^{-} \rightleftharpoons A_s^{m} \tag{1.85}$$

$$AV^{j} + I^{u} + (j + u - m)e^{-} \rightleftharpoons A_s^{m} \tag{1.86}$$

describe the dopant-defect pair assisted recombination of V^k and I^u.

Generally, reactions (1.80), (1.81), (1.82), (1.83), (1.84), (1.85), and (1.86) are fast processes compared to the time scale of diffusion. For these conditions local equilibrium is established. Local equilibrium in the case of the vacancy (1.80), dissociative (1.82), and kick-out mechanisms (1.83) is characterized by

$$\frac{C_{A_s^m}\,C_{V^k}}{C_{AV^j}\,n^{(j-m-k)}} = \frac{C_{A_s^m}^{eq}\,C_{V^k}^{eq}}{C_{AV^j}^{eq}\,(n^{eq})^{(j-m-k)}} \qquad (1.87)$$

$$\frac{C_{A_s^m}}{C_{A_i^q}\,C_{V^k}\,n^{(q+k-m)}} = \frac{C_{A_s^m}^{eq}}{C_{A_i^q}^{eq}\,C_{V^k}^{eq}\,(n^{eq})^{(q+k-m)}} \qquad (1.88)$$

$$\frac{C_{A_s^m}\,C_{I^u}}{C_{A_i^q}\,n^{(q-m-u)}} = \frac{C_{A_s^m}^{eq}\,C_{I^u}^{eq}}{C_{A_i^q}^{eq}\,(n^{eq})^{(q-m-u)}}. \qquad (1.89)$$

Here C_X (C_X^{eq}) represents the (thermal equilibrium) concentration of the defect X with X $\in \{AV^j, A_i^q, A_s^m, I^u, V^k\}$. n (n^{eq}) denotes the (maximum) free electron concentration. Local equilibrium does *not* imply that the concentrations of self-interstitials and vacancies equal their thermal equilibrium values $C_{I^u}^{eq}$ and $C_{V^k}^{eq}$, locally the concentration of the native defects can deviate from thermal equilibrium but the reactions are in dynamic equilibrium. Introducing defect and electron concentrations normalized to their thermal equilibrium values, $\widetilde{C}_X = C_X/C_X^{eq}$ and $\tilde{n} = n/n^{eq}$, Eqs. (1.88) and (1.89) yield

$$\frac{\widetilde{C}_{A_s^m}}{\widetilde{C}_{A_i^q}\,\widetilde{C}_{V^k}\,\tilde{n}^{(q+k-m)}} = \frac{\widetilde{C}_{A_s^m}\,\widetilde{C}_{I^u}}{\widetilde{C}_{A_i^q}\,\tilde{n}^{(q-m-u)}} \qquad (1.90)$$

which simplifies to

$$\frac{1}{\widetilde{C}_{V^k}\widetilde{C}_{I^u}\tilde{n}^{(k+u)}} = 1 \qquad (1.91)$$

and expresses local equilibrium of the Frenkel-pair reaction (1.84). Obviously, this equilibrium state is simultaneously established when the kick-out and dissociative reactions are in local equilibrium.

Reactions (1.80), (1.81), and (1.83) are mathematically equivalent among each other. The mobile dopant-defects AV^j, AI^v, and A_i^q transform to substitutional dopants A_s^m thereby creating a native point defect (either V^k or I^u). On the other hand, reactions (1.82), (1.85), and (1.86) are also mathematically equivalent, since a mobile dopant defect and isolated native defect on the left hand side of the reactions transform to A_s^m. This similarity between the reaction mechanisms indicates that dopant diffusion alone can hardly distinguish between AV^j, AI^v, and A_i^q. However, dopant diffusion in isotope heterostructures can, in principle, differentiate between interstitial foreign-atoms and dopant-defect pairs, because the latter defects also contribute to self-diffusion.

1.6.2.2 Mathematical Formulation of Dopant Diffusion

The formation of substitutional shallow dopants to concentrations exceeding the intrinsic carrier concentration give rise to an electric field that affects the diffusion of charged mobile defects. Accordingly, a drift of charged defects in the electric field must be added to Eq. (1.1), which, in one dimension, reads

$$J_{\mathrm{X}} = -D_{\mathrm{X}} \frac{\partial C_{\mathrm{X}}}{\partial x} + \mu_{\mathrm{X}} C_{\mathrm{X}} \epsilon(x). \tag{1.92}$$

μ and ϵ denote the mobility and the electric field, respectively. Taking into account that the electric field is the derivative of the potential ψ

$$\epsilon(x) = -\frac{\partial \psi(x)}{\partial x} \tag{1.93}$$

and that the free electron concentration is given by

$$n(x) = \mathrm{N_C} \exp\left(\frac{E_f - E_{\mathrm{C}} + e\psi(x)}{kT} \right) \tag{1.94}$$

with e being the elementary charge, the electric field can be written as

$$\epsilon(x) = -\frac{kT}{e} \frac{1}{n(x)} \frac{\partial n(x)}{\partial x}. \tag{1.95}$$

Substituting Eq. (1.95) in Eq. (1.92) and using the Einstein relation

$$\mu = \mathrm{z}\, e \frac{D_{\mathrm{X}}}{kT} \tag{1.96}$$

with the charge state z of the diffusing species, we obtain

$$J_{\mathrm{X}} = -D_{\mathrm{X}} \frac{\partial C_{\mathrm{X}}}{\partial x} - \mathrm{z}\, C_{\mathrm{X}} D_{\mathrm{X}} \frac{1}{n(x)} \frac{\partial n(x)}{\partial x}. \tag{1.97}$$

In the case that the electric field is caused by acceptors, the flux equation reads

$$J_{\mathrm{X}} = -D_{\mathrm{X}} \frac{\partial C_{\mathrm{X}}}{\partial x} + \mathrm{z}\, C_{\mathrm{X}} D_{\mathrm{X}} \frac{1}{p(x)} \frac{\partial p(x)}{\partial x} \tag{1.98}$$

where p denotes the free hole concentration. Additional terms enter Eq. (1.92) when besides chemical and electrical forces also mechanical strain and/or hydrostatic pressure affect the flux of a point defect.

In the following dopant diffusion via the vacancy and dissociative mechanisms (1.80) and (1.82) are treated representative for reactions (1.80) and (1.86) in more detail. Starting from the full differential equation system, useful approximations are presented that describe the relation between the shape of the diffusion profile and the charge states of the point defects involved in the diffusion process. Understanding this relation is very helpful to identify the underlying mechanisms of dopant diffusion.

Dopant Diffusion via the Vacancy Mechanism

The vacancy mechanism is given by reaction (1.80) in a generalized form. Adding the forward and backward rate constants which are denoted k_+ and k_-, respectively, reaction (1.80) reads

$$AV^{\mathrm{j}} + (\mathrm{j}-\mathrm{m}-\mathrm{k})\mathrm{e}^- \underset{k_-}{\overset{k_+}{\rightleftharpoons}} \mathrm{A}_\mathrm{s}^\mathrm{m} + V^\mathrm{k}. \tag{1.99}$$

Taking into account Eq. (1.97), the following three coupled partial differential equations describe the diffusion and reaction of the point defects involved in reaction (1.99)

$$\frac{\partial C_{\mathrm{A_s^m}}}{\partial t} = \frac{\partial}{\partial x}\left(D_{\mathrm{A_s}^\mathrm{m}}\frac{\partial C_{\mathrm{A_s}^\mathrm{m}}}{\partial x} + \mathrm{m}\frac{C_{\mathrm{A_s}^\mathrm{m}} D_{\mathrm{A_s}^\mathrm{m}}}{n(x)}\frac{\partial n(x)}{\partial x}\right) + k_+ C_{\mathrm{AV^j}}\, C_\mathrm{o}\, n^{(\mathrm{j-m-k})} - k_- C_{\mathrm{A_s^m}} C_{V^\mathrm{k}}\, n_\mathrm{in}^{(\mathrm{j-m-k})} \tag{1.100}$$

$$\frac{\partial C_{\mathrm{AV^j}}}{\partial t} = \frac{\partial}{\partial x}\left(D_{\mathrm{AV^j}}\frac{\partial C_{\mathrm{AV^j}}}{\partial x} + \mathrm{j}\frac{C_{\mathrm{AV^j}} D_{\mathrm{AV^j}}}{n(x)}\frac{\partial n(x)}{\partial x}\right) - k_+ C_{\mathrm{AV^j}}\, C_\mathrm{o}\, n^{(\mathrm{j-m-k})} + k_- C_{\mathrm{A_s^m}} C_{V^\mathrm{k}}\, n_\mathrm{in}^{(\mathrm{j-m-k})} \tag{1.101}$$

$$\frac{\partial C_{V^\mathrm{k}}}{\partial t} = \frac{\partial}{\partial x}\left(D_{V^\mathrm{k}}\frac{\partial C_{V^\mathrm{k}}}{\partial x} + \mathrm{k}\frac{C_{V^\mathrm{k}} D_{V^\mathrm{k}}}{n(x)}\frac{\partial n(x)}{\partial x}\right) + k_+ C_{\mathrm{AV^j}}\, C_\mathrm{o}\, n^{(\mathrm{j-m-k})} - k_- C_{\mathrm{A_s^m}} C_{V^\mathrm{k}}\, n_\mathrm{in}^{(\mathrm{j-m-k})} \tag{1.102}$$

where C_X (D_X) with $\mathrm{X} \in \{\mathrm{A_s^m}, \mathrm{A}V^\mathrm{j}, V^\mathrm{k}\}$ is the concentration (diffusion coefficient) of the respective point defect. $D_{\mathrm{A_s}^\mathrm{m}}$ equals the diffusivity of the substitutionally dissolved foreign atom via direct exchange. $C_\mathrm{o}(= 5\times10^{22}\ \mathrm{cm}^{-3}$ for Si) is the number density of substitutional lattice sites. The first term on the right hand side of reactions (1.100), (1.101), and (1.102) describes the change of the concentration of the point defect via diffusion and drift. The second term represents the formation/annihilation of the point defect via reaction (1.99).

Introducing normalized concentrations $\widetilde{C}_{\mathrm{X}} = C_{\mathrm{X}}/C_{\mathrm{X}}^{\mathrm{eq}}$ and $\tilde{n} = n/n^{\mathrm{eq}}$, reduced diffusivities $D_{\mathrm{X}}^{*} = C_{\mathrm{X}}^{\mathrm{eq}} D_{\mathrm{X}}/C_{\mathrm{A_s^m}}^{\mathrm{eq}}$, and the relationship between the rate constants k_{+} and k_{-}

$$\frac{k_{+}}{k_{-}} = \frac{C_{\mathrm{A_s^m}}^{\mathrm{eq}}\, C_{V^{\mathrm{k}}}^{\mathrm{eq}}\, n_{\mathrm{in}}^{(\mathrm{j-m-k})}}{C_{\mathrm{AV^j}}^{\mathrm{eq}}\, C_{\mathrm{o}}\, (n^{\mathrm{eq}})^{(\mathrm{j-m-k})}}, \tag{1.103}$$

due to the law of mass action, the differential equations (1.100), (1.101), and (1.102) read

$$\frac{\partial \widetilde{C}_{\mathrm{A_s^m}}}{\partial t} = \frac{\partial}{\partial x}\left(D_{\mathrm{A_s m}} \frac{\partial \widetilde{C}_{\mathrm{A_s m}}}{\partial x} + \mathrm{m}\frac{\widetilde{C}_{\mathrm{A_s m}} D_{\mathrm{A_s m}}}{\tilde{n}(x)} \frac{\partial \tilde{n}(x)}{\partial x}\right) + k_{-} C_{V^{\mathrm{k}}}^{\mathrm{eq}}\, n_{\mathrm{in}}^{(\mathrm{j-m-k})} \left(\widetilde{C}_{\mathrm{AV^j}}\, \tilde{n}^{(\mathrm{j-m-k})} - \widetilde{C}_{\mathrm{A_s^m}} \widetilde{C}_{V^{\mathrm{k}}}\right) \tag{1.104}$$

$$\frac{C_{\mathrm{AV^j}}^{\mathrm{eq}}}{C_{\mathrm{A_s^m}}^{\mathrm{eq}}} \frac{\partial \widetilde{C}_{\mathrm{AV^j}}}{\partial t} = \frac{\partial}{\partial x}\left(D_{\mathrm{AV^j}}^{*} \frac{\partial \widetilde{C}_{\mathrm{AV^j}}}{\partial x} + \mathrm{j}\frac{\widetilde{C}_{\mathrm{AV^j}} D_{\mathrm{AV^j}}^{*}}{\tilde{n}(x)} \frac{\partial \tilde{n}(x)}{\partial x}\right) - k_{-} C_{V^{\mathrm{k}}}^{\mathrm{eq}}\, n_{\mathrm{in}}^{(\mathrm{j-m-k})} \left(\widetilde{C}_{\mathrm{AV^j}}\, \tilde{n}^{(\mathrm{j-m-k})} - \widetilde{C}_{\mathrm{A_s^m}} \widetilde{C}_{V^{\mathrm{k}}}\right) \tag{1.105}$$

$$\frac{C_{V^{\mathrm{k}}}^{\mathrm{eq}}}{C_{\mathrm{A_s^m}}^{\mathrm{eq}}} \frac{\partial \widetilde{C}_{V^{\mathrm{k}}}}{\partial t} = \frac{\partial}{\partial x}\left(D_{V^{\mathrm{k}}}^{*} \frac{\partial \widetilde{C}_{V^{\mathrm{k}}}}{\partial x} + \mathrm{k}\frac{\widetilde{C}_{V^{\mathrm{k}}} D_{V^{\mathrm{k}}}^{*}}{\tilde{n}(x)} \frac{\partial \tilde{n}(x)}{\partial x}\right) + k_{-} C_{V^{\mathrm{k}}}^{\mathrm{eq}}\, n_{\mathrm{in}}^{(\mathrm{j-m-k})} \left(\widetilde{C}_{\mathrm{AV^j}}\, \tilde{n}^{(\mathrm{j-m-k})} - \widetilde{C}_{\mathrm{A_s^m}} \widetilde{C}_{V^{\mathrm{k}}}\right). \tag{1.106}$$

This system of differential equations describes the diffusion of A via the vacancy mechanism when $\mathrm{A_s^m}$ acts as donor. For $\mathrm{A_s^m}$ being an acceptor, $\tilde{n}$ and $\frac{1}{\tilde{n}}\frac{\partial \tilde{n}}{\partial x}$ are replaced by means of $np = n_{\mathrm{in}}^2$ with $1/\tilde{p}$ and $-\frac{1}{\tilde{p}}\frac{\partial \tilde{p}}{\partial x}$, respectively.

The electron and hole concentrations are related via the charge neutrality equation

$$n = p + \mathrm{m}C_{\mathrm{A_s^m}} + \mathrm{k}C_{V^{\mathrm{k}}} + \mathrm{j}C_{\mathrm{AV^j}} \tag{1.107}$$

with the concentrations of the point defects. For a substitutional donor $\mathrm{A_s^m}$ with $\mathrm{m} \in \{+1, +2, \ldots\}$, we obtain

$$\tilde{n}(x) = \frac{\left(\mathrm{m}\widetilde{C}_{\mathrm{A_s^m}} + \mathrm{j}\frac{C_{\mathrm{AV^j}}^{\mathrm{eq}}}{C_{\mathrm{A_s^m}}^{\mathrm{eq}}}\widetilde{C}_{\mathrm{AV^j}} + \mathrm{k}\frac{C_{V^{\mathrm{k}}}^{\mathrm{eq}}}{C_{\mathrm{A_s^m}}^{\mathrm{eq}}}\widetilde{C}_{V^{\mathrm{k}}}\right) + \sqrt{\left(\mathrm{m}\widetilde{C}_{\mathrm{A_s^m}} + \mathrm{j}\frac{C_{\mathrm{AV^j}}^{\mathrm{eq}}}{C_{\mathrm{A_s^m}}^{\mathrm{eq}}}\widetilde{C}_{\mathrm{AV^j}} + \mathrm{k}\frac{C_{V^{\mathrm{k}}}^{\mathrm{eq}}}{C_{\mathrm{A_s^m}}^{\mathrm{eq}}}\widetilde{C}_{V^{\mathrm{k}}}\right)^2 + 4\tilde{n}_{\mathrm{in}}^2}}{\left(\mathrm{m} + \mathrm{j}\frac{C_{\mathrm{AV^j}}^{\mathrm{eq}}}{C_{\mathrm{A_s^m}}^{\mathrm{eq}}} + \mathrm{k}\frac{C_{V^{\mathrm{k}}}^{\mathrm{eq}}}{C_{\mathrm{A_s^m}}^{\mathrm{eq}}}\right) + \sqrt{\left(\mathrm{m} + \mathrm{j}\frac{C_{\mathrm{AV^j}}^{\mathrm{eq}}}{C_{\mathrm{A_s^m}}^{\mathrm{eq}}} + \mathrm{k}\frac{C_{V^{\mathrm{k}}}^{\mathrm{eq}}}{C_{\mathrm{A_s^m}}^{\mathrm{eq}}}\right)^2 + 4\tilde{n}_{\mathrm{in}}^2}} \tag{1.108}$$

and

$$\frac{1}{\tilde{n}(x)}\frac{\partial \tilde{n}(x)}{\partial x} = \frac{1}{\sqrt{\left(\mathrm{m}\widetilde{C}_{\mathrm{A}_\mathrm{s}^\mathrm{m}} + \mathrm{j}\frac{C^{\mathrm{eq}}_{\mathrm{AVj}}}{C^{\mathrm{eq}}_{\mathrm{A}_\mathrm{s}^\mathrm{m}}}\widetilde{C}_{\mathrm{AVj}} + \mathrm{k}\frac{C^{\mathrm{eq}}_{V\mathrm{k}}}{C^{\mathrm{eq}}_{\mathrm{A}_\mathrm{s}^\mathrm{m}}}\widetilde{C}_{V\mathrm{k}}\right)^2 + 4\tilde{n}_{\mathrm{in}}^2}} \times \frac{\partial\left(\mathrm{m}\widetilde{C}_{\mathrm{A}_\mathrm{s}^\mathrm{m}} + \mathrm{j}\frac{C^{\mathrm{eq}}_{\mathrm{AVj}}}{C^{\mathrm{eq}}_{\mathrm{A}_\mathrm{s}^\mathrm{m}}}\widetilde{C}_{\mathrm{AVj}} + \mathrm{k}\frac{C^{\mathrm{eq}}_{V\mathrm{k}}}{C^{\mathrm{eq}}_{\mathrm{A}_\mathrm{s}^\mathrm{m}}}\widetilde{C}_{V\mathrm{k}}\right)}{\partial x} \tag{1.109}$$

with $\tilde{n}_{\mathrm{in}} = n_{\mathrm{in}}/C^{\mathrm{eq}}_{\mathrm{A}_\mathrm{s}^\mathrm{m}}$. Considering a substitutional acceptor $\mathrm{A}_\mathrm{s}^\mathrm{m}$ with $\mathrm{m} \in \{-1, -2, \ldots\}$ we get

$$\tilde{p}(x) = \frac{\left(-\mathrm{m}\widetilde{C}_{\mathrm{A}_\mathrm{s}^\mathrm{m}} - \mathrm{j}\frac{C^{\mathrm{eq}}_{\mathrm{AVj}}}{C^{\mathrm{eq}}_{\mathrm{A}_\mathrm{s}^\mathrm{m}}}\widetilde{C}_{\mathrm{AVj}} - \mathrm{k}\frac{C^{\mathrm{eq}}_{V\mathrm{k}}}{C^{\mathrm{eq}}_{\mathrm{A}_\mathrm{s}^\mathrm{m}}}\widetilde{C}_{V\mathrm{k}}\right) + \sqrt{\left(-\mathrm{m}\widetilde{C}_{\mathrm{A}_\mathrm{s}^\mathrm{m}} - \mathrm{j}\frac{C^{\mathrm{eq}}_{\mathrm{AVj}}}{C^{\mathrm{eq}}_{\mathrm{A}_\mathrm{s}^\mathrm{m}}}\widetilde{C}_{\mathrm{AVj}} - \mathrm{k}\frac{C^{\mathrm{eq}}_{V\mathrm{k}}}{C^{\mathrm{eq}}_{\mathrm{A}_\mathrm{s}^\mathrm{m}}}\widetilde{C}_{V\mathrm{k}}\right)^2 + 4\tilde{n}_{\mathrm{in}}^2}}{\left(-\mathrm{m} - \mathrm{j}\frac{C^{\mathrm{eq}}_{\mathrm{AVj}}}{C^{\mathrm{eq}}_{\mathrm{A}_\mathrm{s}^\mathrm{m}}} - \mathrm{k}\frac{C^{\mathrm{eq}}_{V\mathrm{k}}}{C^{\mathrm{eq}}_{\mathrm{A}_\mathrm{s}^\mathrm{m}}}\right) + \sqrt{\left(-\mathrm{m} - \mathrm{j}\frac{C^{\mathrm{eq}}_{\mathrm{AVj}}}{C^{\mathrm{eq}}_{\mathrm{A}_\mathrm{s}^\mathrm{m}}} - \mathrm{k}\frac{C^{\mathrm{eq}}_{V\mathrm{k}}}{C^{\mathrm{eq}}_{\mathrm{A}_\mathrm{s}^\mathrm{m}}}\right)^2 + 4\tilde{n}_{\mathrm{in}}^2}} \tag{1.110}$$

and

$$\frac{1}{\tilde{p}(x)}\frac{\partial \tilde{p}(x)}{\partial x} = \frac{1}{\sqrt{\left(-\mathrm{m}\widetilde{C}_{\mathrm{A}_\mathrm{s}^\mathrm{m}} - \mathrm{j}\frac{C^{\mathrm{eq}}_{\mathrm{AVj}}}{C^{\mathrm{eq}}_{\mathrm{A}_\mathrm{s}^\mathrm{m}}}\widetilde{C}_{\mathrm{AVj}} - \mathrm{k}\frac{C^{\mathrm{eq}}_{V\mathrm{k}}}{C^{\mathrm{eq}}_{\mathrm{A}_\mathrm{s}^\mathrm{m}}}\widetilde{C}_{V\mathrm{k}}\right)^2 + 4\tilde{n}_{\mathrm{in}}^2}} \times \frac{\partial\left(-\mathrm{m}\widetilde{C}_{\mathrm{A}_\mathrm{s}^\mathrm{m}} - \mathrm{j}\frac{C^{\mathrm{eq}}_{\mathrm{AVj}}}{C^{\mathrm{eq}}_{\mathrm{A}_\mathrm{s}^\mathrm{m}}}\widetilde{C}_{\mathrm{AVj}} - \mathrm{k}\frac{C^{\mathrm{eq}}_{V\mathrm{k}}}{C^{\mathrm{eq}}_{\mathrm{A}_\mathrm{s}^\mathrm{m}}}\widetilde{C}_{V\mathrm{k}}\right)}{\partial x}. \tag{1.111}$$

Equations (1.104), (1.105), and (1.106) together with Eq. (1.109) (Eq. (1.111)) represent the full set of equations to calculate diffusion profiles of donors (acceptors) $\mathrm{A}_\mathrm{s}^\mathrm{m}$ and other point defects involved in reaction (1.99) for specific initial and boundary conditions.

In the following approximations of the differential equation system (1.104), (1.105), and (1.106) are considered that lead to a single differential equation of the form

$$\frac{\partial C_\mathrm{X}}{\partial t} - \frac{\partial}{\partial x}\left(D_\mathrm{X}\frac{\partial C_\mathrm{X}}{\partial x}\right) = G_\mathrm{X}. \tag{1.112}$$

with $G_\mathrm{X} = 0$. Two different modes of dopant diffusion are distinguished, i.e., the native-defect and foreign-atom controlled modes. Each mode predicts characteristic profiles for $\mathrm{A}_\mathrm{s}^\mathrm{m}$. The characteristic shape of the profiles reflects the charge states of the point defects controlling the diffusion process.

Native-Defect Controlled Dopant Diffusion via the Vacancy Mechanism

The vacancy-controlled mode of dopant diffusion is established when the relationship $D^*_{\mathrm{AV^j}} \gg D^*_{V^\mathrm{k}}$ holds. The higher transport capacity $C^{\mathrm{eq}}_{\mathrm{AV^j}} D_{\mathrm{AV^j}}$ of the dopant-defect pair compared to $C^{\mathrm{eq}}_{V^\mathrm{k}} D_{V^\mathrm{k}}$ leads to an almost homogeneous distribution of the dopant-vacancy pairs after sufficiently long diffusion times, i.e., $\widetilde{C}_{\mathrm{AV^j}} \approx 1$. Accordingly, Eq. (1.105) can be neglected. The contribution of the direct diffusion of $\mathrm{A^m_s}$ to the diffusion of the dopant A is generally small compared to the indirect diffusion via $\mathrm{A}V$ pairs. Taking the difference between Eqs. (1.104) and (1.106) and assuming $C^{\mathrm{eq}}_{V^\mathrm{k}} \ll C^{\mathrm{eq}}_{\mathrm{A^m_s}}$ we obtain

$$\frac{\partial \widetilde{C}_{\mathrm{A^m_s}}}{\partial t} \approx -\frac{\partial}{\partial x}\left(D^*_{V^\mathrm{k}} \frac{\partial \widetilde{C}_{V^\mathrm{k}}}{\partial x} + \mathrm{k} \frac{\widetilde{C}_{V^\mathrm{k}} D^*_{V^\mathrm{k}}}{\tilde{n}(x)} \frac{\partial \tilde{n}(x)}{\partial x} \right). \tag{1.113}$$

For local equilibrium of reaction (1.99),

$$\frac{\widetilde{C}_{\mathrm{A^m_s}} \widetilde{C}_{V^\mathrm{k}}}{\widetilde{C}_{\mathrm{AV^j}}\ \tilde{n}^{(\mathrm{j-m-k})}} = 1, \tag{1.114}$$

and of reaction $\mathrm{A}V^\mathrm{j} + \mathrm{j}e^- \rightleftharpoons \mathrm{A}V^0$,

$$\frac{\widetilde{C}_{\mathrm{AV^0}}}{\widetilde{C}_{\mathrm{AV^j}}\ \tilde{n}^\mathrm{j}} = 1, \tag{1.115}$$

Eqs. (1.114) and (1.115) yield

$$\widetilde{C}_{V^\mathrm{k}} = \frac{\widetilde{C}_{\mathrm{AV^0}} \tilde{n}^{(-\mathrm{m-k})}}{\widetilde{C}_{\mathrm{A^m_s}}} \approx \frac{\tilde{n}^{(-\mathrm{m-k})}}{\widetilde{C}_{\mathrm{A^m_s}}} \tag{1.116}$$

because $\widetilde{C}_{\mathrm{AV^0}} \approx 1$. For high donor concentrations, i.e., $C^{\mathrm{eq}}_{\mathrm{A^m_s}}$ exceeds n_{in}, $\tilde{n} \approx \widetilde{C}_{\mathrm{A^m_s}}$ holds and Eq. (1.113) transforms with Eq. (1.116) to a differential equation of the form

$$\frac{\partial \widetilde{C}_{\mathrm{A^m_s}}}{\partial t} - \frac{\partial}{\partial x} D^{\mathrm{eff}}_{\mathrm{A^m_s}} \frac{\partial \widetilde{C}_{\mathrm{A^m_s}}}{\partial x} = 0 \tag{1.117}$$

with an effective diffusivity of $\mathrm{A^m_s}$ ($\mathrm{m} \in \{+1, +2, \ldots\}$) given by

$$D^{\mathrm{eff}}_{\mathrm{A^m_s}} = (\mathrm{m}+1) D^*_{V^\mathrm{k}} (\widetilde{C}_{\mathrm{A^m_s}})^{-\mathrm{m-k-2}}. \tag{1.118}$$

For high concentrations of acceptors $\mathrm{A^m_s}$ ($\mathrm{m} \in \{-1, -2, \ldots\}$) with $\tilde{p} \approx \widetilde{C}_{\mathrm{A^m_s}}$, Eq. (1.117) with

$$D^{\mathrm{eff}}_{\mathrm{A^m_s}} = (-\mathrm{m}+1) D^*_{V^\mathrm{k}} (\widetilde{C}_{\mathrm{A^m_s}})^{\mathrm{m+k-2}} \tag{1.119}$$

is obtained. Equations (1.118) and (1.119) reveal that in case of dopant diffusion via the vacancy mechanism the effective dopant diffusion coefficient $D^{\mathrm{eff}}_{\mathrm{A_s^m}}$ is interrelated to the reduced diffusion coefficient $D^*_{V^\mathrm{k}} = C^{\mathrm{eq}}_{V^\mathrm{k}} D_{V^\mathrm{k}} / C^{\mathrm{eq}}_{\mathrm{A_s^m}}$ and thus to the transport capacity of V^k. This interrelation reflects the V-controlled mode of dopant diffusion.

Foreign-Atom Controlled Dopant Diffusion via the Vacancy Mechanism

The foreign-atom controlled mode of dopant diffusion is established when the relationship $D^*_{\mathrm{A}V^\mathrm{j}} \ll D^*_{V^\mathrm{k}}$ holds. The higher transport capacity of V^k compared to that of $\mathrm{A}V^\mathrm{j}$ pairs leads to an almost homogeneous distribution of V^k after sufficiently long times, i.e., $\widetilde{C}_{V^\mathrm{k}} \approx 1$. Accordingly, Eq. (1.106) can be neglected. Taking into account that the direct diffusion of $\mathrm{A_s^m}$ is negligible compared to the indirect diffusion of $\mathrm{A_s^m}$ via $\mathrm{A}V$ pairs and that $C^{\mathrm{eq}}_{\mathrm{A}V^\mathrm{j}} \ll C^{\mathrm{eq}}_{\mathrm{A_s^m}}$ holds, the sum of Eqs. (1.104) and (1.105) yields

$$\frac{\partial \widetilde{C}_{\mathrm{A_s^m}}}{\partial t} \approx \frac{\partial}{\partial x}\left(D^*_{\mathrm{A}V^\mathrm{j}} \frac{\partial \widetilde{C}_{\mathrm{A}V^\mathrm{j}}}{\partial x} + \mathrm{j} \frac{\widetilde{C}_{\mathrm{A}V^\mathrm{j}} D^*_{\mathrm{A}V^\mathrm{j}}}{\tilde{n}(x)} \frac{\partial \tilde{n}(x)}{\partial x} \right). \tag{1.120}$$

Assuming local equilibrium of reaction (1.99) and of the reaction $V^\mathrm{k} + \mathrm{k}e^- \rightleftharpoons V^0$, which is expressed by Eq. (1.114) and

$$\frac{\widetilde{C}_{V^0}}{\widetilde{C}_{V^\mathrm{k}}\, \tilde{n}^\mathrm{k}} = 1, \tag{1.121}$$

respectively, one gets

$$\widetilde{C}_{\mathrm{A}V^\mathrm{j}} \approx \widetilde{C}_{\mathrm{A_s^m}} \tilde{n}^{(\mathrm{m-j})} = (\widetilde{C}_{\mathrm{A_s^m}})^{(\mathrm{m-j+1})}. \tag{1.122}$$

For donors $\mathrm{A_s^m}$ ($\mathrm{m} = +1, +2, \ldots$) with $\tilde{n} \approx \widetilde{C}_{\mathrm{A_s^m}}$, Eq. (1.120) transforms to Eq. (1.117) with

$$D^{\mathrm{eff}}_{\mathrm{A_s^m}} = (\mathrm{m}+1) D^*_{\mathrm{A}V^\mathrm{j}} (\widetilde{C}_{\mathrm{A_s^m}})^{\mathrm{m-j}}. \tag{1.123}$$

For acceptors $\mathrm{A_s^m}$ ($\mathrm{m} \in \{-1, -2, \ldots\}$) with $\tilde{p} \approx \widetilde{C}_{\mathrm{A_s^m}}$, the diffusion coefficient in Eq. (1.117) is given by

$$D^{\mathrm{eff}}_{\mathrm{A_s^m}} = (-\mathrm{m}+1) D^*_{\mathrm{A}V^\mathrm{j}} (\widetilde{C}_{\mathrm{A_s^m}})^{-\mathrm{m+j}}. \tag{1.124}$$

Equations (1.123) and (1.124) reveal that in case of the foreign-atom controlled mode of dopant diffusion via the vacancy mechanism the effective dopant diffusion coefficient $D^{\mathrm{eff}}_{\mathrm{A_s^m}}$ is interrelated to the reduced diffusion coefficient $D^*_{\mathrm{A}V^\mathrm{j}} = C^{\mathrm{eq}}_{\mathrm{A}V^\mathrm{j}} D_{\mathrm{A}V^\mathrm{j}} / C^{\mathrm{eq}}_{\mathrm{A_s^m}}$ and thus to the transport capacity of $\mathrm{A}V^j$. This interrelation reflects the $\mathrm{A}V$-controlled mode of dopant diffusion.

1.6.2.3 Dopant Diffusion via the Dissociative Mechanism

The dissociative mechanism is representative for the mechanisms (1.82), (1.85) and (1.86), which all possess a similar reaction scheme. The partial differential equation system for modeling dopant diffusion via the dissociative mechanism is easily translated to the equation system for modeling diffusion via the dopant-defect pair assisted recombination mechanism (1.85) (mechanism (1.86)) by replacing the defect(s) $\mathrm{A_i^q}$ (A_i^q, V^k) with AI^v (AV^j, I^u).

Introducing forward and backward rate constants k_f and k_b, respectively, the dissociative mechanism reads

$$\mathrm{A_i^q} + V^{\mathrm{k}} + (\mathrm{q}+\mathrm{k}-\mathrm{m})e^- \underset{k_b}{\overset{k_f}{\rightleftharpoons}} \mathrm{A_s^m}. \tag{1.125}$$

k_f and k_b are related via the law of mass action according to

$$\frac{k_f}{k_b} = \frac{C^{\mathrm{eq}}_{\mathrm{A_s^m}}\, C'_{\mathrm{o}}\, n_{\mathrm{in}}^{(\mathrm{q+k-m})}}{C^{\mathrm{eq}}_{\mathrm{A_i^q}} C^{\mathrm{eq}}_{V^{\mathrm{k}}} (n^{\mathrm{eq}})^{(\mathrm{q+k-m})}}, \tag{1.126}$$

where C'_{o} is the number density of interstitial sites. Taking into account normalized concentrations and reduced diffusivities, the differential equation system for modeling dopant diffusion via the dissociative mechanism (1.125) is

$$\begin{aligned}\frac{\partial \widetilde{C}_{\mathrm{A_s^m}}}{\partial t} = \frac{\partial}{\partial x}\left(D_{\mathrm{A_s{}^m}}\frac{\partial \widetilde{C}_{\mathrm{A_s{}^m}}}{\partial x} + \mathrm{m}\frac{\widetilde{C}_{\mathrm{A_s{}^m}} D_{\mathrm{A_s{}^m}}}{\tilde{n}(x)}\frac{\partial \tilde{n}(x)}{\partial x}\right) + \\ k_b\, C'_{\mathrm{o}}\, n_{\mathrm{in}}^{(\mathrm{q+k-m})}\left(\widetilde{C}_{\mathrm{A_i^q}}\widetilde{C}_{V^{\mathrm{k}}}\,\tilde{n}^{(\mathrm{q+k-m})} - \widetilde{C}_{\mathrm{A_s^m}}\right)\end{aligned} \tag{1.127}$$

$$\begin{aligned}\frac{C^{\mathrm{eq}}_{\mathrm{A_i^q}}}{C^{\mathrm{eq}}_{\mathrm{A_s^m}}}\frac{\partial \widetilde{C}_{\mathrm{A_i^q}}}{\partial t} = \frac{\partial}{\partial x}\left(D^*_{\mathrm{A_i^q}}\frac{\partial \widetilde{C}_{\mathrm{A_i^q}}}{\partial x} + \mathrm{q}\frac{\widetilde{C}_{\mathrm{A_i^q}} D^*_{\mathrm{A_i^q}}}{\tilde{n}(x)}\frac{\partial \tilde{n}(x)}{\partial x}\right) - \\ k_b\, C'_{\mathrm{o}}\, n_{\mathrm{in}}^{(\mathrm{q+k-m})}\left(\widetilde{C}_{\mathrm{A_i^q}}\widetilde{C}_{V^{\mathrm{k}}}\,\tilde{n}^{(\mathrm{q+k-m})} - \widetilde{C}_{\mathrm{A_s^m}}\right)\end{aligned} \tag{1.128}$$

$$\begin{aligned}\frac{C^{\mathrm{eq}}_{V^{\mathrm{k}}}}{C^{\mathrm{eq}}_{\mathrm{A_s^m}}}\frac{\partial \widetilde{C}_{V^{\mathrm{k}}}}{\partial t} = \frac{\partial}{\partial x}\left(D^*_{V^{\mathrm{k}}}\frac{\partial \widetilde{C}_{V^{\mathrm{k}}}}{\partial x} + \mathrm{k}\frac{\widetilde{C}_{V^{\mathrm{k}}} D^*_{V^{\mathrm{k}}}}{\tilde{n}(x)}\frac{\partial \tilde{n}(x)}{\partial x}\right) - \\ k_b\, C'_{\mathrm{o}}\, n_{\mathrm{in}}^{(\mathrm{q+k-m})}\left(\widetilde{C}_{\mathrm{A_i^q}}\widetilde{C}_{V^{\mathrm{k}}}\,\tilde{n}^{(\mathrm{q+k-m})} - \widetilde{C}_{\mathrm{A_s^m}}\right).\end{aligned} \tag{1.129}$$

The diffusion of a donor $\mathrm{A_s^m}$ (m$\in$ $\{+1, +2, \ldots\}$) is described by Eqs. (1.127), (1.128), and (1.129) and by equations for $\tilde{n}$ and $\frac{1}{\tilde{n}(x)}\frac{\partial \tilde{n}(x)}{\partial x}$ that are similar to

Eqs. (1.108) and (1.109). For an acceptor $\mathrm{A_s^m}$ (m$\in$ $\{-1,-2,\ldots\}$), $\tilde{n}$ and $\frac{1}{\tilde{n}}\frac{\partial \tilde{n}}{\partial x}$ are replaced with $1/\tilde{p}$ and $-\frac{1}{\tilde{p}}\frac{\partial \tilde{p}}{\partial x}$, respectively, and substituted by expressions similar to Eqs. (1.110) and (1.111).

Native-Defect Controlled Dopant Diffusion via the Dissociative Mechanism

The native-defect controlled mode of dopant diffusion holds for $D^*_{\mathrm{A_i^q}} \gg D^*_{V^\mathrm{k}}$. The higher transport capacity $C^{\mathrm{eq}}_{\mathrm{A_i^q}} D_{\mathrm{A_i^q}}$ compared to $C^{\mathrm{eq}}_{V^\mathrm{k}} D_{V^\mathrm{k}}$ leads to $\widetilde{C}_{\mathrm{A_i^q}} \approx 1$ after sufficiently long times. Under this condition we can neglect Eq. (1.128). Taking the sum of Eqs. (1.127) and (1.129) and assuming that the diffusion of $\mathrm{A_s^m}$ via direct exchange is small compared to the indirect diffusion of $\mathrm{A_s^m}$ via $\mathrm{A_i^q}$ and that $C^{\mathrm{eq}}_{V^\mathrm{k}} \ll C^{\mathrm{eq}}_{\mathrm{A_s^m}}$ is fulfilled, we obtain

$$\frac{\partial \widetilde{C}_{\mathrm{A_s^m}}}{\partial t} \approx \frac{\partial}{\partial x}\left(D^*_{V^\mathrm{k}} \frac{\partial \widetilde{C}_{V^\mathrm{k}}}{\partial x} + \mathrm{k}\frac{\widetilde{C}_{V^\mathrm{k}} D^*_{V^\mathrm{k}}}{\tilde{n}(x)}\frac{\partial \tilde{n}(x)}{\partial x}\right). \tag{1.130}$$

Considering local equilibrium of reaction (1.125) and $\mathrm{A_i^q} + \mathrm{q}e^- \rightleftharpoons \mathrm{A_i^0}$,

$$\frac{\widetilde{C}_{\mathrm{A_s^m}}}{\widetilde{C}_{\mathrm{A_i^q}} \widetilde{C}_{V^\mathrm{k}}\ \tilde{n}^{(\mathrm{q+k-m})}} = 1 \tag{1.131}$$

$$\frac{\widetilde{C}_{\mathrm{A_i^0}}}{\widetilde{C}_{\mathrm{A_i^q}}\ \tilde{n}^{\mathrm{q}}} = 1, \tag{1.132}$$

we obtain

$$\widetilde{C}_{V^\mathrm{k}} \approx \widetilde{C}_{\mathrm{A_s^m}}\, \tilde{n}^{(\mathrm{m-k})}. \tag{1.133}$$

For donors $\mathrm{A_s^m}$ (m$\in$ $\{+1,+2,\ldots\}$) with $\tilde{n}(x) \approx \widetilde{C}_{\mathrm{A_s^m}}$, Eq. (1.130) transforms to Eq. (1.117) with

$$D^{\mathrm{eff}}_{\mathrm{A_s^m}} = (\mathrm{m}+1) D^*_{V^\mathrm{k}} (\widetilde{C}_{\mathrm{A_s^m}})^{-\mathrm{k}}. \tag{1.134}$$

For acceptors $\mathrm{A_s^m}$ (m $\in \{-1,-2,\ldots\}$) with $\tilde{p} \approx \widetilde{C}_{\mathrm{A_s^m}}$, Eq. (1.117) with

$$D^{\mathrm{eff}}_{\mathrm{A_s^m}} = (-\mathrm{m}+1) D^*_{V^\mathrm{k}} (\widetilde{C}_{\mathrm{A_s^m}})^{-\mathrm{m}+\mathrm{k}} \tag{1.135}$$

is obtained.

Foreign-Atom Controlled Dopant Diffusion via the Dissociative Mechanism

The foreign-atom controlled mode of dopant diffusion via the dissociative mechanism is established in the case when $D^*_{A_i^q} \ll D^*_{V^k}$ holds. Considering sufficiently long times, thermal equilibrium of V^k is established, i.e., $\widetilde{C}_{V^k} \approx 1$. Thus we can neglect Eq. (1.129). Taking the sum of Eqs. (1.127) and (1.128) and assuming $C^{eq}_{A_i^q} \ll C^{eq}_{A_s^m}$ and $D_{A_s^m} (= D^{ex}_{A_s^m}) \approx 0$, we obtain

$$\frac{\partial \widetilde{C}_{A_s^m}}{\partial t} \approx \frac{\partial}{\partial x}\left(D^*_{A_i^q} \frac{\partial \widetilde{C}_{A_i^q}}{\partial x} + \mathrm{q} \frac{\widetilde{C}_{A_i^q} D^*_{A_i^q}}{\tilde{n}(x)} \frac{\partial \tilde{n}(x)}{\partial x} \right). \tag{1.136}$$

With

$$\widetilde{C}_{A_i^q} \approx \widetilde{C}_{A_s^m}\, \tilde{n}^{\,(\mathrm{m}-\mathrm{q})} \tag{1.137}$$

which results from Eqs. (1.121) and (1.131), Eq. (1.136) reduces to Eq. (1.117) with

$$D^{eff}_{A_s^m} = (\mathrm{m}+1) D^*_{A_i^q} (\widetilde{C}_{A_s^m})^{\mathrm{m}-\mathrm{q}}. \tag{1.138}$$

This effective diffusivity holds for donors A_s^m ($\mathrm{m} \in \{+1, +2, \ldots\}$) with $\tilde{n}(x) \approx \widetilde{C}_{A_s^m}$.

For acceptors A_s^m ($\mathrm{m} \in \{-1, -2, \ldots\}$) with $\tilde{p} \approx \widetilde{C}_{A_s^m}$, $D^{eff}_{A_s^m}$ in Eq. (1.117) is given by

$$D^{eff}_{vA_s^m} = (-\mathrm{m}+1) D^*_{A_i^q} (\widetilde{C}_{A_s^m})^{-\mathrm{m}+\mathrm{q}}. \tag{1.139}$$

Characteristics of Dopant Diffusion in Semiconductors

The effective diffusion coefficients $D^{eff}_{A_s^m}$ for the vacancy (dissociative) mechanism in the native-defect controlled mode given by Eqs. (1.118) and (1.119) (Eqs. (1.134) and (1.135)) reveal that dopant diffusion is affected by the charge states m and k of A_s^m and V^k. On the other hand, within the foreign-atom controlled mode of dopant diffusion via the vacancy mechanism (dissociative mechanism) $D^{eff}_{A_s^m}$ is determined by the charge states m and j (q) of A_s^m and AV^j (A_i^q) (see Eqs. (1.123) and (1.124) (Eqs. (1.138) and (1.139))). Generally, the foreign-atom controlled mode of dopant diffusion via reactions (1.80), (1.81), (1.82) and reactions (1.85) and (1.86) is *not* sensitive to the charge states of the native point defects. The equations derived for $D^{eff}_{A_s^m}$ in the native-defect and foreign-atom controlled diffusion mode confirm expressions for the effective dopant diffusion coefficient given by Gösele [115].

Table 1.2 Effective diffusion coefficients $D^{\mathrm{eff}}_{\mathrm{A_s^m}}$ of the substitutional foreign-atoms $\mathrm{A_s^m}$ obtained on the basis of the vacancy mechanism (1.80) for (a) the native-defect controlled mode (see Eqs. (1.118) and (1.119)) and (b) the foreign-atom controlled mode (see Eqs. (1.123) and (1.124)) of dopant diffusion. The concentration dependence of $D^{\mathrm{eff}}_{\mathrm{A_s^m}}$ determines the shape of the dopant profile. Typical diffusion profiles of an element A are shown in Fig. 1.14

Vacancy mechanism: $AV^{\mathrm{j}} + (\mathrm{j}-\mathrm{m}-\mathrm{k})e^- \rightleftharpoons \mathrm{A_s^m} + V^{\mathrm{k}}$		
With m = 1+(donor)		
Reaction	(a) $D^{\mathrm{eff}}_{\mathrm{A_s^m}}$	(b) $D^{\mathrm{eff}}_{\mathrm{A_s^m}}$
$AV^0 + e^- \rightleftharpoons \mathrm{A_s^+} + V^{2-}$	$2\,D^*_{V^{2-}}\,(\widetilde{C}_{\mathrm{A_s^+}})^{-1}$	$2\,D^*_{AV^0}\,(\widetilde{C}_{\mathrm{A_s^+}})^{1}$
$AV^0 \rightleftharpoons \mathrm{A_s^+} + V^{1-}$	$2\,D^*_{V^-}\,(\widetilde{C}_{\mathrm{A_s^+}})^{-2}$	$2\,D^*_{AV^0}\,(\widetilde{C}_{\mathrm{A_s^+}})^{1}$
$AV^0 - e^- \rightleftharpoons \mathrm{A_s^+} + V^0$	$2\,D^*_{V^0}\,(\widetilde{C}_{\mathrm{A_s^+}})^{-3}$	$2\,D^*_{AV^0}\,(\widetilde{C}_{\mathrm{A_s^+}})^{1}$
$AV^- \rightleftharpoons \mathrm{A_s^+} + V^{2-}$	$2\,D^*_{V^{2-}}\,(\widetilde{C}_{\mathrm{A_s^+}})^{-1}$	$2\,D^*_{AV^-}\,(\widetilde{C}_{\mathrm{A_s^+}})^{2}$
$AV^- - e^- \rightleftharpoons \mathrm{A_s^+} + V^-$	$2\,D^*_{V^-}\,(\widetilde{C}_{\mathrm{A_s^+}})^{-2}$	$2\,D^*_{AV^-}\,(\widetilde{C}_{\mathrm{A_s^+}})^{2}$
$AV^- - 2e^- \rightleftharpoons \mathrm{A_s^+} + V^0$	$2\,D^*_{V^0}\,(\widetilde{C}_{\mathrm{A_s^+}})^{-3}$	$2\,D^*_{AV^-}\,(\widetilde{C}_{\mathrm{A_s^+}})^{2}$
With m = 1−(acceptor)		
Reaction	(a) $D^{\mathrm{eff}}_{\mathrm{A_s^m}}$	(b) $D^{\mathrm{eff}}_{\mathrm{A_s^m}}$
$AV^0 + e^- \rightleftharpoons \mathrm{A_s^-} + V^0$	$2\,D^*_{V^0}\,(\widetilde{C}_{\mathrm{A_s^-}})^{-3}$	$2\,D^*_{AV^0}\,(\widetilde{C}_{\mathrm{A_s^-}})^{1}$
$AV^0 \rightleftharpoons \mathrm{A_s^-} + V^+$	$2\,D^*_{V^+}\,(\widetilde{C}_{\mathrm{A_s^-}})^{-2}$	$2\,D^*_{AV^0}\,(\widetilde{C}_{\mathrm{A_s^-}})^{1}$
$AV^0 - e^- \rightleftharpoons \mathrm{A_s^-} + V^{2+}$	$2\,D^*_{V^{2+}}\,(\widetilde{C}_{\mathrm{A_s^-}})^{-1}$	$2\,D^*_{AV^0}\,(\widetilde{C}_{\mathrm{A_s^-}})^{1}$
$AV^+ + 2e^- \rightleftharpoons \mathrm{A_s^-} + V^0$	$2\,D^*_{V^0}\,(\widetilde{C}_{\mathrm{A_s^-}})^{-3}$	$2\,D^*_{AV^+}\,(\widetilde{C}_{\mathrm{A_s^-}})^{2}$
$AV^+ + e^- \rightleftharpoons \mathrm{A_s^-} + V^+$	$2\,D^*_{V^+}\,(\widetilde{C}_{\mathrm{A_s^-}})^{-2}$	$2\,D^*_{AV^+}\,(\widetilde{C}_{\mathrm{A_s^-}})^{2}$
$AV^+ \rightleftharpoons \mathrm{A_s^-} + V^{2+}$	$2\,D^*_{V^{2+}}\,(\widetilde{C}_{\mathrm{A_s^-}})^{-1}$	$2\,D^*_{AV^+}\,(\widetilde{C}_{\mathrm{A_s^-}})^{2}$

Tables 1.2 and 1.3 summarize the concentration dependence of $D^{\mathrm{eff}}_{\mathrm{A_s^m}}$ for the vacancy mechanism and dissociative mechanism, respectively, taking into account charge states of point defects, that are relevant for dopant diffusion in semiconductors. The charge state m of $\mathrm{A_s^m}$ was chosen 1+ (1−) because singly ionized donors (acceptors) are commonly used for high n-type (p-type) doping of both elemental and compound semiconductors. It is emphasized that the concentration dependence of $D^{\mathrm{eff}}_{\mathrm{A_s^m}}$ deduced for the vacancy mechanism also holds for the interstitialcy (kick-out) mechanism given by reaction (1.81) (reaction (1.83)). $D^{\mathrm{eff}}_{\mathrm{A_s^m}}$ for the interstitialcy (kick-out) mechanism is given by Table 1.2 when AV is changed to AI ($\mathrm{A_i}$) and V is changed to I whereas the charge states of the defects are kept unchanged. On the other hand $D^{\mathrm{eff}}_{\mathrm{A_s^m}}$ deduced for the dissociative mechanism and given in Table 1.3 also reflects $D^{\mathrm{eff}}_{\mathrm{A_s^m}}$ of the dopant-defect pair assisted recombination mechanisms (1.85) and (1.86). This shows that Tables 1.2 and 1.3 are generally applicable to predict the concentration dependence of $D^{\mathrm{eff}}_{\mathrm{A_s^m}}$ for dopant diffusion via one of the mechanisms given by reactions (1.80), (1.81), (1.82), and (1.83) and reactions (1.85) and (1.86).

Table 1.2 reveals that $D^{\mathrm{eff}}_{\mathrm{A_s^m}}$ is proportional to $(C_{\mathrm{A_s^m}})^{\mathrm{r}}$ with $\mathrm{r} \in \{-1, -2, -3\}$ for the native-defect controlled mode of dopant diffusion via the vacancy mechanism. This power dependence is a consequence of the charge states of V^{k} under high n-type or p-type doping. Diffusion profiles of an element A with concentration dependent

Table 1.3 Effective diffusion coefficients $D^{\text{eff}}_{A^m_s}$ of the substitutional foreign-atoms A^m_s obtained on the basis of the dissociative mechanism (1.82) for (a) the native-defect controlled mode (see Eqs. 1.134 and 1.135) and (b) the foreign-atom controlled mode (see Eqs. 1.138 and 1.139) of dopant diffusion. The concentration dependence of $D^{\text{eff}}_{A^m_s}$ determines the shape of the dopant diffusion profile. Typical diffusion profiles of an element A are shown in Fig. 1.14

Dissociative mechanism: $A^q_i + V^k + (q+k-m)e^- \rightleftharpoons A^m_s$		
With m =1+(donor)		
Reaction	(a) $D^{\text{eff}}_{A^m_s}$	(b) $D^{\text{eff}}_{A^m_s}$
$A^0_i + V^{2-} - 3e^- \rightleftharpoons A^+_s$	$2\,D^*_{V^{2-}}\,(\widetilde{C}_{A^+_s})^3$	$2\,D^*_{A^0_i}\,(\widetilde{C}_{A^+_s})^1$
$A^0_i + V^- - 2e^- \rightleftharpoons A^+_s$	$2\,D^*_{V^-}\,(\widetilde{C}_{A^+_s})^2$	$2\,D^*_{A^0_i}\,(\widetilde{C}_{A^+_s})^1$
$A^0_i + V^0 - e^- \rightleftharpoons A^+_s$	$2\,D^*_{V^0}\,(\widetilde{C}_{A^+_s})^1$	$2\,D^*_{A^0_i}\,(\widetilde{C}_{A^+_s})^1$
$A^-_i + V^{2-} - 4e^- \rightleftharpoons A^+_s$	$2\,D^*_{V^{2-}}\,(\widetilde{C}_{A^+_s})^3$	$2\,D^*_{A^-_i}\,(\widetilde{C}_{A^+_s})^2$
$A^-_i + V^- - 3e^- \rightleftharpoons A^+_s$	$2\,D^*_{V^-}\,(\widetilde{C}_{A^+_s})^2$	$2\,D^*_{A^-_i}\,(\widetilde{C}_{A^+_s})^2$
$A^-_i + V^0 - 2e^- \rightleftharpoons A^+_s$	$2\,D^*_{V^0}\,(\widetilde{C}_{A^+_s})^1$	$2\,D^*_{A^-_i}\,(\widetilde{C}_{A^+_s})^2$
With m =1−(acceptor)		
Reaction	(a) $D^{\text{eff}}_{A^m_s}$	(b) $D^{\text{eff}}_{A^m_s}$
$A^0_i + V^0 + e^- \rightleftharpoons A^-_s$	$2\,D^*_{V^0}\,(\widetilde{C}_{A^-_s})^1$	$2\,D^*_{A^0_i}\,(\widetilde{C}_{A^-_s})^1$
$A^0_i + V^+ + 2e^- \rightleftharpoons A^-_s$	$2\,D^*_{V^+}\,(\widetilde{C}_{A^-_s})^2$	$2\,D^*_{A^0_i}\,(\widetilde{C}_{A^-_s})^1$
$A^0_i + V^{2+} + 3e^- \rightleftharpoons A^-_s$	$2\,D^*_{V^{2+}}\,(\widetilde{C}_{A^-_s})^3$	$2\,D^*_{A^0_i}\,(\widetilde{C}_{A^-_s})^1$
$A^+_i + V^0 + 2e^- \rightleftharpoons A^-_s$	$2\,D^*_{V^0}\,(\widetilde{C}_{A^-_s})^1$	$2\,D^*_{A^+_i}\,(\widetilde{C}_{A^-_s})^2$
$A^+_i + V^+ + 3e^- \rightleftharpoons A^-_s$	$2\,D^*_{V^+}\,(\widetilde{C}_{A^-_s})^2$	$2\,D^*_{A^+_i}\,(\widetilde{C}_{A^-_s})^2$
$A^+_i + V^{2+} + 4e^- \rightleftharpoons A^-_s$	$2\,D^*_{V^{2+}}\,(\widetilde{C}_{A^-_s})^3$	$2\,D^*_{A^+_i}\,(\widetilde{C}_{A^-_s})^2$

diffusion coefficients D_A proportional to $(C_A)^r$ with $r \in \{-2, 0, 1, 2\}$ are illustrated in Fig. 1.14. For $r < 0$ the shape of the dopant profile is concave. On the other hand, convex dopant profiles result in case of foreign-atom controlled mode of dopant diffusion via the vacancy mechanism, i.e., $D^{\text{eff}}_{A^m_s}$ is proportional to $(C_{A^m_s})^r$ with $r \in \{1, 2\}$. This power dependence is a consequence of the charge states of A^m_s and AV^j.

In contrast to the vacancy mechanism, the dissociative mechanism predicts box-shaped diffusion profiles for singly ionized donors (acceptors) both for the native defect and foreign-atom controlled diffusion mode in the case that neutral and negatively (positively) charged vacancies are favored under n-type (p-type) conditions (see Table 1.3). More specifically, $D^{\text{eff}}_{A^m_s}$ is proportional to $(C_{A^m_s})^r$ with $r \in \{1, 2, 3\}$ in the case when V^k with $k \in \{0, 1-, 2-\}$ ($k \in \{0, 1+, 2+\}$) mediates donor (acceptor) diffusion. In the foreign-atom controlled mode $D^{\text{eff}}_{A^m_s}$ is also proportional to $(C_{A^m_s})^r$ with $r \in \{1+, 2+\}$ when A^q_i is neutral and singly ionized, respectively. The power dependence results from the difference in the charge states between A^m_s and A^q_i, in the same way as for the vacancy mechanism.

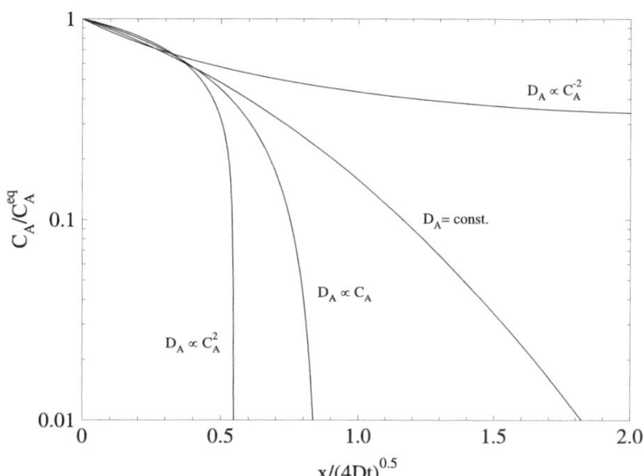

Fig. 1.14 Concentration profiles of an element A normalized by its equilibrium concentration C_A^{eq} versus the normalized penetration depth $x/\sqrt{4D_A t}$. The profiles represent solutions of Eq. (1.112) with $G_X = 0$ and $X = A$ for different concentration dependent diffusion coefficients D_A as indicated

This demonstrates that the shape of dopant profile alone can not tell which diffusion mode is operative.

Nonetheless, the scheme of the concentration dependence of $D^{eff}_{A^m_s}$ given by Tables 1.2 and 1.3 is very helpful for the understanding of dopant profiles in Si, Ge and its alloys. Moreover, the scheme also holds for dopant diffusion in compound semiconductors such as SiC, GaAs, GaSb, ... assuming the diffusion process is restricted to one sublattice (see e.g. [116–125] and references therein). Also charge states of point defects other than those considered in Tables 1.2 and 1.3 may become important. Then the corresponding concentration dependence of $D^{eff}_{A^m_s}$ is given by Eqs. (1.118), (1.119), (1.134) and (1.135) for the native-defect controlled mode and by Eqs. (1.123), (1.124), (1.138), and (1.139) for the foreign-atom controlled diffusion mode.

In order to identify the mechanisms of diffusion, a direct comparison of experimental profiles with numerical solutions of the full partial differential equation system that represent the assumed diffusion model and a comparison of the reduced diffusion coefficients D^*_X with self-diffusion data is required. Generally, several mechanisms can contribute to the diffusion coefficient D_A of a mainly substitutionally dissolved element A. We expect

$$D_A = D_A^{(1)} + D_A^{(2)} + D_A^{(3)} + \ldots \tag{1.140}$$

where the number in the exponent represents contributions due to various indirect diffusion mechanisms. Each diffusion coefficient can be a complicated function of the concentration and diffusion coefficient of the defects involved in the particular reaction (see above). Equation (1.140) shows that it can be difficult to identify the diffusion mechanisms when several mechanisms contribute simultaneously.

As example we consider the diffusion of an element A under isoconcentration condition. Under this condition the diffusion is not affected by internal electric fields and/or chemical concentration gradients and thus always proceeds in thermal equilibrium, that is, the concentrations of V and I are in thermal equilibrium. Accordingly, diffusion under isoconcentration conditions represents the foreign-atom controlled mode. Assuming A_i^q, AV^j, and AI^v of reactions (1.80), (1.81), (1.82), and (1.83) as the possible mobile defects, D_A for isoconcentration conditions is

$$D_A = \frac{C^{eq}_{A_i^q} D_{A_i^q}}{C^{eq}_{A_s^m} + C^{eq}_{A_i^q}} + \frac{C^{eq}_{AV^j} D_{AV^j}}{C^{eq}_{A_s^m} + C^{eq}_{AV^j}} + \frac{C^{eq}_{AI^v} D_{AI^v}}{C^{eq}_{A_s^m} + C^{eq}_{AI^v}}. \tag{1.141}$$

This follows from the differential equation system of the underlying reactions for the foreign-atom controlled diffusion mode (see e.g. Sects. 1.6.1.2 or 1.6.1.3). In the case when A is mainly dissolved on interstitial sites, i.e., $C^{eq}_{A_i^q} \gg C^{eq}_{A_s^m}, C^{eq}_{AV^j}, C^{eq}_{AI^v}$, and $D_{A_i^q} \gg D_{AV^j}, D_{AI^v}$, we get $D_A \approx D_{A_i^q}$ as expected for the diffusion via the direct interstitial mechanism. For a mainly substitutionally dissolved element ($C^{eq}_{A_s^m} \gg C^{eq}_{A_i^q}, C^{eq}_{AV^j}, C^{eq}_{AI^v}$) the diffusivity D_A is given by the sum of the reduced diffusion coefficients $D^*_X = C^{eq}_X D_X / C^{eq}_{A_s^m}$ with $X \in \{A_i^q, AV^j, AI^v\}$.

Equation (1.141) also describes the diffusion of hybrid elements in dislocation-free crystals under isoconcentration conditions [119]. In this case the first term on the right hand side of Eq. (1.141) dominates. In order to identify the mechanisms of diffusion of an element A, complementary diffusion experiments under different experimental condition have to be performed [126]. In this respect also studies on the impact of self-interstitial and vacancy perturbations on self- and dopant diffusion are advantageous. Such experiments were extensively performed with Si (see references in [49]). Oxidation (nitridation) of a bare Si surface is known to inject self-interstitials (vacancies). The results considerably contributed to our present understanding of dopant diffusion in Si (see e.g. [127–136]).

The availability of highly enriched stable isotopes of Si and Ge combined with modern epitaxial deposition techniques such as chemical vapor deposition and molecular beam epitaxy enabled the preparation of isotopically controlled semiconductor layer structures. By means of such isotopically controlled heterostructures not only self-diffusion experiments in Si and Ge could be performed over a wide range of temperatures but also the impact of dopant diffusion on self-diffusion could be investigated directly [137–141]. Modeling of the simultaneous self- and dopant diffusion is a bit more complex than modeling of the separate processes but is a worthwhile effort because more information about the underlying mechanisms and properties of the point defects is obtained than from self- and dopant diffusion experiments studied separately [64, 75].

In the following section typical experimental diffusion profiles of hybrid and dopant atoms in Si are presented that provide strong evidence for different indirect diffusion mechanisms in Si.

1.7 Experimental Diffusion Profiles

Experimental diffusion profiles of Zn in Si [41] are presented in the following as an example on the diffusion behavior of hybrid atoms in Si. Similar profiles are observed for Au [83–86] and Pt [95, 96, 98–100]. In addition to Zn profiles in dislocation-free and highly dislocated Si, also experimental profiles are illustrated that reveal the impact of a reduced Zn vapor pressure on Zn diffusion and the time evolution of Zn diffusion. The Zn diffusion behavior under the different experimental conditions is consistently described on the basis of the kick-out mechanism constituting crucial evidence that Zn in Si mainly diffuses via the kick-out mechanism [41, 94].

As an example on the diffusion behavior of dopants in Si experimental profiles obtained after dopant diffusion in isotopically controlled Si heterostructures are shown [64, 137–139]. These concurrent self- and dopant diffusion experiments reveal the impact of both point-defect reactions and doping on self- and dopant diffusion. Such diffusion studies are very valuable to determine the charge states of the point defects involved in the diffusion process [76].

1.7.1 Diffusion Profiles of Hybrid Atoms

Diffusion of hybrid elements in defect-free and highly dislocated crystals provide the effective diffusion coefficients $D^{\mathrm{eff}}_{V,I}$ and $D^{\mathrm{eff}}_{\mathrm{A}_i}$, respectively, that strongly differ from each other. An impact of the defect structure on the diffusion of hybrid elements via the dissociative and kick-out mechanism is only expected when the relation $C^{\mathrm{eq}}_{\mathrm{A}_i} D_{\mathrm{A}_i} \gg C^{\mathrm{eq}}_{\mathrm{V,I}} D_{\mathrm{V,I}}$ is fulfilled. This condition implies that the transformation of A_i to A_s via reaction (1.26) and (1.27) is controlled by the diffusion of either V, I or A_i. In the case when $C^{\mathrm{eq}}_{\mathrm{A}_i} D_{\mathrm{A}_i} \ll C^{\mathrm{eq}}_{\mathrm{V,I}} D_{\mathrm{V,I}}$ holds, the A_i-A_s transformation is controlled by the diffusion of A_i. Accordingly, the property of dislocations acting as source and sink of native defects will not affect the diffusion of hybrid elements. However, the presence of dislocations can procure additional defect reactions such as trapping of hybrid atoms at dislocations that alters their diffusion behavior [94, 142, 143]. Experiments on the diffusion of Ag and Au in Ge reveal that their diffusion is independent of the defect density, i.e., equal diffusion data are obtained from profiles in dislocation-free and highly dislocated Ge [81]. Even for Zn diffusion in Si experimental conditions can be realized that provide similar diffusion profiles in dislocation-free and highly dislocated Si. Figure 1.15 illustrates $\mathrm{Zn_s}$ concentration profiles $C_{\mathrm{Zn_s}}(x)$ resulting from spreading-resistance (SR) measurements on dislocation-free (profile (+)) and highly dislocated Si samples (profile (×)) that were simultaneously diffused with Zn at 1115 °C for 2880 s [41].

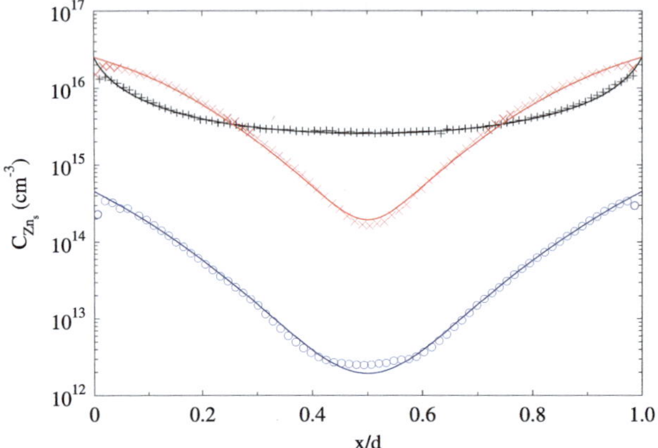

Fig. 1.15 Concentration profiles of substitutional Zn resulting from spreading resistance measurements performed on dislocation-free (+) and highly dislocated (×) Si samples after zinc diffusion at 1115 °C for 2880 s using elemental Zn as diffusion source. Profile (◯) indicates a Zn_s profile in dislocation-free Si measured after Zn diffusion at 1115 °C for 1620 s using $ZnCl_2$ as diffusion source. *Solid lines*: theoretical kick-out profiles simultaneously fitted to the data by optimizing the two essential model parameters $D^*_{Zn_i} = C^{eq}_{Zn_i} D_{Zn_i} / C^{eq}_{Zn_s}$ and $D^*_I = C^{eq}_I D_I / C^{eq}_{Zn_s}$ [41]

The boundary concentration of about $C_{Zn_s}(x = 0) = 2.5 \times 10^{16}\,cm^{-3}$ established at this temperature with a pure Zn source equals the Zn solubility [41, 93]. Also shown in Fig. 1.15 is a Zn_s concentration profile in dislocation-free Si obtained after 1620 s of Zn diffusion at 1115 °C using $ZnCl_2$ as diffusion source (profile: (◯)) [41]. This source yields $C_{Zn_s}(x = 0) = 4.5 \times 10^{14}\,cm^{-3}$ which is a factor 55 smaller than the solubility limit at 1115 °C attained in equilibrium with the elemental vapor source. The shape of the profiles in Fig. 1.15 reveals that the profile (+) in dislocation-free Si is convex whereas the profile (×) in the simultaneously diffused highly dislocated Si is concave. Also the lower profile (◯) has a concave curvature. Obviously, in dislocation-free Si the profile shape changes from convex to concave when the thermodynamic activity of the Zn source is lowered. This shows that the diffusion of Zn in Si is not only sensitive to the defect structure but also to the prevailing Zn vapor pressure (see Sect. 1.6.1.5 and Fig. 1.11). These results together with the time evolution of Zn diffusion in dislocation-free Si which is illustrated by the profiles in Fig. 1.16 provide strong evidence that Zn in Si mainly diffuses via the kick-out mechanism. Only this mechanism yields a consistent description of all profiles in Figs. 1.15 and 1.16 as illustrated by the solid lines [41]. The profiles of Zn_i and I corresponding to the calculated Zn_s profiles of Fig. 1.16 are shown in Figs. 1.17 and 1.18, respectively.

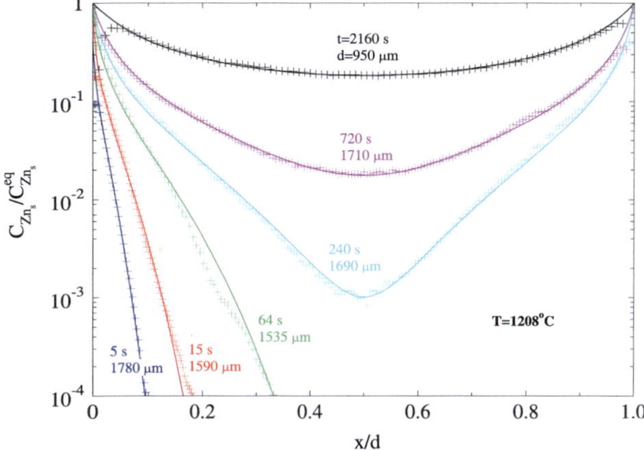

Fig. 1.16 Concentration profiles of substitutional Zn_s measured by means of the spreading resistance technique after Zn diffusion at 1208 °C for times listed in the figure. The penetration depth x is normalized to the thickness d of the particular sample given in the figure. *Solid lines* show calculated profiles based on the kick-out model with one set of parameters [41]

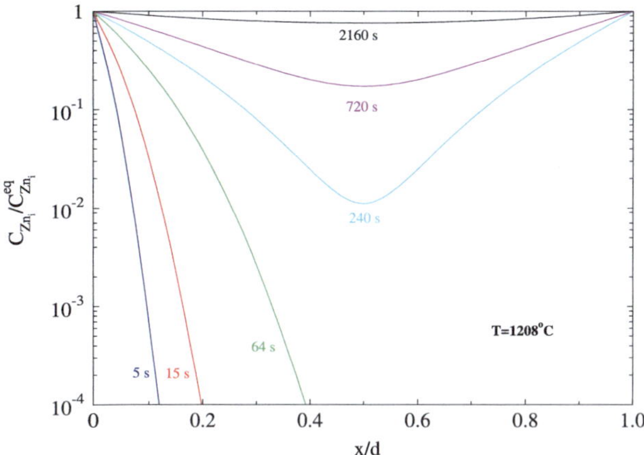

Fig. 1.17 Calculated normalized concentration profiles of interstitial Zn_i corresponding to the calculated Zn_s profiles shown in Fig. 1.16

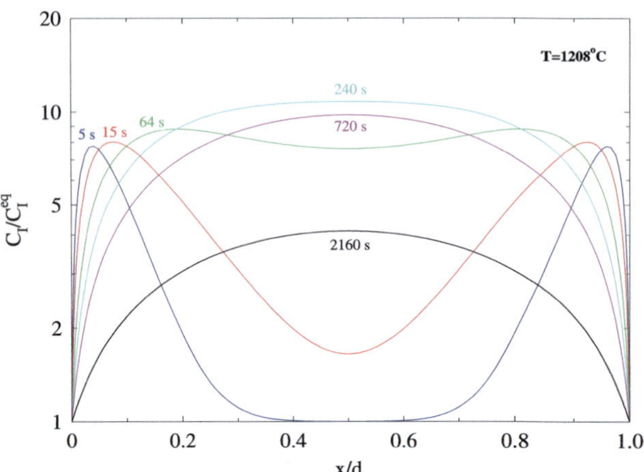

Fig. 1.18 Calculated normalized concentration profiles of self-interstitials *I* corresponding to the calculated Zn_s profiles shown in Fig. 1.16

1.7.2 Diffusion Profiles of Dopant Atoms

Numerous experiments on dopant diffusion in Si were performed over the past decades to determine the atomic mechanisms of diffusion [49]. The studies were motivated by the demand to control the diffusion and activation of dopants in the fabrication of Si-based electronic devices. Dopant diffusion in semiconductors can be rather complex since doping affects the position of the Fermi level and thus the formation of charged defects [114]. Moreover, dopant diffusion can induce native defect concentrations deviating from thermal equilibrium. The impact of the doping level on the shape of dopant profiles in Si is e.g. impressively demonstrated by the P diffusion experiments reported by Yoshida et al. [144, 145]. Most recently performed experiments on dopant diffusion in Si isotope structures not only confirm the concentration dependence of dopant diffusion but also reveal additional insight on the mechanisms of dopant diffusion and on the charge states of the native point defects involved [64, 76].

Concentration profiles of ^{11}B, ^{75}As, and ^{31}P and the corresponding ^{30}Si profiles measured with SIMS after diffusion annealing are illustrated in Fig. 1.19. Figure 1.19a shows the As-implanted Si isotope structure that reflects exemplary also B- and P-implanted isotope structures. The reduced concentration of ^{30}Si within the enriched ^{28}Si layers compared to the adjacent ^{nat}Si layers gives rise to a strong gradient at the ^{nat}Si/^{28}Si interface. Accordingly, the isotopically modulated structure is ideally suited for studying self-diffusion and the impact of dopant diffusion on self-diffusion. The dopant diffuses from the top implanted amorphous Si layer into the isotope structure. All Si profiles obtained after concurrent self- and dopant diffusion reveal a depth-dependent broadening. The concentration profiles of P and Si within the topmost ion-implanted amorphous Si layer are missing in Fig. 1.19d because this layer was removed prior to SIMS profiling (see Ref. [64]). The depth-dependent broadening of the Si isotope structure observed after dopant diffusion is due to charged point defects whose thermal equilibrium concentration depends on

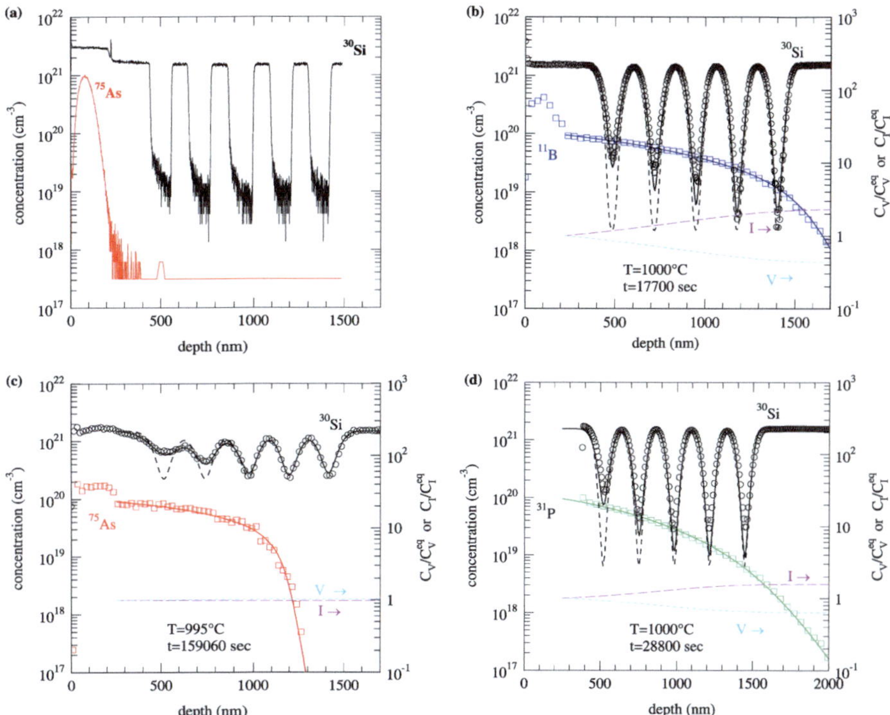

Fig. 1.19 Concentration profiles of ^{30}Si of a ^{nat}Si/^{28}Si isotope multilayer structures with a top amorphous Si layer (about 250 nm *thick*) implanted with either boron (^{11}B), arsenic (^{75}As), and phosphorous (^{31}P) and subsequently annealed at the temperatures and times indicated. The profiles were measured by means of secondary ion mass spectrometry. (**a**): ^{30}Si and ^{75}As profiles of the As-implanted Si isotope multilayer structure. (**b**), (**c**), and (**d**) illustrate, respectively, the ^{11}B, ^{75}As, and ^{31}P diffusion profiles and corresponding self-diffusion profiles. The *solid lines* in (**b**)–(**d**) are best fits to the experimental dopant and self-atom profiles obtained on the basis of indirect diffusion mechanisms [64]. The *lower dashed lines* show the corresponding concentrations $C_{V,I}$ of V (*short-dashed line*) and I (*long-dashed line*) normalized to their thermal equilibrium concentration $C_{V,I}^{eq}$ (see right ordinate). The diffusion of B and P leads to a super- and undersaturation of I and V, respectively, whereas As does not induce any significant deviation from the equilibrium concentration of I and V. The *upper thin dashed lines* in (**b**), (**c**), and (**d**) show ^{30}Si profiles that are expected for self-diffusion under electronic intrinsic and thermal equilibrium conditions. The top amorphous layer of the P-diffused Si isotope structure was removed by chemical-mechanical polishing before SIMS profiling

the position of the Fermi level [76]. As a consequence self-diffusion is altered by doping. Moreover dopant diffusion can lead to concentrations of native point defects that deviate from thermal equilibrium. This, in particular, holds for the native-defect controlled mode of dopant diffusion. These two effects of doping and dopant diffusion on self-diffusion are unraveled by the simultaneous self- and dopant-atom diffusion experiments. These experiments also provide information about possible contributions of AV and AI pairs to self-diffusion. Whether such pairs contribute significantly to self-diffusion depends on the magnitude of the transport coefficients $C^{eq}_{AV}D_{AV}$ and $C^{eq}_{AI}D_{AI}$ compared to those of the native defects and on the correlation factors for self-diffusion via AV and AI. A comprehensive mathematical treatment of the simultaneous diffusion of self- and dopant atoms and of the impact of the Fermi level on the concentration of charged point defects is given in Refs. [64, 76].

Figure 1.19b shows that the Si profile associated with the in-diffusion of B reveals a faster Si diffusion for high B concentrations close to the amorphous/crystalline interface than at the B diffusion front. Comparison of the Si profile obtained by in-diffusion of B with that expected for electronically intrinsic and thermal equilibrium conditions reveals a higher Si diffusivity under B in-diffusion even at the deepest ^{28}Si/^{nat}Si interface. This points to a supersaturation of native point defects established by B diffusion. Additional results on B diffusion in Si isotope structures obtained at different temperatures, on enhanced (retarded) B diffusion under I (V) injection [129, 132] and of isoconcentration diffusion experiments [133] reveal the charge states of the mobile B-related defect and of the native point defects involved in B diffusion. Altogether, B diffusion in Si under various experimental conditions is consistently described by the following reactions [64]

$$\mathrm{B_i^0} \rightleftharpoons \mathrm{B_s^-} + I^{\mathrm{u}} + (1-\mathrm{u})\mathrm{h} \tag{1.142}$$

$$\mathrm{B_i^0} + V^{\mathrm{k}} \rightleftharpoons \mathrm{B_s^-} + (1+\mathrm{k})\mathrm{h} \tag{1.143}$$

with neutral (I^0), singly and doubly positively (I^+, I^{2+}) charged self-interstitials and neutral (V^0), singly positive (V^+), and singly and doubly negative (V^-, V^{2-}) vacancies. h denotes a hole to satisfy charge neutrality of the reactions. The dissociative mechanism assures local equilibrium between I and V after sufficient long diffusion times. For a more detailed discussion on B diffusion in Si the interested reader is referred to Refs. [64, 105, 146, 147].

Si profiles obtained by As diffusion (see Fig. 1.19c) show an enhanced Si diffusion within the topmost isotope layers. Here the As doping concentration exceeds the intrinsic carrier concentration. The Si profile at the deepest interface is accurately described with the diffusion coefficient for intrinsic and thermal equilibrium conditions [64]. This and the shape of the As diffusion profiles points to a foreign-atom controlled mode of dopant diffusion. Indeed available experimental results on As diffusion in Si, which comprise As diffusion experiments under native defect injection [68, 132], under isoconcentration conditions [148], and on the simultaneous self- and dopant-atom diffusion [64], are accurately described by the defect reactions

$$\mathrm{As}V^{\mathrm{j}} \rightleftharpoons \mathrm{As_s^+} + V^{\mathrm{k}} + (1+\mathrm{k}-\mathrm{j})\mathrm{e}^- \tag{1.144}$$

$$\mathrm{As}I^0 \rightleftharpoons \mathrm{As_s^+} + I^{\mathrm{u}} + (1+\mathrm{u})\mathrm{e}^- \tag{1.145}$$

$$\mathrm{As}V^{\mathrm{j}} + I^{\mathrm{u}} \rightleftharpoons \mathrm{As_s^+} + (1-\mathrm{j}-\mathrm{u})\mathrm{e}^- \tag{1.146}$$

$$\mathrm{As}I^0 + V^{\mathrm{k}} \rightleftharpoons \mathrm{As_s^+} + (1-\mathrm{k})\mathrm{e}^-. \tag{1.147}$$

Various charge states for V ($\in \{V^+, V^0, V^-, V^{2-}\}$ and I ($\in \{I^{2+}, I^+, I^0, I^-\}$) were assumed as well as neutral and singly negatively charged dopant-vacancy pairs AsV^0 and AsV^-. The parameters deduced from modeling As diffusion reveal a foreign-atom controlled mode of diffusion, i.e., the equilibrium concentration of V and I is not disturbed by the diffusion of As. This is also demonstrated by the normalized concentration of V and I shown in Fig. 1.19c. A comprehensive discussion on As diffusion in Si is given in Ref. [64].

Self-diffusion profiles associated with the diffusion of P also indicate an enhanced self-atom diffusion at the topmost natural and isotopically enriched layers (see Fig. 1.19d)). On the other hand the diffusion of Si at the deepest ^{28}Si/^{nat}Si interface is described with a diffusion coefficient that is close to the value determined for intrinsic and thermal equilibrium conditions. This was a surprising result because the pronounced tail of P profiles which are clearly at variance with the box-shaped As profiles were generally assumed to be due to a supersaturation of Si self-interstitials [146, 149]. The broadening of the Si profiles demonstrates that the I supersaturation due to P diffusion is significantly lower than expected on the basis of the previous considered P diffusion models. Taking into account additional results on P diffusion in Si obtained under different experimental conditions that comprise diffusion studies under intrinsic and extrinsic doping [150, 151], under defect injection [129, 132, 152], and under isoconcentration conditions [150], the following reactions were determined to accurately describe P diffusion in Si [64]

$$\mathrm{P}V^- \rightleftharpoons \mathrm{P_s^+} + V^{\mathrm{k}} + (2+\mathrm{k})\mathrm{e}^- \tag{1.148}$$

$$\mathrm{P}V^- + I^{\mathrm{u}} \rightleftharpoons \mathrm{P_s^+} + (2-\mathrm{u})\mathrm{e}^- \tag{1.149}$$

$$\mathrm{P_i^q} \rightleftharpoons \mathrm{P_s^+} + I^{\mathrm{u}} + (1+\mathrm{u}-\mathrm{q})\mathrm{e}^- \tag{1.150}$$

$$\mathrm{P_i^q} + V^{\mathrm{k}} \rightleftharpoons \mathrm{P_s^+} + (1-\mathrm{k}-\mathrm{q})\mathrm{e}^-. \tag{1.151}$$

$\mathrm{P_i^q}$ exists as neutral and singly positively charged defect. The charge states of V^{k} and I^{u} are the same as those considered in reactions (1.142), (1.143), (1.144), (1.145), (1.146), and (1.147) for modeling B and As diffusion.

Successful modeling of the simultaneous P and Si diffusion requires a contribution of singly positively charged mobile P defects to P diffusion. Generally, the extended tail of the P profiles compared to the much steeper As profile at the diffusion front was explained with an I-supersaturation established during P diffusion. Experiments on P diffusion in isotope multilayer structures clearly showed that this interpretation can not be correct. The supersaturation required to explain extended P profiles would predict an enhanced self-diffusion in the tail region that is not observed experimentally. In order to unravel this discrepancy, a mobile P species was proposed whose effective diffusivity is concentration independent. According to Eqs. (1.123) and (1.138), the charge difference between substitutional $\mathrm{P_s^+}$ and the mobile P must be zero. Therefore a singly positively charged interstitial phosphorus defect $\mathrm{P_i^+}$ is considered in reactions (1.150) and (1.151).

1.8 Concluding Remarks

Direct and indirect mechanisms of diffusion in semiconductors are discussed with particular emphasizes on the charge state of the point defects involved. Continuum theoretical calculations of atomic diffusion in semiconductors based on differential equations are presented that offer a robust and fast approach to describe experimental diffusion profiles. With this approach not only the diffusion of hybrid foreign-atoms that occupy both interstitial and substitutional lattice sites but also the impact of doping on self- and foreign-atom diffusion can be treated. Computer simulations reveal the significance of the various model parameters on the shape of foreign-atom profiles. In many cases the diffusion process is mainly determined by the transport coefficients $C_X^{eq} D_X$ of the point defects X involved in the diffusion process. Basically two different diffusion modes, the foreign-atom and native-defect controlled modes, can be distinguished. These modes are characterized by specific relations between the experimentally apparent effective diffusion coefficient of the foreign atom and the native point defect that controls the formation of the substitutional dopant A_s^m. This relation is crucial for the understanding of dopant diffusion in semiconductors.

With the availability of isotopically controlled semiconductor multilayer structures, advanced diffusion studies became possible that provide more detailed insight on the diffusion mechanisms and properties of defects involved than earlier studies [64, 76, 140]. The advantage of isotopically controlled semiconductors for diffusion studies is demonstrated by experiments and simulations of the simultaneous diffusion of self- and foreign-atoms in isotope multilayer structures. These experiments enable to determine the charge states of native point defects and help to distinguish between doping effects, i.e., Fermi level effect, and effects due to defect reactions on self- and dopant diffusion. Together with diffusion studies under various experimental conditions a consistent set of defect reactions are derived that accurately describe the diffusion of the most common dopants boron, phosphorous, and arsenic in Si.

With the availability of high-purity, isotopically enriched Si [153] sophisticated diffusion studies can be performed with well designed layered structures. Moreover, the computing capacity of today's personal computers is sufficient to solve the complex diffusion equations within acceptable times. Although, numerous diffusion studies have been performed with silicon, basic questions concerning the properties of self-interstitials and vacancies are still unsolved [65–67].

Today the continuous down-scaling of CMOS (complementary metal-oxide-semiconductors) devices reached 2D/3D device structures in nanometer size. Associated with this high level of integration is the need to control the dopant distribution and its activation [154]. Further miniaturization is intimately connected with an improved understanding of the properties and interaction of atomic defects. This interaction concerns the interaction of defects in the volume but also the interaction of defects with surfaces and interfaces since the surface/interface-to-volume ratio increases with decreasing size of electronic devices. The properties of point defects in bulk Si, i.e., their type, mobility, interaction and electronic behavior are well understood to some extent, thanks to the progress achieved in spectroscopic and diffusion studies over the past decades and recently by means of isotopically controlled Si structures (see e.g. [49, 50, 64, 147, 155, 156]). However, the properties of dopants and self-atoms determined for bulk materials are not necessarily applicable to nanoscale Si structures because in small dimensions the thermodynamics and kinetics of point defects can be different from that in the bulk [157, 158]. Surface and interface states can affect the behavior of point defects in the whole nanostructure due to the high surface/interface-to-volume ratio.

Finite size, surface, and interface effects on atomic transport in nanosized semiconductors are technologically of fundamental significance to control and engineer the electronic properties of nanoelectronic devices and to successfully implement novel transistor concepts (see e.g. Ref. [159]). Atomistic calculations are increasingly used to predict defect types, defect interactions, their stability, mobility, and electronic properties. Finally, however, the relevance of these calculations needs to be verified experimentally.

The characterization of atomic defects and their interaction with other point defects, surfaces and interfaces in nanosized materials is very challenging. Not only experiments under well-defined conditions with two (2D) and/or three dimensional (3D) nanostructures must be performed but also appropriate analyses techniques are necessary to resolve nanosized 2D/3D structures and their properties.

References

1. Ammon, W.: Defects in Monocrystalline silicon. In: Kasap, S., Capper, P. (eds.) Springer Handbook of Electronic and Photonic Materials, pp. 101–120. Springer, New York (2007). doi:10.1007/978-0-387-29185-7_5
2. Car, R., Kelly, P.J., Oshiyama, A., Pantelides, S.T.: Microscopic theory of atomic diffusion mechanisms in silicon. Phys. Rev. Lett. **52**, 1814 (1984). doi:10.1103/PhysRevLett.52.1814
3. Bar-Yam, Y., Joannopoulos, J.D.: Barrier to migration of the silicon self-interstitial. Phys. Rev. Lett. **52**, 1129 (1984). doi:10.1103/PhysRevLett.52.1129
4. Blöchl, P.E., Smargiassi, E., Car, R., Laks, D.B., Andreoni, W., Pantelides, S.T.: First-principles calculations of self-diffusion constants in silicon. Phys. Rev. Lett. **70**, 2435 (1993). doi:10.1103/PhysRevLett.70.2435
5. Clark, S.J., Ackland, G.J.: Ab initio calculations of the self-interstitial in silicon. Phys. Rev. B. **56**, 47 (1997). doi:10.1103/PhysRevB.56.47
6. Sadigh, B., Lenosky, Th.J., Theiss, S.K., Caturla, M.-J., de la Rubia, T.D., Foad, M.A.: Mechanism of boron diffusion in silicon: an ab initio and Kinetic Monte Carlo study. Phys. Rev. Lett. **83**, 4341 (1999). doi:10.1103/PhysRevLett.83.4341
7. Leung, W.-K., Needs, R.J., Rajagopal, G., Itoh, S., Ihara, S.: Calculations of silicon self-interstitial defects. Phys. Rev. Lett. **83**, 2351 (1999). doi:10.1103/PhysRevLett.83.2351
8. Windl, W., Bunea, M.M., Stumpf, R., Dunham, S.T., Masquelier, M.P.: First-principles study of boron diffusion in silicon. Phys. Rev. Lett. **83**, 4345 (1999). doi:10.1103/PhysRevLett.83.4345
9. Jeong, J.-W., Oshiyama, A.: Atomic and electronic structures of a Boron impurity and its diffusion pathways in crystalline Si. Phys. Rev. B. **64**, 235204 (2001). doi:10.1103/PhysRevB.64.235204

10. Liu, X.-Y., Windl, W., Beardmore, K.M., Masquelier, M.P.: First-principles study of phosphorus diffusion in silicon: interstitial- and vacancy-mediated diffusion mechanisms. Appl. Phys. Lett. **82**, 1839 (2003). doi:10.1063/1.1562342
11. Al-Mushadani, O.K., Needs, R.J.: Free-energy calculations of intrinsic point defects in silicon. Phys. Rev. B **68**, 235205 (2003). doi:10.1103/PhysRevB.68.235205
12. Lopez, G.M., Fiorentini, V.: Structure, energetics, and extrinsic levels of small self-interstitial clusters in silicon. Phys. Rev. B **69**, 155206 (2004). doi:10.1103/PhysRevB.69.155206
13. El-Mellouhi, F., Mousseau, N., Ordejón, P.: Sampling the diffusion paths of a neutral vacancy in silicon with quantum mechanical calculations. Phys. Rev. B **70**, 205202 (2004). doi:10.1103/PhysRevB.70.205202
14. Jones, R., Briddon, P.R.: The ab initio cluster method and the dynamics of defects in semiconductors. In: Willardson, R.K., Weber, E.R., Stavola, M. (eds.) Identification of Defects in Semiconductors. Semiconductors and Semimetals. Academic Press, San Diego (1998)
15. Coutinho, J.: Density functional modeling of defects and impurities in silicon materials. In: Yoshida, Y., Langouche, G. (eds.) Defects and Impurities in Silicon Materials: An Introduction to Atomic-Level Silicon Engineering. Springer, Tokyo (2016)
16. Song, E.G., Kim, E., Lee, Y.H., Hwang, Y.G.: Fully relaxed point defects in crystalline silicon. Phys. Rev. B **48**, 1486 (1993). doi:10.1103/PhysRevB.48.1486
17. Tang, M., Colombo, L., Zhu, J., de la Rubia, T.D.: Intrinsic point defects in crystalline silicon: tight-binding molecular dynamics studiesof self-diffusion, interstitial-vacancy recombination, and formation volumes. Phys. Rev. B **55**, 14279 (1997). doi:10.1103/PhysRevB.55.14279
18. Alippi, P., Colombo, L., Ruggerone, P., Sieck, A., Seifert, G., Frauenheim, Th.: Atomic-scale characterization of boron diffusion in silicon. Phys. Rev. B **64**, 075207 (2001). doi:10.1103/PhysRevB.64.075207
19. Jääskeläinen, A., Colombo, L., Nieminen, R.: Silicon self-diffusion constants by tight-binding molecular dynamics. Phys. Rev. B **64**, 233203 (2001). doi:10.1103/PhysRevB.64.233203
20. Schober, H.R.: Extended interstitials in silicon and germanium. Phys. Rev. B **39**, 13013(R) (1989). doi:10.1103/PhysRevB.39.13013
21. Maroudas, D., Brown, R.A.: Atomistic calculation of the self-interstitial diffusivity in silicon. Appl. Phys. Lett. **62**, 172 (1993). doi:10.1063/1.109361
22. Maroudas, D., Brown, R.A.: Calculation of thermodynamic and transport properties of intrinsic point defects in silicon. Phys. Rev. B **47**, 15562 (1993). doi:10.1103/PhysRevB.47.15562
23. Posselt, M., Gao, F., Zwicker, D.: Atomistic study of the migration of di- and tri-interstitials in silicon. Phys. Rev. B **71**, 245202 (2005). doi:10.1103/PhysRevB.71.245202
24. Watkins, G.D.: EPR and ENDOR studies of defects in semiconductors. In: Willardson, R.K., Weber, E.R., Stavola, M. (eds.) Identification of Defects in Semiconductors. Semiconductors and Semimetals. Academic Press, San Diego (1998)
25. Spaeth, J.-M.: Magneto-optical and electrical detection of paramagnetic resonance in semiconductors. In: Willardson, R.K., Weber, E.R., Stavola, M. (eds.) Identification of Defects in Semiconductors. Semiconductors and Semimetals. Academic Press, San Diego (1998)
26. Weil, J.A., Bolton, J.R.: Electron Paramagnetic Resonance: Elementary Theory and Practical Applications, 2nd edn. Wiley, Hoboken (2007)
27. Perkowitz, S.: Optical Characterization of Semiconductors: Infrared, Raman, and Photoluminescence Spectroscopy. Elsevier Science, Burlington (2012)
28. Lang, D.V.: Deep-level transient spectroscopy: a new method to characterize traps in semiconductors. J. Appl. Phys. **45**, 3023 (1974). doi:10.1063/1.1663719
29. Wichert, Th., Recknagel, E.: Perturbed angular correlation. In: Microscopic Methods in Metals. Topics in Current Physics, vol. 40, pp. 317–364. Springer, Berlin/Heidelberg (1986). doi:10.1007/978-3-642-46571-0_11
30. Siegel, R.W.: Positron annihilation spectroscopy. Ann. Rev. Mater. Sci. **10**, 393–425 (1980). doi:10.1146/annurev.ms.10.080180.002141

31. Saarinen, K., Hautojärvi, P., Corbel, C.: Positron annihilation spectroscopy of defects in semiconductors. In: Willardson, R.K., Weber, E.R., Stavola, M. (eds.) Identification of Defects in Semiconductors. Semiconductors and Semimetals. Academic Press, San Diego (1998)
32. Krause-Rehberg, R., Leipner, H.S.: Positron Annihilation in Semiconductors: Defect Studies. Springer, Berlin/Heidelberg (1999)
33. Greenwood, N.N., Gibb, T.C.: Mössbauer Spectroscopy. Chapman and Hall, London (1971). doi:10.1007/978-94-009-5697-1
34. Würschum, R., Bauer, W., Maier, K., Seeger, A., Schaefer, H.-E.: Defects in semiconductors after electron irradiation or in high-temperature thermal equilibrium, as studied by positron annihilation. J. Phys.: Condens. Matter **1**, SA33–SA48 (1989). doi:10.1088/0953-8984/1/SA/005
35. Langouche, G., Yoshida, Y.: Nuclear methods to study defects and impurities in Si materials. In: Yoshida, Y., Langouche, G. (eds.) Defects and Impurities in Silicon Materials: An Introduction to Atomic-Level Silicon Engineering. Springer, Tokyo (2016).
36. Ueki, T., Itsumi, M., Takeda, T.: Octahedral void defects observed in the bulk of Czochralski silicon. Appl. Phys. Lett. **70**, 1248 (1997). doi:10.1063/1.118543
37. Nishimura, M., Yamaguchi, Y., Nakamura, K., Jablonski, J., Wantanabe, M.: The role of oxygen impurities in the formation of grown-in laser scattering tomography defects in silicon single crystals. In: Claeys, C.L., Rai-Choudhury, P., Watanabe, M., Stallhofer, P., Dawson, H.J. (eds.) High Purity Silicon V, Proceedings Volume 98-13, pp. 188–199. The Electrochemical Society, Pennington (1998)
38. Abe, T., Kato, Y.: The effects of polishing damage and oxygen concentration on gate oxide integrity in silicon crystals. Jpn. J. Appl. Phys. **32**, 1879–1883 (1993). doi:10.1143/JJAP.32.1879
39. Sugra, H., Abe, H., Koya, H., Yoshimi, T., Suzuki, I., Yoshioka, H., Kagawa, N.: Proceedings of the 2nd Symposium on Physics and Chemistry of SiO_2 and Si-SiO_2 Interfaces, vol. 2, p. 279 (1993)
40. Ammon, W.v., Ehlert, A., Lambert, U., Gräf, D., Brohl, M., Wagner, P.: Gate oxide related bulk properties of oxygen doped floating zone and Czochralski silicon. In: Huff, H.R., Bergholz, W., Sumino, K. (eds.) Semiconductor Silicon/1994: Proceedings of the Seventh International Symposium on Silicon Materials Science and Technology. Electrochemical Society Proceedings vol. 94–10, pp. 136–147, Pennington (1994)
41. Bracht, H., Stolwijk, N.A., Mehrer, H.: Properties of intrinsic point defects in silicon determined by zinc diffusion experiments under nonequilibrium conditions. Phys. Rev. B **52**, 16542–16560 (1995). doi:10.1103/PhysRevB.52.16542
42. Edelin, G., Mathiot, D.: A model for the determination of the defect concentrations in III-V compounds. Philos. Mag. B **42**, 95–110 (1980). doi:10.1080/01418638008225641
43. Tan, T.Y.: Point defect thermal equilibria in GaAs. Mater. Sci. Eng. B **10**, 227–239 (1991). doi:10.1016/0921-5107(91)90130-N
44. Tan, T.Y., You, H.-M., Gösele, U.M.: Thermal equilibrium concentrations and effects of negatively charged Ga vacancies in n-type GaAs. Appl. Phys. A **56**, pp. 249–258 (1993). doi:10.1007/BF00539483
45. More generally, the diffusion flux is proportional to the gradient of the chemical potential (see e.g. [46]). For ideal solid solutions such as doped semiconductors the chemical potential is proportional to the concentration gradient. Accordingly, dopant diffusion in semiconductors is generally described in terms of concentration gradients. On the other hand, in case of e.g. binary diffusion couples the gradient in the chemical potential is considered for the treatment of the interdiffusion process
46. Mehrer, H.: Diffusion in Solids: Fundamentals, Methods, Materials, Diffusion-Controlled Processes. Springer Series in Solid-State Sciences. Springer, Berlin/New York (2007).
47. Crank, J.: The Mathematics of Diffusion. Oxford Science Publications/Clarendon Press, Oxford (1979)

48. Philibert, J.M., Atom Movements – Diffusion and Mass Transport in Solids, Hors Collection, EDP Sciences. Editions de Physique, Les Ulis (2012)
49. Beke, D.L., Landolt, H., Börnstein, R. (eds.): Diffusion in Semiconductors and Non-Metallic Solids. New Series, Group III, vol. 33A. Springer, Berlin (1998)
50. Schulz, M., Landolt, H., Börnstein, R. (eds.): Impurities and Defects in Group IV Elements, IV-IV and III-V Compounds. New Series, Group III, vol. 41A2, Part α. Springer, Berlin (2002)
51. Compaan, K., Haven, Y.: Correlation factors for diffusion in solids. Trans. Faraday Soc. **52**, 786–801 (1956). doi:10.1039/TF9565200786
52. Frank, W., Gösele, U., Mehrer, H., Seeger, A.: In: Murch, G.E., Nowick, A.S. (eds.) Diffusion in Crystalline Solids. Academic Press, New York (1984)
53. Landolt, H., Börnstein, R.: In: Mehrer, H. (ed.) Diffusion in Solids Metals and Alloys, New Series III, vol. 26. Springer, Berlin (1990)
54. Heumann, Th.: In: Ilschner, B. (ed.) Diffusion in Metallen, Werkstoff-Forschung und -Technik Bd., vol. 10. Springer, Berlin (1992)
55. Compaan, K., Haven, Y.: Correlation factors for diffusion in solids. Part 2. – indirect interstitial mechanism. Trans. Faraday Soc. **54**, 1498–1508 (1958). doi:10.1039/TF9585401498
56. Posselt, M., Gao, F., Bracht, H.: Correlation between self-diffusion in Si and the migration mechanisms of vacancies and self-interstitials: an atomistic study. Phys. Rev. B **78**, 035208 (2008). doi:10.1103/PhysRevB.78.035208
57. Chen, R., Dunham, S.T.: Correlation factors for interstitial-mediated self-diffusion in the diamond lattice: kinetic lattice Monte Carlo approach. Phys. Rev. B **83**, 134124, (2011). doi:10.1103/PhysRevB.83.134124
58. Bracht, H., Haller, E.E., Clark-Phelps, R.: Silicon self-diffusion in isotope heterostructures. Phys. Rev. Lett. **81**, 393 (1998). doi:10.1103/PhysRevLett.81.393
59. Ural, A., Griffin, P.B., Plummer, J.D.: Self-diffusion in silicon: similarity between the properties of native point defects. Phys. Rev. Lett. **83**, 3454 (1999). doi:10.1103/PhysRevLett.83.3454
60. Bracht, H., Haller, E.E.: Comment on "Self-Diffusion in silicon: similarity between the properties of native point defects". Phys. Rev. Lett. **85**, 4835 (2000). doi:10.1103/PhysRevLett.85.4835
61. Ural, A., Griffin, P.B., Plummer, J.D.: Ural, griffin, and plummer reply. Phys. Rev. Lett. **85**, 4836 (2000). doi:10.1103/PhysRevLett.85.4836
62. Aid, S.R., Sakaguchi, T., Toyonaga, K., Nakabayashi, Y., Matumoto, S., Sakuraba, M., Shimamune, Y., Hashiba, Y., Murota, J., Wada, K., Abe, T.: Si self-diffusivity using isotopically pure ^{30}Si epitaxial layers. Mater. Sci. Eng. B **114–115**, 330 (2004). doi:10.1016/j.mseb.2004.07.055
63. Shimizu, Y., Uematsu, M., Itoh, K.M.: Experimental evidence of the vacancy-mediated silicon self-diffusion in single-crystalline silicon. Phys. Rev. Lett. **98**, 095901 (2007). doi:10.1103/PhysRevLett.98.095901
64. Bracht, H., Silvestri, H.H., Sharp, I.D., Haller, E.E.: Self- and foreign-atom diffusion in semiconductor isotope heterostructures. II. Experimental results for silicon. Phys. Rev. B **75**, 035211 (2007). doi:10.1103/PhysRevB.75.035211
65. Kube, R., Bracht, H., Hüger, E., Schmidt, H., Lundsgaard Hansen, J., Nylandsted Larsen, A., Ager III, J.W., Haller, E.E., Geue, T., Stahn, J.: Contributions of vacancies and self-interstitials to self-diffusion in silicon under thermal equilibrium and nonequilibrium conditions. Phys. Rev. B. **88**, 085206 (2013). doi:10.1103/PhysRevB.88.085206
66. Suezawa, M., Iijima, Y., Yonenaga, I.: Comment on "Contributions of vacancies and self-interstitials to self-diffusion in silicon under thermal equilibrium and nonequilibrium conditions". Phys. Rev. B **90**, 117201 (2014). doi:10.1103/PhysRevB.90.117201
67. Kube, R., Bracht, H., Hüger, E., Schmidt, H., Lundsgaard Hansen, J., Nylandsted Larsen, A., Ager III, J.W., Haller, E.E., Geue, T., Stahn, J., Uematsu, M., Itoh, K.M.: Reply to "Comment on 'Contributions of vacancies and self-interstitials to self-diffusion in silicon under thermal equilibrium and nonequilibrium conditions'". Phys. Rev. B **90**, 117202 (2014). doi:10.1103/PhysRevB.90.117202

68. Fahey, P.M., Griffin, P.B., Plummer, J.D.: Point defects and dopant diffusion in silicon. Rev. Mod. Phys. **61**, 289 (1989). doi:10.1103/RevModPhys.61.289
69. Watkins, G.D., Corbett, J.W.: Defects in irradiated silicon: electron paramagnetic resonance and electron-nuclear double resonance of the Si-E center. Phys. Rev. **134**, A1359 (1964). doi:10.1103/PhysRev.134.A1359
70. Manning, J.R.: Correlation factors for impurity diffusion. bcc, Diamond, and fcc Structures. Phys. Rev. **136**, A1758 (1964). doi:10.1103/PhysRev.136.A1758
71. Hu, S.M.: On interaction potential, correlation factor, vacancy mobility, and activation energy of impurity diffusion in diamond lattice. Phys. Stat. Sol. B **60**, 595 (1973). doi:10.1002/pssb.2220600215
72. Dunham, S.T., Wu, C.D.: Atomistic models of vacancy-mediated diffusion in silicon. J. Appl. Phys. **78**, 2362 (1995). doi:10.1063/1.360156
73. Brotzmann, S., Bracht, H.: Intrinsic and extrinsic diffusion of phosphorus, arsenic, and antimony in germanium. J. Appl. Phys. **103**, 033508 (2008). doi:10.1063/1.2837103
74. Chroneos, A., Bracht, H., Grimes, R.W., Uberuaga, B.P.: Vacancy-mediated dopant diffusion activation enthalpies for germanium. Appl. Phys. Lett. **92**, 172103 (2008). doi:10.1063/1.2918842
75. Brotzmann, S., Bracht, H., Lundsgaard Hansen, J., Nylandsted Larsen, A., Simoen, E., Haller, E.E., Christensen, J.S., Werner, P.: Diffusion and defect reactions between donors, C, and vacancies in Ge. I. Experimental results. Phys. Rev. B **77**, 235207 (2008). doi:10.1103/PhysRevB.77.235207
76. Bracht, H.: Self- and foreign-atom diffusion in semiconductor isotope heterostructures. I. Continuum theoretical calculations. Phys. Rev. B **75**, 035210 (2007). doi:10.1103/PhysRevB.75.035210
77. Frank, F.C., Turnbull, D.: Mechanism of diffusion of copper in germanium. Phys. Rev. **104**, 617 (1956). doi:10.1103/PhysRev.104.617
78. Stolwijk, N.A., Frank, W., Hölzl, J., Pearton, S.J., Haller, E.E.: Diffusion and solubility of copper in germanium. Appl. Phys. **57**, 5211 (1985). doi: 10.1063/1.335259
79. Bracht, H.: Copper related diffusion phenomena in germanium and silicon. Mater. Sci. Semicond. Process. **7**, 113 (2004). doi:10.1016/j.mssp.2004.06.001
80. Equation (4) in Ref. [79] should read $D^{\mathrm{eff}}_{\mathrm{Cu}(1)} = 7.8 \times 10^{-5} \exp\left(-\frac{0.084\,\mathrm{eV}}{k_{\mathrm{B}}T}\right)\ \mathrm{cm}^2\mathrm{s}^{-1}$
81. Bracht, H., Stolwijk, N.A., Mehrer, H.: Diffusion and solubility of copper, silver, and gold in germanium. Phys. Rev. B **43**, 14465 (1991). doi:10.1103/PhysRevB.43.14465
82. Gösele, U., Frank, W., Seeger, A.: Mechanism and kinetics of the diffusion of gold in silicon. Appl. Phys. **23**, 361 (1980). doi:10.1007/BF00903217
83. Stolwijk, N.A., Schuster, B., Hölzl, J., Mehrer, H., Frank, W.: Diffusion and solubility of gold in silicon. Physica B+C **116**, 335 (1983). doi:10.1016/0378-4363(83)90271-1
84. Morehead, F., Stolwijk, N.A., Meyberg, W., Gösele, U.: Self-interstitial and vacancy contributions to silicon self-diffusion determined from the diffusion of gold in silicon. Appl. Phys. Lett. **42**, 690 (1983). doi:10.1063/1.94074
85. Stolwijk, N.A., Schuster, B., Hölzl, J.: Diffusion of gold in silicon studied by means of neutron-activation analysis and spreading-resistance measurements. Appl. Phys. A **33**, 133 (1984). doi:10.1007/BF00617619
86. Stolwijk, N.A., Hölzl, J., Frank, W., Weber, E.R., Mehrer, H.: Diffusion of gold in dislocation-free or highly dislocated silicon measured by the spreading-resistance technique. Appl. Phys. A. **39**, 37 (1986). doi:10.1007/BF01177162
87. Hauber, J., Stolwijk, N.A., Tapfer, L., Mehrer, H., Frank, W.: U- and W-shaped diffusion profiles of gold in silicon. J. Phys. C; Solid State Phys. **19**, 5817 (1986). doi:10.1088/0022-3719/19/29/007
88. Stolwijk, N.A., Hölzl, J., Frank, W., Hauber, J., Mehrer, H.: Decoration of defects in silicon with gold, and related subjects. Phys. Stat. Sol. (A) **104**, 225 (1987). doi:10.1002/pssa.2211040117

89. Föll, H., Gösele, U., Kolbesen, B.O.: Microdefects in silicon and their relation to point defects. J. Cryst. Growth **52**, 907 (1981). doi:10.1016/0022-0248(81)90397-3
90. Meyberg, W., Frank, W., Seeger, A., Peretti, H.A., Mondino, M.A.: The migration of interstitials to immobile vacancies and dislocations, with application to plastically deformed tantalum. Cryst. Lattice Defects Amorph. Mater. **10**, 1 (1983)
91. Bracht, H., Overhof, H.: Kinetics of interstitial-substitutional exchange of Zn, Pt, and Au in Si: experimental results and theoretical calculations. Phys. Stat. Sol. (A) **158**, 47 (1996). doi:10.1002/pssa.2211580107
92. Perret, M., Stolwijk, N.A., Cohausz, L.: Kick-out diffusion of zinc in silicon at 1262 K. J. Phys.: Condens. Matter **1**, 6347 (1989). doi:10.1088/0953-8984/1/36/004
93. Grünebaum, D., Czekalla, Th., Stolwijk, N.A., Mehrer, H., Yonenaga, I., Sumino, K.: Diffusion and solubility of zinc in dislocation-free and plastically deformed silicon crystals. Appl. Phys. A **53**, 65 (1991). doi:10.1007/BF00323437
94. Bracht, H., Stolwijk, N.A., Yonenage, I., Mehrer, H.: Interstitial-substitutional diffusion kinetics and dislocation-induced trapping of zinc in plastically deformed silicon. Phys. Stat. Sol. (A) **137**, 499 (1993). doi:10.1002/pssa.2211370220
95. Hauber, J., Frank, W., Stolwijk, N.A.: Diffusion and solubility of platinum in silicon. Mater. Sci. Forum **38–41**, 707 (1989). doi:10.4028/www.scientific.net/MSF.38-41.707
96. Zimmermann, H., Ryssel, H.: Observation of inverse U-shaped profiles after platinum diffusion in silicon. Appl. Phys. Lett. **59**, 1209 (1991). doi:10.1063/1.105505
97. Zimmermann, H., Ryssel, H.: Gold and platinum diffusion: the key to the understanding of intrinsic point defect behavior in silicon. Appl. Phys. A **55**, 121 (1992). doi:10.1007/BF00334210
98. Zimmermann, H., Ryssel, H.: The modeling of platinum diffusion in silicon under non-equilibrium conditions. J. Electrochem. Soc. **139**, 256 (1992). doi:10.1149/1.2069180
99. Lerch, W., Stolwijk, N.A., Mehrer, H., Poisson, Ch.: Diffusion of platinum into dislocated and non-dislocated silicon. Semicond. Sci. Technol. **10**, 1257 (1995). doi:10.1088/0268-1242/10/9/009
100. Mantovani, S., Nava, F., Nobili, S., Ottaviani, G.: In-diffusion of Pt in Si from the PtSi/Si interface. Phys. Rev. B **33**, 5536 (1986). doi:10.1103/PhysRevB.33.5536
101. Giese, A., Bracht, H., Stolwijk, N.A., Walton, J.T.: Out-diffusion of Zn from Si: a method to study vacancy properties in Si. J. Appl. Phys. **83**, 8062 (1998). doi:10.1063/1.367900
102. Giese, A., Bracht, H., Stolwijk, N.A., Baither, D.: Microscopic defects in silicon induced by zinc out-diffusion. Mater. Sci. Eng. B **71**, 160 (2000). doi:10.1016/S0921-5107(99)00367-0
103. Gösele, U., Frank, W., Seeger, A.: An entropy barrier against vacancy-interstitial recombination in silicon. Solid State Commun. **45**, 31 (1983). doi:10.1016/0038-1098(83)90878-5
104. Cowern, N.E.B., Janssen, K.T.F., van de Walle, G.F.A., Gravesteijn, D.J.: Impurity diffusion via an intermediate species: the B-Si system. Phys. Rev. Lett. **65**, 2434 (1990). doi:10.1103/PhysRevLett.65.2434
105. Cowern, N.E.B., van de Walle, G.F.A., Gravesteijn, D.J., Vriezema, C.J.: Experiments on atomic-scale mechanisms of diffusion. Phys. Rev. Lett. **67**, 212 (1991). doi:10.1103/PhysRevLett.67.212
106. Gossmann, H.-J., Gilmer, G.H., Rafferty, C.S., Unterwald, F.C., Boone, T., Poate, J.M., Luftman, H.S., Frank, W.: Determination of Si self-interstitial diffusivities from the oxidation-enhanced diffusion in B doping-superlattices: the influence of the marker layers. J. Appl. Phys. **77**, 1948 (1995). doi:10.1063/1.358828
107. Stolk, P.A., Gossmann, H.-J., Eaglesham, D.J., Jacobson, D.C., Rafferty, C.S., Gilmer, G.H., Jaraíz, M., Poate, J.M., Luftman, H.S., Haynes, T.E.: Physical mechanisms of transient enhanced dopant diffusion in ion-implanted silicon. J. Appl. Phys. **81**, 6031 (1997). doi:10.1063/1.364452
108. Gossmann, H.-J.: Dopants and intrinsic point-defects during Si device processing. In: Huff, H.R., Gösele, U., Tsuya, H. (eds.) Silicon Materials Science and Technology. Electrochemical Society Proceedings, vol. 98-1, p. 884, Pennington (1998)

109. Ural, A., Griffin, P.B., Plummer, J.D.: Fractional contributions of microscopic diffusion mechanisms for common dopants and self-diffusion in silicon. J. Appl. Phys. **85**, 6440 (1999). doi:10.1063/1.370285
110. Cowern, N., Rafferty, C.: Enhanced diffusion in silicon processing. MRS Bull. **25**, 39 (2000). doi:10.1557/mrs2000.97
111. Bracht, H., Stolwijk, N.A., Mehrer, H., Yonenaga, I.: Short-time diffusion of zinc in silicon for the study of intrinsic point defects. Appl. Phys. Lett. **59**, 3559 (1991). doi:10.1063/1.106393
112. Stolwijk, N.A., Grünebaum, D., Perret, M., Brohl, M.: Zinc and sulfur in silicon, experimental evidence for kick-out diffusion behavior. Mater. Sci. Forum **38–41**, 701 (1989). doi:10.4028/www.scientific.net/MSF.38-41.701
113. Rollert, F., Stolwijk, N.A., Mehrer, H.: Diffusion of sulfur-35 into silicon using an elemental vapor source. Appl. Phys. Lett. **63**, 506 (1993). doi:10.1063/1.109987
114. Shockley, W., Moll, J.L.: Solubility of Flaws in Heavily-Doped Semiconductors. Phys. Rev. **119**, 1480 (1960). doi:10.1103/PhysRev.119.1480
115. Gösele, U.M.: Fast diffusion in semiconductors. Ann. Rev. Mater. Sci. **18**, 257 (1988). doi:10.1146/annurev.ms.18.080188.001353
116. Bracht, H., Stolwijk, N.A., Laube, M., Pensl, G.: Diffusion of boron in silicon carbide: evidence for the kick-out mechanism. Appl. Phys. Lett. **77**, 3188 (2000). doi:10.1063/1.1325390
117. Rüschenschmidt, K., Bracht, H., Laube, M., Stolwijk, N.A., Pensl, G.: Diffusion of boron in silicon carbide. Physica B **308–310**, 734 (2001). doi:10.1016/S0921-4526(01)00889-4
118. Rüschenschmidt, K., Bracht, H., Stolwijk, N.A., Laube, M., Pensl, G., Brandes, G.R.: Self-diffusion in isotopically enriched silicon carbide and its correlation with dopant diffusion. J. Appl. Phys. **96**, 1458 (2004). doi:10.1063/1.1766101
119. Yu, S., Tan, T.Y., Gösele, U.: Diffusion mechanism of chromium in GaAs. J. Appl. Phys. **70**, 4827 (1991). doi:10.1063/1.349049
120. Uematsu, M., Wada, K., Gösele, U.: Non-equilibrium point defect phenomena influencing beryllium and zinc diffusion in GaAs and related compounds. Appl. Phys. A **55**, 301 (1992). doi:10.1007/BF00324076
121. Uematsu, M., Werner, P., Schultz, M., Tan, T.Y., Gösele, U.M.: Sulfur diffusion and the interstitial contribution to arsenic self-diffusion in GaAs. Appl. Phys. Lett. **67**, 2863 (1995). doi:10.1063/1.114810
122. Tan, T.Y.: Point defects and diffusion mechanisms pertinent to the Ga sublattice of GaAs. Mater. Chem. Phys. **40**, 245 (1995). doi:10.1016/0254-0584(95)01488-8
123. Bösker, G., Stolwijk, N.A., Thordson, J.V., Södervall, U., Andersson, T.G.: Diffusion of nitrogen from a buried doping layer in gallium arsenide revealing the prominent role of as interstitials. Phys. Rev. Lett. **81**, 3443 (1998). doi:10.1103/PhysRevLett.81.3443
124. Bracht, H., Brotzmann, S.: Zinc diffusion in gallium arsenide and the properties of gallium interstitials. Phys. Rev. B **71**, 115216 (2005). doi:10.1103/PhysRevB.71.115216
125. Sunder, K., Bracht, H., Nicols, S.P., Haller, E.E.: Zinc and gallium diffusion in gallium antimonide. Phys. Rev. B **75**, 245210 (2007). doi:10.1103/PhysRevB.75.245210
126. Ural, A., Griffin, P.B., Plummer, J.D.: Atomic-scale diffusion mechanisms via intermediate species. Phys. Rev. B **65**, 134303 (2002). doi:10.1103/PhysRevB.65.134303
127. Mizuo, S., Higuchi, H.: Retardation of Sb Diffusion in Si during Thermal Oxidation. Jpn. J. Appl. Phys. **20**, 739 (1981). doi:10.1143/JJAP.20.739
128. Mizuo, S., Higuchi, H.: Effects of Oxidation on Aluminum Diffusion in Silicon. Jpn. J. Appl. Phys. **21**, 56 (1982). doi:10.1143/JJAP.21.56
129. Mizuo, S., Kusaka, T., Shintani, A., Nanba, M., Higuchi, H.: Effect of Si and SiO2 thermal nitridation on impurity diffusion and oxidation induced stacking fault size in Si. J. Appl. Phys. **54**, 3860 (1983). doi:10.1063/1.332611
130. Matsumoto, S., Ishikawa, Y., Niimi, T.: Oxidation enhanced and concentration dependent diffusions of dopants in silicon. J. Appl. Phys. **54**, 5049 (1983). doi:10.1063/1.332776
131. Ishikawa, Y., Tomisato, M., Honma, H., Matsumoto, S., Niimi, T.: The retarded diffusion of arsenic in silicon by thermal oxidation in extrinsic conditions. J. Electrochem. Soc. **130**, 2109 (1983). doi:10.1149/1.2119532

132. Fahey, P., Barbuscia, G., Moslehi, M., Dutton, R.W.: Kinetics of thermal nitridation processes in the study of dopant diffusion mechanisms in silicon. Appl. Phys. Lett. **46**, 784 (1985). doi:10.1063/1.95909
133. Miyake, M.: Oxidation-enhanced diffusion of ion-implanted boron in silicon in extrinsic conditions. J. Appl. Phys. **57**, 1861 (1985). doi:10.1063/1.334416
134. Ishikawa, Y., Nakamichi, I., Matsumoto, S., Niimi, T.: The effect of thermal oxidation of silicon on boron diffusion in extrinsic conditions. Jpn. J. Appl. Phys. **26**, 1602 (1987). doi:10.1143/JJAP.26.1602
135. Giles, M.D.: Extrinsic transient diffusion in silicon. Appl. Phys. Lett. **58**, 2399 (1991). doi:10.1063/1.104883
136. Gossmann, H.-J., Haynes, T.E., Stolk, P.A., Jacobson, D.C., Gilmer, G.H., Poate, J.M., Luftman, H.S., Mogi, T.K., Thompson, M.O.: The interstitial fraction of diffusivity of common dopants in Si. Appl. Phys. Lett. **71**, 3862 (1997). doi:10.1063/1.120527
137. Sharp, I.D., Bracht, H.A., Silvestri, H.H., Nicols, S.P., Beeman, J.W., Hansen, J., Nylandsted Larsen, A., Haller, E.E.: Self- and dopant diffusion in extrinsic boron doped isotopically controlled silicon multilayer structures. Mater. Res. Soc. Symp. Proc. **719**, F13.11 (2002). doi:10.1557/PROC-719-F13.11
138. Silvestri, H.H., Sharp, I.D., Bracht, H.A., Nicols, S.P., Beeman, J.W., Hansen, J., Nylandsted Larsen, A., Haller, E.E.: Dopant and Self-Diffusion in Extrinsic n-Type Silicon Isotopically Controlled Heterostructures. Mater. Res. Soc. Symp. Proc. **719**, F13.10 (2002). doi:10.1557/PROC-719-F13.10
139. Silvestri, H.H., Bracht, H., Sharp, I.D., Lundsgaard Hansen, J., Nylandsted Larsen, A., Haller, E.E.: Simultaneous phosphorus and Si self-diffusion in extrinsic, isotopically controlled silicon heterostructures. Mater. Res. Soc. Symp. Proc. **810**, C3.3. (2004) C3.3. doi:10.1557/PROC-810-C3.3
140. Bracht, H., Silvestri, H.H., Haller, E.E.: Advanced diffusion studies with isotopically controlled materials. Solid State Commun. **133**, 727 (2005). doi:10.1016/j.ssc.2004.12.024
141. Bracht, H., Diffusion mediated by doping and radiation-induced point defects. Physica B **376–377**, 11 (2006). doi:10.1016/j.physb.2005.12.006
142. Bracht, H., Rodriguez Schachtrup, A., Yonenaga, I.: Segregation of gold at dislocations confirmed by gold diffusion into highly dislocated silicon. Mater. Sci. Forum **258–263**, 1783 (1997)
143. Rodriguez, A., Bracht, H., Yonenaga, I.: Impact of high B concentrations and high dislocation densities on Au diffusion in Si. J. Appl. Phys. **95**, 7841 (2004). doi:10.1063/1.1751235
144. Yoshida, M., Arai, E., Nakamura, H., Terunuma, Y.: Excess vacancy generation mechanism at phosphorus diffusion into silicon. J. Appl. Phys. **45**, 1498 (1974). doi:10.1063/1.1663450
145. Yoshida, M.: Numerical solution of phosphorus diffusion equation in silicon. Jpn. J. Appl. Phys. **18**, 479 (1979). doi:10.1143/JJAP.18.479
146. Uematsu, M.: Simulation of boron, phosphorus, and arsenic diffusion in silicon based on an integrated diffusion model, and the anomalous phosphorus diffusion mechanism. J. Appl. Phys. **82**, 2228 (1997). doi:10.1063/1.366030
147. Mirabella, S., De Salvador, D., Napolitani, E., Bruno, E., Priolo, F.: Mechanisms of boron diffusion in silicon and germanium. J. Appl. Phys. **113**, 031101 (2013). doi:10.1063/1.4763353
148. Masters, B.J., Fairfield, J.M.: Arsenic isoconcentration diffusion studies in silicon. J. Appl. Phys. **40**, 2390 (1969). doi:10.1063/1.1658001
149. Yoshida, M., Tanaka, S.: Simulation of phosphorus diffusion profiles with different phosphorus surface concentration at the same diffusion temperature in silicon. Jpn. J. Appl. Phys. **41**, 5493 (2002). doi:10.1143/JJAP.41.5493
150. Makris, J.S., Masters, B.J.: Phosphorus isoconcentration diffusion studies in silicon. J. Electrochem. Soc. **120**, 1252 (1973). doi:10.1149/1.2403672
151. Fair, R.B., Tsai, J.C.C.: A Quantitative model for the diffusion of phosphorus in silicon and the emitter dip effect. J. Electrochem. Soc. **124**, 1107 (1977). doi:10.1149/1.2133492
152. Tan, T.Y., Gösele, U.: Point defects, diffusion processes, and swirl defect formation in silicon. Appl. Phys. A **37**, 1 (1985). doi:10.1007/BF00617863

153. Ager III, J.W., Beeman, J.W., Hansen, W.L., Haller, E.E., Sharp, I.D., Liao, C., Yang, A., Thewalt, M.L.W., Riemann, H.: High-purity, isotopically enriched bulk silicon. J. Electrochem. Soc. **152**, G448 (2005). doi:10.1149/1.1901674
154. see International Technology Roadmap for Semiconductors. http://www.itrs.net
155. Thewalt, M.L.W.: Spectroscopy of excitons and shallow impurities in isotopically enriched silicon-electronic properties beyond the virtual crystal approximation. Solid State Commun. **133**, 715 (2005). doi:10.1016/j.ssc.2004.12.023
156. Steger, M., Yang, A., Sekiguchi, T., Saeedi, K., Thewalt, M.L.W., Henry, M.O., Johnston, K., Riemann, H., Abrosimov, N.V., Churbanov, M.F., Gusev, A.V., Kaliteevskii, A.K., Godisov, O.N., Becker, P., Pohl, H.-J.: Photoluminescence of deep defects involving transition metals in Si: new insights from highly enriched 28Si. J. Appl. Phys. **110**, 081301 (2011). doi:10.1063/1.3651774
157. Gleiter, H.: Nanostructured materials: basic concepts and microstructure. Acta Mater. **48**, 1 (2000). doi:10.1016/S1359-6454(99)00285-2
158. Gleiter, H., Weissmüller, J., Wollersheim, O., Würschum, R.: Nanocrystalline materials: a way to solids with tunable electronic structures and properties. Acta mater. **49**, 737 (2001). doi:10.1016/S1359-6454(00)00221-4
159. Colinge, J.-P., Lee, C.-W., Afzalian, A., Akhavan, N.D., Yan, R., Ferain, I., Razavi, P., O'Neill, B., Blake, A., White, M., Kelleher, A.-M., McCarthy, B., Murphy, R.: Nanowire transistors without junctions. Nat. Nanotechnol. **5**, 225 (2010). doi:10.1038/nnano.2010.15

Chapter 2
Density Functional Modeling of Defects and Impurities in Silicon Materials

José Coutinho

Abstract This is a contribution targeted for early scientists, from both academia and industry, providing the grounds for defect modeling in silicon materials using density functional methods. It starts with a revision of the theoretical framework and tools, including relevant approximations such as the treatment of the exchange-correlation interactions and the use of pseudopotentials. It then describes how to step up from total energies, electron densities and Kohn-Sham states to the actual defect observables. Particular emphasis is given to the calculation of spectroscopic observables such as electrical levels, local vibrational modes, spin densities, migration barriers, and defect response to uniaxial stress. Each of these techniques is accompanied by examples of defect calculations. It is shown how these results can be crucial in unraveling a detailed picture of many complexes, including substitutional and interstitial impurities, dopants, transition metals, carbon, oxygen or hydrogen. In the last section we take a look at some developments in modeling defects in silicon nanostructures. While holding promising optical and magnetic properties, nano-silicon presents many challenges, particularly with regard to defect control, doping and electrical transport.

Keywords Density functional theory • Defects in silicon • Modeling • Spectroscopy

2.1 Introduction

Silicon has been the material of choice for applications in consumer electronics during the last four decades. This is must owed to a coalition of several factors like abundance, doping flexibility, the possibility to fabricate clean and sharp interfaces with a stable oxide, not to mention other valuable attributes such as low toxicity and environmental friendship.

J. Coutinho (✉)
Department of Physics and I3N, University of Aveiro, Campus Santiago, 3810-193 Aveiro, Portugal
e-mail: jose.coutinho@ua.pt

Y. Yoshida, G. Langouche (eds.), *Defects and Impurities in Silicon Materials*, Lecture Notes in Physics 916, DOI 10.1007/978-4-431-55800-2_2

Understanding defects in silicon has paved the way for its success. Thanks to a combination of applied, experimental and theoretical research on defects, silicon is the most pure material ever known. However, as structure dimensions in micro- and nano-electronics are scaled down to achieve economic benefit and functionality, stringent control of statistical fluctuations of dopants and defects is becoming a key ingredient in device design [1]. Defect and impurity issues have also been a matter of great interest in material growth industries. The recent developments in purification methods to upgrade metallurgical Si directly to solar-Si stand as a prominent example where the importance of defect studies is huge. Although avoiding chlorosilanes, hydrochloric acid and being less energy hungry than the standard Siemens process, the resulting material is dirtier and less perfect [2]. These are just two examples of defect-related problems in Si materials that need our attention today.

Density functional theory (DFT) [3, 4] has played an enormous role along this road and holds equally great expectations for the future of defect modeling. DFT was groundbreaking by fundamentally changing the object of study in electronic structure calculations – Instead of looking for the electronic wavefunction of defects in solids, which in the simplest case depends on the coordinates of all electrons, DFT is able to describe the all-electron problem by means of the electron density, i.e. a function of solely three space coordinates, and this is all done without adjusting parameters [5]. Combined with several advances and techniques like the use of efficient and accurate pseudopotentials [6, 7], order-N diagonalization algorithms [8], real-space mesh implementations [9], basis set filtration [10] or the increasing availability of computational power, DFT has empowered us to carry out accurate calculations of the electronic structure of atomistic models of thousands of atoms with increasing complexity.

Although we may find in the literature several in-depth reviews about modeling defects in semiconductors by means of first-principles methods (see for instance [11] and references therein), it is also true that novice researchers, both in academia and industry, often face severe difficulties in making the most out of a myriad of DFT packages available (many of them for free!). Such difficulties range from syntax issues to the validity of the approximations to be used, but often are related to the process of bridging theory to the observations, i.e. how to calculate defect-related observables from the output data of a DFT code. This very last sentence states the main goal of the present contribution, and this is done with the help of several exemplifying calculations made for defects in silicon materials, including vacancy-related complexes, transition metal impurities, hydrogen and oxygen.

The chapter starts with a brief revision of the density functional theoretical framework and tools, including a visit to the Hartree-Fock method and several approximations like the use of pseudopotentials, the choice of basis functions and boundary conditions. It then describes how to make use of total energies, electron densities and Kohn-Sham states to arrive at the observed properties of defects. Particular emphasis is given to the calculation of electrical levels, local vibrational modes, migration barriers, and defect response to uniaxial stress. All these topics are accompanied by examples of calculations, as well as references to the literature for those who may want to deepen their understanding on specific issues.

2.2 Theoretical Framework

2.2.1 The Many-Body Problem

In the absence of external fields, the Schrödinger equation for a non-relativistic and stationary problem involving a set of N_e electrons and N_n atomic nuclei can be written as

$$\hat{H}\Psi = E\Psi. \tag{2.1}$$

Here, the Hamiltonian $\hat{H}$ contains the usual kinetic and potential terms,[1]

$$\begin{aligned}\hat{H} &= -\frac{1}{2}\sum_{i}^{N_\mathrm{e}}\nabla_i^2 - \sum_{\alpha}^{N_\mathrm{n}}\frac{1}{2M_\alpha}\nabla_\alpha^2 \\ &+ \frac{1}{2}\sum_{\substack{i,j=1\\ i\neq j}}^{N_\mathrm{e}}\frac{1}{|\mathbf{r}_i-\mathbf{r}_j|} - \sum_{i,\alpha=1}^{N_\mathrm{e},N_\mathrm{n}}\frac{Z_\alpha}{|\mathbf{r}_i-\mathbf{R}_\alpha|} + \frac{1}{2}\sum_{\substack{\alpha,\beta=1\\ \alpha\neq\beta}}^{N_\mathrm{n}}\frac{Z_\alpha Z_\beta}{|\mathbf{R}_\alpha-\mathbf{R}_\beta|}, \quad (2.2)\\ &= \hat{T}_\mathrm{e} + \hat{T}_\mathrm{n} + \hat{V}_\mathrm{ee} + \hat{V}_\mathrm{en} + \hat{V}_\mathrm{nn} \quad (2.3)\end{aligned}$$

where M_α, Z_α and $\mathbf{R}_\alpha$ represent the mass, charge and location of the α-th nucleus, and $\mathbf{r}_i$ the i-th electron coordinate. The many-body wavefunction Ψ from Eq. 2.1 is therefore a function of N_n nuclei coordinates, plus N_e electron space- and spin-coordinates, $\mathbf{r}_i$ and s_i respectively,

$$\Psi \equiv \Psi(\mathbf{r}_1, s_1, \ldots, \mathbf{r}_{N_\mathrm{e}}, s_{N_\mathrm{e}}; \mathbf{R}_1, \ldots, \mathbf{R}_{N_\mathrm{n}}). \tag{2.4}$$

Disregarding any analytical solution for Eq. 2.1 (only known for problems such as the hydrogen and He^+ atoms [12]), the fact that Ψ depends on $3N_\mathrm{n} + 4\sum_\alpha Z_\alpha$ scalar variables makes the simplest of the problems intractable, even when the fastest computers are put at our disposal. This is the time where bold approximations have to come into play.

[1]All quantities are expressed in atomic units (unless otherwise specified). In this system of units, $\hbar$, e, m and $4\pi\epsilon_0$ are taken to be unity, where e, m and ϵ_0 are the electron charge, electron mass, and the permittivity of vacuum respectively. The unit of length is 0.529 Å, and the unit of energy is 1 Hr = 27.211 eV.

2.2.2 Born-Oppenheimer Approximation

By noting that the electron mass is ~2000 times lighter than that of a neutron or proton, we expect electrons to respond *instantaneously* to atomic motion. In other words, the shape of an electron-only wavefunction will adapt adiabatically to nuclear motion.[2] The idea proposed by Born and Oppenheimer [13], was that the many-body wavefunction Ψ in Eq. 2.4 could be approximated to first order by the ansatz

$$\Psi(\mathbf{r};\mathbf{R}) = \psi_{\mathbf{R}}(\mathbf{r})\,\phi(\mathbf{R}), \tag{2.5}$$

where ψ and ϕ are independent electronic and nuclear wavefunctions, with $\mathbf{r}$ and $\mathbf{R}$ representing all electronic and nuclear degrees of freedom, respectively. The subscripted general coordinate in $\psi_{\mathbf{R}}$ stresses the fact that the electronic wavefunction depends on $\mathbf{R}$ in a parametric way.

Let us for the moment ignore any eventual magnetic interactions and drop the spin quantum numbers. By combining Eqs. 2.5 and 2.3, one obtains

$$\begin{aligned}\hat{H}\psi_{\mathbf{R}}(\mathbf{r})\phi(\mathbf{R}) &= (E_{\rm n}+E_{\rm e})\,\psi_{\mathbf{R}}(\mathbf{r})\phi(\mathbf{R})\\ &\quad -\sum_{\alpha}^{N_{\rm n}}\frac{1}{2M_\alpha}\left[\phi(\mathbf{R})\nabla_\alpha^2\psi_{\mathbf{R}}(\mathbf{r}) + 2\,\boldsymbol{\nabla}_\alpha\psi_{\mathbf{R}}(\mathbf{r})\cdot\boldsymbol{\nabla}_\alpha\phi(\mathbf{R})\right],\end{aligned} \tag{2.6}$$

with $E_{\rm n}$ and $E_{\rm e}$ being eigenvalues from the separate nuclear and electronic problems respectively,

$$\left(\hat{T}_{\rm n} + \hat{V}_{\rm nn}\right)\phi = E_{\rm n}\phi \tag{2.7}$$

$$\left(\hat{T}_{\rm e} + \hat{V}_{\rm ee} + \hat{V}_{\rm en}\right)\psi = E_{\rm e}\psi. \tag{2.8}$$

Equations 2.7 and 2.8 include all kinetic ($\hat{T}$) and potential ($\hat{V}$) terms in the many-body Hamiltonian (Eq. 2.3). Subscripts 'n' and 'e' label the nuclear and electronic terms respectively. Now, assuming $M_\alpha \gg 1$, the energy is $E \approx E_{\rm e} + E_{\rm n}$ and any stationary state of the many-body system is obtained by finding ψ for a set of fixed atomic positions.

It is worth noting that when electron-nuclear coupling is strong, the Born-Oppenheimer approximation breaks down. This is the case for Jahn-Teller systems, where $\boldsymbol{\nabla}_\alpha\psi$ is large enough as to impose a nuclear distortion due mixing of the nuclear and electronic wavefunctions. The Jahn-Teller effect demands a proper treatment of the vibronic coupling at the unstable partially occupied manifold, and

---

[2] Hydrogen-containing problems are somewhat special, in the sense that electron-phonon coupling effects can be strong and in some cases have to be accounted for explicitly.

that is beyond the capability of static electronic structure calculations. In spite of these limitation, DFT along with the Born-Oppenheimer approximation has been able to match the experimentally observed JT-relaxed structures of many defects in Si, the single vacancy being a prototypical example [14–16].

2.2.3 Hartree-Fock Method

The Hartree-Fock (HF) method [17] uses a variational approach in which the wave function of the many-electron system has the form of an anti-symmetrized product of one-electron wavefunctions (therefore taking into account the Pauli exclusion principle in a natural way). These are elegantly written in the form of Slater determinants, [18]

$$\Psi_{\mathrm{SD}}(\mathbf{x}_1,\ldots,\mathbf{x}_N) = \frac{1}{\sqrt{N!}} \det \begin{bmatrix} \psi_1(\mathbf{x}_1) & \cdots & \psi_N(\mathbf{x}_1) \\ \vdots & \ddots & \vdots \\ \psi_1(\mathbf{x}_N) & \cdots & \psi_N(\mathbf{x}_N) \end{bmatrix}, \tag{2.9}$$

where $\mathbf{x}_i \equiv (\mathbf{r}_i, s_i)$ stores space- and spin-coordinates of the i-th electron, ψ_i are one-electron *spin-orbitals*, which we assume to be orthogonal, and $N \equiv N_\mathrm{e}$ is the number of electrons. The wavefunctions Ψ_{SD} carry the exchange properties of a many-body wavefunction. The determinantal form ensures that exchanging two electron coordinates affects Ψ_{SD} by a factor of -1, while $\Psi_{\mathrm{SD}} = 0$ if two electrons with the same coordinate are present.

Considering Ψ_{SD} as a solution to the Hamiltonian in Eq. 2.1, and considering electronic terms only, one arrives at an expectation value for the total energy,

$$E = \sum_i^N H_i + \frac{1}{2} \sum_{i,j=1}^N \left(J_{ij} - K_{ij} \right), \tag{2.10}$$

where H_i are one-electron integrals

$$H_i = \int \mathrm{d}\mathbf{x}\, \psi_i^*(\mathbf{x}) \left[-\frac{1}{2}\nabla_i^2 - \sum_{\alpha=1}^{N_\mathrm{n}} \frac{Z_\alpha}{|\mathbf{r}_i - \mathbf{R}_\alpha|} \right] \psi_i(\mathbf{x}), \tag{2.11}$$

and J_{ij} and K_{ij} are the well-known Hartree and exchange integrals, respectively. These are two-electron integrals,

$$J_{ij} = \iint \mathrm{d}\mathbf{x}\, \mathrm{d}\mathbf{x}'\, \psi_i(\mathbf{x}) \psi_i^*(\mathbf{x}) \frac{1}{|\mathbf{r} - \mathbf{r}'|} \psi_j(\mathbf{x}') \psi_j^*(\mathbf{x}') \tag{2.12}$$

$$K_{ij} = \iint \mathrm{d}\mathbf{x}\, \mathrm{d}\mathbf{x}'\, \psi_i(\mathbf{x}) \psi_j^*(\mathbf{x}) \frac{1}{|\mathbf{r} - \mathbf{r}'|} \psi_i(\mathbf{x}') \psi_j^*(\mathbf{x}'), \tag{2.13}$$

which consider volume integration and spin summation, and account for electron-electron Coulomb repulsion and exchange interactions due to spin-correlation effects, respectively. Minimization of Eq. 2.10 subject to the orthogonality condition $\langle\psi_i|\psi_j\rangle = \int \mathrm{d}\mathbf{r}\, \psi_i\psi_j^* = \delta_{ij}$, results in the Hartree-Fock equations [19],

$$\hat{F}\,\psi_i(\mathbf{x}) = \sum_{j=1}^{N} \epsilon_{ij}\,\psi_j(\mathbf{x}), \tag{2.14}$$

where the Fock operator $\hat{F}$ is defined as

$$\hat{F} = \hat{h} + \hat{j} - \hat{k} \tag{2.15}$$

with $\langle\psi_i|\hat{h}|\psi_i\rangle = H_i$ (see Eq. 2.11), and

$$\hat{j}(\mathbf{x})f(\mathbf{x}) = \sum_{k=1}^{N} \int \mathrm{d}\mathbf{x}'\, \psi_k^*(\mathbf{x}')\psi_k(\mathbf{x}') \frac{1}{|\mathbf{r}-\mathbf{r}'|} f(\mathbf{x}) \tag{2.16}$$

$$\hat{k}(\mathbf{x})f(\mathbf{x}) = \sum_{k=1}^{N} \int \mathrm{d}\mathbf{x}'\, \psi_k^*(\mathbf{x}')f(\mathbf{x}') \frac{1}{|\mathbf{r}-\mathbf{r}'|} \psi_k(\mathbf{x}) \tag{2.17}$$

for an arbitrary function $f(\mathbf{x})$. The matrix elements ϵ_{ij} consist of Lagrange multipliers from the minimization procedure. Diagonal elements are obtained after integrating the Fock operator, i.e.,

$$\epsilon_i \equiv \epsilon_{ii} = \langle\psi_i|\hat{F}|\psi_i\rangle = H_i + \sum_{j=1}^{N} \left(J_{ij} - K_{ij}\right), \tag{2.18}$$

which after summation, and according to Eq. 2.10, gives the total energy as

$$E = \sum_{i=1}^{N} \epsilon_i - \frac{1}{2}\sum_{i,j=1}^{N} \left(J_{ij} - K_{ij}\right). \tag{2.19}$$

Usually, in a computer implementation of the HF method, the spin-orbitals ψ's are expanded as a linear combination of M atomic orbitals $\phi_i(\mathbf{x})$,

$$\psi_j(\mathbf{x}) = \sum_{i}^{M} c_{ij}\phi_i(\mathbf{x}). \tag{2.20}$$

The Fock equation then takes a matrix form, and we end up with a generalized eigenvalue problem known as Roothaan equation [17],

$$\mathbf{F}\cdot\mathbf{c}_j = \epsilon_j\,\mathbf{S}\cdot\mathbf{c}_j, \tag{2.21}$$

where S is a $M \times M$ overlap matrix $S_{ij} = \langle\phi_i|\phi_j\rangle$. The above equation is solved in a self-consistent manner, until the wavefunctions (i.e. the c_{ij} coefficients) converge and become consistent with the one-electron and total energies. The physical meaning of the eigenvalues from the Roothaan equation is provided by the Koopmans' theorem [20],

Theorem 1 (Koopmans' theorem) *Assuming that the eigenstates* $\mathbf{c}_j$ *do not vary after removal of one electron from the system, the ionization energy* I_m *of the m-th electron is given by* $I_m = -\epsilon_m$.

Density functional theory also has a Koopman's theorem analogue which is more restricted in the sense that it can does not apply to all eigenstates. It links the first vertical ionization energy (without atomic relaxation) of a many-electron system to the negative of the highest occupied Kohn-Sham state ($I = -\epsilon_{\mathrm{max}}$) [21].

The Hartree-Fock method has one major short-coming – the wavefunction does not incorporate electron-correlation, that is, the spin-orbitals do not have a functional dependence on $\mathbf{r}_i - \mathbf{r}_j$. This problem can be solved by expressing Ψ as a linear combination of Slater determinants, known as Configuration Interaction (CI) method. Unfortunately, this approach brings a serious computational cost as the number of determinants grows rapidly with the size of the problem. Also the evaluation of J_{ij} and K_{ij} matrix elements is a demanding computational task – its CPU-time scales with the fourth power of the basis size. In fact, even with help of modern supercomputers, only systems with a few tens of atoms can be studied.

2.2.4 Density-Functional Theory

As we mentioned above, application of the all-electron HF method to extended systems is computationally prohibitive. The non-local character of both $\hat{j}$ and $\hat{k}$ operators (their values at $\mathbf{r}$ depend on another coordinate $\mathbf{r}'$), complicates the evaluation of the J_{ij} and K_{ij} matrix elements. Most of today's state-of-the-art electronic structure calculations of defects in solids use density-functional theory [3, 4]. This method has been extensively reviewed in the past and we divert the most interested readers to several dedicated reports available in the literature [5, 22–24]. Accordingly, the total energy $E \equiv E[n(\mathbf{r})]$ is a functional of the electron density $n(\mathbf{r})$ (as opposed to the HF method where E is directly connected to the wavefunction). It is therefore proposed that,

$$E[n] = F[n] + \int \mathrm{d}^3\mathbf{r}\, v_{\mathrm{ext}}(\mathbf{r})\, n(\mathbf{r}), \tag{2.22}$$

with v_{ext} representing an external potential to which the electrons are subjected. This includes electron-ion interactions and others, such as applied electrical and magnetic fields. The functional F is universal (system-independent), and accounts for the electronic kinetic energy, as well as electronic Coulomb and exchange correlation interactions [3].

Theorem 2 (First Hohenberg-Kohn theorem) *The external potential is determined, within a trivial additive constant, by the electron density* $n(\mathbf{r})$.

According to DFT, the number of electrons, the external potential, and therefore the ground-state along with all its properties are uniquely defined by the electron density [3].

Theorem 3 (Second Hohenberg-Kohn theorem) *For a trial density* $\tilde{n}(\mathbf{r})$*, such that* $\tilde{n}(\mathbf{r}) \geq 0$ *and* $\int \mathrm{d}\mathbf{r}\, \tilde{n}(\mathbf{r}) = N$*, the ground state energy* E_0 *is*

$$E_0 \leq E[\tilde{n}]. \tag{2.23}$$

The previous expression illustrates the variational character of DFT, i.e.,

$$E_0 = \min_{\tilde{n}} E[\tilde{n}]. \tag{2.24}$$

The striking feature of this method, is that despite the huge simplification of adopting the electron density as the variational variable, no approximations are made. Still as in HF, the exchange-correlation contributions expressed by the functional F have a non-local character.

2.2.4.1 Kohn-Sham Equations

In DFT, we make use of the following set of independent-particle equations, [4]:

$$\left[-\frac{1}{2}\nabla^2 - \sum_{\alpha} \frac{Z_\alpha}{|\mathbf{r}-\mathbf{R}_\alpha|} + \int \frac{n(\mathbf{r}')}{|\mathbf{r}-\mathbf{r}'|}\mathrm{d}^3\mathbf{r}' + \frac{\delta E_{\mathrm{xc}}[n]}{\delta n(\mathbf{r})}\right]\psi_\lambda(\mathbf{r}) = \epsilon_\lambda \psi_\lambda(\mathbf{r}). \tag{2.25}$$

with the density n obtained by summing up all occupied spin-orbital states,

$$n(\mathbf{r}) = \sum_{\lambda \in \mathrm{occ}} |\psi_\lambda(\mathbf{r})|^2 . \tag{2.26}$$

The sum of the three potential terms in Eq. 2.25 is usually referred to as the *effective potential*, $v_{\mathrm{eff}}(\mathbf{r})$, and comprises the external potential imposed by the ions, the Hartree potential, and the exchange-correlation potential which lumps together all remaining many-body effects. One main pillar of DFT is the existence of an universal E_{xc} density functional (depending only on the electron density n of the system), that leads to the knowledge of the *exact* ground-state total energy.

The total energy functional for a many-electron system is given by [23],

$$E[n] = \sum_{\lambda \in \mathrm{occ}} \epsilon_\lambda - J[n] + E_{\mathrm{xc}}[n] - \int \mathrm{d}^3\mathbf{r}\, V_{\mathrm{xc}}[n]\, n(\mathbf{r}). \tag{2.27}$$

In Eq. 2.27, the Hartree functional (J) and the exchange-correlation potential (V_{xc}) are written as

$$J[n(\mathbf{r})] = \frac{1}{2}\int \mathrm{d}^3\mathbf{r} \int \mathrm{d}^3\mathbf{r}' \, \frac{n(\mathbf{r})n(\mathbf{r}')}{|\mathbf{r}-\mathbf{r}'|} \tag{2.28}$$

$$V_{\text{xc}}[n(\mathbf{r})] = \frac{\partial E_{\text{xc}}}{\partial n}. \tag{2.29}$$

Unfortunately the exact form of E_{xc} is unknown, although several approximations are currently available. In fact, improvements to this functional is still one of today's most active research areas.

Equations 2.25 and 2.26 – known as *Kohn-Sham equations* [4], are solved in a self-consistent manner as shown schematically in Fig. 2.1. We first choose a trial charge density, often a superposition of atomic density data (atomic wave functions, atomic electron densities or electrostatic potentials), and calculate the effective potential. Then we move on and solve the Kohn-Sham single-electron equations. The resulting eigenvectors are fed into 2.26 to build a new charge-density, which is the input to calculate a new effective potential. This process (*self-consistency cycle*) keeps going until the charge-density, eigenvalues, total energy, or a combination of these, do not show an appreciable change between consecutive steps (usually implemented by imposing certain tolerance limits δ for these changes).

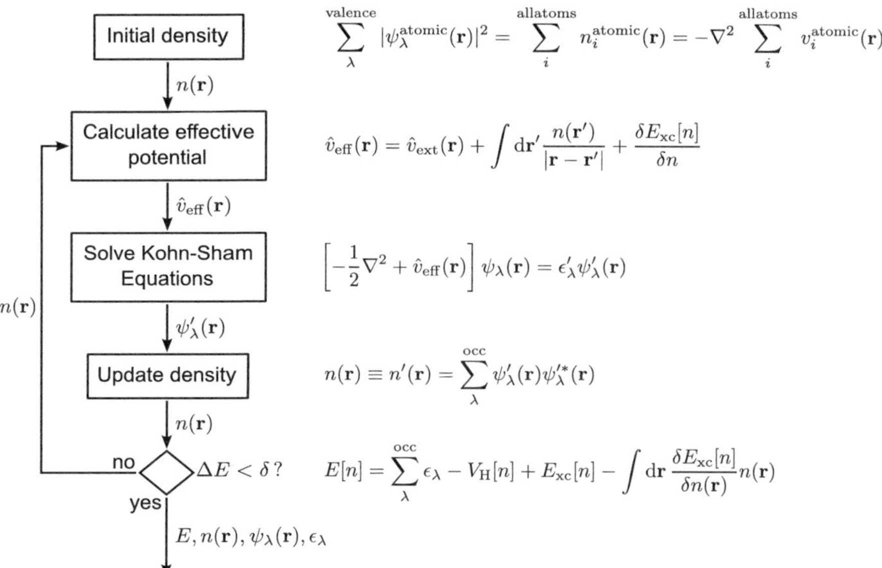

Fig. 2.1 Simplified algorithm of the self-consistent cycle in a density functional code

The above formalism neglects any spin-dependence. The extension of the Kohn-Sham scheme to spin-polarized systems was first proposed by assuming a local treatment of $E_{\rm xc}$ [25, 26]. Here the electron density is made up of spin-up and spin-down contributions, $n(\mathbf{r}) = n^{\uparrow}(\mathbf{r}) + n^{\downarrow}(\mathbf{r})$. This will be discussed further ahead. Note that we have now one-electron spin-orbitals, and therefore the total electron density is

$$n(\mathbf{r}) = \sum_{\lambda,s\in{\rm occ}} |\psi_{\lambda,s}(\mathbf{r})|^2 . \tag{2.30}$$

2.2.4.2 The Exchange-Correlation Functional

In DFT, the exchange-correlation energy is accounted for by the $E_{\rm xc}[n]$ functional, which is unknown and non-local. Several methods have been proposed to overcome this problem. A common approach is to adopt the local density approximation (LDA), or local spin-density approximation (LSDA) for non-zero spin systems [4, 25, 27], where it is assumed that the exchange-correlation energy is a local functional of the density (up and down components) and can be expressed with help of an exchange-correlation energy density, $\epsilon_{\rm xc}(n)$,

$$E_{\rm xc}[n^{\uparrow}, n^{\downarrow}] = \int {\rm d}^3\mathbf{r}\, \epsilon_{\rm xc}(n^{\uparrow}, n^{\downarrow})\, n(\mathbf{r}).$$

Usually this is separated into exchange and correlation components, where within LSDA notation,

$$E_{\rm xc}[n^{\uparrow}, n^{\downarrow}] = E_{\rm x}[n^{\uparrow}, n^{\downarrow}] + E_{\rm c}[n^{\uparrow}, n^{\downarrow}]. \tag{2.31}$$

If we consider the exchange contribution from the homogeneous electron gas we arrive at [25],

$$E_{\rm x}[n^{\uparrow}, n^{\downarrow}] = -\frac{3}{4}\left(\frac{3}{\pi}\right)^{1/3} \int {\rm d}^3\mathbf{r} \left(n^{\uparrow 4/3} + n^{\downarrow 4/3}\right). \tag{2.32}$$

The correlation part is however more complicated. In the high-density regime, perturbation theory yields one expression [27], while in the low-density regime, a Green function quantum Monte Carlo method yields another [28, 29]. A simple parameterized functional form is then fitted to the numerical results. Several parameterizations are available, for instance those from Perdew and Zunger [27], Vosko, Wilk and Nusair [30], and Perdew and Wang [31].

It is remarkable how such a simplistic approach has been so successful in the description of the many-body effects in many semiconductors including Si. As noted by Hood and co-workers [32], this is mostly due to a fortunate cancelation of errors in the calculated potential. This is effect is shown in Fig. 2.2, which represents local

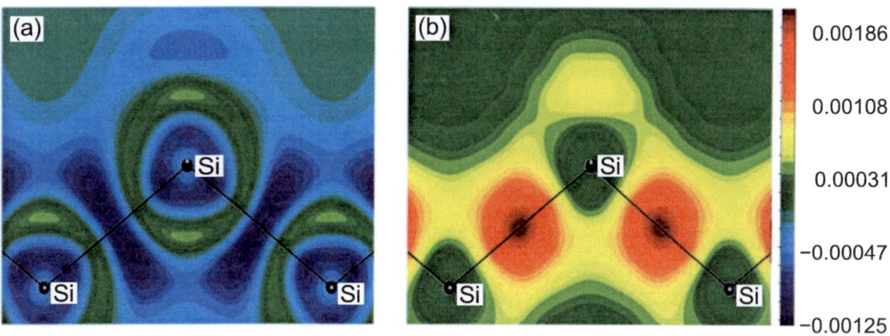

Fig. 2.2 Contour plots in the (110) plane of a Si crystal passing through atoms for (**a**) $e_{\rm x}^{\rm VMC}(\mathbf{r}) - e_{\rm x}^{\rm LDA}(\mathbf{r})$, and (**b**) $e_{\rm c}^{\rm VMC}(\mathbf{r}) - e_{\rm c}^{\rm LDA}(\mathbf{r})$. The atoms and bonds in the (110) plane are schematically represented. The contours are in atomic units (Reprinted with permission from Ref. [32]. ©1998 American Physical Society)

differences of the exchange energy density ($e_{\rm x}$) and the correlation energy density ($e_{\rm c}$) between LDA calculations and highly accurate Variational quantum Monte Carlo (VMC) calculations in bulk Si. It is clear that within LDA, the exchange contribution is overestimated almost everywhere, whereas the correlation part is underestimated almost everywhere.

First-order expansions of $E_{\rm xc}$ in the density have also been developed. This is on the basis of the generalized gradient approximation (GGA), and includes terms depending on ∇n [33, 34]. This is still a local functional, although it depends on the gradient of the density at every point in space (hence being often referred as *semi-local*),

$$E_{\rm xc}[n^{\uparrow}, n^{\downarrow}] = \int {\rm d}^3\mathbf{r}\, \epsilon_{\rm xc}(n^{\uparrow}, n^{\downarrow}, \nabla n^{\uparrow}, \nabla n^{\downarrow})\, n(\mathbf{r}),$$

and the exchange correlation potential

$$V_{\rm xc}[n(\mathbf{r})] = \frac{\partial E_{\rm xc}}{\partial n} - \nabla \cdot \frac{\partial E_{\rm xc}}{\partial(\nabla n(\mathbf{r}))},$$

is calculated numerically.

Usually, the use of GGA exchange-correlation functionals lead to superior results compared to LDA calculations, especially when the electron density varies *quickly* in space. This is often the case of non-screened problems such as molecules and other vacuum surrounded problems. For *infinite* solids, and in particular for group-IV semiconductors, gradient corrected functionals do not bring relevant advantages over LDA, except perhaps when dealing with defects that include species with strongly correlated 3d-, 4d- or f-shell electrons, where even GGA shows insufficiencies. In such cases, a successful method has been the LDA+U (or GGA+U) functional, where a Hartree-Fock-like treatment of the on-site interactions is carried out for electrons with angular momentum $l = 2$ or $l = 3$ on specific atoms, and uses plain LDA (or GGA) to describe all other interactions in the system [35].

2.2.5 *Pseudopotentials*

It is well known that the standard properties of molecules or solids are governed by valence electrons. Electrons in core-states behave much like those in isolated atoms, creating an effective potential that screens the nuclei to the outer shells – this is basically a *pseudopotential*. The idea behind the pseudopotential method is to split the all-electron Coulomb operator into two separate parts, i.e., one that takes care of the inner core-states, and the other dealing with valence states. This allows us to drop matrix elements regarding the core-electrons, and focus only on few remaining ones. Detailed information on the pseudopotential method has been extensively reported [36–40].

Two main reasons justify this approach, (i) a full electron potential results in large total energies, therefore leading to large errors when comparing similar systems, (ii) the number of basis functions required to fit all states (especially those at the core) would become prohibitively large, and affordable expansions would carry catastrophic consequences to the accuracy of the results.

Obviously, being an approximation, the use of pseudopotentials has several caveats. Some assumptions are made, namely (i) the *frozen core* approximation which assumes the core states to be unperturbed when the atom is transferred between different environments, and (ii) the *small core* approximation where one assumes a negligible overlap between core and valence states.

2.2.5.1 Basic Formulation

Transferability *measures* the quality of a pseudopotential. That is, a pseudopotential will be transferable if it can reproduce the properties of the valence electrons in a wide rage of different problems. To accomplish this, the pseudopotential must be constructed by stripping off the valence electrons from the atom, leaving behind an ionic-pseudopotential [38]. Within the density-functional framework,

$$\hat{V}^{\mathrm{ps}}_{\mathrm{ion}}(\mathbf{r}) = \hat{V}^{\mathrm{ps}}(\mathbf{r}) - \int \mathrm{d}^3\mathbf{r}' \, \frac{n^{\mathrm{ps}}(\mathbf{r})}{|\mathbf{r} - \mathbf{r}'|} - \hat{V}_{\mathrm{xc}}[n(\mathbf{r})], \tag{2.33}$$

with

$$n^{\mathrm{ps}}(\mathbf{r}) = \sum_{\lambda} \left|\psi^{\mathrm{ps}}_{\lambda}(\mathbf{r})\right|^2 \tag{2.34}$$

and

$$\hat{V}^{\text{ps}}(\mathbf{r}) = V^{\text{ps}}_{\text{loc}}(\mathbf{r}) + \sum_{i,j} d_{ij} \left|p_i\right\rangle \left\langle p_j\right| \tag{2.35}$$

is the pseudopotential of the neutral atom that depends on a local term and a non-local contribution. The latter carries several species-specific matrix elements (d_{ij}), and operators (p_i) that project the electronic states on to specific angular/radial states i.

Note that in 2.33 we not only subtract a classic Coulomb term, but also all exchange-correlation interactions $\hat{V}_{\text{xc}}$ between the valence electrons [40]. Pseudopotentials are usually formulated in order to suit a particular type of basis functions underlying the pseudo-wavefunction ($\psi_{\lambda}^{\text{ps}}$) and electron density expansion. For instance, the pseudopotentials proposed by Bachelet, Hamann and Schlüter [41], or those by Hartwigsen, Goedecker and Hutter [7], are based on a Gaussian formulation, and therefore are particularly efficient for codes that use atomic-like local-basis functions. These pseudopotentials can be particularly *hard* in the sense that they account for strongly varying potentials (that lead to high kinetic energy states), making the expansion of the valence easy, usually with help of about 10–30 atomic-like functions per atom. Conversely the norm-conserving pseudopotential scheme proposed by Troullier and Martins [42], the ultra-soft pseudopotentials by Vanderbilt [43] and the Blöchl's projector augmented wave (PAW) method by Kresse and Joubert [44], stand as the most attractive for plane-wave basis codes as they tend to minimize the energy cutoff needed to accurately account for the potential, the density, and therefore the energy.

2.2.6 Boundary Conditions

Ideally, one would like to model a single defect in an infinite solid. Since this not practical, one is forced to approximate the solid host to a chunk of material. Two alternative routes are usually followed by setting appropriate boundary conditions to the Hamiltonian, namely finite and periodic boundary conditions. With finite boundary conditions the defective solid is treated as a vacuum-surrounded crystallite with a few hundred atoms (or more, depending on the available computational power). While attractive from the computational point of view (all Hamiltonian matrix elements are constructed in real-space), the surface of the crystallite has to be somehow passivated, otherwise surface states would interact with the defect states. Conversely, with periodic boundary conditions we are free of surface-related problems, but as opposed to finite systems, this approach suffers from the artificial periodicity of defects across the lattice. A well known problem arises when we want to study a charged defect – within a periodic approach we end up with a lattice of charged defects, with severe spurious long-range Coulomb effects.

Despite these shortcomings, the use of periodic boundary conditions (usually referred to as the supercell approach) has been the most popular method for the study of defects in solids.

2.2.6.1 The Supercell Method

Let us assume a crystalline material with primitive lattice vectors $\mathbf{a}_i$, with $i = 1, 2$ and 3, primitive cell volume v_0, and a set of n_a atoms per cell located at $\mathbf{r}_\alpha$ ($\alpha = 1, \ldots, n_a$). The Bravais lattice points are then given by $\mathbf{l_n} = \sum_i n_i \mathbf{a}_i$. Reciprocal-space unit vectors $\mathbf{G}_i$ are implicitly defined as

$$\mathbf{G}_i = 2\pi \frac{\mathbf{a}_j \times \mathbf{a}_k}{\mathbf{a}_i \cdot (\mathbf{a}_j \times \mathbf{a}_k)}. \tag{2.36}$$

The linear transformation

$$\mathbf{A}_j = \sum_{i=1}^{3} W_{ij} \mathbf{a}_i, \tag{2.37}$$

with W_{ij} elements being integers, defines a *supercell* with lattice vectors $\mathbf{A}_j$, volume $\Omega = v_0 \left|\det(W)\right|$, and respective Bravais lattice $\mathbf{L_n} = \sum_i n_i \mathbf{A}_i$ for all integers n_i. Analogously, respective reciprocal space lattice vectors $\mathbf{g}_j$ are given by

$$\mathbf{g}_j = \sum_{i=1}^{3} W_{ij}^{-1} \mathbf{G}_i \tag{2.38}$$

with W_{ij}^{-1} representing the elements of the inverse matrix W^{-1}. The supercell holds $N_a = n_a \left|\det(W)\right|$ atoms, and their location $\mathbf{R}_\alpha$ ($\alpha = 1, \ldots, N_a$) can be defined by any set of N_a independent atomic positions,

$$\mathbf{R}_\alpha = \mathbf{l_n} + \mathbf{r}_\beta, \text{ provided that } \mathbf{R}_\alpha - \mathbf{R}_{\alpha'} \neq \mathbf{L_m}. \tag{2.39}$$

In the previous expression, β can take any value between $1 \leq \beta \leq n_a$, and the index $\mathbf{n}$ can be any integer vector. The important point is that any vector separating a pair of atoms (within the supercell) cannot match a (super)lattice vector $\mathbf{L_m}$.

Inserting a defect into the supercell has important consequences. Firstly we lose the internal translational symmetry and the basis vectors $\mathbf{A}_i$ now define the unit cell. Although the goal is to model a defect in a crystalline environment, the outer periodicity of the supercell introduces defect images located on each cell. Note that it does not matter where we place the defect – for example, displacing a substitutional impurity by $\mathbf{l_n}$ implies that all defect images will be equally displaced. However, if the supercell is not adequate in terms of size and shape, spurious effects can take place in the form of defect-image coupling through elastic and Coulomb interactions.

The shape of the supercell may have peculiar implications regarding the symmetry of the problem, and by this way its choice may be relevant when we are interested in studying electronic levels that are degenerate. The space-group of the supercell has equal or less symmetry than the underlying crystalline unit cell. For example, suppose we construct an hexagonal supercell of perfect crystalline Si and introduce a substitutional defect. In a real Si sample a substitutional defect would have T_d symmetry, whereas in our model it will have S_6 symmetry (imposed by the hexagonal lattice of the supercell). We know from group theory that tetrahedral centers hold single-, double- and triple-degenerate states, whereas a trigonal one holds single- and double-degenerancy only. Such fundamental difference could have important consequences to the results, especially if involving partially occupied multiplets (like the Jahn-Teller effect). In such cases the use of a cubic supercell are usually preferable.

2.2.7 *Brillouin-Zone Sampling*

Within the supercell method, the calculation of physical quantities (for example the total energy, charge density, or the density of states), require integrations over the Brillouin-zone (BZ). The integrand function $f(\mathbf{k})$ is periodic across the reciprocal space, and usually has no simple analytic form. To avoid numerical integration over a dense mesh, several schemes were developed. As proposed by Baldereschi [45], Chadi and Choen [46], one or a set of N *special* $\mathbf{k}_i$-points can be used to obtain the average $\bar{f}$ over the BZ, where

$$\bar{f} = \frac{\Omega}{(2\pi)^3}\int_{\mathrm{BZ}} \mathrm{d}^3\mathbf{k} f(\mathbf{k}) \approx \sum_i^N w_i f(\mathbf{k}_i), \tag{2.40}$$

with w_i being normalized weighting factors. Despite the convenience of this method, it provided no obvious way of checking for convergence, and ultimately for the quality of the calculations. A clearer scheme was proposed by Monkhorst and Pack (MP) for cubic cells [47], and latter generalized for lower symmetry lattices [48]. Besides its simplicity, the method allowed convergence to be verified. Accordingly, the MP-scheme defines a sequence of running variables,

$$u_i = (2i - I - 1)/2I, \qquad (i = 1, \ldots I) \tag{2.41}$$

$$u_j = (2j - J - 1)/2J, \qquad (i = 1, \ldots J) \tag{2.42}$$

$$u_k = (2k - K - 1)/2K, \qquad (i = 1, \ldots K), \tag{2.43}$$

with the integers I, J and $K \geq 1$ defining a $I \times J \times K$ uniform grid of $\mathbf{k}_{ijk}$ points in the BZ,

$$\mathbf{k}_{ijk} = u_i\mathbf{g}_1 + u_j\mathbf{g}_2 + u_k\mathbf{g}_3, \tag{2.44}$$

where $\mathbf{g}_1, \ldots, \mathbf{g}_3$ are the reciprocal lattice vectors. We note that if F, G and H are point-groups for the supercell with a defect, respective Brillouin-zone and the Wigner-Seitz cell (constructed from the lattice of the supercell) respectively, with respective orders (number of symmetry operations) g_F, g_G and g_H, then $F \subseteq G \subseteq H$. Further, for non-relativistic calculations we also know that the BZ has time-reversal symmetry, so that $G = H \times \mathrm{i}$ where 'i' is the inversion operation.

The MP scheme, uses the following recipe:

1. Define a uniform grid $\mathbf{k}_{ijk}$ for a given set of I, J and K integers (usually referred to as MP-$I \times J \times K$ grid);
2. For each $\mathbf{k}_{ijk}$ generate a star of vectors using all symmetry operations from the Wigner-Seitz group H;
3. Find all irreducible (symmetry inequivalent) $\mathbf{k}$ vectors within the BZ point group G. This is usually carried out by folding each star of points with the application of all symmetry operations in G, making sure that all folded $\mathbf{k}$-points lie inside the BZ. The resulting points are the special $\mathbf{k}$-points;
4. Each special $\mathbf{k}_n$-point has a weighting factor w_n that relates to the representativity of $\mathbf{k}_n$ in the BZ. Being g_n the order of its site symmetry in within the BZ,

$$w_n = g_G/g_n. \tag{2.45}$$

For example, for a defect (in a supercell) with $F = C_{2v}$ point-group, the BZ has $G = H \times \mathrm{i} = D_{2d}$ symmetry. If the principal axis of the defect is along the z-axis, a $\mathbf{k}$ vector $(0, 0, 1/4)$ (with C_{2v} site symmetry) has a weighting factor $w = 8/4 = 2$, corresponding to the equivalent $(0, 0, 1/4)$ and $(0, 0, -1/4)$ vectors.

2.2.8 Basis Functions

In a one-electron description of a periodic system we need two quantum-numbers to label electronic states, $\psi_{\mathbf{k}\lambda}$, namely a Bloch wave vector $\mathbf{k}$ and a one-electron band index λ. Bloch theorem states that the values of $\psi_{\mathbf{k}\lambda}$ at any two points in real space separated by a lattice vector differ by a phase factor only, that is

$$\psi_{\mathbf{k}\lambda}(\mathbf{r} + \mathbf{L_n}) = \psi_{\mathbf{k}\lambda}(\mathbf{r}) \exp(\mathrm{i}\,\mathbf{k} \cdot \mathbf{L_n}), \tag{2.46}$$

meaning that $\psi_{\mathbf{k}\lambda}$ can be written as

$$\psi_{\mathbf{k}\lambda}(\mathbf{r}) = u_{\mathbf{k}\lambda}(\mathbf{r}) \exp(\mathrm{i}\,\mathbf{k} \cdot \mathbf{r}), \tag{2.47}$$

with $u_{\mathbf{k}\lambda}(\mathbf{r}+\mathbf{L_n}) = u_{\mathbf{k}\lambda}(\mathbf{r})$ being any function with the periodicity of the lattice.

In the so-called plane-wave codes $u_{\mathbf{k}\lambda}$ is constructed with help of a Fourier series,

$$u_{\mathbf{k}\lambda}(\mathbf{r}) = \frac{1}{\sqrt{\Omega}} \sum_{\mathbf{g}} c_{\mathbf{gk}\lambda} \exp\left(\mathrm{i}\,\mathbf{g}\cdot\mathbf{r}\right),$$

$$\psi_{\mathbf{k}\lambda}(\mathbf{r}) = \frac{1}{\sqrt{\Omega}} \sum_{\mathbf{g}} c_{\mathbf{gk}\lambda} \exp\left[\mathrm{i}\,(\mathbf{g}+\mathbf{k})\cdot\mathbf{r}\right],$$

where the coefficients $c_{\mathbf{gk}\lambda}$ are varied in order minimize the total energy, and only plane waves with a kinetic energy lower than a user-specified cut-off $E_{\mathrm{cut}} = |\mathbf{g}_{\mathrm{cut}} + \mathbf{k}|^2/2$ are included in the summation (over reciprocal lattice vectors $\mathbf{g}$). This type of basis functions are particularly convenient and robust. They are a natural choice for periodic problems, they are not *biased* (in the sense that there are not regions within the cell with more functions than others) and are easy to handle with respect to convergence checking – all we have to do is to increase E_{cut} until we are satisfied with the resulting accuracy.

Alternatively, we may express $u_{\mathbf{k}\lambda}$ in Eq. 2.47 with help of atomic-like functions. For instance, we may define a set of Bloch basis functions $B_{\mathbf{k}i}(\mathbf{r})$, built from a set of Cartesian-Gaussian functions ϕ_i centered at atomic positions $\mathbf{R}_i$, periodically repeated over N_L lattice vectors $\mathbf{L_n}$,

$$B_{\mathbf{k}i}(\mathbf{r}) = \frac{1}{\sqrt{N_L}} \sum_{\mathbf{L_n}} \phi_i(\mathbf{r}-\mathbf{R}_i-\mathbf{L_n}) \exp\left[\mathrm{i}\,\mathbf{k}\cdot(\mathbf{r}-\mathbf{L_n})\right]. \tag{2.48}$$

The localized orbitals are given by

$$\phi_i(\mathbf{r}) = (x-R_{ix})^{l_1}(y-R_{iy})^{l_2}(z-R_{iz})^{l_3} \exp\left[-a_i(\mathbf{r}-\mathbf{R}_i)^2\right], \tag{2.49}$$

with s-, p-, d-, or f-like orbitals corresponding to $\sum_k l_k = 0, 1, 2$ or 3 for $l_k \geq 0$, and a_i being a suitably chosen decay coefficient. The Kohn-Sham orbitals $\psi_{\mathbf{k}\lambda}$ are then expanded over $B_{\mathbf{k}i}$ basis functions as

$$\psi_{\mathbf{k}\lambda}(\mathbf{r}) = \sum_i c_{i\mathbf{k}\lambda} B_{i\mathbf{k}}(\mathbf{r}) \tag{2.50}$$

Advantages of using Gaussian-like functions include the fact that many integrals can be hardcoded analytically, and in contrast to oscillating basis functions, they can be made as to quickly vanish away from $\mathbf{R}_i$. This makes them *computationally cheaper* in the description of highly localized states, and more suitable to implement in order-N methods. On the other hand, they are not orthogonal and over-completeness can lead to numerical instabilities.

2.3 Calculation of Defect Observables

Density functional theory provides a ground-state methodology to calculate the electronic structure of molecules and solids. By definition, excited states are not accessible, and strictly speaking, Kohn-Sham eigenstates are not one-electron wavefunctions. They represent a set of basis functions that build up the electron density $n(\mathbf{r})$. This is the price we have to pay for writing the total energy simply as a functional of the density. Nevertheless, and despite its limitations, the theory is exact and many observables can be calculated from total energies and electron densities. Among these we have defect structures, solubilities, electrical levels, vibrational modes, elastic properties, migration barriers, and many others. This section explains how several defect-related observables are obtained, and the experiments to which they can be compared.

2.3.1 *Structure of Solids and Defects*

The final quality of any defect study strongly relies on the capability of the method to reproduce the properties of the solid material hosting the defect. This includes structural and elastic properties. The equilibrium structure of a crystal is often obtained with help of an equation of state (EOS). A particularly popular EOS is that proposed by Birch following the earlier work of Murnaghan [49]. Accordingly the internal energy (energy within a control volume v) of a solid is given by the Birch-Murnaghan equation,

$$E(v) = E_0 + \frac{9\,v_0\,B}{16}\left\{\left[\left(\frac{v0}{v}\right)^{\frac{2}{3}} - 1\right]^3 B' + \left[\left(\frac{v0}{v}\right)^{\frac{2}{3}} - 1\right]^2 \left[6 - 4\left(\frac{v0}{v}\right)^{\frac{2}{3}}\right]\right\}, \tag{2.51}$$

where B and B' are respectively the bulk modulus and its pressure derivative under equilibrium conditions, corresponding to $v \equiv v_0$ and minimum internal energy E_0. In practice, the calculation is carried out by considering the unit cell of interest and assuming a series of values around v_0 for the volume. Total energies are then calculated from first-principles, and all four parameters in $E(v)$ are found by means of a fitting procedure.

Figure 2.3 reports the calculation of the EOS for the case of crystalline Si using LDA and GGA to account for the exchange-correlation energy (left and right curves, respectively). The results are also compared to the experimental EOS (dashed line) constructed from the experimental lattice constant, B and B' values [50]. From the figure it becomes clear that the LDA and GGA under-estimate and over-estimate the equilibrium lattice parameter of Si. This follows from a well known propensity of LDA and GGA in predicting binding energies of molecules (or cohesive energies

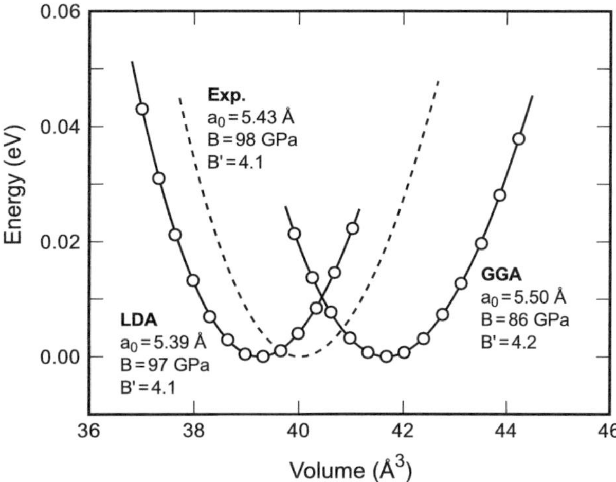

Fig. 2.3 Total energy *(circles)* of a Si unit cell (2 Si atoms) as a function of cell volume. Calculations using the local density approximation (LDA on the left) and the generalized gradient approximation (GGA on the right) to the exchange-correlation potential are compared to the experimental equation of state (*dashed line* at the *center*)

of solids) in excess and deficit with respect to the experiments, respectively. While the LDA under-estimates the lattice constant by about 0.7 %, GGA over-estimates it by about 1.3 %. These deviations are small enough to allow reliable defect calculations. Nevertheless, it is important to note that all calculations (including defective supercells) should always consider the calculated a_0 value (and not the experimental one). This will minimize artificial strain effects with impact on the elastic defect/host coupling (provided that large enough supercells are used).

Now that we have a proper pristine supercell structure, the next step in studying a defect is finding its ground state structure. To that end we need to come up an initial guess (both in terms of chemical species and geometry) by setting-up a defective supercell. Among all components of the input data of any atomistic first-principles package, the initial coordinates/species file is the one that needs most human intervention. Other input components essentially relate to calculation specifications and several approximations to be assumed, like the method to calculate the exchange-correlation energy, the pseudopotentials to be used, or the special **k**-points to sample the Brillouin zone.

Having the input files ready, the calculation protocol usually proceeds through (1) the evaluation of the self-consistent electron (spin) density and energy, (2) calculation of the internal forces acting on individual nuclei, (3) displacement of the nuclei along the forces, and start all over again (until the resulting structure/energy/forces fall below some user-specified tolerance limits). Once a self-consistent $n(\mathbf{r})$ is known, the force $\mathbf{F}_\alpha$ acting on atom α can be obtained from $-\nabla E$, where E is the total energy. In practice, forces are calculated analytically in line with the Hellmann-Feynman theorem [51, 52] that states the following,

Theorem 4 (Hellmann-Feynman theorem) *Let ζ be a parameter in the Hamiltonian and Ψ_ζ an eigenstate of $\hat{H}$. Then*

$$\frac{\partial E}{\partial \zeta} = \langle \Psi | \frac{\partial \hat{H}}{\partial \zeta} | \Psi \rangle . \tag{2.52}$$

This means that the forces can be directly calculated from the self-consistent wavefunction coefficients $c_{i\mathbf{k}\lambda}$. Hence, there is no need for the evaluation of numerical derivatives of the energy and the computation is carried out rather efficiently.

An example of the calculation of the ground state structure of interstitial carbon (C_i) in silicon is shown in Fig. 2.4. C_i defects are usually observed in low-temperature *e*-irradiated carbon-rich Si, upon generation of mobile Si interstitials which are readily trapped by substitutional carbon defects [53–55]. The figure represents both the inner (SCF electronic structure) and outer (atomic structure) relaxation cycles with help of graphs of the total energy change during each SCF cycle and the force acting on the carbon atom. Total energy change and force tolerance limits are respectively 10^{-7} and 10^{-4} in atomic units.

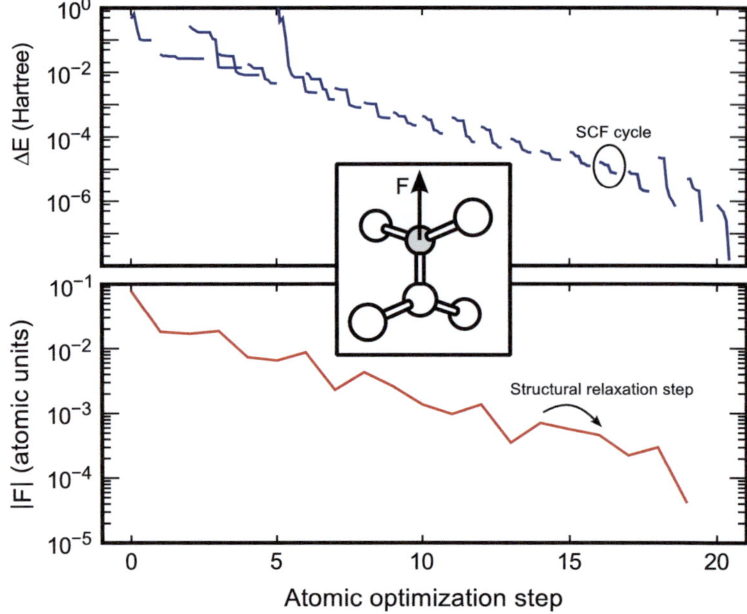

Fig. 2.4 SCF electronic structure relaxation (*above*) and atomic structure relaxation (*below*) of an interstitial carbon defect in a 64 Si atom supercell. In the upper graph, each line segment represents the total energy change during a SCF cycle (set of steps) for each structure along the (outer) atomic relaxation cycle. The force on the carbon atom along the symmetric C-Si axis (shown in the *inset*) is represented in lower graph

As shown in the inset, the defect consists on a C-Si unit aligned along the [001] crystallographic axis, replacing a Si atom in the lattice. This results in a C_{2v} symmetric structure where both C and Si atoms are three-fold coordinated. From early local density functional calculations (in the neutral charge state), we know that the axial C-Si bond (1.73 Å long) is slightly shorter and stronger than the other two C-Si bonds (1.82 Å long) [54]. This explains the observed ordering of the two C-related local vibrational modes connected to this defect at 930 and 921 cm^{-1} [56]. Accordingly the higher frequency mode results from the in-axis vibration of the shorter and stiffer C-Si unit, whereas the 921-cm^{-1} mode relates to the off-axes motion of the C atoms along the bonds with its other two Si neighbors.

2.3.2 Electronic Structure of Defects

We know that the $\psi_{\mathbf{k}\lambda}(\mathbf{r}) = |c_{1\mathbf{k}\lambda}, \ldots, c_{N\mathbf{k}\lambda}\rangle$ Kohn-Sham states from Eq. 2.50 are just eigenvectors within the eigenspace of possible $c_{i\mathbf{k}\lambda}$ coefficients, and that only $\sum_{\mathbf{k},\lambda=\text{occ}} |\psi_{\mathbf{k}\lambda}(\mathbf{r})|^2$ has physical significance. However, it is also acknowledged that $\psi_{\mathbf{k}\lambda}(\mathbf{r})$'s carry much resemblance with true one-electron states (like those in the Hartree-Foch method). This allows us to understand the electronic structure of defects by inspecting the extent, shape and symmetry of the Kohn-Sham states, in particular those within the band gap, whose occupancy can vary as a result of doping, light exposure, carrier trapping, or during many kinds of spectroscopic experiments. Usually, gap states of defects are grouped as deep or shallow, depending on how "far" in the energy scale they are from the host band edges. This is by no means a unique classification scheme, where perhaps another useful way is by comparing the degree of mixing with the crystalline states – an empty state with a strong resemblance with the valence band top (i.e. spanning many unit cells and localized on Si-Si bonds) is likely to be shallow acceptor, whereas a state that is highly localized on a foreign atomic species will most likely give rise to a transition state deep in the gap. Let us start with a description of the shallow type of states.

Substitution of a host Si atom by a singly-ionized group-V species (like phosphorous), leaves all neighboring Si atoms saturated, and the band-gap free of deep states. However, the excess of positive charge at the substitutional defect creates an attractive potential for electrons. This potential is asymptotically Coulombic and slowly varying, meaning that any electron moving in this potential spends most of its time at remote regions from the positive center. The properties of the extra electron (needed to neutralize the system), are then analogous to those in the hydrogen problem. However, as the potential is strongly screened by the medium, the binding energy of this electron to P^{+} is of the order of 50 meV in Si (as opposed to the ionization energy of 13.6 eV for the free H atom), i.e., at room temperature the electron is effectively donated to the Si host to become a n-type free carrier. Analogously, a substitutional B atom is prone to grab one electron to complete four bonds with all its Si neighbors. This effectively takes place by partial oxidation of many neighboring Si atoms, creating an extensive volume of positive net charge in the crystal, i.e., a shallow hole state.

The above shallow defects are normally referred to as dopants because they are deliberately introduced in order to *calibrate* the electrical conductivity of Si in specific parts of devices. Besides the ground state a few tens of eV below the band edges, they usually have several excited states observed in the infra-red spectrum. These transitions are analogous to the Lyman series ($1s \rightarrow 2p$, $1s \rightarrow 3p$, ...) and are accurately described by effective-mass theory (EMT), early developed by Walter Kohn [58, 59], and nicely reviewed by Stoneham [39], Ramdas and Rodriguez [60]. EMT starts with a description of the bulk band structure, especially the valence band top or conduction band bottom states. These bands are represented in red in Fig. 2.5, which shows that Si is an indirect gap semiconductor, where the valence band top is at the center of the BZ ($\mathbf{k} = \Gamma$) and the conduction band minima lie at $\mathbf{k} = \mathbf{k}_{\mathrm{m}}$ along the Δ direction and close to $\mathbf{k} = X$. There are six equivalent conduction band minima along the $\langle 100 \rangle$ direction family, each with transverse and longitudinal effective masses $m_{\mathrm{t}} = 0.19\, m_{\mathrm{e}}$ and $m_{\mathrm{l}} = 0.92\, m_{\mathrm{e}}$, where m_{e} is the free-electron mass.

Let us first assume that the conduction band minimum of Si is isotropic with a single (diagonal) effective mass m_{e}, and that a P^{+} ion replaces a host Si atom. All four valence electrons of P may be assumed to form sp^3 hybrid orbitals with its four Si neighbors. The excess of positive charge at the impurity then generates a screened Coulomb potential

$$V_{\mathrm{i}} = -1/\epsilon r, \tag{2.53}$$

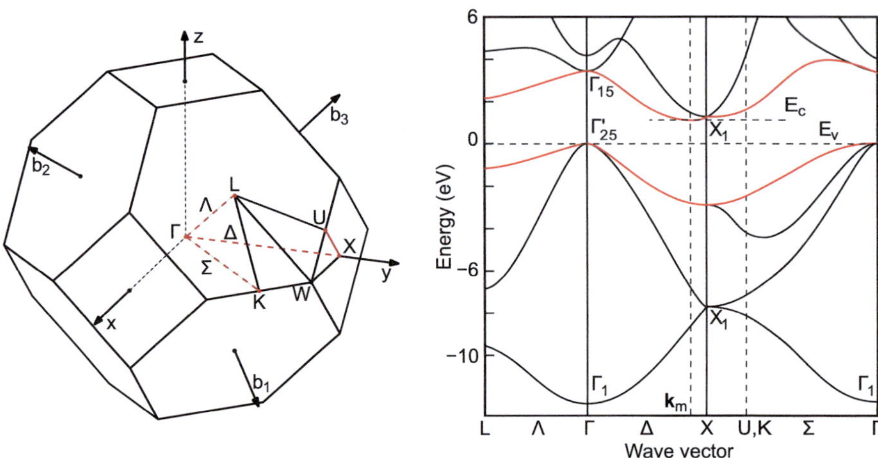

Fig. 2.5 Brillouin zone and electronic band structure of Si along some high symmetry paths in **k**-space shown as red segments. High symmetry points, directions and electronic states are labeled according to the usual notation. Reciprocal space lattice vectors ($\mathbf{b}_1$, $\mathbf{b}_2$ and $\mathbf{b}_3$) and Cartesian directions are also shown. Relativistic effects like spin-orbit coupling are not considered in the band structure (Adapted with permission from Ref. [57]. ©1974 American Physical Society)

within the crystal with a static dielectric constant ϵ. Although this equation is not valid at the immediate neighborhood of the impurity (referred to as central-cell region), it is a very good approximation for $r \gg a_0$. If the ionized electron is captured by P^+, it will move with effective-mass m_e in the hydrogen-like potential V_i, and similarly to the hydrogen atom, a series of bound states are anticipated with energy,

$$E_n = -m_e/2\epsilon^2 n^2, \tag{2.54}$$

some of which (usually for $n = 1$ and 2) are located just below the conduction band minimum. A more rigorous treatment of the problem account for the six-fold degeneracy of the conduction band minima, and the *hydrogenic* wavefunction should be expanded in terms of Bloch functions of conduction band minima $\psi_{\mathbf{k}_m\mathrm{CB}}$,

$$\Psi(\mathbf{r}) = \sum_{j=1}^{6} \alpha_j F_j(\mathbf{r}) \psi_{\mathbf{k}^j_m\mathrm{CB}}(\mathbf{r}). \tag{2.55}$$

Here the index j runs over the six equivalent conduction-band minima, the coefficients α_j are evaluated numerically (usually disregarding any redundancy from time-reversal and defect symmetry), and F_j is a hydrogen-like envelope function, which is also a solution of the following hydrogenic Hamiltonian,

$$\left[-\frac{\hbar^2}{2m_t}\left(\frac{\partial^2}{\partial x^2}+\frac{\partial^2}{\partial y^2}\right)-\frac{\hbar^2}{2m_l}\frac{\partial^2}{\partial z^2}+V_i(r)\right]F_j(\mathbf{r}) = E\,F_j(\mathbf{r}). \tag{2.56}$$

In Eq. 2.56 the anisotropy of the conduction band minima is taken into account, which has the effect of splitting polarized p states into p_0 and $p_\pm$ components. Overall, every s, p_0 and $p_\pm$ states are respectively six-, six- and twelve-fold degenerate for an impurity with T_d symmetry. However, due to inter-valley coupling, this symmetry is lowered, where a maximum of three-fold degeneracy is allowed in Si. Accordingly, s and p_0 states unfold into $a_1 + e + t_2$, whereas $p_\pm$ states split into $2t_1 + 2t_2$ [60]. For an s state, this splitting leads to coefficients $\boldsymbol{\alpha}$ that can be deduced by inspection,

$$\begin{aligned}
\boldsymbol{\alpha}^{a_1} &= (1,\ 1,\ 1,\ 1,\ 1,\ 1)/\sqrt{6}\\
\boldsymbol{\alpha}^{e'} &= (-1,\ -1,\ -1,\ -1,\ 2,\ 2)/\sqrt{12}\\
\boldsymbol{\alpha}^{e''} &= (1,\ 1,\ -1,\ -1,\ 0,\ 0)/2\\
\boldsymbol{\alpha}^{t_2'} &= (-1,\ 1,\ 0,\ 0,\ 0,\ 0)/\sqrt{2}\\
\boldsymbol{\alpha}^{t_2''} &= (0,\ 0,\ -1,\ 1,\ 0,\ 0)/\sqrt{2}\\
\boldsymbol{\alpha}^{t_2'''} &= (0,\ 0,\ 0,\ 0,\ -1,\ 1)/\sqrt{2}.
\end{aligned}$$

We note that only the wavefunction of the fully symmetric s-like a_1-state shows non-zero amplitude at the origin, making it the ground state for an attractive center like P^+. Therefore, unlike the excited states, its binding energy is strongly dependent on the chemical nature of the defect. This is the case of common dopants like substitutional phosphorus or arsenic, as demonstrated by early EPR experiments [61, 62]. However, group theory is unable by itself to establish the ordering of the splitting, and other arrangements are possible. One example is interstitial Li in Si, where the ground state was attribute to a five-fold degenerate $1s(e + t_2)$ state, 1.8 meV below the $1s(a_1)$ state [63].

Now we look at deep centers. The interstitial iron impurity in Si is a classical example, and perhaps the most relevant contaminant in solar Si due to its plethora of contamination sources and huge impact to the carrier life-time performance in p-type material. The interstitial Fe atom occupies the tetrahedral site of the Si lattice as represented by the red spot at the center of Fig. 2.6b. From deep-level transient spectroscopy and Hall-effect measurements it has been established that the defect has a donor state at 0.39 eV above the top of the valence band (E_v) [64, 65]. In p-type Si it is positively charged and captures minority carriers (electrons) rather efficiently. The more interested readers are diverted to Ref. [66] for further reading about many experimental findings related to this center.

Detailed density-functional calculations of iron and other transition metals in Si can be found in the literature (see for instance the work of the Estreicher group

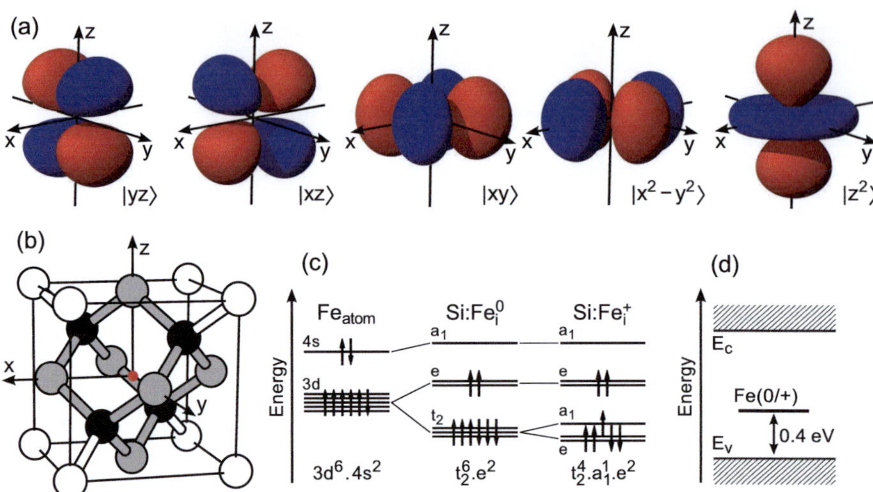

Fig. 2.6 Electronic structure analysis of an interstitial iron impurity in silicon. (**a**) Wavefunction isosurfaces of the $m = -2, \ldots, +2$ atomic d-shell components. (**b**) Conventional unit cell of crystalline Si with a tetrahedral interstitial site (where the impurity is located) represented by a *red dot* at the center. First and second neighboring Si atoms are shown as *black* and *gray balls*, respectively. (**c**) One-electron structure (3d and 4s shells) of atomic Fe, as well as interstitial Fe in Si in the neutral and positive charge states. (**d**) Donor level of an interstitial Fe impurity in Si at about $E_v + 0.4$ eV

[67, 68] and references therein). The results are summarized in Fig. 2.6 which shows the typical shape of the 3d orbital components, the tetrahedral interstitial location of the impurity in the lattice, the electronic structure of Fe_i in Si, and the location of its donor level within the gap. According to the early interpretation of Ludwig and Woodbury [69], the local crystal field of the Si host lifts the degeneracy of the 3d manifold orbitals of Fe_{atom} into doublet (e) and triplet (t_2) states in Si:Fe_i^0. This is depicted in Fig. 2.6c. The t_2 triplet state in Fe_i^0, with lobes along xy, xz and yz directions, are lower in energy as they overlap the positively charged (electron depleted) regions localized at the four closest back-bonding Si atoms (represented as black balls in Fig. 2.6b). Another factor promoting the higher stability of t_2 (in comparison to e) is the fact that all t_2 components are nodal along the x-, y- and z-axis. This avoids overlap with the high electron density regions of the Si second neighbors (represented as gray balls in Fig. 2.6b). Conversely, the e states ($|x^2 - y^2\rangle$ and $|z^2\rangle$) rise in energy as they couple to the octahedral-shaped electron density with maxima localized on bonds to second-nearest Si neighbors (along x, y and z directions), represented as gray bonds in Fig. 2.6b.

The calculations predict that Fe_i^0 and Fe_i^+ have an electronic structure with occupancy $t_2^6 \cdot e^2$ and $t_2^5 \cdot e^2$, corresponding to total spin states $S = 1$ and $S = 3/2$, respectively [67, 70–72]. These results account rather well to the observed electron paramagnetic resonance data [73]. The $S = 0$ state of Fe_i^0 was found to be metastable by 0.74 eV. We note that in the perfect tetrahedral Fe_i^+ defect, the electron occupancy among all t_2 components is asymmetric and the atomic structure is prone to change via Jahn-Teller distortion. While this effect is beyond the scope of Born-Oppenheimer approximated density functional theory, we depict in Fig. 2.6c a possible splitting pattern of t_2 into $e + a_1$ (also with $S = 3/2$), that corresponds to a small trigonal distortion. Finally, we note that ionization of Fe_i^+ into Fe_i^{++} (promoting an electron to the conduction band bottom) has been predicted to be at least as costly as the band gap energy. Analogously, hole emission from Fe_i^0 with formation of Fe_i^- (plus a free hole in the valence band top) was also anticipated to cost more than the Si band gap energy [67]. This means that interstitial iron has only a donor level in the gap, calculated at $E_v + 0.28$ eV [67], which is to be compared to $E_v + 0.39$ eV from DLTS measurements [64, 65]. These results lead us to the transition-state diagram depicted in Fig. 2.6d. Details about the calculation of electrical levels will be provided further ahead.

Unambiguous evidence for the presence of considerable concentrations of substitutional Fe in silicon has been provided by Mössbauer experiments. In addition to interstitial Fe, substitutional Fe, Fe clusters and $FeSi_2$ precipitates have been directly observed at 1273K on ^{57}Fe-doped Czochralski, floating-zone and Si-on-Insulator wafers at high temperatures [74]. Other experiments where substitutional iron has been detected involve ion-implants. Here, the Mössbauer experiments make use of implanted ^{57}Mn mother isotopes, with a half-life of about 90 s, which decay into the 14.4 keV exited state of ^{57}Fe. Detection of the resulting 14.4 eV γ-rays (and respective shifts) is then used to measure the Mössbauer effect [75–79]. The typical room-temperature Mössbauer spectra of ^{57}Mn/^{57}Fe implanted Si shows two singlets with isomer shifts around 0.8 and −0.03 mm/s (against the α-Fe reference) [78, 80]. These were assigned to interstitial and substitutional Fe species, respectively.

The calculation of isomer shifts (and other hyperfine interactions between nuclei and surrounding electrons) involves an explicit treatment of the core-electrons in the Hamiltonian. The use of pseudopotentials is unsuitable and the frozen core approximation breaks down. Unfortunately, the number of plane waves to describe the electronic states is determined by the smallest length scales that are to be described in real space. Since near the nuclei the wave function of states like 3d, 4s, etc. shows steep behavior, a typical solid state problem would need over 10 million plane waves, which is way beyond the capability of the fastest supercomputers. This problem may be surmounted with the use of the augmented plane-wave method. Accordingly, in regions close to the nuclei, the wavefunctions are described with help of local orbitals like spherical harmonics, whereas within the interstitial regions they are expressed in plane-waves. For a detailed description of full-potential (all-electron) methods based on augmented plane-waves the reader is diverted to Refs. [81] and [82].

Isomer shifts of interstitial and substitutional Fe impurities in Si were calculated in the past by means of the augmented spherical wave method within the local-spin density approximation [83]. The resulting shifts were 0.89 and 0.13 mm/s relatively to α-iron for Fe_i and Fe_s, respectively. These mach reasonably well the experiments, although a revisit to this problem using an improved basis, a more accurate exchange-correlation treatment and a larger supercell is highly desirable. We note though that state-of-the-art linear augmented plane-wave calculations of substitutional Fe on the carbon site in SiC and interstitial Fe surrounded by four Si neighbors in SiC, resulted in isomer shifts of −0.55 and 0.49 mm/s [84], respectively, not far from the data in pure Si.

Early deep level transient spectroscopy studies of electron-irradiated Cz-Si samples contaminated with Fe, indicated a clear interaction between Fe interstitials and three prominent radiation-induced defects, namely the vacancy-oxygen (VO), divacancy (V_2) and phosphorous-vacancy (PV) complexes [85, 86]. In fact, an electron trap at $E_c - 0.36$ eV was assigned to a FeVO$(-/0)$ level based on a clear correlation between the growth rate of this trap and the decreasing rate of the VO$(-/0)$ trap at $E_c - 0.17$ eV [85]. Despite many efforts, there were no indications of the formation of traps that could be connected to substitutional Fe, most probably due to the fact that the concentration of impurities other than Fe_i acting as traps for vacancies (such as oxygen, carbon and dopants) were incommensurate with the iron contamination level. More recently, Tang and co-workers [87, 88] reported the observation of three hole traps at 0.25, 0.29 and 0.34 eV above the valence band edge, and connected the two latter signals to an Fe impurity bound to a divacancy. These observation came in support of previous first-principles calculations [89] that predicted a donor level at $E_v + 0.25$ eV and an acceptor level at $E_c - 0.73$ eV for two stable FeV_2 structures.

Substitutional Fe in silicon has also been investigated theoretically. The details of the electronic structure can be described by means of the Watkins vacancy model [90] depicted in Fig. 2.7. The picture basically shows a resonance between the t_2^6

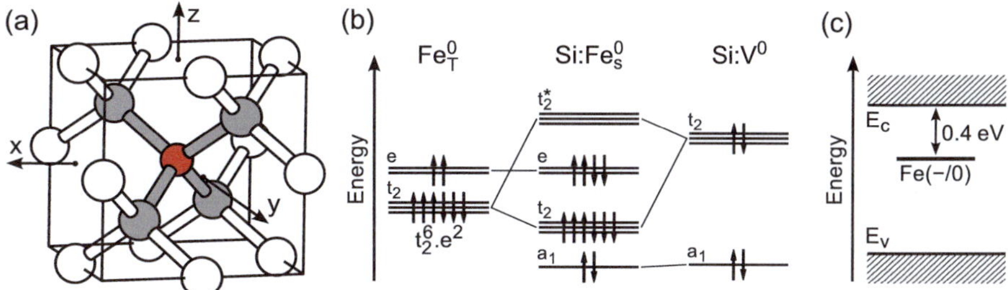

Fig. 2.7 Electronic structure analysis of a substitutional iron impurity in silicon. (**a**) Conventional unit cell of crystalline Si centered at the tetrahedral substitutional site. The Fe impurity is represented by the *red* atom at the center. First neighboring Si atoms and bonds are shown in *gray*. (**b**) One-electron structure detailing the resonance between d-state derivatives ($t_2 + e$) of Fe and those of the single vacancy in Si ($a_1 + t_2$). (**c**) Calculated acceptor level of a substitutional Fe defect in Si at about $E_c - 0.4\,\text{eV}$ [68]

states of the Fe atom [derived from the splitting $d^8 \rightarrow t_2^6 + e^2$ due to the local tetrahedral crystal field] with the t_2^2 states of the vacancy [arising from the analogous splitting of four dangling bonds sharing four electrons, $4(sp3)^1 \rightarrow a_1^2 + t_2^2$, located at the site where the Fe atom sits and shown as gray bonds in Fig. 2.7a]. The result is the formation of a fully occupied t_2^6 bonding state (low in energy) and also the formation of an empty t_2^* anti-bonding counter-part (high in energy).

Now the relevant question is what is the location of these states (particularly when their occupancy is changed) within the silicon band structure? While early calculations based on the Green's function method anticipated an inert Fe_s with no levels in the gap [71, 91], i.e., the e and t_2^* states were predicted to lie below E_v and above E_c, respectively, more recent density functional calculations placed the t_2^* state well within the gap, and calculated an acceptor level in the upper half of the gap, either at $E_c - 0.41\,\text{eV}$ (as depicted in Fig. 2.7c) or $E_c - 0.29\,\text{eV}$, depending on the specifics of the calculations, such as the type of pseudopotentials and basis functions employed [89]. These results challenge both theorists and experimentalists, suggesting that further studies should be carried out.

2.3.3 *Adiabatic Mechanisms: Reactions, Migration, ...*

To calculate the activation energy of a mechanism such a chemical reaction, migration step, or simply the reorientation of a defect, we have to find the energy and structure of a saddle point lying somewhere along the minimum energy path (MEP) between reactants and products (initial and final structures) [92]. This can be achieved in a number of ways, but in general it is done by relaxing the atoms subject to some constraints. For instance, Fig. 2.8 depicts the energy barrier separating two equivalent ground state structures of a neutral divacancy in silicon

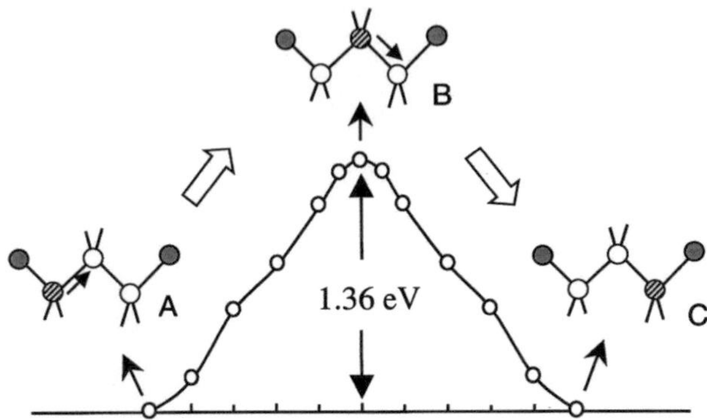

Fig. 2.8 Total energy variation while a neighboring Si atom (as indicated by the *hatched circle*) jumps across the neutral-state divacancy (*open circles*) along a ⟨110⟩ direction (Reprinted with permission from Ref. [93]. ©2002 American Physical Society)

[93]. The defect comprises two adjacent vacancies shown as white circles, and the calculations were carried out using the supercell plane-wave method. The total energy variation was calculated while a neighboring Si atom jumps across the vacant sites (from the left to the right as shown in the figure), enforcing discrete and equal x and y coordinates for the moving Si atom, while its z coordinate was free to change. The number of constrains should be always kept as low as possible, and in this case the moving Si atom is enforced to lie on the plane of the figure. The remaining atoms in the simulation cell were fully relaxed. The calculated barrier is 1.36 eV high, comparing rather well with the experimental figure of 1.4 eV obtained from the recovery kinetics of the stress-induced dichroism measured by electron paramagnetic resonance [94].

For some simple defects, symmetry constraints can help us to find saddle point structures and respective barriers. Let us represent the atomic structure at the saddle point by $\mathbf{Q}_{\mathrm{SP}}$, and assume that the number of symmetry operations that leave the structure unchanged (order of its point group) is g_{SP}. Analogously, the number of symmetry operations and structure after perturbing $\mathbf{Q}_{\mathrm{SP}}$ by $\delta\mathbf{Q}$ are represented by g and $\mathbf{Q} = \mathbf{Q}_{\mathrm{SP}} + \delta\mathbf{Q}$, respectively. If $g_{\mathrm{SP}} > g$ for any $\delta\mathbf{Q}$, then $\mathbf{Q}_{\mathrm{SP}}$ is an extreme configuration within the potential energy surface, i.e. it is either a minimum, maximum or a saddle-point structure. Now, how easy it will be to distinguish between these three possibilities depends on the defect at hand. All we have to do is to start by guessing a structure for $\mathbf{Q}_{\mathrm{SP}}$ finding its relaxed structure without breaking the symmetry. Then we apply several $\delta\mathbf{Q}$ perturbations followed by structural relaxation to inspect the surroundings of the potential energy surface.

Let us use the mechanism of Fig. 2.8 as an example. We note that at the saddle point ($\mathbf{Q}_{\mathrm{SP}} = B$) the divacancy has C_{2v} symmetry ($g = 4$). We start by relaxing that structure keeping the symmetry (symmetrizing the forces during every relaxation step – most density functional packages available can do this). Then, by applying a small displacement $\delta\mathbf{Q}$ to the Si atom at the center towards the left or right direction in the figure, we lower the symmetry of the defect to C_{1h} ($g = 2$). After relaxing the perturbed structure we end up either at structures A or C (depending on the initial displacement direction). This result implies that structure B is either a maximum or a saddle point. By applying further displacements we could infer more precisely the nature of structure B, for instance by displacing the central Si atom along the direction perpendicular to the figure. We note that during these calculations, defects may have close structures with different point group symmetries, and therefore special care should be taken when setting up the special **k**-points for Brillouin zone sampling.

Another example of the use of symmetry constraints to estimate migration barriers of defects is the case of interstitial iron. The diffusion of interstitial iron in silicon is thought to be charge-state-dependent, with the height of the migration barrier of Fe_i^+ (about 0.69 eV) being lower than that of Fe_i^0 (about 0.84 eV) [66, 95, 96]. This is currently however a matter of debate, since the measurements are highly sensitive to trapping events between iron and other impurities/defects like p-type dopants.

Iron in Si is close to an isolated atom sitting at the tetrahedral interstitial (T) site, and therefore has little covalent overlap with its Si neighbors. This impurity diffuses along a straight line from one T site to the next across the saddle point at the hexagonal (H) interstitial site with D_{3d} point-group symmetry [97]. All we have to do is to relax the structure of the impurity, subject to D_{3d}-symmetrized forces, and compare the resulting energy to that of the ground-state tetrahedral defect. An important point to make here is that since the time between consecutive migration steps is much shorter than the spin-lattice relaxation time [98], the total spin must be conserved along the path, and at the saddle point it must be identical to that at the ground state. According to Estreicher et al. [97], such calculations using the LDA give a migration barrier height of 0.82 eV and 0.88 for Fe_i^+ and Fe_i^0, respectively, although they mach better the experiments by using a variety of flavors of the GGA to the exchange-correlation potential, with resulting barriers of 0.69 eV and 0.86 for Fe_i^+ and Fe_i^0, respectively [97].

Perhaps the most popular recipe among the community to find reaction paths and solid state processes at the atomic scale is the nudged elastic band (NEB) method [99]. It is a robust algorithm that works when the initial and final states of the mechanism are known. The method is here described with help of Fig. 2.9. It basically involves the relaxation of an initial pathway (shown as a set of structures along a straight line between initial and final structures) to a minimum-energy path (MEP) between reactants and products (shown as a set of structures along the red segments forming an approximately curved path). Both end-structures are kept fixed during the relaxation process, and the MEP determined from the NEB can be complex and involve degrees of freedom that are not anticipated.

Accordingly the initial reaction path is represented by a set of N images, and the NEB force on image i (with $2 \leq i \leq N - 1$) is

$$\mathbf{F}_i^{\mathrm{NEB}} = \mathbf{F}_i^{\mathrm{s}\|} + \mathbf{F}_i^{\mathrm{g}\perp}, \tag{2.57}$$

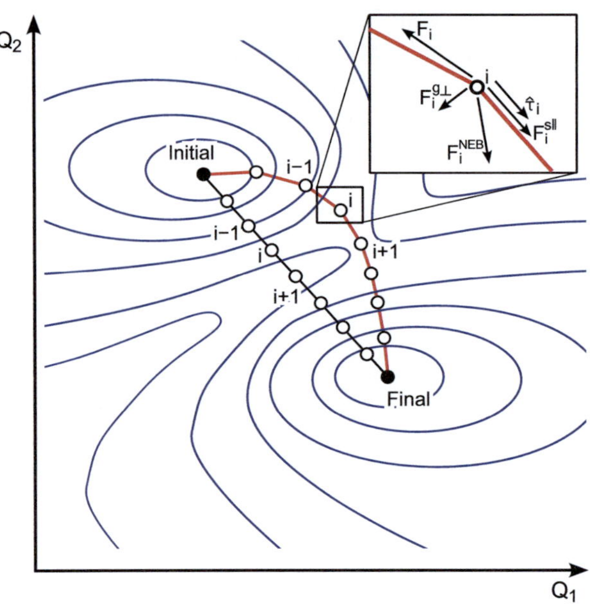

Fig. 2.9 Schematic potential energy surface across two general coordinates Q_1 and Q_2 showing a nudged elastic band along with respective end-structures (*close circles*), intermediate moving structures (*open circles*) and respective forces. The initial elastic band is set to be linear between initial and final structures. After full relaxation the elastic band passes through (or close to) the saddle point structure. This is represented by the curved *red* line

where $\mathbf{F}_i^{\mathrm{g}\perp}$ is the force due to the gradient of the potential perpendicular to the path, and $\mathbf{F}_i^{\mathrm{s}\|}$ is the spring force parallel to the path. $\mathbf{F}_i^{\mathrm{g}\perp}$ is calculated from the first-principles force $\mathbf{F}_i^{\mathrm{g}}$ acting on the atoms of each image as

$$\mathbf{F}_i^{\mathrm{g}\perp} = \mathbf{F}_i^{\mathrm{g}} - \left(\mathbf{F}_i^{\mathrm{g}} \cdot \hat{\boldsymbol{\tau}}_i\right) \hat{\boldsymbol{\tau}}_i, \tag{2.58}$$

where $\hat{\boldsymbol{\tau}}_i$ is an *up-winding* unitary tangent pointing towards the adjacent image with the higher potential energy. Spring forces $\mathbf{F}_i^{\mathrm{s}\|}$ are added to avoid images to collapse together as well as to keep them well separated,

$$\mathbf{F}_i^{\mathrm{s}\|} = k\left(|\delta\mathbf{Q}_+| - |\delta\mathbf{Q}_-|\right) \hat{\boldsymbol{\tau}}_i, \tag{2.59}$$

with k being a spring constant, and the distance to the adjacent images are,

$$\delta\mathbf{Q}_+ = \mathbf{Q}_{i+1} - \mathbf{Q}_i, \tag{2.60}$$

$$\delta\mathbf{Q}_- = \mathbf{Q}_i - \mathbf{Q}_{i-1}, \tag{2.61}$$

were $\mathbf{Q}_i$ represents the Cartesian positions of the atoms in the i-th image. Further refinements to this method, like driving the image with the highest energy up to the saddle point, or including the lattice vectors into the optimization process can also be carried out and were reported elsewhere [99].

An example of a defect-related mechanism studied using the NEB method is the transformation of the bistable trivacancy complex in Si [100]. This is one of the most important radiation products in heavy particle irradiated Si at room temperature [101]. Right after irradiation the observed defect has the structure depicted in Fig. 2.10a. We refer to this structure as part-of-hexagonal-ring (PHR) because it is obtained after removing three adjacent Si atoms from an hexagonal ring in the Si lattice (dashed atoms). The V_3[PHR] structure has Si dangling bonds that are responsible for several deep donor and acceptor levels observed during deep-level transient spectroscopy measurements [101]. However, storage for a few weeks of

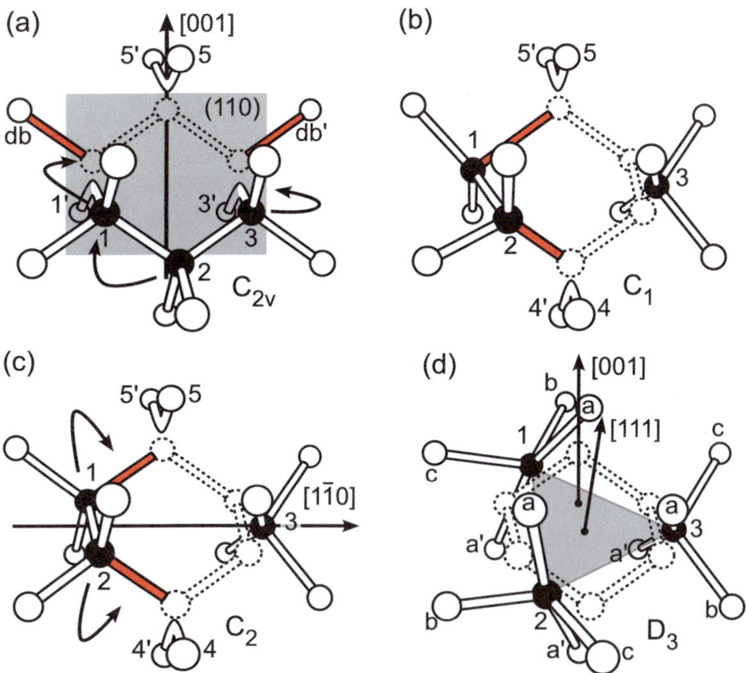

Fig. 2.10 Atomic structure of stable and metastable V_3 complexes in silicon along with high symmetry crystallographic directions and planes. The PHR→FFC transformation mechanism is depicted by the *arrows* indicating the motion of three core Si atoms (1, 2, and 3) from substitutional sites in V_3[PHR] (**a**) to interstitial sites in V_3[FFC] (**d**). These are shown in *black* for a better perception of their placement with respect to the vacant sites (drawn with a *dashed line*). Intermediate structures along the mechanism are shown in (**b**) and (**c**). Broken bonds are shown as *solid red sticks*, reconstructed radical pairs Si_i-$Si_{i'}$ are shown as *banana-bonds*, and Si atoms that are slightly perturbed (or unperturbed) from their lattice sites are shown in *white* (Reprinted with permission from Ref. [100]. ©2012 American Physical Society)

the irradiated Si samples at room temperature or shorter anneals in the temperature range 50–100 °C resulted in the transformation of V_3[PHR] related deep levels into a shallow electron trap at 75 meV bellow the conduction band bottom. The concentration of the PHR defects could be fully restored by application of a forward bias injection at $T \approx 300$ K. The shallow defect, now known to have a structure as represented in Fig. 2.10d, is the lowest energy structure in the neutral charge state and has all Si atoms four-fold coordinated (FFC).

The NEB method was applied to study the MEP for the transformation V_3[PHR]→V_3[FFC]. The saddle point structure lies somewhere between structures in Fig. 2.10a, b, and the barrier height was estimated to be 1.15 eV and 1.14 eV for neutral and negatively charged V_3 complexes, respectively [100]. These figures match very well their respective experimental counterparts of 1.16 and 1.15 eV obtained from transformation kinetics measurements (by monitoring the concentration of electron traps related to V_3[PHR] and V_3[FFC] during anneals) [101].

2.3.4 Formation Energies and Electronic Levels

The chemical potential μ_X of a material constituent X (where X stands for an electron or a chemical species) is the derivative of the Gibbs free energy $G = E + PV - TS$ (for a given phase) with respect to their number n_X around the equilibrium state [92, 102],

$$\mu_X = \partial G / \partial n_X. \tag{2.62}$$

Under such conditions μ_X must be the same in all phases in contact, so it can be considered as the free energy per particle. E is the internal energy and the PV term can be neglected as pressure (P) and volume (V) changes involved in solid-state processes are usually small. For the sake of simplicity we will also drop the entropy term, although this could be important at high temperatures [103]. Hence the formation energy E_f of a crystal with a neutral defect can be expressed as

$$E_\mathrm{f} = E_\mathrm{D} - \sum_X n_X \mu_X, \tag{2.63}$$

where E_D is the internal energy of the defective crystal (in practice a defective supercell) made of n_X atoms/electrons. E_D is directly obtained from the calculated total energy ($E[n]$ in Eq. 2.27) of the relaxed supercell. Chemical potentials are usually considered as the energy of X in a standard thermodynamic state, and can be easily calculated from first-principles as well. For instance, for the study interstitial oxygen complexes in Si, one may assume that the reference states of Si and O atoms are in bulk Si and α-SiO_2 (quartz). Hence, μ_Si is calculated from the total energy per Si atom in a bulk calculation, whereas $\mu_\mathrm{O} = [E(\alpha\text{-SiO}_2) - \mu_\mathrm{Si}]/2$, where $E(\alpha\text{-SiO}_2)$ is the total energy per SiO_2 stoichiometric unit from a bulk quartz calculation.

As a first example, let us look at the thermodynamics of titanium impurities in Si. Ti is among the most feared recombination centers in solar Si. The ground state at the tetrahedral interstitial site (Ti_i) can adopt up to four charge states with transitions at two deep donor levels at $E(0/+) = E_c - 0.27$ eV and $E(+/++) = E_v + 0.26$ eV, and at an acceptor level at $E(-/0) = E_c - 0.09$ eV (see Refs. [35, 68] and references therein). Of special relevance is the $Ti_i(0/+)$ level, that showing a large capture cross section for minority carriers in p-type Si, can cause more than 60 % loss to the power conversion efficiency in poly-crystalline Si solar cells for a Ti concentration of 10^{14} cm^{-3} [104]. The substitutional form o titanium impurities (Ti_s) in Si has never been unambiguously connected to a spectroscopic fingerprint, although the above mentioned acceptor level has also been assigned to $Ti_s(-/0)$ [105]. In order to shed light into this controversy, Markevich and co-workers [35] calculated the formation energy of Ti centers in Si. Chemical potentials of Si and Ti species were calculated from crystalline Si and $TiSi_2$ (C49 phase), respectively. The formation energy of neutral Ti_i was therefore calculated against the energy of a Ti atom in a $TiSi_2$ precipitate. The resulting figure was $E_f = 2.96$ eV, in excellent agreement with the 3.05 eV measured from depth-profile DLTS data on $TiSi_2$/Si structures [106].

In the same work, the formation energy of neutral substitutional Ti was 3.82 eV [35]. The solubility of Ti in Si is $n_0 \exp(-E_f/k_B T)$ at a specific temperature T, where n_0 is the density of available sites (5×10^{22} cm^{-3} for Ti_s and half of that for Ti_i), and k_B is the Boltzmann constant. For neutral Ti (in intrinsic Si) this gives $[Ti_i] = 3.6 \times 10^{13}$ cm^{-3} and $[Ti_s] = 1.9 \times 10^{11}$ cm^{-3} at the melting point ($T \sim 1400$ °C), meaning that no more than ~0.5 % of dispersed Ti will be incorporated at substitutional sites.

Now when defects possess gap states, i.e. if they can be found on different charge states depending on the Fermi level, the formation energy of a defect must account for the chemical potential of electrons μ_e. A natural choice is to assume that the defect can exchange electrons *stored* in a reservoir with energy $\mu_e = E_v + E_F$, where E_v is the top of the valence band and $0 \leq E_F \leq E_g$ is the Fermi level of the material which is just above 0 for p-type material or just below the band-gap energy for n-type material. The formation energy $E_f^q(E_F)$ of the crystal containing a defect in a charge state q (with $q = 0, \pm 1, \pm 2, \ldots$) is a function of the Fermi energy and depends parametrically on q,

$$E_f^q(E_F) = E_D^q - \sum_X n_X \mu_X + q(E_v^q + E_F) \tag{2.64}$$

This equation allows us to estimate several thermodynamic properties as a function of the Fermi level, including solubilities, defect concentrations, occupancy levels, among other quantities [107–109]. For example, the location of a $(q/q+1)$ level is given by the position of the Fermi-level for which charge the states q and $q+1$ have the same formation energy. For reasons that will be made clear below, the valence band top also depends parametrically on q.

Equation 2.64 is accompanied by a few caveats. The formation energy of a charged defect depends on the position of the valence-band top E_v. Due to finite size effects, the location of E_v in a defective supercell depends on the charge state of the defect and differs from that of the bulk supercell. Not to mention that in some cases, mixing between crystalline and defect states can be strong enough as to make impossible any distinction between their character. A method to estimate the offset of the band structure in the defective supercell with respect to a bulk reference was proposed by Van de Walle and Martin [110]. Accordingly we simply have to calculate the difference between average electrostatic potentials taken across equivalent volumes of bulk-like regions in the defective cell ($\bar{V}_\mathrm{D}$) and in the pristine bulk supercell ($\bar{V}_\mathrm{B}$), respectively. Assuming that the valence-band top reference is the highest occupied state from a bulk calculation at $\mathbf{k} = \Gamma$, i.e. $\epsilon_{\max,\Gamma}$, then

$$E_\mathrm{v}^q = \epsilon_{\max,\Gamma} + \bar{V}_\mathrm{D}^q - \bar{V}_\mathrm{B}, \tag{2.65}$$

Another difficulty with the calculation of the formation energy of charged defects arrives from image-charge spurious interactions when using periodic boundary conditions. It is easy to show that the Coulomb potential of a three-dimensional infinite array of identical charges is infinite. To circumvent this problem, when we charge a defect by means of removing or adding electrons to the unit cell (or more precisely by changing the occupancy of a localized state), we also have to add a *counter-charge* of opposite sign, uniformly distributed across the cell (often referred to as *jellium*). However, the total energy refers now to a periodic arrangement of localized charges immersed in a compensating charged background, and not an isolated charged defect embedded in a neutral solid. The simplest recipe to mitigate this problem is known as the Madelung correction as proposed by Makov and Payne [111]. The formation energy then becomes,

$$E_\mathrm{f}^q = E_D^q - \sum_X n_X \mu_X + q\left[\epsilon_{\max,\Gamma} + \bar{V}_\mathrm{D}^q - \bar{V}_\mathrm{B} + E_\mathrm{F}\right] + \alpha_\mathrm{M}\frac{q^2}{\epsilon_\mathrm{s} L}, \tag{2.66}$$

where α_M is the Madelung constant for the appropriate lattice with characteristic length L and ϵ_s the macroscopic static dielectric constant of the host material ($\epsilon_\mathrm{s} = 11.7$ for Si). The formulation of the Madelung correction assumes that the problem consists on a lattice of perfect point-like charges (immersed in a jellium of opposite charge), which can be far from the case being studied, for instance for shallow centers. Despite this correction being often negligible in Si supercells with a few hundreds of atoms (about 80 meV for positively or negatively charged center in a 512 atom cubic supercell), for highly charged systems ($|q| > 2$) its importance becomes significant. Further, more sophisticated methods to account for these interactions are available, which consider the specific localization and shape of the charged states under scrutiny [112].

An alternative method to calculate electrical levels of defects is the marker method [113]. This can also be combined with Eq. 2.63 to obtain formation energies of charged defects avoiding the calculation of corrections to the valence band top and image-charges [100]. The marker method is particularly suitable to estimate the energy of electronic transitions like those observed in DLTS or optical experiments. Accordingly, we compare the ionization potential $I_{\rm D}(q/q+1) = E_{\rm D}^{q} - E_{\rm D}^{q+1}$ of the defect under scrutiny, to that of a defect that is well established experimentally, usually referred to as *Marker*, i.e. $I_{\rm M}(q/q+1) = E_{\rm M}^{q} - E_{\rm M}^{q+1}$ [114]. The procedure is better understood with help of Fig. 2.11. The marker method basically assumes that the calculated offset between ionization potentials of a defect and a marker $\delta_{\rm calc} = I_{\rm D}(q/q+1) - I_{\rm M}(q/q+1)$ is equal to the analogous offset between the experimentally measured levels of those defects $\delta_{\rm exp} = E_{\rm D}(q/q+1) - E_{\rm M}(q/q+1)$, that is

$$E_{\rm D}(q/q+1) = I_{\rm D}(q/q+1) - I_{M}(q/q+1) + E_{\rm M}(q/q+1), \tag{2.67}$$

Note that we can also write,

$$\begin{aligned} E_{\rm D}(q/q+1) - E_{\rm v} = {} & I_{\rm D}(q/q+1) - I_{M}(q/q+1) + \\ & + E_{\rm M}(q/q+1) - E_{\rm v}, \end{aligned} \tag{2.68}$$

or

$$\begin{aligned} E_{\rm c} - E_{\rm D}(q/q+1) = {} & I_{\rm D}(q/q+1) - I_{M}(q/q+1) + \\ & + E_{\rm c} - E_{\rm M}(q/q+1), \end{aligned} \tag{2.69}$$

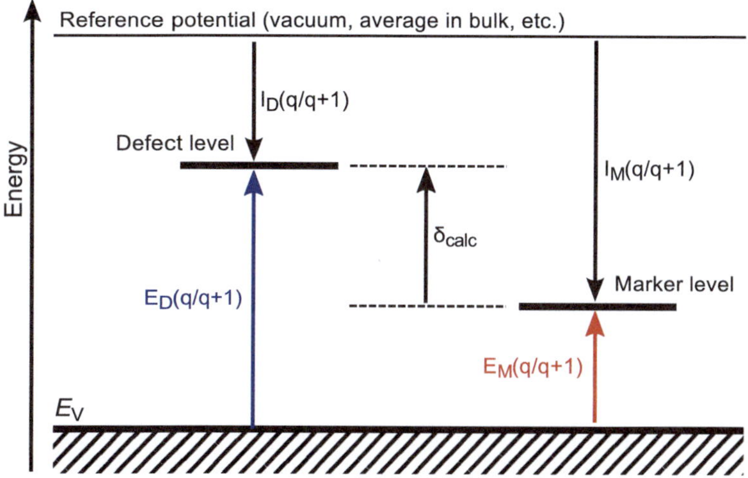

Fig. 2.11 Ionization energy diagram describing the marker method. The defect level with respect to the valence band top is the quantity to be calculated. From first principles calculations we can only obtain the level $I_{\rm D}(q/q+1)$ with respect to a reference potential. The problem is solved by introducing a *marker level* whole location with respect to $E_{\rm v}$ is known in *red*

where we should use the equation that suits better the purpose of comparison with the experimental data, that is, if the marker level has been measured with respect to the valence band top (*ex.* a hole trap in p-type material) we should use Eq. 2.68. If on the other hand the experimental data refers to a level with respect to the conduction band bottom (*ex.* an electron trap in n-type material) the Eq. 2.69 is the one to use. This is clearly a semi-empirical approach as the difference between the last two terms in Eqs. 2.68 and 2.69 comes from an experimental observation. The method assumes that the valence band top and defect-image interaction (including charge) corrections are similar for both the marker and the defect under scrutiny. Hence, the quality of the results will be greater the greater the similarity between the localization and shape of the donor (or acceptor) states of both defects.

Unfortunately many defects have no particularly obvious marker to be compared with, and often $I_{\mathrm{M}}(q/q+1)$ in Eq. 2.67 is replaced by $I_{\mathrm{bulk}}(q/q+1)$ and consequently $E_{\mathrm{M}}(q/q+1) = E_{\mathrm{v}}$ for $q \geq 0$. For $q < 0$ we either use $E_{\mathrm{M}}(q/q+1) = E_{\mathrm{c}}$ or $E_{\mathrm{M}}(q/q+1) = E_{\mathrm{v}} + E_{\mathrm{g}}$ (whatever is the most convenient) with E_{g} being the band gap of silicon. The subscript "bulk" tells that the calculation is carried out using a defect-free supercell [115]. This particular *flavor* of the marker method is rather appealing as it is parameter free if we avoid the introduction of an $\mathrm{E_g}$ value from experiments.

As an example of the calculation of defect formation energies as a function of the Fermi energy, including their electrical levels, let us look at the thermodynamics of titanium-hydrogen complexes in silicon [116, 117]. The passivation of Ti by H has been revisited recently by Kolkovsky et al. [105, 118] and Leonard et al. [119] by means of conventional DLTS high-resolution Laplace DLTS. A total of four levels ascribed to several Ti-H defects were reported, namely E40' at $E_{\mathrm{c}}-0.80$ meV, E170 at $E_{\mathrm{c}} - 0.34$ eV, E170' at $E_{\mathrm{c}} - 0.38$ eV and a deeper level at $E_{\mathrm{c}} - 0.57$ eV detected by conventional DLTS and previously labeled E(260) [120]. The superposition of E170 and E170' peaks was also assigned to the E(175) band of a previous work by Jost and Weber [120]. From depth profile data there was a clear indication that E170' involved more H atoms than both E170 and E40' which showed similar decay profile slopes into the sample. Lately, Leonard et al. [119] employed conventional DLTS and capacitance-voltage (C-V) profiling measurements in n-type Si. They reported three bands related to the formation of TiH_n complexes. Two of the bands, labeled $E(175)$ and $E(270)$, were consistent with the measurements of Jost and Weber [120], and corresponded to electron traps with activations energies of 0.34 eV and 0.56 eV, respectively. The third level, labeled $E(45)$, was located at $E_{\mathrm{c}}-75$ meV, being consistent with the E40' peak [105]. By inspecting concentration depth profile data, the authors concluded that all three levels should be related to different $\mathrm{TiH_n}$ complexes. Analogously to Ref. [120], the total concentration of $\mathrm{TiH_n}$ electron traps did not add up to the concentration of $\mathrm{Ti_i}$ impurities present before the H treatment, demonstrating that complete passivation of Ti could be attained.

Several structures were investigated for $\mathrm{Ti_iH}_n$ complexes in Si (with $n = 1, \ldots, 4$) and its was concluded that H atoms prefer to bind directly to the Ti metal at anti-bonding *ab*-sites as depicted in Fig. 2.12a. By comparing $I(q/q+1)$ values for defective and bulk supercells we arrived at the electrical levels reported in Table 2.1.

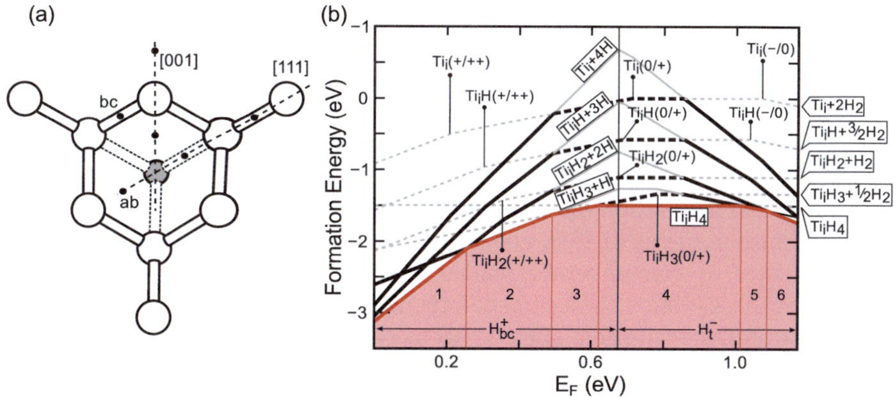

Fig. 2.12 (**a**) Schematic representation of some sites (*black dots*) neighboring a Ti_i impurity (*gray atom* at the center) which can be occupied by H to form Ti_iH_n complexes. Particularly stable 'ab' and 'bc' sites are labeled accordingly. (**b**) Formation energies of $Ti_iH_n + (4-n)H$ (*solid lines*) and $Ti_iH_n + 1/2(4-n)H_2$ (*dashed lines*) as a function of the Fermi energy with $0 \le n \le 4$. H and H_2 stand for dissolved atomic (either H^+_{bc} or H^-_t) and molecular hydrogen in Si, respectively. *Thick lines* represent phase diagrams of individual Ti_iH_n complexes. The lowest energy set of *thick (red) lines* represent the combined ground state for different regions of the Fermi energy, namely: *1* – $Ti_iH_2^{++} + 2H^+_{bc}$, *2* – $Ti_iH_3^+ + H^+_{bc}$, *3* – $Ti_iH_3^+ + 1/2H_2$, *4* – Ti_iH_4, *5* – $Ti_iH_3 + H^-_t$ and *6* – $Ti_iH_2 + 2H^-_t$ (Reprinted with permission from Ref. [116]. ©2014 AIP Publishing LLC)

Table 2.1 Calculated electrical levels of Ti_iH_n complexes in Si [116]. Ti_iH_4 was predicted to be electrically inert. Experimental levels are shown within braces and were taken from Refs. [105] and [119]. Electrical levels are divided into electron traps (data rows 1 and 2) and hole traps (data rows 3 and 4). Donor levels calculated with respect to the conduction band bottom, $E_c - E(0/+)$, were obtained from $E(0/+) - E_v$ values considering a band gap of 1.17 eV. All values are in eV

	Ti_i	Ti_iH	Ti_iH_2	Ti_iH_3
$E_c - E(-/0)$	0.10(0.07)	0.13(0.080)		
$E_c - E(0/+)$	0.46(0.27)	0.48(0.34)	0.50(0.56)	0.39(0.38)
$E(0/+) - E_v$	0.71	0.69	0.67	0.77
$E(+/++) - E_v$	0.22(0.26)	0.30	0.36	

Clearly, gradual hydrogenation of Ti leads to a progressive removal of electronic states from the Ti_i ion and to the creation of Ti-H bonding states within the valence band.

Bearing in mind that (i) E40', E170' and E(270) traps should be related to transitions from distinct complexes [118, 119], and that (ii) E40' and E170 should be related to transitions from defects with identical number of H atoms [118], we conclude that: (1) At least three Ti_iH_n complexes are electrically active. (2) The E40' level measured at $E_c - 0.80$ meV is the shallowest electron traps, and from Table 2.1, $Ti_iH(-/0)$ seems to be the obvious candidate. Further support to this assignment comes from the concentration profile of E40' indicating that this trap is the one that penetrates deeper from the surface, consistent with being related to a complex with a low number of hydrogen atoms. (3) Concentration profiles of $E(270)$ at $E_c - 0.56$ eV and E170' at $E_c - 0.38$ eV decrease faster, suggesting that they involve a larger number of hydrogen atoms. Their relative positioning in the gap suggest that they are related to (0/+) transitions of Ti_iH_2 and Ti_iH_3 calculated at $E_c - 0.50$ eV and $E_c - 0.39$ eV, respectively. The $Ti_iH(0/+)$ level calculated at $E_c - 0.48$ eV was tentatively assigned to E170 at $E_c - 0.34$ eV based on Ref. [118] which indicates that it should involve the same number H atoms as E40'.

The formation energy of Ti_iH_n complexes in Si as a function of the Fermi energy was calculated by combining the marker method reported above with Eq. 2.63. First we obtain E_f values for neutral defects, while formation energies for charged defects are obtained with help of the levels calculated from the ionization energies. Chemical potentials of Si, H and Ti species were taken from energies of bulk Si, H_2 molecules in Si, and Ti_i impurities in Si. Figure 2.12b summarizes the results from which we draw two main conclusions. (1) Reactions involving molecular hydrogen like $Ti_iH_n + {}^1\!/\!{}_2H_2 \rightarrow Ti_iH_{n+1}$ (downward transitions between dashed lines) and atomic hydrogen like $Ti_iH_n + H \rightarrow Ti_iH_{n+1}$ (downward transitions between solid lines) are exothermic under intrinsic and moderately doped n-type conditions. (2) The maximum number of H atoms that can decorate a Ti ion depends on the Fermi level. In p-type material and probably in heavily n-doped Si these reactions are endothermic when involving complexes with larger number of H atoms.

2.3.5 *Local Vibrational Modes of Defects*

Local vibrational modes (LVMs) and their frequencies are quantities which can be accurately predicted by density-functional methods. These can be directly compared to infra-red absorption, Raman scattering or photoluminescence data. The adiabatic (or Born-Oppenheimer) approximation is usually assumed, in the sense that the nuclei are regarded as point-like masses surrounded by electrons that adiabatically follow their vibrational movement. The harmonic approximation is also commonly accepted, according to which the total energy E of a system composed of N atoms is expanded to second order as a function of the atomic displacements [39, 121–123],

$$E = E_0 + \frac{1}{2}\sum_{i,j}^{N} \mathbf{u}_i \cdot \mathbf{K}_{i,j} \cdot \mathbf{u}_j, \tag{2.70}$$

where E_0 is the static internal energy at the ground state, $\mathbf{u}_i$ is the displacement vector of atom i, and $\mathbf{K}_{i,j}$ is a second-rank tensor whose elements $K_{ia,jb}$ are referred to as force-constants along Cartesian direction pairs a and b. They are calculated from the second derivatives of the total energy with respect to atomic motion around the equilibrium structure,

$$K_{ia,jb} = \frac{\partial^2 E}{\partial u_{ia} \partial u_{jb}}. \tag{2.71}$$

The equation of motion of an atom i is then

$$M_i \frac{\partial^2 \mathbf{u}_i}{\partial t^2} = -\nabla_i E = -\sum_j^N \mathbf{K}_{i,j} \cdot \mathbf{u}_j. \tag{2.72}$$

Assuming an harmonic solution to $\mathbf{u}_i(t)$, one arrives at an eigenvalue problem,

$$\mathbf{D} \cdot \mathbf{U} = \omega^2 \mathbf{U}, \tag{2.73}$$

where the $3N \times 3N$ matrix $\mathbf{D}$ is known as *dynamical matrix* [121],

$$D_{ia,jb} = \frac{1}{\sqrt{M_i M_j}} K_{ia,jb}. \tag{2.74}$$

Its respective $3N$ eigenvalues ω^2 are the square frequencies associated with the $3N$ normal modes $\mathbf{U}$ that represent the relative displacements of all atoms for a particular LVM.

For periodic systems we have to include a lattice (cell) index l into the formulation. Hence the i-th atom in the l-th unit cell has a displacement represented by $\mathbf{u}_{li}$, and its interaction with the j-th atom in the m-th unit cell is described by a force constant tensor $\mathbf{K}_{li,mj}$. A solution to Eq. 2.72 assumes now the form of a traveling wave, and the dynamical matrix elements for the unit cell at the origin (labelled as 0) are now given by,

$$D_{ia,jb}(\mathbf{k}) = \frac{1}{\sqrt{M_i M_j}} \sum_l K_{0ia,ljb} \exp\left[\mathrm{i}\,\mathbf{k} \cdot (\mathbf{r}_{lj} - \mathbf{r}_{0i})\right], \tag{2.75}$$

where a phase factor that depends on the equilibrium positions of atoms is introduced. Eigenvectors and eigenvalues from $\mathbf{D}(\mathbf{k})$ are now complex and have a $\mathbf{k}$-dependence to account for dispersion effects. In practice, either Eq. 2.74 or Eq. 2.75 with $\mathbf{k} = \Gamma$ are used to calculate local vibrational modes, with the second derivatives $K_{ia,jb}$ usually obtained directly from the density-functional code. This involves either moving all atom-pairs along Cartesian directions, or calculating the second derivatives of the total energy analytically. For localized modes, usually only a few atom shells around the defect are necessary to include in $\mathbf{D}$. This follows from the fact that in these modes only the neighboring atoms have considerable amplitude on $\mathbf{U}$. However, modes which lie close to the crystal vibrational states are better described if additional neighbors are included.

Of special interest in the identification of defects by means of LVM infrared absorption spectroscopy, is the fact that the vibrational frequency of a mode localized on a defect is inversely proportional to the square-root of the reduced mass of that mode, $\mu^{-1/2}$. This leads to observation of shifts in the frequencies, when one or more chemical species at the core of the defect are replaced by heavier or lighter isotopes. Let us look for instance at the interstitial carbon-oxygen complex (C_iO_i)

in Si. This defect is a product of the interaction between mobile interstitial carbon centers and interstitial oxygen in oxygen-rich Si samples. It is a stable defect with an annealing temperature around 350–450 °C.

Early density functional modeling found that the oxygen atom in C_iO_i was strongly affected by carbon and was not two-fold coordinated as depicted in Fig. 2.13a, but rather, was bonded to three Si atoms as illustrated in Fig. 2.13b [125]. This effect follows from the large electronegativity of carbon, that encourages electron transfer from the Si_2 radical in Fig. 2.13a. This leaves the Si_2 dangling bond orbital empty and prone to accept two electrons donated by a lone pair orbital on oxygen, and finally to the strong structural relaxation of the defect with the formation of the additional Si_2-O bond.

Unambiguous evidence for the above picture comes from Fourier transform infra-red (FTIR) spectroscopy [124]. The defect gives rise to three prominent LVM absorption bands at 1116.3, 865.9 and 742.8 cm^{-1}. These figures come from samples where most C and O species where ^{12}C and ^{16}O, respectively. Additional bands closer to the Raman frequency edge of Si were also assigned to C_iO_i but they are not relevant for the present discussion. When Si samples are enriched with ^{13}C or ^{18}O isotopes, the spectra change dramatically as shown by the peak frequencies and related isotope-shifts reported in Table 2.2.

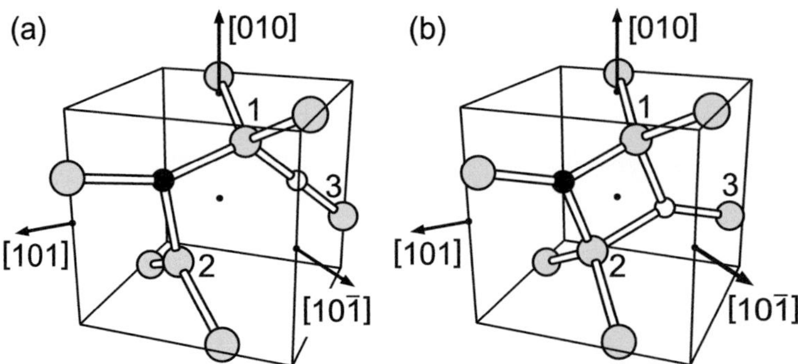

Fig. 2.13 Atomic models for the C_iO_i complex in silicon. (**a**) Two-fold coordinated oxygen model and (**b**) three-fold coordinated oxygen model. *Gray, black, white atoms* are Si, C, O species. Crystallographic axes and principal directions of the B tensor (see text in the next section) are also shown (Reprinted with permission from Ref. [124]. ©2001 American Physical Society)

Table 2.2 Observed and Calculated local vibrational mode frequencies (cm^{-1}), for C_iO_i in silicon [124]. The first two columns report absolute frequencies, whereas columns three onwards give their downward isotopic shifts

$^{12}C_i^{16}O_i$		$^{13}C_i^{16}O_i$		$^{12}C_i^{18}O_i$	
Exp.	Calc.	Exp.	Calc.	Exp.	Calc.
1116.3	1137.6	36.4	37.7	1.0	1.0
865.9	876.1	23.9	24.5	0.15	0.1
742.8	759.6	0.5	0.3	33.4	35.6

Of special interest is the oxygen related mode (that shows the large ^{18}O-shift) measured at 746.2 cm^{-1} and closely reproduced at 760 cm^{-1}. This frequency is well below the asymmetric stretch mode from the two-fold coordinated interstitial oxygen center at 1136 cm^{-1}. This is a characteristic of the weakened bonds of over-coordinated oxygen relative to those in a two-fold coordinated Si-O-Si structure. The two higher frequency modes are mostly localized on the C atom, as shown by the large shift measured after ^{13}C enrichment of the samples. All observed and calculated isotope-shifts are also very similar in magnitude, providing compelling evidence for the assignment of the model represented in Fig. 2.13b.

2.3.6 Defect Response to Uniaxial Stress

An important characteristic of a defect with lower symmetry than that of the host crystal, is the stress-energy tensor or piezospectroscopic tensor **B**. In the absence of any imposed external stress, defects are found equally distributed over all equivalent orientations allowed by the symmetry of the crystal. However, applying a compressive stress along a particular direction increases the energy of those defects which also exert a compressive stress along that direction, and conversely decreases the energy of those which impose a tensile stress.

Let us suppose that we have a volume V_0 of a cubic crystal under strain, containing N defects, N_i of which are aligned along a particular direction i. Within the elastic regime, the total energy is expanded to second order as

$$E = E_0 + NE_\mathrm{f} + \sum_i N_i\,\mathbf{B}^{(i)} \cdot \boldsymbol{\epsilon} + \frac{V_0}{2}\boldsymbol{\epsilon} \cdot \mathbf{C} \cdot \boldsymbol{\epsilon}, \tag{2.76}$$

subject to the condition $N = \sum_i N_i$. The first term is the total energy of the perfect crystal in the absence of external strain $\boldsymbol{\epsilon}$. The second term is the formation energy of all defects. The last term is the strain energy stored in the crystal calculated with help of the elastic constants C_{klmn}, and finally, the term linear in $\boldsymbol{\epsilon}$ represents the response of defects to strain $\boldsymbol{\epsilon}$. This response is accounted for by means of a traceless second rank tensor $\mathbf{B}^{(i)}$ whose elements are

$$B^{(i)}_{lm} = \frac{\partial E^{(i)}_\mathrm{f}}{\partial \epsilon_{lm}}, \tag{2.77}$$

representing the variation of the formation energy $E^{(i)}_\mathrm{f}$ of a defect oriented along i, with respect to all strain components.

Now, if all possible defect orientations are equally distributed ($N_i = N_j$ for every i and j) it can be shown that summing $\mathbf{B}^{(i)}$ over all orientations results in a diagonal tensor, and therefore

$$\sum_i N_i\,\mathbf{B}^{(i)} \cdot \boldsymbol{\epsilon} = \frac{N}{3}\,\mathrm{Tr}\,\mathbf{B}\,(\epsilon_{11} + \epsilon_{22} + \epsilon_{33}). \tag{2.78}$$

Since at the equilibrium $\partial E/\partial \epsilon_{ij} = 0$, we conclude that $\mathrm{Tr}\,\mathbf{B} = 0$.

Let us now assume that a specific defect has r inequivalent orientations with respect to an external strain. In a simplistic picture this could be $r = 2$, standing for defects either aligned along or perpendicular to a strain direction. In a silicon crystal $r = 1$ for a tetrahedral center, but can be up $r = 24$ for a C_1-symmetry defect. For samples where the density of defects is low and defect-defect interactions are negligible, under equilibrium conditions the probability p_i of finding a defect aligned along i is

$$p_i = \exp\left(-\beta\, \mathbf{B}^{(i)} \cdot \boldsymbol{\epsilon}\right) / \sum_j^r \exp\left(-\beta\, \mathbf{B}^{(j)} \cdot \boldsymbol{\epsilon}\right), \tag{2.79}$$

with $\beta = 1/k_B T$, and hence for two different alignments we have,

$$\frac{p_i}{p_j} = \exp\left[-\beta\left(\mathbf{B}^{(i)} - \mathbf{B}^{(j)}\right) \cdot \boldsymbol{\epsilon}\right], \text{ subject to } \sum_i^r p_i = 1. \tag{2.80}$$

Both Eqs. 2.79 and 2.80 are used by experimentalists to obtain the B_{lm} components by measuring $\ln(p_i/p_j)$ as a function of T and $\boldsymbol{\epsilon}$.

There are two types of measurements: stress-alignment and stress-splitting. In stress-alignment experiments, the sample is kept under uniaxial stress at several temperatures for some time, allowing defects to reorient and find their equilibrium distribution under those conditions (cf. Eq. 2.79). Measurements of p_i/p_j population fractions are carried out by means of an orientation-sensitive technique such as electron paramagnetic resonance or an optical technique employing a polarized light source (dichroism). These measurements often need to be carried out at low temperatures, and to that end defects are quenched by quickly lowering the sample temperature.

In stress-splitting measurements the populations of defects are not modified. Instead, a transition between two states (electronic, vibrational, etc.) is measured as a function of stress. The stress-induced change in the transition is then expected to vary as

$$\Delta E = (\mathbf{B}_{\text{final}} - \mathbf{B}_{\text{initial}}) \cdot \boldsymbol{\epsilon},$$

where **B**-tensors of the final and initial states have to be taken into account, and the strain is related to the applied stress by means of the compliance constants.

The calculation of the stress tensor of a defect is usually simplified by noting that not all B_{lm} components are independent, namely

1. $B_{11} + B_{22} + B_{33} = 0$. This condition follow from the fact that in the absence of stress, all defects are randomly aligned, and the volume derivative of the energy of the defective crystal must vanish at the equilibrium;
2. The **B** matrix is symmetric ($B_{lm} = B_{ml}$);

3. $\mathbf{B}$ is symmetric with respect to any symmetry operation $\mathbf{R}$ of the point-group of the defect, $\mathbf{R}^{-1} \cdot \mathbf{B} \cdot \mathbf{R} = \mathbf{B}$.

The recipe to calculate the $\mathbf{B}$-tensor of a defect involves the following steps,

1. A volumetric (isotropic) relaxation of the defective supercell. This step will enforce the $\mathrm{Tr}\,\mathbf{B} = 0$ condition and can be done by simply relaxing the lattice constant of the defective supercell.
2. Identify the independent Cartesian components of $\mathbf{B}$ to be calculated by solving $\mathrm{Tr}\,\mathbf{B} = 0$, $B_{lm} = B_{ml}$ and $\mathbf{R}^{-1} \cdot \mathbf{B} \cdot \mathbf{R} = \mathbf{B}$ for all symmetry operations $\mathbf{R}$ of the space group of the supercell. Here we should also consider those operations involving fractional lattice translations in non-symmorphic space groups. For example, for a trigonal defect in Si, diagonal and off-diagonal elements are $B_{ll} = 0$ and $B_{lm} = B$, respectively, i.e. there is a single independent component to be found.
3. Construct a set of strain deformation matrices $\boldsymbol{\epsilon}$ (to be applied to atom positions $\mathbf{r}$ and lattice vectors of the supercell $\mathbf{A}_1$, $\mathbf{A}_2$ and $\mathbf{A}_3$) along the directions of the non-zero elements of $\mathbf{B}$. Returning to the previous example of a trigonal defect in Si, we would need a single strain tensor with at least one non-zero off-diagonal element, say $\epsilon_{lm} = \gamma$ for $l \neq m$, and $\epsilon_{ll} = 0$. To second order, the resulting change in the energy of the supercell is

$$\Delta E(\gamma) = \sum_{lm} B_{lm}\epsilon_{lm} = 12B\gamma + c\gamma^2 \tag{2.81}$$

where the generalized elastic constant c accounts for the quadratic dependence of ΔE on strain.
4. For each strain matrix from step 3, chose a few γ values around zero, say ± 0.02, ± 0.01 and 0.00, apply the respective deformations to the atoms and lattice – for instance, new atom positions will be at $\mathbf{r}' = \mathbf{r}\,(\mathbf{1} + \boldsymbol{\epsilon})$, with $\mathbf{1}$ being a 3 by 3 identity matrix – and then calculate the respective total energies. Finally we have to fit Eq. 2.81 to the data. The generalized elastic constant c is not a relevant quantity in the present context and can be discarded after each fit.

We may chose not to relax the volume of the supercell by skipping step 1. I this case an extra volumetric component will be present in a non-traceless stress-energy tensor $\mathbf{B}'$. Using Eqs. 2.78 and 2.76, and assuming that a population of randomly aligned defects impose a volumetric distortion to a cubic material, i.e. $\epsilon_{ll} = \epsilon$ and the bulk modulus is $K = (C_{1111} + 2\,C_{1122})/3$, we now have

$$E = E_0 + NE_f + \frac{9}{2} V_0 K \epsilon^2 + N\,\mathrm{Tr}\,\mathbf{B}'\,\epsilon, \tag{2.82}$$

To find the volume change per defect $\Delta V/N = (V - V_0)/N$ we simply solve $\partial E/\partial\epsilon = 0$, to find that

$$\frac{\Delta V}{N} = -\frac{1}{3K}\,\mathrm{Tr}\,\mathbf{B}'. \tag{2.83}$$

Comparing Eqs. 2.76 and 2.82, we readily conclude that

$$\mathbf{B} = \mathbf{B}' - \frac{1}{3}\,\mathrm{Tr}\,\mathbf{B}'. \tag{2.84}$$

Equations 2.83 and 2.84 allow us to calculate the anisotropic stress-energy tensor elements B_{lm}, as well as to predict the volume relaxation of the crystal for a specific defect density.

We now look at the vacancy-oxygen-hydrogen (VOH) complex in Si as an example of the calculation of the stress-energy tensor. VOH is an electrically active complex with acceptor and donor levels at $E_c - 0.31\,\text{eV}$ and $E_v + 0.28\,\text{eV}$, respectively (see Ref. [126] and references therein for further details). It is a monoclinic-I (C_{1h}) symmetric defect, with a (110) mirror plane, where three among the four Si dangling bonds of the vacancy are saturated by Si-O-Si and Si-H units (see structure inset in Fig. 2.14a). This leaves a single Si radical that is responsible for the electrical activity.

A combined experimental and theoretical investigation of the uniaxial stress response of VOH in Si was reported by Coutinho et al. [126]. They employed high-resolution Laplace deep level transient spectroscopy (LDLTS) in combination with uniaxial stress, and total-energy density-functional calculations to produce a detailed model consistent with the measured dynamical, electrical, and piezospectroscopic data. For the sake of simplicity we will restrict our analysis to the [100] stress direction. Accordingly, under stress the VOH(−/0) signal in LDLTS splits into two peaks with relative intensities 1:2 and their variation with stress is shown in Fig. 2.14a as blue and red circles, respectively. From the difference of the slopes we obtain $\Delta E = 21$ meV for a stress of 1.0 GPa.

A C_{1h} symmetric defect in Si has 12 possible orientations and they are labelled with help of letter pairs (a, b, c, d). Each pair can have two different orientations (say *left* and *right*) and for an arbitrary stress direction, *ab* can in principle be distinguishable from *ba*. When left and right orientations are indistinguishable with respect to the applied stress, we simply make use of over-lined labels. For instance, when orientations *cd* and *dc* are equivalent with respect to the applied stress, we use $\overline{cd}$ to represent both orientations. In practical terms, for the VOH complex we can assume that the first letter of the pair refers to the Si atom that binds to H, whereas the second letter is the three-fold coordinated Si radical. According to Fig. 2.14b, the *cd* structure has a Si_a-O-Si_b unit, a Si_c-H bond and a Si_d dangling bond. From the calculated **B**-tensors for neutral and negatively charged VOH complexes, we arrived at the configuration coordinate diagram shown in Fig. 2.14c. It indicates that under [100] stress the VOH(−/0) signal splits into two transitions. Specifically, we have a lower energy transition with faster emission rate (shown as a red arrow) that shifts as $-32 + 24 = -8$ meV/GPa involving 8 degenerate orientations ($\overline{ab}$, $\overline{ad}$, $\overline{bc}$ and $\overline{bd}$), and a higher energy and slower transition (blue arrow) with emission rate varying as $+64 - 48 = 16$ meV/GPa involving a total of four degenerate orientations ($\overline{ab}$ and $\overline{cd}$). The net effect is a shift between peaks of $\Delta E = 16 + 8 = 24$ meV/GPa, which nicely accounts for the observed 21 meV/GPa. Further, from the populations

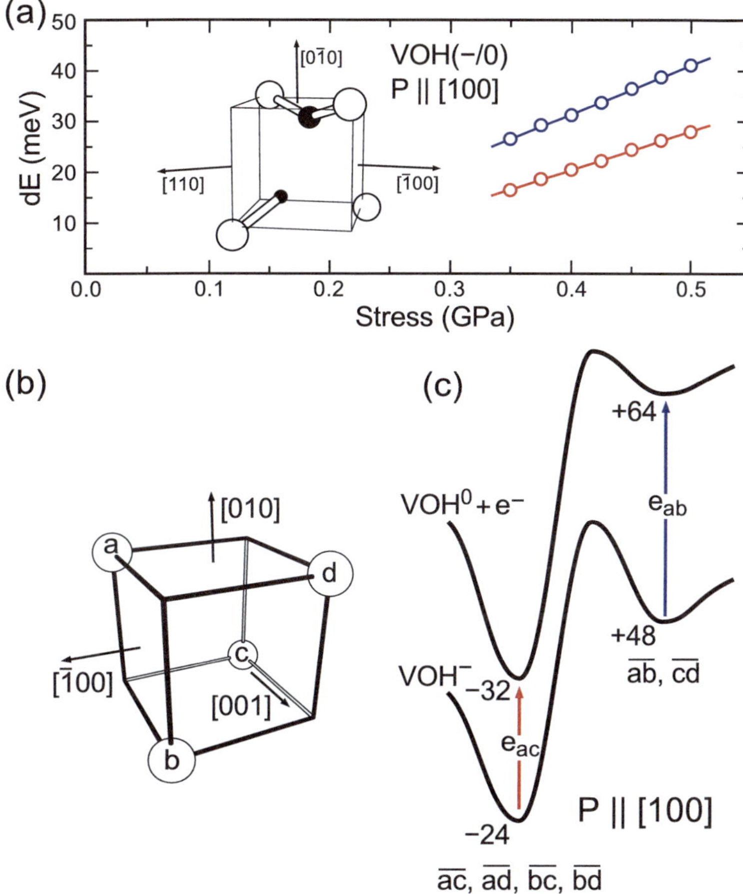

Fig. 2.14 (**a**) Energy shifts versus applied stress for the split lines of the VOH(−/0) level under stress along the ⟨100⟩ direction. A positive slope means that the energy level approaches the conduction band under stress. The net splitting of both lines is 21 meV/GPa. The structure of VOH is shown in the *inset*. Large and small black spheres represent O and H atoms. Si atoms are shown in white. (**b**) Labeling scheme for the 12 different orientations of a C_{1h}-symmetry defect in a cubic crystal (see text). (**c**) Configuration coordinate diagrams for VOH^0 and VOH$^-$ defects under 1 GPa uniaxial stresses along [100] (Reprinted with permission from Ref. [126]. ©2003 American Physical Society 2003)

of the two transitions, theory predicts a 4:8 population ratio, which is the same as the observed 1:2 intensity ratio. A similar analysis was done for stress directions other than [100] and we direct the most interested readers to Ref. [126] and references therein for further details.

2.4 Defects in Silicon Nanostructures

In the last decade, density functional computer simulations of the quantum-mechanics of electrons and atomic nuclei have had an increasingly important impact on materials science. Besides assisting researchers in the fundamental understanding of many effects, DFT has also paved the way towards improving materials design and future technologies. Efficient algorithms for solving the Kohn-Sham equations have emerged and implemented with increasing sophistication, tremendously boosting the applicability of DFT methods. With this, new doors are opening to innovative research on materials across physics, chemistry, materials science, surface science, and nanotechnology.

Nanostructures with precisely controlled size and shape can be assembled into more complex structures to form new meta-solids, including ordered superlattices [127]. These materials can be explored for various applications, including sensors, transistors, data storage, solar cells, and thermoelectrics. Si nanostructures are currently breaking new ground for applications ranging from thin-film transistors and thermoelectric power devices to light emitters and fluorescent tags for biomedical applications [128–131]. Presently, high-quality (low defect content) Si nanostructures can be produced in very large quantities, at low cost, and with relatively narrow size distribution by means of several methods [132, 133]. Photovoltaic devices based on Si nanostructures are particularly exciting because of their size-tunable optical absorption edge and the potential for multi-exciton generation [128]. Some of the unique features of Si nanostructures have been investigated in detail for some years now. However, many problems still lay ahead and many questions remain unanswered. In particular, the effect of quantum confinement on the energy levels and band structure, the exceptionally high quantum efficiencies in photon-to-exciton conversion or the role of defects, are topics currently under active research.

The usual protocol to be followed during a DFT study involves three distinct steps: (1) translation of the problem to a computable atomistic model, (2) computation of the required physico-chemical observables and quantities, and (3) validation of the simulation results by confrontation with measurements. In this section I will address step 1 (problem translation) since this is the one that differs mostly from a study of defects in a crystalline solid. This is explored for the case of thin films of Si nanocrystals (free-standing and embedded in a host material) and nanowires.

2.4.1 Freestanding and Particulate Nanostructures

Freestanding Si nanocrystals (Si-NCs) are attractive from a processing perspective. Liquid dispersions of NCs (NC inks) can be used for low-cost, print-type deposition of semiconductors, as opposed to the conventional and more costly vacuum deposition methods. Besides allowing for patterning devices onto large-area and flexible substrates, the freestanding nature of the NCs open the possibility for controlling the specifics of the NC surface, enabling a great deal of flexibility in selecting and controlling the inter-NC medium.

Atomistic-scale density functional calculations of NC solids have been hampered for several reasons. Possibly the most important one stems from the sheer size of the problem. Progress has had to rely on bold simplifications, especially with regard to the size, geometry and boundary conditions of the model. A common approach is to limit the problem to a localized region of interest, either by (i) using finite boundary conditions [134, 135], or by (ii) adopting a periodic slab-, wire-, or particle-in-a-box approach [136–138], hoping that within such a limited region the model will be able to adequately capture the electronic local density of states. In the latter case, it is also common to impose a sufficiently large vacuum space to isolate the nanostructure periodic replicas, and by this way to mitigate unwanted periodicity effects [139]. The methods just described usually aim at studying a single nanostructure, eventually interacting with dopants, adsorbants, surfactants, and other agents. Unfortunately, the resemblance that such models bear to a NC solid, regarding the chemical environment of atoms, steric hindrance, or long-range electrostatics, may however be unsatisfactory.

A possible approach to model a NC solid is to pack NC arrays with minimal vacuum fraction by using appropriate periodic boundary conditions [140]. This approach has been shown to lead to reasonable and realistic properties. For instance, while the calculated energy gap of a vacuum-surrounded Si NC with a diameter of 2 nm (3 nm) is about 3.1 eV (2.5 eV), the energy gap of a face-centered cubic (fcc) superlattice of identical NCs is only approximately 2.0 eV (1.8 eV). This results from Coulomb screening effects in the calculation of isolated and charged NC^+ and NC^- (to obtain the quasi-particle gap), which in the case of a packed superlattice are effectively stronger than in a particle-in-a-box-like calculation. This is very significant if we want to understand the electronic structure of defects and impurities in a NC solid.

An example is depicted in Fig. 2.15a that shows a close-packed Si NC superlattice with F_4-TCNQ molecules at interstitial sites [140]. The electronic structure calculations suggest that F_4-TCNQ molecules within the SL interstitialcies introduce empty electronic states that bridge and overlap neighboring NCs, providing extra electronic connectivity to the NC network.

Inspection of the Kohn-Sham states showed that although F_4-TCNQ is able to polarize the NC surface (by becoming fractionally charged with about one fifth of an electron from the NC surface), the LUMO of the doped system is heavily localized on the molecule. That means that traveling electrons can be trapped at this empty state. However, after being trapped they will readily feel a strong Coulomb repulsion from lower-energy molecular electrons, and the propensity to hop into the LUMO of adjacent NCs will also be high. This picture is supported by the calculated deep acceptor levels shown in Fig. 2.16a and by the sudden increase of the conductivity of F_4-TCNQ-doped films with concentrations of about one molecule per NC.

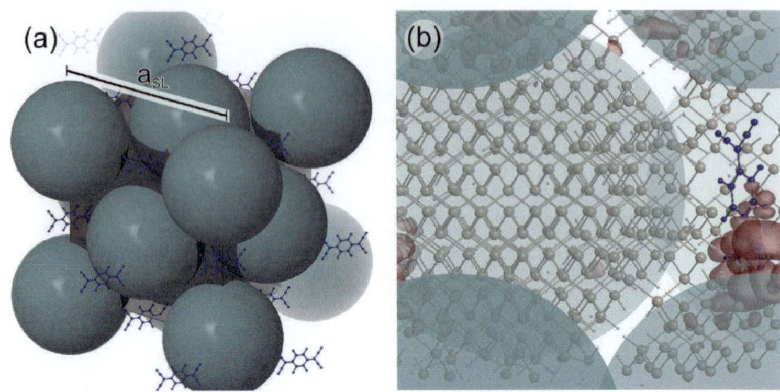

Fig. 2.15 (**a**) Three-dimensional view of an fcc superlattice of Si NCs (*large solid spheres*) doped with F_4-TCNQ molecules (*blue*). (**b**) Electron density isosurface (*red*) of the HOMO level overlapping both the F_4-TCNQ molecule and an adjacent NC. *Circles* enclosing individual NCs are drawn for eye guidance purposes (Reprinted with permission from Ref. [140]. ©2014 American Chemical Society)

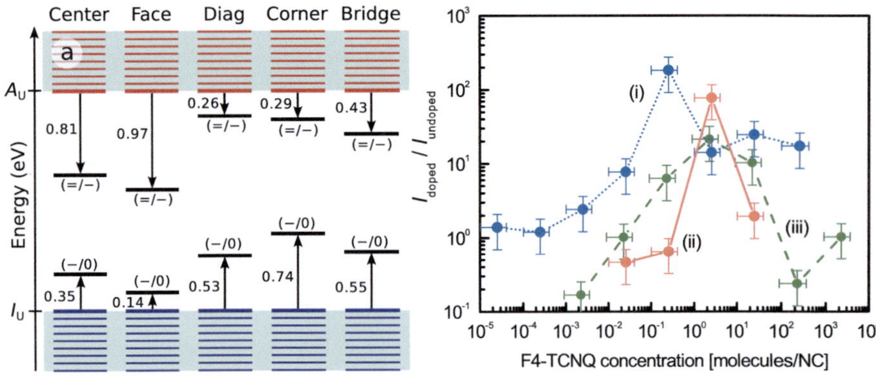

Fig. 2.16 (**a**) Calculated $(-/0)$ and $(=/-)$ acceptor levels (*black horizontal lines*) of F4-TCNQ molecules in a Si NC superlattice. *Arrows* are labeled with energy differences with respect to the ionization potential (I_U) and electron affinity (A_U) energies of the undoped SL. SL band energies are represented as *blue* (bottom) and *red* (top) *horizontal lines*. (**b**) Evolution of the current ratio $I_{doped}/I_{undoped}$ at $V = +30$ V as a function of F_4-TCNQ concentration for films of Si NCs with (i) diameter $d_{NC} = 4.2$ nm, (ii) $d_{NC} = 15.7$ nm, and (iii) $d_{NC} = 17$ nm (Reprinted with permission from Ref. [140]. ©2014 American Chemical Society)

2.4.2 Embedded Nanostructures

Charge and energy transfer between adjacent NCs is a fundamental problem with huge impact on applications such as batteries, memories and photovoltaics. For instance, multi-exciton generation has been proposed to take place after conversion of a high energy excitation localized on a cluster of adjacent Si NCs, to a set of low-energy excitons [141]. While the role of the size, shape, surface decoration, and embedding medium on the optoelectronic properties of Si NCs has been investigated to some extent, the study of the effects of NC-NC interplay has received less attention.

The work of Guerra and Ossicini [142] addresses some of these issues by looking at the strain-induced changes on the optical properties of Si NCs, imposed both by a SiO_2 host and by short-range NC-NC interactions. They show that below a certain NC-NC separation threshold, inter-NC strain forces play an important role in their optical properties. In the context of this chapter, the method used to construct the embedded model is particularly interesting. Basically the authors made use of a SiO_2 cubic phase known as β-cristobalite, which can be constructed simply by starting with a perfect Si unit cell or supercell, an placing an oxygen atom right at every Si-Si bond-center site (obviously accompanied by an increase in the lattice constant to end up with a ~ 1.6 Å Si-O bond length). They started with a $Si_{512}O_{1014}$ cubic cell, removed a certain number of O atoms within two close (non-overlapping) spherical regions, and relaxed the atomic structure to *heal* the dangling bonds. The result is depicted on the left-hand side of Fig. 2.17. The resulting structure builds up a considerable tensile strain, and its impact can be inspected either by allowing the lattice parameters of the supercell to relax, or as proposed in Ref. [142] by de-embedding the NCs and passivating any remaining Si- and O-radicals.

Another interesting method to produce SiO_2-embedded Si NCs has been inspired in the bond-switching method of Wooten, Winer, and Weaire [143]. It starts with the

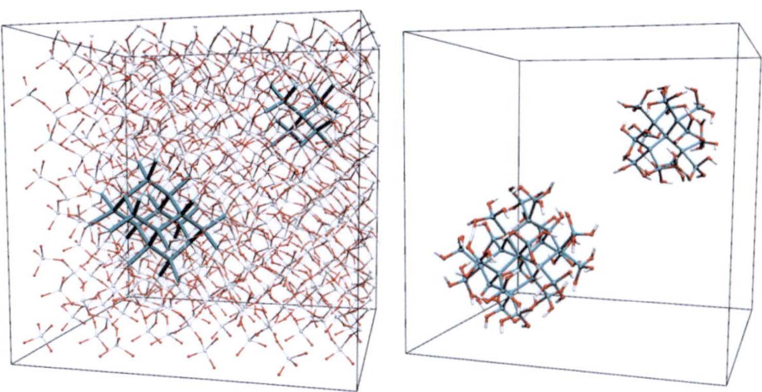

Fig. 2.17 Si_{17}+Si_{32} embedded in amorphous SiO_2 (*left*) and analogous vacuum-surrounded OH-terminated NCs pair (*right*) at a surface-to-surface distance $d = 0.8$ nm. Si, O, and H atoms are represented in *cyan*, *red*, and *white*, respectively. Si atoms of the SiO2 are represented in *white* for clarity. The simulation box, displayed by *black lines*, has a side of 2.78 (*left*) and 2.60 nm (*right*) (Reprinted with permission from Ref. [142]. ©2013 American Physical Society 2013)

cristobalite structure of SiO2 (like above), but before relaxing the atomic structure, a set of bond-switching operations over Si-O-Si linear structures is carried out. This recipe provides more realistic models embedded in amorphous SiO_2 (see for example Ref. [144]).

2.4.3 *Doping of Si Nanostructures*

Deliberate introduction of extrinsic species into semiconductors lies at the heart of microelectronics. A prototypical example is the replacement of a few lattice sites in a billion in crystalline Si by phosphorous or boron, to confer electrical conductivity to an otherwise poor insulator. Analogously, doping of nanocrystal (NC) solids [145, 146] and metamaterials made of wave-function tunable nanostructures [147, 148] is expected to leverage exciting and disruptive technologies.

Doping of nanocrystals, in particular those made of Si, has been already reported. For instance, the introduction of phosphorous into Si NCs has been demonstrated by means of electron paramagnetic resonance and ^{31}P hyperfine-splitting data [149, 150]. The measurements show that the size of the NCs has a striking influence on the localization of the P donor state for NC diameters smaller than $d \sim 15$ nm. It suggests that substitutional P, which in bulk silicon is a shallow donor with ionization energy of only 44 meV, becomes a deep donor in small nanostructures [151, 152]. In fact, in nanowires and nanocrystals with $d \sim 2$ nm the donor electron was predicted to be tightly bound to the P^+ ion by about 0.5 eV and over 1.5 eV, respectively [136, 153–155]. Hence, for Si-NCs which are interesting from the confinement and optical perspectives, electrical doping is anticipated to be hardly achievable.

A promising route for electrical doping of small nanostructures was proposed for the case of nanowires by Amato and co-workers [156]. Accordingly, the production of a one-dimensional free-carrier gas was suggested to take place in type-II offsetted Ge-Si (or Si-Ge) core-shell nanowires suitably doped with phosphorous or boron. For instance, in Ge-Si core-shell nanowires the replacement of a Ge core atom by phosphorous leads to a semi-occupied HOMO level mostly localized on the Si shell. Such an electron-transfer mechanism can be ascribed to the electronic band offset across a Si/Ge interface. This is depicted in Fig. 2.18a, showing that the conduction band at the core is higher in energy than at the shell. Hence donated electrons tend to accumulate at the Si shell. Conversely, Fig. 2.18b shows that a boron atom at the Si core of a Si-Ge core-shell nanowire will produce a free hole that will relax to the valence band top of the Ge shell. Although having no parallel in the case of zero-dimensional core-shell nanocrystals (cs-NCs), this principle could be explored for heterostructures made of adjacent NCs.

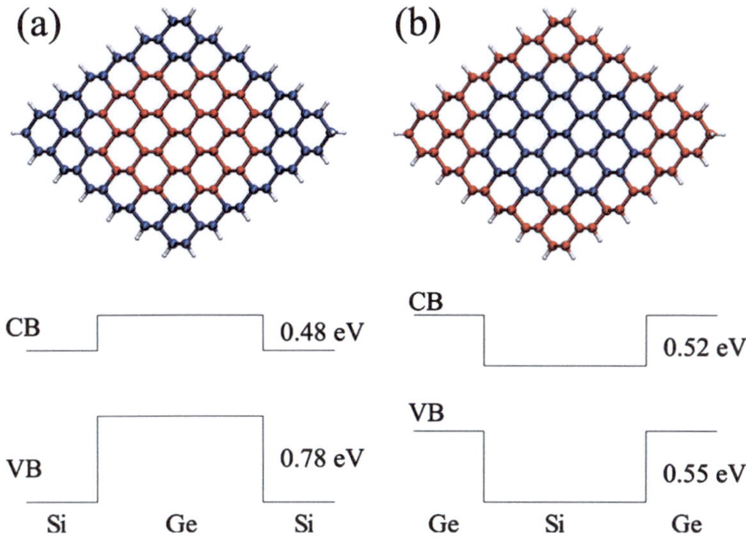

Fig. 2.18 Cross-section view and calculated band alignment (band gap not in scale) of (**a**) Ge-core/Si-shell nanowires and (**b**) Si-core/Ge-shell nanowires. Ge and Si atoms are shown as *red* and *blue spheres*, respectively; *white spheres* represent H atoms passivating the surface (Reprinted with permission from Ref. [156]. ©2011 American Chemical Society)

2.5 Summary

In this chapter we described the details of the calculation of several observables related to defects in Si materials by means of density functional methods. A major concern was to make the subject accessible to early-stage researchers, especially to post-graduate students.

We started with a general overview of the theoretical framework, from the Hartree-Fock method to density functional theory. It is noted that commonly used recipes to treat the exchange-correlation interaction among electrons, for instance the local density approximation or the generalized gradient approximation suffer from the well-known *band-gap underestimation* problem. Perhaps this is the most serious source of errors affecting the calculations and it is especially severe for the case of electrical levels. More sophisticated approximations to the exchange-correlation energy have been made available, and those incorporating a portion of exact exchange from Hartree-Fock theory are becoming rather popular. However, although these methods are promising for studying molecules and defects in supercells with a few hundred atoms, their use in nano-sized problems (with thousands of atoms) is still intractable.

We described the process of making use of the basic output from density functional codes, namely total energies, Kohn-Sham eigenstates and electron densities, to obtain several defect properties. These include defect structures, migration barriers and respective mechanisms, solubilities, binding energies, electrical levels and local vibrational modes. These can be compared to a multitude of spectroscopic techniques, like deep-level transient spectroscopy, optical absorption, photoluminescence, electron paramagnetic resonance, among others. The response of defects to uniaxial stress (as monitored by some spectroscopic method), is a powerful route to extract insightful information about the symmetry, as well as the nature of the defect-crystal elastic coupling. We showed how to calculate these properties, and how to compare them with stress-splitting or stress-alignment experiments.

The chapter ends with a few selected problems on Si nanostructures, namely nanocrystals and nanowires. Metamaterials made of such wave-function tunable nanostructures are expected to leverage exciting and disruptive technologies like quantum computing, solotronics, third-generation photovoltaics, thermoelectrics, biological tagging, to mention but a few. The examples presented in this chapter demonstrate that density functional methods are able to address this class of problems, and will play a major role in the understanding and development of future Si-based technologies.

References

1. Bukhori, M., Roy, S., Asenov, A.: Simulation of statistical aspects of charge trapping and related degradation in bulk mosfets in the presence of random discrete dopants. IEEE Trans. Electron Devices **57**(4), 795–803 (2010). doi:10.1109/TED.2010.2041859
2. Safarian, J., Tranell, G., Tangstad, M.: Processes for upgrading metallurgical grade silicon to solar grade silicon. Energy Procedia **20**, 88–97 (2012). doi:10.1016/j.egypro.2012.03.011
3. Hohenberg, P., Kohn, W.: Inhomogeneous electron gas. Phys. Rev. **136**, B864–B871 (1964)
4. Kohn, W., Sham, L.J.: Self-consistent equations including exchange and correlation effects. Phys. Rev. **140**, A1133–A1138 (1965)
5. Jones, R.O., Gunnarsson, O.: The density functional formalism, its applications and prospects. Rev. Mod. Phys. **61**, 689–746 (1989). doi:10.1103/RevModPhys.61.689
6. Blöchl, P.E.: Projector augmented-wave method. Phys. Rev. B **50**, 17953–17979 (1994). doi:10.1103/PhysRevB.50.17953
7. Hartwigsen, C., Goedecker, S., Hutter, J.: Relativistic separable dual-space Gaussian pseudopotentials from H to Rn. Phys. Rev. B **58**, 3641–3662 (1998). doi:10.1103/PhysRevB.58.3641
8. Goedecker, S.: Linear scaling electronic structure methods. Rev. Mod. Phys. **71**, 1085–1123 (1999). doi:10.1103/RevModPhys.71.1085
9. Beck, T.L.: Real-space mesh techniques in density-functional theory. Rev. Mod. Phys. **72**, 1041–1080 (2000). doi:10.1103/RevModPhys.72.1041
10. Rayson, M.J., Briddon, P.R.: Highly efficient method for kohn-sham density functional calculations of 500–10000 atom systems. Phys. Rev. B **80**, 205104 (2009). doi:10.1103/PhysRevB.80.205104
11. Freysoldt, C., Grabowski, B., Hickel, T., Neugebauer, J., Kresse, G., Janotti, A., Van de Walle, C.G.: First-principles calculations for point defects in solids. Rev. Mod. Phys. **86**, 253–305 (2014). doi:10.1103/RevModPhys.86.253
12. Schiff, L.I.: Quantum Mechanics. McGraw-Hill, Singapore (1955)
13. Born, M., Oppenheimer, R.: Zur quantentheorie der molekeln. Annalen der Physik **389**, 457–484 (1927). doi:10.1002/andp.19273892002
14. Watkins, G.D.: Deep Centers in Semiconductors, chap. 3. Gordon and Breach, New York (1986)

15. Probert, M.I.J., Payne, M.C.: Improving the convergence of defect calculations in supercells: an ab initio study of the neutral silicon vacancy. Phys. Rev. B **67**, 075204 (2003). doi:10.1103/PhysRevB.67.075204
16. Wright, A.F.: Phys. Rev. B **74**, 165116 (2006). doi:10.1103/PhysRevB.74.165116
17. Roothaan, C.C.J.: New developments in molecular orbital theory. Rev. Mod. Phys. **23**, 69–89 (1951). doi:10.1103/RevModPhys.23.69
18. Slater, J.C.: The theory of complex spectra. Phys. Rev. **34**, 1293–1322 (1929). doi:10.1103/PhysRev.34.1293
19. Thijssen, J.M.: Computational Physics. Cambridge University Press, Cambridge (1999)
20. Koopmans, T.: Über die zuordnung von wellenfunktionen und eigenwerten zu den einzelnen elektronen eines atoms. Physica **1**(1–6), 104–113 (1934). doi:10.1016/S0031-8914(34)90011-2
21. Perdew, J.P., Levy, M.: Comment on "significance of the highest occupied kohn-sham eigenvalue". Phys. Rev. B **56**, 16021–16028 (1997). doi:10.1103/PhysRevB.56.16021
22. Lundqvist, S., March, N.H. (eds.): Theory of the Inhomogeneous Electron Gas. Springer, New York (1983). doi:10.1007/978-1-4899-0415-7
23. Parr, R.G., Yang, W.: Density-Functional Theory of Atoms and Molecules. The International Series of Monographs on Chemistry. Oxford University Press, New York (1989)
24. Argaman, N., Makov, G.: Density functional theory: an introduction. Am. J. Phys. **68**(1), 69–79 (2000). doi:10.1119/1.19375
25. von Barth, U., Hedin, L.: A local exchange-correlation potential for the spin polarized case. J. Phys. C: Solid State Phys. **5**(13), 1629–1642 (1972). doi:10.1088/0022-3719/5/13/012
26. Rajagopal, A.K., Callaway, J.: Inhomogeneous electron gas. Phys. Rev. B **7**, 1912–1919 (1973). doi:10.1103/PhysRevB.7.1912
27. Perdew, J.P., Zunger, A.: Self-interaction correction to density-functional approximations for many-electron systems. Phys. Rev. B **23**, 5048–5079 (1981). doi:10.1103/PhysRevB.23.5048
28. Ceperley, D.: Ground state of the fermion one-component plasma: a monte carlo study in two and three dimensions. Phys. Rev. B **18**, 3126–3138 (1978). doi:10.1103/PhysRevB.18.3126
29. Ceperley, D.M., Alder, B.J.: Ground state of the electron gas by a stochastic method. Phys. Rev. Lett. **45**, 566–569 (1980). doi:10.1103/PhysRevLett.45.566
30. Vosko, S.H., Wilk, L., Nusair, M.: Accurate spin-dependent electron liquid correlation energies for local spin density calculations: a critical analysis. Can. J. Phys. **58**(8), 1200–1211 (1980). doi:10.1139/p80-159
31. Perdew, J.P., Wang, Y.: Accurate and simple analytic representation of the electron-gas correlation energy. Phys. Rev. B **45**, 13244–13249 (1992). doi:10.1103/PhysRevB.45.13244
32. Hood, R., Chou, M., Williamson, A., Rajagopal, G., Needs, R.: Exchange and correlation in silicon. Phys. Rev. B **57**, 8972–8982 (1998). doi:10.1103/PhysRevB.57.8972
33. Perdew, J.P., Burke, K., Ernzerhof, M.: Generalized gradient approximation made simple. Phys. Rev. Lett. **77**, 3865–3868 (1996). doi:10.1103/PhysRevLett.77.3865
34. Perdew, J.P., Burke, K., Wang, Y.: Generalized gradient approximation for the exchange correlation hole of a manyelectron system. Phys. Rev. B **54**, 16533–16539 (1996). doi:10.1103/PhysRevB.54.16533
35. Markevich, V.P., Leonard, S., Peaker, A.R., Hamilton, B., Marinopoulos, A.G., Coutinho, J.: Titanium in silicon: lattice positions and electronic properties. Appl. Phys. Lett. **104**(15), 152105 (2014). doi:10.1063/1.4871702
36. Harrison, W.A.: Pseudopotentials in the theory of metals. Pseudopotentials in the Theory of Metals. W. A. Benjamim, New York (1966)
37. Brust, D.: The pseudopotential method and the single-particle electronic excitation spectra of crystals. In: Methods in Computational Physics, vol. 8, p. 33. Academic Press, New York (1968)
38. Heine, V.: The pseudopotential concept. In: Ehrenreich, H., Seitz, F., Turnbull, D. (eds.) Solid State Physics vol. 24, pp. 1–36. Academic Press (1970). doi:10.1016/S0081-1947(08)60069-7
39. Stoneham, A.M.: Theory of Defects in Solids. Oxford University Press, London (1975)

40. Pickett, W.E.: Pseudopotential methods in condensed matter applications. Comput. Phys. Rep. **9**(3), 115–197 (1989). doi:10.1016/0167-7977(89)90002-6
41. Bachelet, G.B., Hamann, D.R., Schlüter, M.: Pseudopotentials that work: from h to pu. Phys. Rev. B **26**, 4199–4228 (1982). doi:10.1103/PhysRevB.26.4199
42. Troullier, N., Martins, J.L.: Efficient pseudopotentials for plane-wave calculations. Phys. Rev. B **43**, 1993–2006 (1991). doi:10.1103/PhysRevB.43.1993
43. Vanderbilt, D.: Soft self-consistent pseudopotentials in a generalized eigenvalue formalism. Phys. Rev. B **41**, 7892–7895 (1990). doi:10.1103/PhysRevB.41.7892
44. Kresse, G., Joubert, D.: From ultrasoft pseudopotentials to the projector augmented-wave method. Phys. Rev. B **59**, 1758–1775 (1999). doi:10.1103/PhysRevB.59.1758
45. Baldereschi, A.: Mean-value point in the brillouin zone. Phys. Rev. B **7**, 5212–5215 (1973). doi:10.1103/PhysRevB.7.5212
46. Chadi, D.J., Cohen, M.L.: Electronic structure of $Hg_{1-x}Cd_x$ Te alloys and charge density calculations using representative k points. Phys. Rev. B **7**, 692–699 (1973). doi:10.1103/PhysRevB.7.692
47. Monkhorst, H.J., Pack, J.D.: Special points for brillouinzone integrations. Phys. Rev. B **13**, 5188–5192 (1976). doi:10.1103/PhysRevB.13.5188
48. Pack, J.D., Monkhorst, H.J.: "Special points for brillouinzone integrations"—a reply. Phys. Rev. B **16**, 1748–1749 (1977). doi:10.1103/PhysRevB.16.1748
49. Birch, F.: Finite elastic strain of cubic crystals. Phys. Rev. **71**, 809–824 (1947). doi:10.1103/PhysRev.71.809
50. Martienssen, W., Warlimont, H. (eds.): Springer Handbook of Condensed Matter and Materials Data. Springer, Berlin/Heidelberg (2005). doi:10.1007/3-540-30437-1
51. Feynman, R.P.: Forces in molecules. Phys. Rev. **56**, 340–343 (1939). doi:10.1103/PhysRev.56.340
52. Epstein, S.T., Hurley, A.C., Wyatt, R.E., Parr, R.G.: Integrated and integral hellmann–feynman formulas. J. Chem. Phys. **47**(4), 1275–1286 (1967). doi:10.1063/1.1712080
53. Watkins, G.D., Brower, K.L.: Epr observation of the isolated interstitial carbon atom in silicon. Phys. Rev. Lett. **36**, 1329–1332 (1976). doi:10.1103/PhysRevLett.36.1329
54. Leary, P., Jones, R., Öberg, S., Torres, V.J.B.: Dynamic properties of interstitial carbon and carbon carbon pair defects in silicon. Phys. Rev. B **55**, 2188–2194 (1997). doi:10.1103/PhysRevB.55.2188
55. Zirkelbach, F., Stritzker, B., Nordlund, K., Lindner, J.K.N., Schmidt, W.G., Rauls, E.: Defects in carbon implanted silicon calculated by classical potentials and first-principles methods. Phys. Rev. B **82**, 094110 (2010). doi:10.1103/PhysRevB.82.094110
56. Bean, A.R., Newman, R.C.: Low temperature electron irradiation of silicon containing carbon. Solid State Commun. **8**(3), 175–177 (1970). doi:10.1016/0038-1098(70)90074-8
57. Chelikowsky, J.R., Cohen, M.L.: Electronic structure of silicon. Phys. Rev. B **10**, 5095–5107 (1974). doi:10.1103/PhysRevB.10.5095
58. Kohn, W.: Phys. Rev. **98**, 915 (1955)
59. Kohn, W.: Solid State Phys. **5**, 257 (1957)
60. Ramdas, A.K., Rodriguez, S.: Spectroscopy of the solidstate analogues of the hydrogen atom: donors and acceptors in semiconductors. Rep. Prog. Phys. **44**, 1297–1387 (1981). doi:10.1088/0034-4885/44/12/002
61. Fletcher, R.C., Yager, W.A., Pearson, G.L., Holden, A.N., Read, W.T., Merritt, F.R.: Spin resonance of donors in silicon. Phys. Rev. **94**, 1392–1393 (1954). doi:10.1103/PhysRev.94.1392.2
62. Feher, G.: Electron spin resonance experiments on donors in silicon. i. electronic structure of donors by the electron nuclear double resonance technique. Phys. Rev. **114**, 1219–1244 (1959). doi:10.1103/PhysRev.114.1219
63. Watkins, G.D., Ham, F.S.: Electron paramagnetic resonance studies of a system with orbital degeneracy: the lithium donor in silicon. Phys. Rev. B **1**, 4071–4098 (1970). doi:10.1103/PhysRevB.1.4071
64. Wünstel, K., Wagner, P.: Interstitial iron and ironacceptor pairs in silicon. Appl. Phys. A **27**(4), 207–212 (1982). doi:10.1007/BF00619081

65. Awadelkarim, O.O., Monemar, B.: A study of ironrelated centers in heavily boron-doped silicon by deep-level transient spectroscopy. J. Appl. Phys. **64**(11), 6306–6310 (1988). doi:10.1063/1.342090
66. Istratov, A.A., Hieslmair, H., Weber, E.R.: Iron and its complexes in silicon. Appl. Phys. A **69**(1), 13–44 (1999). doi:10.1007/s003390050968
67. Sanati, M., Szwacki, N., Estreicher, S.: Interstitial fe in si and its interactions with hydrogen and shallow dopants. Phys. Rev. B **76**:125,204, 125204 (2007). doi:10.1103/PhysRevB.76.125204
68. Backlund, D.J., Estreicher, S.K.: Ti, fe, and ni in si and their interactions with the vacancy and the a center: a theoretical study. Phys. Rev. B **81**, 235213 (2010). doi:10.1103/PhysRevB.81.235213
69. Ludwig, G., Woodbury, H.: Electronic structure of transition metal ions in a tetrahedral lattice. Phys. Rev. Lett. **5**, 98–100 (1960). doi:10.1103/PhysRevLett.5.98
70. Katayama-Yoshida, H., Zunger, A.: Localization and magnetism of an interstitial iron impurity in silicon. Phys. Rev. Lett. **53**, 1256–1259 (1984). doi:10.1103/PhysRevLett.53.1256
71. Beeler, F., Andersen, O., Scheffler, M.: Theoretical evidence for low-spin ground states of early interstitial and late substitutional 3d transition-metal ions in silicon. Phys. Rev. Lett. **55**, 1498–1501 (1985). doi:10.1103/PhysRevLett.55.1498
72. Katayama-Yoshida, H., Zunger, A.: Calculation of the spin-polarized electronic structure of an interstitial iron impurity in silicon. Phys. Rev. B **31**, 7877–7899 (1985). doi:10.1103/PhysRevB.31.7877
73. Woodbury, H., Ludwig, G.: Spin resonance of transition metals in silicon. Phys. Rev. **117**, 102–108 (1960). doi:10.1103/PhysRev.117.102
74. Yoshida, Y., Ogawa, S., Arikawa, K.: Direct observation of substitutional Fe atoms in Si and SOI wafers at 1273 k. Physica B: Condensed Matter **340–342**, 605–608 (2003). doi:10.1016/j.physb.2003.09.125
75. Yoshida, Y., Kobayashi, Y., Hayakawa, K., Yukihira, K., Shimura, F., Yoshida, A., Diao, X., Ogawa, H., Yano, Y., Ambe, F.: In-beam Mössbauer study of interstitial and substitutional 57mn/57fe jumps in si. Defect Diffus. Forum **194–199**, 611–616 (2001). doi:10.4028/www.scientific.net/DDF.194-199.611
76. Gunnlaugsson, H.P., Weyer, G., Dietrich, M., collaboration, I., Fanciulli, M., Bharuth-Ram, K., Sielemann, R.: Charge state dependence of the diffusivity of interstitial Fe in silicon detected by mössbauer spectroscopy. Appl. Phys. Lett. **80**(15), 2657–2659 (2002). doi:10.1063/1.1469216
77. Wahl, U., Correia, J., Rita, E., Araújo, J., Soares, J.: Lattice sites of implanted fe in si. Phys. Rev. B **72**, 014115 (2005). doi:10.1103/PhysRevB.72.014115
78. Yoshida, Y., Kobayashi, Y., Hayakawa, K., Yukihira, K., Yoshida, A., Ueno, H., Shimura, F., Ambe, F.: In situ observation of substitutional and interstitial fe atoms in si after gev-implantation: an in-beam mössbauer study. Physica B: Condens. Matter **376–377**, 69–72 (2006). doi: 10.1016/j.physb.2005.12.019
79. Yoshida, Y., Suzuki, K., Kobayashi, Y., Nagatomo, T., Akiyama, Y., Yukihira, K., Hayakawa, K., Ueno, H., Yoshimi, A., Nagae, D., Asahi, K., Langouche, G.: 57fe charge states in mc-si solar cells under light illumination after gev-implantation of 57Mn. Hyperfine Interact. **204**(1–3), 133–137 (2012). doi:10.1007/s10751-011-0424-3
80. Weyer, G., Burchard, A., Fanciulli, M., Fedoseyev, V., Gunnlaugsson, H., Mishin, V., Sielemann, R.: The electronic configuration of substitutional fe in silicon. Physica B: Condens. Matter **273–274**, 363–366 (1999). doi:10.1016/S0921-4526(99)00478-0
81. Andersen, O.K.: Linear methods in band theory. Phys. Rev. B **12**, 3060–3083 (1975). doi:10.1103/PhysRevB.12.3060
82. Schwarz, K., Blaha, P., Madsen, G.: Electronic structure calculations of solids using the {WIEN2k} package for material sciences. Comput. Phys. Commun. **147**, 71–76 (2002). doi:10.1016/S0010-4655(02)00206-0
83. Kübler, J., Kumm, A.E., Overhof, H., Schwalback, O., Hartick, M., Kankeleit, E., Keck, B., Wende, L., Seilemann, R.: Isomer-shift of interstitial and substitutional iron in silicon and germanium. Z. Phys. B **92**, 155–162 (1993). doi: 10.1007/BF01312171

84. Elzain, M., Al-Harthi, S., Gismelseed, A., Al-Rawas, A., Yousif, A., Widatallah, H., Al-Barwani, M.: The magnetic and hyperfine properties of iron in silicon carbide. Hyperfine Interact. **226**(1–3), 281–287 (2014). doi: 10.1007/s10751-013-0946-y
85. You, Z., Gong, M., Chen, J., Corbett, J.W.: Iron-vacancy-oxygen complex in silicon. J. Appl. Phys. **63**(2), 324–326 (1988). doi:10.1063/1.340297
86. Komarov, B.: Special features of radiation-defect annealing in silicon p-n structures: the role of fe impurity atoms. Semiconductors **38**(9), 1041–1046 (2004). doi:10.1134/1.1797482
87. Tang, C.K., Vines, L., Svensson, B.G., Monakhov, E.V.: Deep level transient spectroscopy on proton-irradiated Fecontaminated p-type silicon. Physica Status Solidi (C) **9**(10–11), 1992–1995 (2012). doi:10.1002/pssc.201200163
88. Tang, C.K., Vines, L., Markevich, V.P., Svensson, B.G., Monakhov, E.V.: Divacancy-iron complexes in silicon. J. Appl. Phys. **113**(4), 044503 (2013). doi:10.1063/1.4788695
89. Estreicher, S., Sanati, M., Gonzalez Szwacki, N.: Iron in silicon: interactions with radiation defects, carbon, and oxygen. Phys. Rev. B **77**, 125214 (2008). doi:10.1103/PhysRevB.77.125214
90. Watkins, G.D.: Deep Centers in Semiconductors, 2nd edn., chap. 3, p. 177. Gordon and Breach Science Publishers, New York (1992)
91. Zunger, A., Lindefelt, U.: Theory of substitutional and interstitial $3d$ impurities in silicon. Phys. Rev. B **26**, 5989–5992 (1982). doi:10.1103/PhysRevB.26.5989
92. Flynn, C.P.: Point Defects and Diffusion. Oxford University Press, Glasgow (1972)
93. Hwang, G.S., Goddard III, W.A.: Diffusion and dissociation of neutral divacancies in crystalline silicon. Phys. Rev. B **65**, 233205 (2002). doi:10.1103/PhysRevB.65.233205
94. Watkins, G.D., Corbett, J.W.: Defects in irradiated silicon: electron paramagnetic resonance of the divacancy. Phys. Rev. **138**, A543–A555 (1965). doi:10.1103/PhysRev.138.A543
95. Heiser, T., Mesli, A.: Charge-state-dependent diffusion and carrier-emission-limited drift of iron in silicon. Phys. Rev. Lett. **68**, 978–981 (1992). doi:10.1103/PhysRevLett.68.978
96. Takahashi, H., Suezawa, M., Sumino, K.: Charge-state-dependent activation energy for diffusion of iron in silicon. Phys. Rev. B **46**, 1882–1885 (1992). doi:10.1103/PhysRevB.46.1882
97. Estreicher, S.K., Backlund, D.J., Carbogno, C., Scheffler, M.: Activation energies for diffusion of defects in silicon: the role of the exchange correlation functional. Angewandte Chemie International Edition **50**(43), 10221–10225 (2011). doi:10.1002/anie.201100733
98. Katayama-Yoshida, H., Hamada, N.: Superhyperfine interaction and spin-lattice relaxation of an interstitial iron impurity in silicon. Phys. Rev. B **35**, 407–410 (1987). doi:10.1103/PhysRevB.35.407
99. Sheppard, D., Xiao, P., Chemelewski, W., Johnson, D.D., Henkelman, G.: A generalized solidstate nudged elastic band method. J. Chem. Phys. **136**, 074103 (2012)
100. Coutinho, J., Markevich, V.P., Peaker, A.R., Hamilton, B., Lastovskii, S.B., Murin, L.I., Svensson, B.J., Rayson, M.J., Briddon, P.R.: Electronic and dynamical properties of the silicon trivacancy. Phys. Rev. B **86**, 174101 (2012). doi:10.1103/PhysRevB.86.174101
101. Markevich, V.P., Peaker, A.R., Lastovskii, S.B., Murin, L.I., Coutinho, J., Torres, V.J.B., Briddon, P.R., Dobaczewski, L., Monakhov, E.V., Svensson, B.G.: Trivacancy and trivacancy-oxygen complexes in silicon: experiments and ab initio modeling. Phys. Rev. B **80**, 235207 (2009). doi:10.1103/PhysRevB.80.235207
102. Reif, F.: Fundamentals of Statistical and Thermal Physics. McGraw-Hill, New York (1965)
103. Estreicher, S.K., Sanati, M., West, D., Ruymgaart, F.: Thermodynamics of impurities in semiconductors. Phys. Rev. B **70**, 125209 (2004). doi:10.1103/PhysRevB.70.125209
104. Rohatgi, A., Davis, J.R., Hopkins, R.H., Rai-Choudhury, P., McMullin, P.G., McCormick, J.R.: Solid-State Electron. **23**, 415–422 (1980). doi:10.1016/0038-1101(80)90076-3
105. Kolkovsky, V., Scheffler, L., Weber, J.: A re-examination of the interstitial ti levels in si. Phys. Stat. Sol. (C) **9**(10–11), 1996–1999 (2012). doi:10.1002/pssc.201200141
106. Hocine, S., Mathiot, D.: Titanium diffusion in silicon. Appl. Phys. Lett. **53**(14), 1269–1271 (1988). doi:10.1063/1.100446

107. Baraff, G.A., Schlüter, M.: Electronic structure, total energies, and abundances of the elementary point defects in gaas. Phys. Rev. Lett. **55**, 1327–1330 (1985). doi:10.1103/PhysRevLett.55.1327
108. Qian, G.X., Martin, R.M., Chadi, D.J.: Firstprinciples study of the atomic reconstructions and energies of ga- and as-stabilized gaas(100) surfaces. Phys. Rev. B **38**, 7649–7663 (1988). doi:10.1103/PhysRevB.38.7649
109. Northrup, J.E.: Energetics of gaas island formation on si(100). Phys. Rev. Lett. **62**, 2487–2490 (1989). doi:10.1103/PhysRevLett.62.2487
110. Van de Walle, C.G., Martin, R.M.: Theoretical study of band offsets at semiconductor interfaces. Phys. Rev. B **35**, 8154–8165 (1987). doi:10.1103/PhysRevB.35.8154
111. Makov, G., Payne, M.C.: Periodic boundary conditions in ab initio calculations. Phys. Rev. B **51**, 4014–4022 (1995). doi:10.1103/PhysRevB.51.4014
112. Komsa, H.P., Rantala, T., Pasquarello, A.: Comparison between various finite-size supercell correction schemes for charged defect calculations. Physica B: Condens. Matter **407**(15), 3063–3067 (2012). doi:10.1016/j.physb.2011.08.028
113. Goss, J.P., Shaw, M.J., Briddon, P.R.: Marker-method calculations for electrical levels using gaussian-orbital basis sets. In: Drabold, D.A., Estreicher, S.K. (eds.) Theory of Defects in Semiconductors. Topics in Applied Physics, vol. 104, pp. 69–94. Springer, Berlin (2007). doi:10.1007/11690320_4
114. Resende, A., Jones, R., Öberg, S., Briddon, P.R.: Calculations of electrical levels of deep centers: application to auh and ag-h defects in silicon. Phys. Rev. Lett. **82**, 2111–2114 (1999). doi:10.1103/PhysRevLett.82.2111
115. Jeong, J.W., Oshiyama, A.: Atomic and electronic structures of a boron impurity and its diffusion pathways in crystalline si. Phys. Rev. B **64**, 235204 (2001). doi:10.1103/PhysRevB.64.235204
116. Santos, P., Coutinho, J., Torres, V.J.B., Rayson, M.J., Briddon, P.R.: Hydrogen passivation of titanium impurities in silicon: effect of doping conditions. Appl. Phys. Lett. **105**(3), 032108 (2014). doi:10.1063/1.4891575
117. Backlund, D.J., Estreicher, S.K.: Structural, electrical, and vibrational properties of ti-h and ni-h complexes in si. Phys. Rev. B **82**, 155208 (2010). doi:10.1103/PhysRevB.82.155208
118. Kolkovsky, V., Scheffler, L., Weber, J.: Transition metals (ti and co) in silicon and their complexes with hydrogen: a lapplace dlts study. Physica B **439**, 24–28 (2014). doi:10.1016/j.physb.2013.11.005
119. Leonard, S., Markevich, V.P., Peaker, A.R., Hamilton, B.: Passivation of titanium by hydrogen in silicon. Appl. Phys. Lett. **103**(13), 132103 (2013). doi:10.1063/1.4822329
120. Jost, W., Weber, J.: Titanium-hydrogen defects in silicon. Phys. Rev. B **54**, R11038–R11041 (1996). doi:10.1103/PhysRevB.54.R11038
121. Born, M., Huang, K.: Dynamical Theory of Crystal Lattices. Oxford University Press, Oxford (1954)
122. Maradudin, A.A., Montroll, E.W., Weiss, G.H.: Lattice Dynamics in the Harmonic Approximation. Academic, New York (1963)
123. Srivastava, G.P. (ed.): The Physics of Phonons. Adam Hilger, Bristol (1990)
124. Coutinho, J., Jones, R., Briddon, P.R., Öberg, S., Murin, L.I., Markevich, V.P., Lindström, J.L.: Interstitial carbon-oxygen center and hydrogen related shallow thermal donors in Si. Phys. Rev. B **65**, 014109 (2001). doi:10.1103/PhysRevB.65.014109
125. Jones, R., Öberg, S.: Oxygen frustration and the interstitial carbon-oxygen complex in si. Phys. Rev. Lett. **68**, 86–89 (1992). doi:10.1103/PhysRevLett.68.86
126. Coutinho, J., Andersen, O., Dobaczewski, L., Bonde Nielsen, K., Peaker, A.R., Jones, R., Öberg, S., Briddon, P.R.: Effect of stress on the energy levels of the vacancy-oxygen-hydrogen complex in si. Phys. Rev. B **68**, 184106 (2003). doi:10.1103/PhysRevB.68.184106
127. Panthani, M.G., Korgel, B.A.: Nanocrystals for electronics. Annu. Rev. Chem. Biomol. Eng. **3**(1), 287–311 (2012). doi:10.1146/annurev-chembioeng-062011-081040
128. Beard, M.C., Knutsen, K.P., Yu, P., Luther, J.M., Song, Q., Metzger, W.K., Ellingson, R.J., Nozik, A.J.: Multiple exciton generation in colloidal silicon nanocrystals. Nano Lett. **7**(8), 2506–2512 (2007). doi:10.1021/nl071486l

129. Holman, Z.C., Liu, C.Y., Kortshagen, U.R.: Germanium and silicon nanocrystal thin-film field-effect transistors from solution. Nano Lett. **10**, 2661–2666 (2010). doi:10.1021/nl101413d
130. Cheng, K.Y., Anthony, R., Kortshagen, U.R., Holmes, R.J.: High-efficiency silicon nanocrystal light-emitting devices. Nano Lett. **11**(5), 1952–1956 (2011). doi:10.1021/nl2001692
131. Kessler, V., Gautam, D., Hülser, T., Spree, M., Theissmann, R., Winterer, M., Wiggers, H., Schierning, G., Schmechel, R.: Thermoelectric properties of nanocrystalline silicon from a scaled-up synthesis plant. Adv. Eng. Mater. **15**(5), 379–385 (2013). doi:10.1002/adem.201200233
132. Gupta, A., Swihart, M.T., Wiggers, H.: Luminescent colloidal dispersion of silicon quantum dots from microwave plasma synthesis: Exploring the photoluminescence behavior across the visible spectrum. Adv. Funct. Mater. **19**(5), 696–703 (2009). doi:10.1002/adfm.200801548
133. Hessel, C.M., Reid, D., Panthani, M.G., Rasch, M.R., Goodfellow, B.W., Wei, J., Fujii, H., Akhavan, V., Korgel, B.A.: Synthesis of ligand-stabilized silicon nanocrystals with size-dependent photoluminescence spanning visible to near-infrared wavelengths. Chem. Mater. **24**(2), 393–401 (2012). doi:10.1021/cm2032866
134. Carvalho, A., Coutinho, J., Barroso, M., Silva, E.L., Öberg, S., Rayson, M., Briddon, P.R.: Electronic structure modification of si nanocrystals with f_4-tcnq. Phys. Rev. B **84**, 125437 (2011). doi:10.1103/PhysRevB.84.125437
135. Chelikowsky, J.R., Alemany, M.M.G., Chan, T.L., Dalpian, G.M.: Computational studies of doped nanostructures. Rep. Progress Phys. **74**, 046501 (2011). doi:10.1088/0034-4885/74/4/046501
136. Cantele, G., Degoli, E., Luppi, E., Magri, R., Ninno, D., Iadonisi, G., Ossicini, S.: First-principles study of n- and p-doped silicon nanoclusters. Phys. Rev. B **72**, 113303 (2005). doi:10.1103/PhysRevB.72.113303
137. Du, M.H., Erwin, S.C., Efros, A.L.: Trapped-dopant model of doping in semiconductor nanocrystals. Nano Lett. **8**, 2878–2882 (2008). doi:10.1021/nl8016169
138. Chan, T.L., Chelikowsky, J.R.: Controlling diffusion of lithium in silicon nanostructures. Nano Lett. **10**(3), 821–825 (2010). doi:10.1021/nl903183n
139. Gali, A., Vörös, M., Rocca, D., Zimanyi, G.T., Galli, G.: High-energy excitations in silicon nanoparticles. Nano Lett. **9**(11), 3780–3785 (2009). doi:10.1021/nl901970u
140. Pereira, R.N., Coutinho, J., Niesar, S., Oliveira, T.A., Aigner, W., Wiggers, H., Rayson, M.J., Briddon, P.R., Brandt, M.S., Stutzmann, M.: Resonant electronic coupling enabled by small molecules in nanocrystal solids. Nano Lett. **14**(7), 3817–3826 (2014). doi:10.1021/nl500932q
141. Timmerman, D., Izeddin, I., Stallinga, P., Yassievich, I.N., Gregorkiewicz, T.: Spaceseparated quantum cutting with silicon nanocrystals for photovoltaic applications. Nat. Photon. **2**(2), 105–109 (2008). doi:10.1038/nphoton.2007.279
142. Guerra, R., Ossicini, S.: Role of strain in interacting silicon nanoclusters. Phys. Rev. B **87**, 165441 (2013). doi:10.1103/PhysRevB.87.165441
143. Wooten, F., Winer, K., Weaire, D.: Computer generation of structural models of amorphous si and ge. Phys. Rev. Lett. **54**, 1392–1395 (1985). doi:10.1103/PhysRevLett.54.1392
144. Seino, K., Bechstedt, F., Kroll, P.: Tunneling of electrons between si nanocrystals embedded in a sio2 matrix. Phys. Rev. B **86**, 075312 (2012). doi:10.1103/PhysRevB.86.075312
145. Erwin, S.C., Zu, L., Haftel, M.I., Efros, A.L., Kennedy, T.A., Norris, D.J.: Doping semiconductor nanocrystals. Nature **436**, 91–94 (2005). doi:10.1038/nature03832
146. Norris, D.J., Efros, A.L., Erwin, S.C.: Doped nanocrystals. Science **319**, 1776–1779 (2008). doi:10.1126/science.1143802
147. Yu, Y., Bosoy, C.A., Hessel, C.M., Smilgies, D.M., Korgel, B.A.: Silicon nanocrystal superlattices. Chem. Phys. Chem. **14**(1), 84–87 (2013). doi:10.1002/cphc.201200738
148. Yu, Y., Bosoy, C.A., Smilgies, D.M., Korgel, B.A.: Self-assembly and thermal stability of binary superlattices of gold and silicon nanocrystals. J. Phys. Chem. Lett. **4**, 3677–3682 (2013). doi:10.1021/jz401964s
149. Fujii, M., Mimura, A., Hayashi, S., Yamamoto, Y., Murakami, K.: Hyperfine structure of the electron spin resonance of phosphorus-doped si nanocrystals. Phys. Rev. Lett. **89**, 206805 (2002). doi:10.1103/PhysRevLett.89.206805

150. Pereira, R.N., Stegner, A.R., Andlauer, T., Klein, K., Wiggers, H., Brandt, M.S., Stutzmann, M.: Dielectric screening versus quantum confinement of phosphorus donors in silicon nanocrystals investigated by magnetic resonance. Phys. Rev. B **79**, 161304 (2009). doi:10.1103/PhysRevB.79.161304
151. Bjork, M.T., Schmid, H., Knoch, J., Riel, H., Riess, W.: Donor deactivation in silicon nanostructures. Nat. Nanotechnol. **4**(2), 103–107 (2009). doi:10.1038/nnano.2008.400
152. Pierre, M., Wacquez, R., Jehl, X., Sanquer, M., Vinet, M., Cueto, O.: Single-donor ionization energies in a nanoscale cmos channel. Nat. Nanotechnol. **5**(2), 133–137 (2010). doi:10.1038/nnano.2009.373
153. Melnikov, D.V., Chelikowsky, J.R.: Quantum confinement in phosphorus-doped silicon nanocrystals. Phys. Rev. Lett. **92**, 046802 (2004). doi:10.1103/PhysRevLett.92.046802
154. Diarra, M., Niquet, Y.M., Delerue, C., Allan, G.: Ionization energy of donor and acceptor impurities in semiconductor nanowires: Importance of dielectric confinement. Phys. Rev. B **75**, 045301 (2007). doi:10.1103/PhysRevB.75.045301
155. Ni, Z., Pi, X., Yang, D.: Doping si nanocrystals embedded in sio2 with p in the framework of density functional theory with p in the framework of density functional theory. Phys. Rev. B **89**, 035312 (2014). doi:10.1103/PhysRevB.89.035312
156. Amato, M., Ossicini, S., Rurali, R.: Band-offset driven efficiency of the doping of sige core-shell nanowires. Nano Lett. **11**(2), 594–598 (2011). doi:10.1021/nl103621s

Chapter 3
Electrical and Optical Defect Evaluation Techniques for Electronic and Solar Grade Silicon

Anthony R. Peaker and Vladimir P. Markevich

Abstract In this chapter we review techniques which characterize and quantify the properties of defects and impurities in silicon materials and devices in terms of their effect on free carriers and their recombination- generation behavior. In particular we explore the application of Deep Level Transient Spectroscopy (DLTS) and its many variants to electronic and solar grade silicon. The physics of carrier recombination at deep level defects is presented and the various methodologies to measure defect parameters related to Shockley-Read-Hall generation-recombination kinetics are discussed. The use of high resolution Laplace DLTS is presented and techniques for determining characteristics of minority carrier traps are explained. Methods to measure minority carrier lifetime and relate these measurements to the defect concentrations and properties are considered. Optical measurements to study defects are presented including optical absorption, Raman techniques and photoluminescence. The status and limitations of such techniques for qualification of silicon material are discussed.

3.1 Introduction

Central to the success of semiconductor technology is the ability to control the electrical properties of silicon material. The role of intentionally added dopants to produce the correct concentrations of holes and electrons is well established. However the presence of unintentional impurities and structural defects in the lattice can have a dramatic effect on the electronic properties of silicon primarily through recombination-generation processes which degrade the lifetime of minority carriers in the silicon. The effect of this lifetime degradation on characteristics of

A.R. Peaker (✉) • V.P. Markevich
Photon Science Institute, University of Manchester, Manchester M13 9PL, UK
e-mail: a.peaker@manchester.ac.uk

Y. Yoshida, G. Langouche (eds.), *Defects and Impurities in Silicon Materials*, Lecture Notes in Physics 916, DOI 10.1007/978-4-431-55800-2_3

most devices is very important. For example, recombination lifetime determines producing high efficiency solar cells, generation lifetime determines the refresh time of DRAM and precise control of lifetime is critical for many power devices.

In high quality electronic grade silicon material destined to make integrated circuits the concentration of unwanted impurities is extremely low, e.g. $< 10^{10}$ cm^{-3}. Indeed they are often undetectable by normal techniques and the material has a near perfect crystal lattice with each atom aligned in a cubic structure. However during processing defects are generated by the displacement of atoms caused by ion implantation or impurities are introduced as contamination at some technological stages. In contrast, silicon which is aimed at the solar cell market is much less pure and often contains many structural defects. Solar silicon is by necessity a low cost product whereas electronic grade material is considerably more expensive per unit area but is a relatively small contributor to the cost of an integrated circuit.

Usually solar and electronic grade materials are made by different or modified growth processes. Solar silicon is typically cast from the melt into large crucibles and allowed to cool solidifying from the bottom and growing as multi-crystalline ingots. Electronic grade material is usually produced by pulling from a melt (Czochralski – Cz) or sometimes by refining using a molten zone which is passed along the ingot (Float Zone – FZ). Both these methods benefit from the fact that a high degree of segregation of metallic impurities into the molten region occurs reducing the metallic impurities in the solid by a factor of $\sim 10^4$. In the case of cast solar silicon the segregation merely redistributes the impurities into the last grown region and as the solid silicon remains in contact with the crucible at high temperature, impurities can diffuse from the crucible lining into the finished ingot.

In the case of Cz silicon high concentrations of oxygen ($\sim 8 \times 10^{17}$ cm^{-3}) are incorporated into the ingot. This has a beneficial effect of strengthening the silicon and normally the oxygen is not electrically active. However it can precipitate creating extended defects and during heat treatments at T>400 °C so creating extended defects and forming electrically active centres.

The variety of silicon material and the many processes used in its fabrication into devices make the task of defect evaluation quite complex. In this paper we look at the effects of defects on electrical and optical properties of silicon and in particular their effect on carrier lifetime. In Fig. 3.1 a map of the recombination lifetime shows local variations due to extended defects (slip lines at the edge), oxygen precipitation (rings near center) and areas of contamination with metals. This Cz slice was grown 20 years ago and suffered careless handling during cutting and polishing. Top quality Cz and FZ slices today show a very uniform lifetime which can be of the order of several milliseconds.

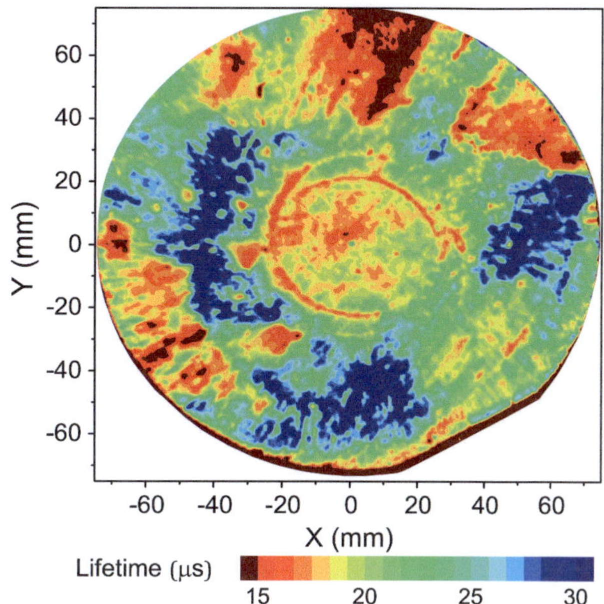

Fig. 3.1 Lifetime map of a degraded n-type Cz-Si wafer taken with the use of a Semilab WT 2000PVN instrument working in microwave photo-conductance decay regime

3.2 Recombination-Generation Processes in Silicon

3.2.1 Carrier Lifetime

The term lifetime is used to describe two different processes. These are recombination lifetime, which is essentially how long an excess minority carrier survives in the silicon, and generation lifetime, which is the time required to produce an excess minority carrier. Generation lifetime is related to leakage currents in devices while recombination lifetime affects for example the efficiency of solar cells and gain of power bipolar transistors. In the case where these processes are associated with defects, the generation- recombination processes are referred to as the Shockley-Read-Hall (SRH) [1–3] process. Although in this chapter we are concerned with the effect of defects on carrier lifetime, it is important to put this in the context of other recombination processes which may affect the recombination lifetime. The measured lifetime is a result of the combination of all recombination processes, typically SRH, Auger and radiative recombination, as described in the equation below:

$$\tau_r = \frac{1}{\tau_{SRH}^{-1} + \tau_{rad}^{-1} + \tau_{Auger}^{-1}} \tag{3.1}$$

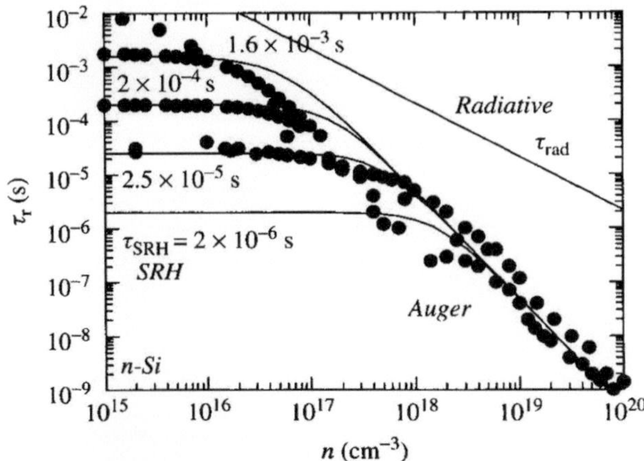

Fig. 3.2 Effective lifetime in silicon resulting from band to band radiative and Auger recombination together with various values of experimentally observed SRH lifetime (Reproduced with permission from Ref. [3])

In direct gap materials the band to band radiative recombination is very important and its rate increases linearly with carrier concentration. As can be seen in Fig. 3.2 it is not significant in silicon and this is true for all indirect gap materials.

Band to band Auger recombination results from two carriers recombining and the energy being removed by a third carrier. The recombination rate increases as the square of carrier concentrations so it becomes important at high carrier densities. The lifetime limited by the Auger recombination in many cases can be described by the following equation:

$$\begin{aligned}\tau_{\text{Auger}} &= \frac{1}{C_p\left(p_0^2 + 2p_0\Delta n + \Delta n^2\right) + C_n\left(n_0^2 + 2n_0\Delta n + \Delta n^2\right)} \\ &\approx \frac{1}{C_p\left(p_0^2 + 2p_0\Delta n + \Delta^2 n\right)}\end{aligned} \tag{3.2}$$

where C_p and C_n are the Auger recombination coefficients, for holes and electrons, respectively. The approximation can be used for p-type material. For silicon $C_p \approx 10^{-31}$ and $C_n \approx 2.8 \times 10^{-31}$ cm^6 s^{-1} [4]. This gives the lifetimes shown in Fig. 3.2 where the Auger lifetime is the line with a slope of two between the radiative and SRH lifetimes. A more accurate quantitative description of the Auger recombination rate in silicon as a function of the dopant density and the carrier injection level has been published recently [5]. Defect related Auger processes also exist [6] but are rarely considered in silicon.

3.2.2 Shockley-Read-Hall (SRH) Statistics

The carrier generation-recombination processes associated with defects are usually modeled with Shockley-Read-Hall (SRH) statistics [1, 2]. This is a simple kinetic model which can be used to calculate carrier capture and emission rates based on the defect energy level in the band gap and a parameter referred to as the carrier capture cross section. In general these parameters can be measured experimentally and hence the SRH model can be used to calculate recombination and generation rates and further the effect that particular defects have on the carrier lifetimes. In practice, measurement of the capture cross section is sometimes very difficult and so for this and other reasons which will be discussed later the application of the SRH model is not always successful.

Figure 3.3 shows an overview of the principal charge exchange mechanisms involving defect energy levels within the band gap which have a greater ionization energy than conventional shallow dopants. Because of this they are often referred to as deep states. The arrows denote the direction of electron transfer while the symbol C represents the capture rate and e the emission rate. The subscript (p) indicates a hole process and (n) an electron process. (A) represents carrier generation, (B) electron trapping and re-emission, (C) hole trapping and re-emission and (D) recombination. The symbols e_{n} and e_{p} represent electron and hole emission rates with units of s^{-1} while c_n and c_p represent the electron and hole capture coefficients with units of $\mathrm{cm}^3\ \mathrm{s}^{-1}$ (some authors use c_n and c_p to represent capture rates which have units of s^{-1}. We use the symbol C_{cn} to represent this quantity, i.e., $C_{cn} = nc_n$). c_n and c_p are often written in terms of capture cross sections σ_n and σ_p which have units of cm^2

$$c_n = \sigma_n v_{\mathrm{th}} \tag{3.3}$$

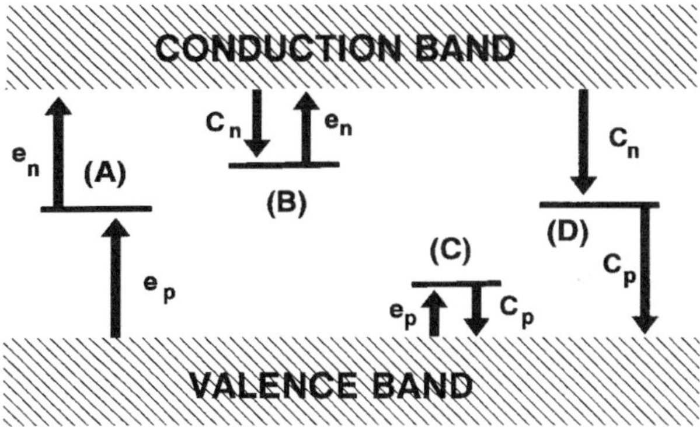

Fig. 3.3 Principal charge exchange mechanisms associated with deep states

where v_{th} is the thermal velocity of electrons. Defining n_T as the number of traps occupied by electrons, the number of empty traps is $N_T - n_T$. At equilibrium,

$$\left(c_n n + e_p\right)\left(N_T - n_T\right) = \left(e_n + c_p p\right) n_T \tag{3.4}$$

This gives a steady state value for the occupation factor f_T for the level, defined by $f_T = n_T/N_T$:

$$f_T = \frac{c_n n + e_p}{e_n + c_p p + c_n n + e_p} \tag{3.5}$$

3.3 Quantifying the Properties of Defects

3.3.1 Measurement of the Concentrations of Shallow Donors and Acceptors

Although intentionally added donors and acceptors are often not thought of as defects in the sense of materials physics they are perturbations of the lattice and so in that sense are defects. Knowing their concentration and distribution is obviously of crucial importance in a device context and a necessary prerequisite to some of the defect measurements described later in this chapter. However these techniques are well described in many texts (e.g. Schroder Chaps. 1 and 2 [3]) and so all we provide here is an overview of the main techniques to put them in context.

3.3.1.1 Resistivity Based Methods

These techniques are commonly used methods of estimating the donor and acceptor concentration in silicon. They measure the resistivity (ρ) either averaging over a large volume, (e.g. four point probe methods using the standard method SEMI MF84-0312), or locally (e.g. spreading resistance using the standard method SEMI MF525-0312 and MF672-0412) from which under certain conditions the donor or acceptor concentration can be calculated. The basic relationship on which all such measurements are based is given below:

$$\rho = \frac{1}{q\left(n\mu_n + p\mu_p\right)} \tag{3.6}$$

Usually in silicon one carrier type predominates sufficiently that compensation can be neglected. This results in the mobility μ being a well defined function of the carrier concentration and so the number of carriers can be determined from the resistivity. The common donors and acceptors in silicon (P, As, B) are quite shallow so that at room temperature in the absence of compensation the number of carriers is almost equal to the number of dopant atoms. However in some types of silicon the no compensation approximation may not be true, for example in upgraded metallurgical silicon (umg-Si) used for solar cells. In these circumstances the number of ionized dopant atoms will be greater than the carrier concentration and the mobility will be less than expected for electronic grade silicon.

3.3.1.2 Hall Effect Measurements

Hall Effect measurements determine directly the number of carriers, so in conjunction with resistivity measurements can also measure mobility. This is a valuable attribute in very highly doped material particularly implanted regions. The technique uses a sample with four contacts in a magnetic field. Although widely used for III-V materials it is not extensively used for silicon. Temperature dependent Hall measurements can be used to determine the activation energy of dopants [3].

3.3.1.3 Capacitance-Voltage Techniques

Capacitance-Voltage (CV) measurements are very widely used in various forms to measure ionized dopant concentration (in most cases equal to the carrier concentration) and are an essential prerequisite to DLTS measurement of deep defects. The technique relies on the fact that the depletion width changes as a function of applied voltage. It measures the change of capacitance of a depletion region as a function of applied bias. It can be applied to Schottky diodes fabricated on unprocessed wafers, to p-n junctions and to MOS capacitances. Structures for CV measurement are invariably included in integrated circuit test patterns.

We choose a p-type Schottky barrier (metal semiconductor contact) to illustrate the concepts since this structure is much used as an experimental probe on semiconductor layers. However, almost identical arguments can be applied to n^+p junctions, where the n-type layer is very heavily doped compared to the p-type region.

Capacitance is measured in the small signal high frequency regime. A small ac sinusoidal voltage is imposed on any quiescent bias, V_T, and the measured capacitance is:

$$C = -\frac{dQ}{dV} \tag{3.7}$$

where Q is the semiconductor charge and V is the junction voltage. The measurement frequency is typically 1 MHz and the ac voltage amplitude is typically $10\ \text{mV} \leq V_{osc} \leq 100\ \text{mV}$. The measuring instrument senses the ac current flow resulting from the oscillator voltage and from the complex impedance extracts the capacitance. It is assumed that the depletion approximation holds good, ie, the electron density at the edge of the depletion layer ($X = W$) rises from zero to its bulk value over a distance which is negligible. The diode is reversed biased (a positive potential is applied to the metal contact on the p-type substrate) so that the total voltage across the structure is

$$V = V_R + V_d \tag{3.8}$$

where V_{R} is the applied voltage and V_{d} is the built in or diffusion potential.

$$C = -\frac{dQ}{dV} = \frac{\varepsilon_0 \varepsilon_s A}{W} = qAN(W)\frac{dW}{dV} \tag{3.9}$$

The only way that Q can vary with applied voltage is by expansion or contraction of depletion layer. During the incremental reverse bias change due to the sinusoidal ac test voltage the depletion edge moves out by some increment δW adding to the barrier depletion charge $qN_{\mathrm{D}}\delta W$. During the incremental change of the sinusoid towards forward bias, the depletion edge moves in reducing the depletion charge by the same increment. This is used as the basis for measuring N_A^- (or N_D^+).

$$N_A(W) = -\frac{C^3}{q\varepsilon_0\varepsilon_s A^2\,(dC/dV)} \tag{3.10}$$

$$N_A(W) = \frac{2}{q\varepsilon_0\varepsilon_s A^2\,[d\,(1/C^2)\,/dV]} \tag{3.11}$$

Plotting $1/C^2$ against V_{R} produces a straight line provided N_{A} is uniform. The slope relates to N_A^- and the intercept on the voltage axis gives V_{d}, the diffusion potential of the barrier. In fact, the value of N_A^- in the equation is strictly the value of N_A^- at the depletion edge associated with value of W corresponding to the total potential drop V_{T} and this fact can be used to obtain carrier profiles from *C-V* measurements. An example of such a carrier concentration profile derived from C-V measurements on a Schottky diode on p-type Si is shown in Fig. 3.4.

Quite obviously it is only possible to use the equation over the region of the semiconductor through which the depletion region can be swept. In practice, this is limited to a minimum value near the zero bias point and to a maximum value at which the leakage current interferes with the measurement of capacitance. In the ideal case this is close to the breakdown voltage of the semiconductor. This depends on the doping level. For example, if we consider Si doped to 10^{17} cm^{-3}, the minimum depth that can be sampled at zero bias is 0.1 μm. Breakdown occurs at 12 V reverse bias which is equivalent to a depletion depth of 0.4 μm.

The only way to increase the measurement depth is to physically remove material from the surface of the semiconductor. This is most conveniently done in an electrochemical cell in which the electrolyte acts as both the etchant and the "metal" contact. As long as the electrolyte is fairly concentrated (>0.1 M), there is negligible field penetration into the solution and the interface behaves as a one-sided junction. This is the basis of the electrochemical C-V profiling technique [7].

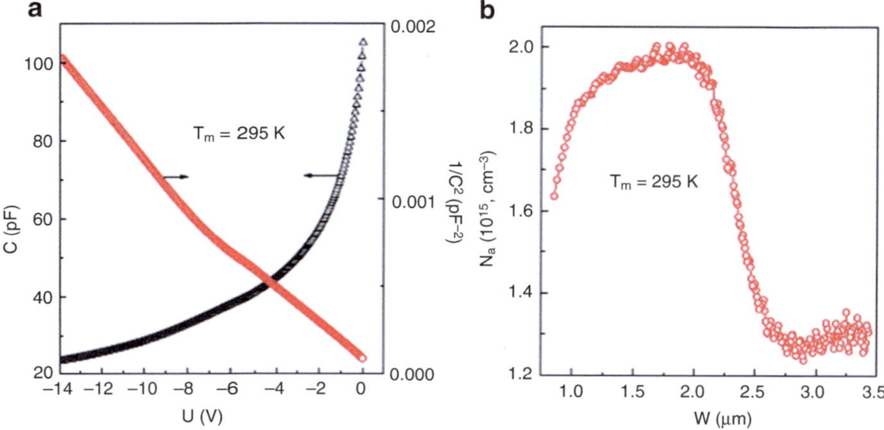

Fig. 3.4 Plots of (**a**) capacitance (C) and $1/C^2$ values versus applied voltage and (**b**) depth profile of concentration of uncompensated shallow acceptors (N_a) for a Schottky diode on p-type Si. The decrease in the N_a in the range from 2 to 3.5 μm is associated with the passivation of electrical activity of boron atoms with hydrogen which was introduced into the sample by wet chemical etching and then driven deeper to the bulk by annealing at 150 °C for 20 min with the applied reverse bias of 8 V

3.3.2 Basic Principles of Measurements of Deep Level Defect Parameters

Equation (3.5) is the basis of most deep level measurements, which usually utilize processes B and C in Fig. 3.3. In all cases of practical importance the concentration of electrically active defects is very small compared to the host lattice, in silicon typically $<10^{13}$ defects cm^{-3}. Experimental techniques for the measurement of the electric parameters of deep states have a very high detectivity and can be regarded as having three stages. Typically these are:

1. The occupancy of the deep electrical state is set to an equilibrium state;
2. The occupancy is perturbed;
3. The subsequent change in occupancy is measured as a function of time, directly or indirectly, as the state returns to its initial equilibrium condition.

It is variants of these three stages which distinguish the methods of measurement. For example, in deep level transient spectroscopy (DLTS) the occupancy is often set by applying zero bias to the normally reverse biased depletion region. The shift of the Fermi level will tend to result in states becoming occupied by majority carriers. The diode is then restored to reverse bias and, at a sufficiently high temperature, the majority carriers are thermally emitted from the deep state. This sequence provides a transient perturbation of the occupancy which is usually monitored by observing the small signal capacitance change of the diode. The transient capacitance is then processed to extract magnitude and time constant. Originally this was done using a simple analogue signal processing technique, today the processing is done digitally and several different methodologies for this exist.

In general we force a change in f_T by changing the trap environment thermally, optically or simply by expanding or collapsing the depletion region so as to shift the Fermi level at the trap location. This latter method is the most commonly used and can provide a sea of carriers equal to the majority carrier concentration to fill the trap. Normally we try to establish an initial condition of $f_T = 1$ and then observe the small signal capacitance change ΔC due to the change in the depletion region depth as $f_T \rightarrow 0$. Usually this will be a time dependent exponential function so that for the emission of electrons in n-type material or holes in p-type (majority carrier emission)

$$\Delta C(t) = \Delta C_{(\text{total})}\left(1 - e^{-t/\tau}\right) \tag{3.12}$$

and for the emission of holes in n-type and the emission of electrons in p-type (minority carrier emission)

$$\Delta C(t) = \Delta C_{(\text{total})}\left(e^{-t/\tau}\right) \tag{3.13}$$

It is the time constant τ that provides information which we can use to investigate the principal properties of the deep state. The next section considers the cases of most interest; thermal emission and carrier capture.

3.3.3 Thermal Emission of Carriers

By far the most common measurements involve using thermal energy to change the occupancy of the center. If we consider the case where a defect is electron occupied and is emitting its electrons to the conduction band, then the thermal emission rate is proportional to a Boltzmann factor:

$$e_n = A_n \exp\left(-E_{\text{a}}/\text{k}T\right) \tag{3.14}$$

where A_n is a coefficient related to the trap-semiconductor system and E_{a} is the activation or trap energy. This form of relationship means that the emission rate changes very rapidly with temperature. If the log of e_n, or more usually for reasons discussed below, $\log(e_n/T^2)$ is plotted against $1/T$, E_{a} can be obtained from the gradient of the line. The combination of E_{a} and A_{n} provide a convenient fingerprint for the defect. An example of such an "Arrhenius plot" for three defect levels is shown in Fig. 3.5.

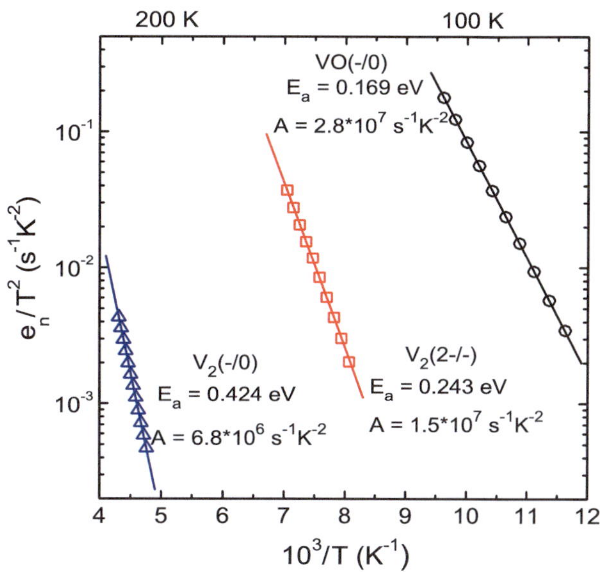

Fig. 3.5 Arrhenius plot of the thermal emission of electrons from three common defects in implanted or irradiated silicon. These are the vacancy oxygen complex and two charge states of the divacancy. The energy is derived from the slope of the plot and A_n from the intercept with the *vertical axis* at $\mathrm{T} = \infty$

Equation (3.14) relating emission rate to temperature is often written in the form:

$$e_n = g v_{\mathrm{th}} N_{\mathrm{c}} \sigma_n exp\left(-E_{\mathrm{a}}/\mathrm{k}T\right), \tag{3.15}$$

where N_{c} is the density of states in the conduction band and g is the degeneracy of the electron state, the other terms are as previously defined. This relationship, derived by applying the principle of detailed balance to the emission and capture processes, can only be applied for near equilibrium conditions for a single reversible process. If the capture proceeds through an intermediate state or states, Eq. (3.6) is erroneous, and there is some evidence that such capture processes do occur. It should be noted that (v_{th} N_{c}) varies as T^2, and, therefore, log e_n/T^2 against $1/T$ is usually plotted. More importantly both σ and E_{T} (the real trap energy depth) are, in general, functions of temperature. As will be discussed later, it is possible to measure σ directly and determine its temperature dependence. Hence if σ increases with increasing temperature with an activation energy, E_σ, then E_{D} (the thermal energy of the state) is taken to be $E_{\mathrm{a}}-E_\sigma$ where E_{a} is the slope of the T^2 corrected Arrhenius plot. This makes the tacit assumption that any dependency of σ on T can be described by an exponential relationship. Only if this dependence is due to an energetic barrier which can be surmounted thermally is this accurate.

In practice, the physical interpretation of E_{a} is of concern primarily when comparing energies obtained from optical measurements with the thermal energies and most importantly in the derivation of capture cross sections from the emission data.

3.3.4 Capture Cross Sections

The process of a defect capturing a carrier is conceptually more straightforward than emission. We merely assume that the defect has a sphere of influence and ascribe it a cross sectional area so that any carrier intercepting that area will be 'captured'. If we take an empty defect state and allow it to capture electrons we can observe the capture process by monitoring the change in occupancy of the state and so derive a capture cross-section. This process will proceed with a rate C_{cn} given by:

$$C_{cn} = \sigma_n v_{\text{th}} n \tag{3.16}$$

This gives rise to an exponential transient in occupancy with a time constant equal to ${C_{cn}}^{-1}$. There are several experimental techniques available for this measurement, two of which are widely used: DLTS with variable filling pulse width which measures majority carrier capture cross sections and minority carrier transient spectroscopy (MCTS) which measures minority carrier capture cross sections. These techniques will be described later. The measurement of capture cross sections enables the SRH model to be applied and although the measurement is normally carried out in an environment in which only one type of carrier is present, the results have been successfully used to calculate minority carrier lifetimes.

The recombination process in Fig. 3.3 occurs in bulk material (not in the depletion region) and we can see that two processes are involved. These are (a) the capture of a hole and (b) the capture of an electron. Equation (3.16) demonstrates that the capture rate depends on the number of carriers present. Electrons have been considered but the case for holes is identical if the hole parameters are substituted for the electron values. In the bulk of p-type material the hole concentration is high and so unless $\sigma_p << \sigma_n$ the hole capture rate will be faster than electron capture, i.e., the state will tend to be hole occupied. Consequently, as we need both processes to occur for recombination, the rate limiting step will be minority carrier capture, in this case, electron capture. Therefore, for low excitation levels (e.g. in p-type material where $\Delta n << p$) we can write a very simple expression for the minority carrier lifetime:

$$\tau_{\min} = (\sigma_n v_{\text{th}} N_{\text{T}})^{-1} \tag{3.17}$$

A similar expression can be written for n-type material depending on the hole capture cross section and concentration.

3.3.5 Comparing Capacitance and Current Measurements

Although the change in occupancy of the deep states in semiconductors can be detected in many ways this is almost always done by a measurement of depletion capacitance or (less often) by the detection of the current produced by the emitted carriers. In a few techniques, changes in conductance of a MOSFET or MESFET are measured.

For the space charge methods, some comments can be made contrasting current and capacitance:

1. In general, capacitance techniques are favored because in the case of a one sided junction (Schottky barriers, $p^{+}n$ and $n^{+}p$ junctions) a distinction can be drawn between hole and electron emission simply by observing the sign of the capacitance change.
2. However, the magnitude of capacitance change induced by unit charge depends on its spatial location within the depletion region. Charge on the junction plane has no effect on the capacitance while charge introduced at the depletion region edge has a maximum effect. In the case of current measurements, a released majority carrier makes an equal contribution to the current irrespective of the location of the trap from which it originated.
3. Perhaps the most confusing feature of current techniques is that fundamentally the peak height for a given trap concentration is a function of the emission rate: this is not the case for capacitance. Apart from the need to take this into account when calculating trap concentrations, it also causes a shift of the peak (in temperature scanned experiments) to higher temperatures. This is often a source of misinterpretation in the literature particularly in comparison of trap 'fingerprints' obtained by different methods.

3.3.6 Transient Space Charge Methods ... DLTS and Related Techniques

Although there are previous publications on transient measurements of deep states the first truly comprehensive experimental description was published by Sah et al. in 1970 [8]. Figure 3.6 illustrates the general principle of these measurements which shows capacitance transients which by the polarity of the signal distinguish between minority and majority carrier emission. The major development which enabled these measurement to be widely used was DLTS which is essentially an analogue signal processing technique first described by Lang in 1974 [9]. The voltage applied to the sample across the space charge region is shown at the top of the diagram, the middle drawing shows the distinction between majority and minority carriers in capacitance measurements and the lower diagram illustrates the more rapid emission at higher temperatures, i.e. an increase in the emission rate as the temperature increases as described by Eqs. (3.14) and (3.15). In Lang's technique two samples of the transients' magnitudes are taken at t_1 and t_2 and the difference is used to plot a graph of this signal against temperature. The technique produces a sequence of peaks. At its simplest, each peak represents a deep state.

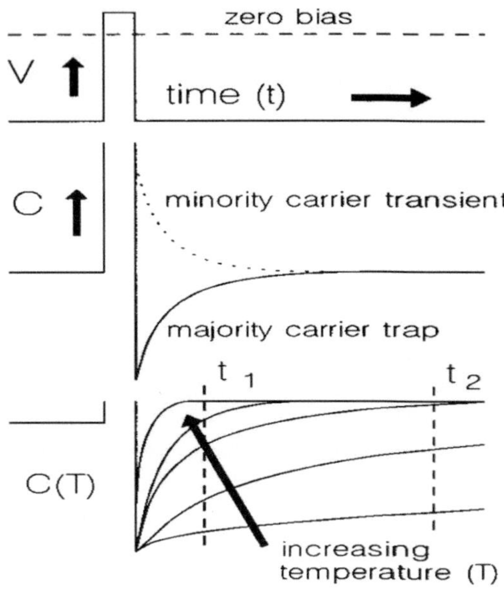

Fig. 3.6 Principle of capacitance transient measurements (see text)

In Fig. 3.6 it can be seen from the voltage plot at the top of the diagram that the device is switched into forward bias for a short period of time (typically 1 ms). This collapses the depletion region and allows the trap to fill from the majority carrier population. When the voltage is switched into reverse bias again the captured majority carriers have reduced the net space charge and the capacitance is getting lower. If the temperature is sufficiently high the majority carriers are emitted and the net space charge increases which we detect as an increase in capacitance. If we apply a 'fill' pulse which forward biases the p-n junction to such an extent that a substantial amount of current flows, minority carriers are injected. Their concentration will always be less than that of the majority carriers but if the minority carrier capture cross-section is comparable with or larger than the majority carrier cross-section then the minority carriers will be captured at the defect and a transient of the opposite sense may be observed as shown by the dotted line.

In DLTS the exponential change of capacitance, the emission transient, is sampled at two times, t_1 and t_2, which are typically in the range 0.1 ms–1 s. The amplitude difference between the two samples provides the DLTS signal which is then plotted against the sample temperature. The reasoning behind this is shown in the lower part of Fig. 3.6. The times t_1 and t_2 are the times after the end of the filling pulse. If the exponential is very slow (as at the bottom of the figure) almost no difference between the values of the capacitances, ΔC, exists. If the exponential is very fast (as at the top) again no difference exists. At intermediate values of time constant ΔC is non-zero and for given values of t_1 and t_2 there is a time constant that will give a maximum output. Because of the Boltzmann factor the time constant changes rapidly with temperature so if we alter the sample temperature the plot of ΔC against T will peak when the emission rate matches the system time constant, i.e. when:

$$e_n^{-1} = \tau = (t_1 - t_2) \ln (t_1/t_2) \,. \tag{3.18}$$

Traps with different emission rates will produce peaks at different temperatures and several different trap species will produce a "spectrum". To produce an activation plot the temperature must be scanned at different values of t_1 and t_2, or different 'rate windows' in DLTS jargon. The Arrhenius plot shown previously (Fig. 3.5) is derived from the positions of the peaks at the various rate windows.

The variation of emission rate gives rise to the peak width and so the peak width could be used to derive the activation energy. However the peak width is not exclusively a function of the trap parameters. It depends on the DLTS system, in particular the t_2/t_1 ratio and the sampling width and so these must also be taken into account. In general, this method is not favored due to the narrow temperature range sampled and the uncertainties associated with the degree of broadening due to other interfering physical effects. It is not only the position of the peaks that provides information; their magnitudes are directly related to the amplitudes of the emission transients and hence the concentrations of the deep states.

Figure 3.7 shows a DLTS plot of electron irradiated n-type Cz silicon using a p^+n diode as a test sample. Identical results would be obtained if a Schottky diode had been fabricated and used as the test sample. The detectivity of DLTS can be very high but in the case of capacitance DLTS depends linearly on the carrier concentration. In general it is almost always possible to detect and analyze results from a defect if $N_T > 10^{-4}n$. In favorable conditions with a very good system $N_T > 10^{-6}n$ can be approached. This translates into a detectivity of 10^{10} cm^{-3}

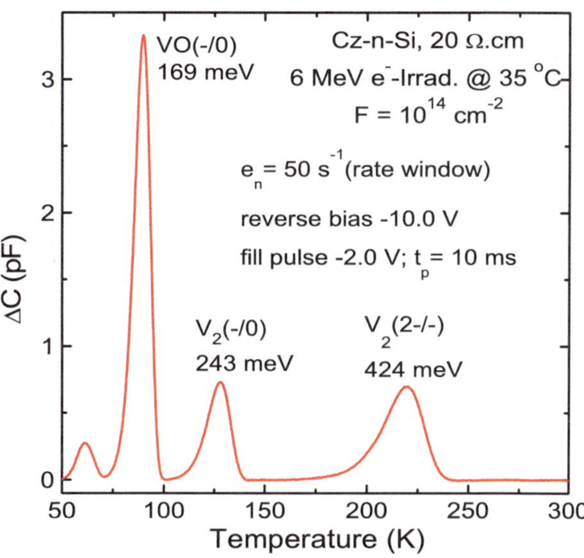

Fig. 3.7 DLTS spectrum of electron irradiated Cz silicon showing the signals from vacancy oxygen defect and the two charge states of the divacancy

defects in typical 20 Ω cm material using an average DLTS system and sample. This is much better than normal chemical techniques can achieve.

3.3.7 High Resolution Laplace DLTS

One of the major problems in relation to DLTS is the lack of resolution. The peaks tend to be very broad and this is a major factor in ascribing definitive identities to thermal emission fingerprints. The origin of the broadening is manifold. Lang's box car technique is extremely crude in signal processing terms and many improvements have been proposed. These developments have been reviewed by Istratov [10].

The subject is complex and if the signal processing is essentially based on filters the system devised by Lang is a good compromise between resolution and noise. However from a mathematical viewpoint, perhaps the most difficult issue is the fact that the zero base line of the exponential decay is not known. Recently significant advances have been made in the practical implementation of high resolution systems by employing more advanced techniques. Because this work is based on the inverse Laplace transform, they have become known generically as "Laplace transform DLTS" [11].

Truly remarkable improvements in resolution have been achieved often approaching an order of magnitude. However Dobaczewski et al. [12] point out that the signal processing issue is only one of a number of factors in achieving reduced line widths of the DLTS signal. The Laplace DLTS methods are isothermal techniques, analyzing the range of rates at a specific temperature and provide a spectral plot of a processed capacitance signal against emission rate rather than against temperature. In order to produce an Arrhenius plot, the experiment is repeated at a number of temperatures. This eliminates the line broadening due to the shift of parameters with temperature which is significant when instrumental broadening is eliminated by the use of the inverse Laplace transform.

Under good experimental conditions Laplace DLTS provides an order of magnitude higher energy resolution than conventional DLTS techniques. An example of this is shown in Fig. 3.8 which shows DLTS and LDLTS spectra of hydrogenated silicon containing gold. The conventional DLTS spectrum is shown as an inset at the top right of the figure. The broad peak centered at 260 K is attributed to electron emission from the gold acceptor and from G4 which is a gold-hydrogen complex. The linewidth of the DLTS peak showed little deviation from an ideal point defect but using Laplace DLTS (shown as the main spectrum) the gold acceptor level and the gold-hydrogen level G4 are clearly separated.

LDLTS has had a profound effect on electrical defect spectroscopy enabling the effect of external probes, such as uniaxial stress, and internal perturbations, such as the proximity of atoms isovalent with the host (e.g. the case of SiGe), to be quantified in terms of electronic behavior. Laplace DLTS provides a synergy with other techniques that was difficult or impossible to achieve previously.

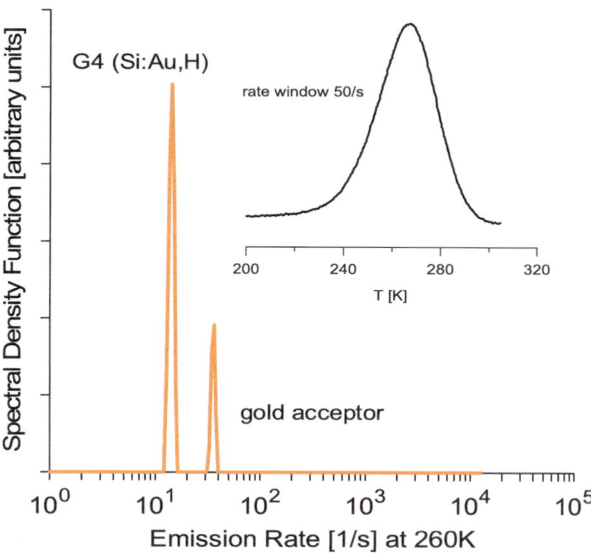

Fig. 3.8 Comparison of a conventional DLTS plot of the electron emission signal from gold and a gold-hydrogen complex in silicon with the isothermal LDLTS spectra which clearly separates the two signals which are indistinguishable in conventional DLTS

A comprehensive review of the technique and examples of results obtained are given elsewhere [13].

3.3.8 Minority Carrier Processes

All the methods considered so far have been based on the change of occupancy of a deep level state by shifting the Fermi level to achieve majority carrier occupancy and then examining the majority carrier emission. It is quite feasible to undertake similar experiments but filling with minority carriers. This is very important because at low excitation densities it is the capture of minority carriers which is the rate limiting process in recombination.

Lang [9] describes such a method in his early papers on DLTS and in this case, minority carrier occupancy was achieved by passing forward current through a p-n junction. It is also possible to perturb the occupancy optically and a number of techniques have been used to this end. Optical methods can be used to excite carriers across the gap and create minority carriers. In this case, the wavelengths of the exciting light must be slightly above band gap. Alternatively, sub band gap light can be used to directly change the occupancy of the state according to its optical cross section. This latter technique is referred to as "optical DLTS" or ODLTS and is rarely used for silicon. Much more important is minority carrier transient spectroscopy or MCTS.

The first application of the generation of minority carriers by above band gap light as a technique for manipulating the occupancy of deep states was described by Hamilton et al. [14] and developed into minority carrier transient spectroscopy

Fig. 3.9 The concept of minority carrier generation using a pulse of above band gap light

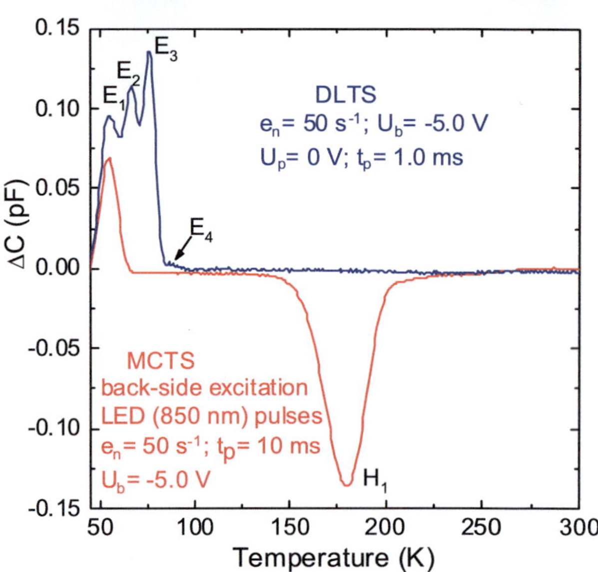

Fig. 3.10 DLTS and MCTS spectra of an n-type Cz silicon sample containing a high concentration of thermal double donors. MCTS shows a negative peak due to minority carrier emission

by Brunwin et al. [15]. Light can be applied through a semi-transparent Schottky diode i.e. from the front of the sample, or from the backside. If the slice is thick compared to the minority carrier diffusion length then thinning is necessary as shown in Fig. 3.9 to allow a sufficient flux of minority carriers to reach the depletion region. The disadvantage of excitation from the front is that majority carriers are also created in the depletion region and these play a part in the capture kinetics. If excitation is from the back this effect is much reduced as the depletion field rejects majority carriers and so if no light is absorbed in the depletion region the carrier flux is exclusively minority carriers.

Figure 3.10 compares DLTS and MCTS spectra for an n-type Cz silicon sample with Au Schottky diodes. The MCTS spectrum has been recorded with the use of backside optical excitation pulses from a light emitting diode with wavelength of 850 nm.

3.3.9 DLTS Measurements of Iron in Silicon

Iron is a common contaminant in silicon processing, an issue which is discussed in detail in Chap. 9 of this book. In addition it is almost always present in solar grade silicon either originating from the feedstock or introduced during processing. DLTS is a very valuable technique for quantifying iron in silicon and in this section we will use it as an illustration of the DLTS technique. However it is important to emphasize that DLTS can only detect electrically active iron. Iron which is present as precipitates is not seen so that the concentrations determined by "chemical" techniques such as Secondary Ion Mass Spectrometry (SIMS) or neutron activation analysis (NAA) may be much greater than the electrically active fraction. This is of course true of all techniques which rely on the measurement of electrical activity such as carrier lifetime methods discussed in Sect. 3.4.3 of this chapter.

In equilibrium iron occupies a tetrahedral interstitial site in the silicon lattice although iron on other sites has been observed in significant concentrations, including substitutional iron, an issue which is discussed in Chap. 8 of this book.

Interstitial iron is quite mobile and in p-type Si is a deep donor carrying a positive charge, Fe_i^+. In consequence the interstitial iron is coulombically attracted to the negatively charged acceptors (most commonly boron) and forms an electrically neutral pair:

$$Fe_i^+ + B_s^- \leftrightarrow (Fe_i\, B_s)^{\circ}$$

The binding energy between the Fe and B is not strong (~0.65 eV) so the pair can be dissociated thermally at quite low temperatures (reaction pushed to the left). Typically at 200 °C with a boron concentration $<10^{16}$ cm^{-3} almost all the iron is present in the unpaired interstitial state. Starting from the room temperature equilibrium state (Fe_iB_s pairs) the dissociation reaction at 200 °C with $[B] = 10^{16}$ cm^{-3} takes about 10 min to be almost complete, i.e., all the iron is in the interstitial state Fe_i. The capture and emission of a hole represented by the transition $Fe_i^{+/0}$ can be easily observed in DLTS with a hole binding energy of 0.39 eV enabling $[Fe_i]$ to be quantified. The iron boron pair can itself capture and release a hole with a binding energy of 0.1 eV, observable in DLTS as the $Fe_iB_s^{+/0}$ related signal. If either Fe_i or B_s becomes neutral the attractive force binding the pair ceases to exist and the iron diffuses away. In consequence the above reaction is reversible ... moving to the left at room temperature if the Fermi level is below the boron acceptor level at $E_v + 0.045$ eV or above the interstitial iron donor level at $E_v + 0.39$ eV. The latter condition will occur in a depletion region or when sufficient minority carriers are injected into the p-type material for example by exposure to light of a wavelength which is absorbed by the silicon. The mechanism of the iron boron reactions has been studied in detail and is now well understood with an extensive literature, e.g. [16–18].

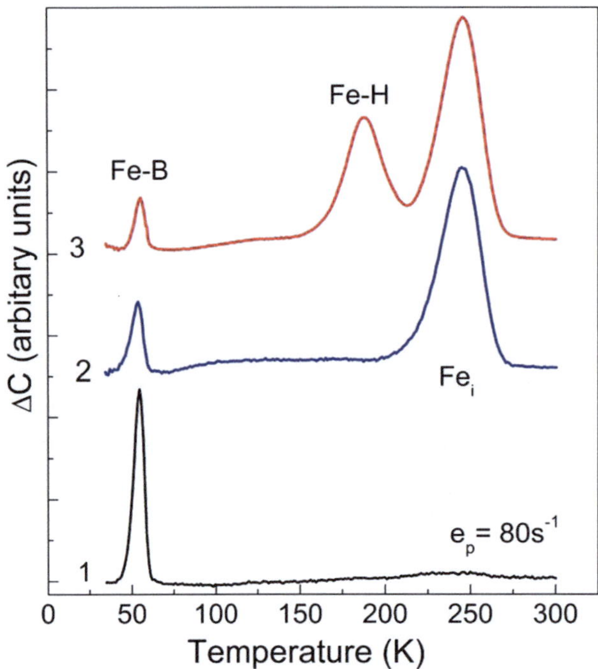

Fig. 3.11 DLTS spectra ($e_p = 80\ s^{-1}$) of iron doped p-type silicon [B] = ~10^{15} cm^{-3}, [Fe] = ~10^{13} cm^{-3}: (1) after storage at room temperature with all the iron present as FeB pairs, (2) after a 175 °C for 30 min anneal of material from sample 1. This dissociates some of the FeB pairs producing isolated interstitial iron, (3) after reverse bias annealing at 110 °C for 30 min material from sample 1. In this case Fe_i has formed and some of this has reacted with hydrogen in the silicon to form FeH pairs. Both reactions are completely reversible with no iron loss (Adapted with permission from Leonard et al. [19])

Figure 3.11 shows some DLTS spectra related to the reactions discussed above. Starting with the equilibrium condition in p-type silicon line 1 shows the DLTS spectrum in which almost all of the iron is paired with boron. After annealing at 175 °C for 30 mins, much of the iron is present as the isolated interstitial due to the dissociation of the iron boron pairs (line 2). If the Fermi level is shifted towards mid gap by the application of a reverse bias to the Schottky diode structure used for the DLTS measurement the dissociation reaction proceeds more rapidly. Plot 3 of Fig. 3.11 shows a DLTS spectrum after a reverse bias anneal at 110 °C for 30 mins. In this case hydrogen diffusing from the surface has reacted with the interstitial iron in the depletion region to create an additional DLTS peak due to an iron-hydrogen complex [19]. This complex is not stable in p-type silicon but can be observed in n-type or, as in this case, within the depletion region.

Laplace DLTS signatures of these three iron related defects in p-type silicon (FeB, Fe_i and FeH) are shown in Fig. 3.12. LDLTS provides a much more precise fingerprint of deep level defects which can give us insight into the defect structure.

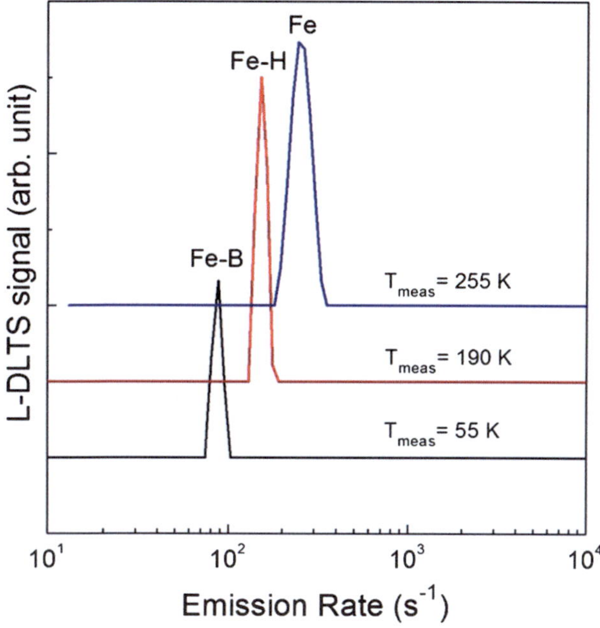

Fig. 3.12 Laplace DLTS spectra of the three defects shown in Fig. 3.11. Note the different measurement temperatures

The FeB and FeH LDLTS hole emission spectra have very narrow line widths indicating that the lattice locations of the Fe and B and of the Fe and H are fixed in a unique or electronically equivalent configuration. A slight broadening is evident in the case of the Fe_i defect. Some of this may be due to the higher measurement temperature used for the Fe_i spectrum but it is also possible that it is an indicator that Fe_i exists in slightly different lattice configurations with displacements from the ideal tetrahedral interstitial site. These issues are discussed in more detail in Chap. 8 of this book.

The iron and the iron boron pair are both important recombination centers but with different characteristics. Consequently the transformation of FeB to Fe_i produces a change in minority carrier lifetime. This led Zoth and Bergholz [20] to propose using a lifetime measurement technique (surface photovoltage) before and after FeB dissociation as an alternative to DLTS for the detection of iron in p-type silicon. This approach has the advantage that the procedure does not need Schottky diode fabrication, is fast and has high detectivity. This method is now used extensively for the detection of iron in combination with various lifetime measurement methods and is discussed in more detail in Sect. 3.4.6 of this chapter.

3.4 Measurement of Carrier Lifetime

3.4.1 Generation Lifetime

The concept of generation lifetime applies where there are few carriers, for example in a depletion region. It is the average time taken in the creation of an electron hole pair. It is strongly temperature dependent and is of major technological importance in DRAMs where it largely defines the refresh time.

The most common method to measure the generation lifetime is via a pulsed MOS capacitor. This is often referred to the Zerbst method but many variants exist most of which have been reviewed by Kang and Schroder [21]. The technique is based on applying a bias to the MOS capacitor which puts the capacitor into deep depletion and then observing the capacitance as a function of time. The change of capacitance results from the thermal generation of carriers. Traditionally this transient is then analyzed by plotting the slope of the transient (the generation rate) against $C_{inv}/C-1$ where C_{inv} is the capacitance of the MOS biased into inversion. The latter term is proportional to depletion width. The generation lifetime is derived from the slope of the Zerbst plot and the intercept with the vertical axis at $C_{inv}/C-1=0$ is the surface recombination velocity.

3.4.2 Recombination Lifetime

Recombination lifetime is an important parameter in qualifying silicon and in process control. In solar cells technology it is the key measurement in assessing the quality of solar silicon. In power devices the minority carrier lifetime is often engineered to obtain minimum forward drop while still having fast enough turn off time. However there are very few other devices today which depend directly on recombination lifetime. Despite this carrier lifetime measurements are a key tool in the semiconductor industry because it is a remarkably sensitive method to determine contamination and processing defects which may affect devices in ways other than the direct impact of lifetime. In the industry lifetime measurement are attractive because they can be very quick and contactless. Most other techniques have much lower sensitivity (e.g. chemical methods) or require the formation of a device e.g. DLTS.

3.4.3 Photoconductance Decay

Photoconductance decay is a long established technique for determining recombination lifetime and very widely used today. Many variants of the measurement exist and commercial systems are sold both for single point measurements and mapping

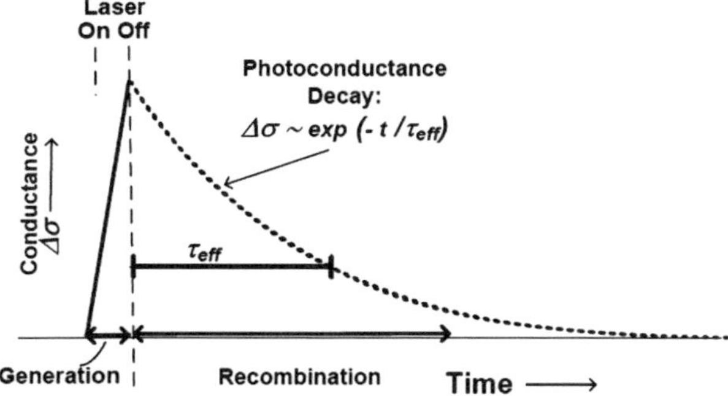

Fig. 3.13 Photoconductivity as a function of time in a recombination lifetime measurement. Excess carriers are generated in the first phase of the measurement by an above band gap LASER which is then turned off and the decay of excess carriers measured

wafers or ingots. The principle of the method is to excite carriers optically in the semiconductor and then measure the change of conductivity $\Delta\sigma$ as a function of time. The increase in conductivity is due to the optical generation of electron hole pairs $(\Delta n + \Delta p)$ by the light pulse,

$$\Delta\sigma = q\left(\Delta n\mu_n + \Delta p\mu_p\right). \tag{3.19}$$

The decay of the photoconductivity depends on the recombination lifetime. In the ideal case the decay is exponential so determining the time constant of the conductivity decay gives the lifetime as shown in Fig. 3.13. In practice there are several recombination processes (e.g. surface and bulk) and often the lifetime is a function of excitation density so the carrier lifetime changes during the decay process.

A key issue is how to measure the change of conductivity. Early systems used contacted samples [22] but this has been largely displaced by contactless measurements using microwaves or inductive coupling because of the ease of operation. In addition contactless measurements can in some cases provide the possibility of mapping rather than taking a single spot measurement. There is a substantial literature on these techniques which although quite simple in principle are susceptible to artifacts which necessitate careful equipment design.

In the case of microwave techniques the measurement is confined to the distance the microwaves penetrate the sample referred to as the "skin depth". This depends on the frequency and the resistivity. Most instruments using this approach operate at around 10 GHz which provides a skin depth of ~0.5 mm in 1 Ω·cm Si and ~3 mm in 20 Ω·cm Si in consequence $\Delta\sigma$ is measured for all or most of the thickness of a slice at these resistivities. Convoluted with this is what depth is excited by the incident light. Typical laser wavelengths lie between 900 nm and 1 μm which give extinction depths of 46 and 215 μm respectively. The kinetics of the generated carriers are quite complex [23] and can be very significantly influenced by surface recombination (S). The effect is summarized for a slice of thickness d with recombination velocities at the front and back face of S by:

$$\tau_s = d/2S \tag{3.20}$$

which decreases the measured value of lifetime according to:

$$\tau_{\text{meas}} = \frac{1}{\tau_s^{-1} + \tau_{\text{bulk}}^{-1}} \tag{3.21}$$

The classic approach is to try to reduce the surface recombination velocity to negligible proportions and many recipes have been published to try to achieve this. Among the more successful approaches are immersion in an ethanolic iodine solution usually preceded by an HF dip [24, 25] and a methanolic solution of quinhydrone [26].

An alternative to passivation by chemical solutions is to deposit a dielectric (usually SiO_2 or Al_2O_3) in a way that is known to produce an extremely low density of interface states and hence a low interface recombination. The effect of the surface on the measurement is particularly dramatic on long lifetime material. Commercial instruments are available which can measure recombination lifetimes in the range 10 ns–30 ms using a LASER focused to a 0.5 mm spot. The slice can be scanned to provide maps of the type shown elsewhere in this chapter. A standard test procedure has been published by SEMI [27].

3.4.4 *Quasi-Steady State Photoconductance*

In the procedure described above the excess carrier density changes during the measurement. This can lead to artifacts and does not provide a good basis for measuring the lifetime as a function of excitation density. True steady state techniques such as surface photo voltage overcome this problem [28] but quasi-steady-state methods have found favor in recent years. In the case of the microwave photo-conductive decay the sample is irradiated at a constant level by a second light source as shown in Fig. 3.14.

An alternative approach to the Quasi-steady state measurement, which is very widely used, is to excite the sample with a flash lamp as described by Sinton and Cuevas [29]. The flash lamp provides a relatively slowly varying intensity of illumination. As the decay is much longer than the minority carrier lifetime the "steady state" condition is maintained. The excitation is over an area of a few cm^2 and the change of conductance detected by eddy current measurements and inductive coupling. This approach is not suitable for lifetime mapping but tends to give a more reliable measure of the bulk lifetime of ingots than its microwave counterpart. This is because the lower frequency (used for the inductive coupling) samples deeper into the silicon. With appropriate filtering of the flash lamp the

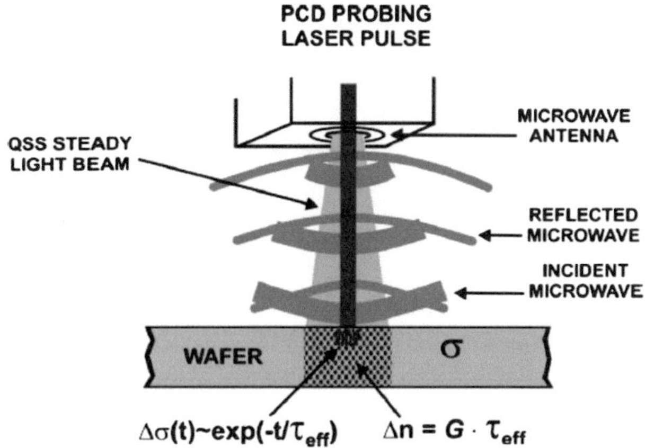

Fig. 3.14 Schematic of a Quasi Stteady State microwave photoconductive decay measurement showing the steady state light beam and the pulsing LASER

system can be made less critically dependent on surface preparation than the microwave technique particularly when measuring ingots.

When measuring solar silicon and particularly multi-crystalline material anomalous long values of recombination lifetime are observed which do not represent the true lifetime of the material. This can be attributed to carrier trapping and is now well understood and procedures adopted to avoid such misleading results [30, 31].

3.4.5 Photoluminescence Based Techniques for Lifetime Measurement

Photoluminescence has become an important method of carrier lifetime assessment for solar silicon. It is not susceptible to the trapping effects described in the previous section but there are fundamental problems in extracting absolute values of carrier lifetime from photoluminescence. Commercial instruments are available for photoluminescence measurements which consist of illuminating a silicon slice with above band gap light and detecting the luminescence with a detector array often a charge coupled device (CCD). The attractiveness of the technique, in addition to it being contactless, as in the previous cases, is its speed ... a 5 inch solar slice can be mapped in seconds compared to the 15–30 min or so required for microwave photoconductive decay methods. Luminescence intensity taken at room temperature depends on many factors besides lifetime so additional measurements are required to establish absolute measures of lifetime. These can be spot measurements based on the techniques described previously or modulated photoluminescence measurements [32–34].

3.4.6 *Lifetime Techniques for the Assessment of Iron Contamination in p-type Silicon*

As discussed in Sect. 3.3.9 of this chapter a change in lifetime occurs when FeB pairs are dissociated. This dissociation is used as a test for iron because although both Fe_i and FeB are important recombination centers Fe_i has more impact on the lifetime than FeB at low excitation densities. Consequently a measure of the change of lifetime on dissociation of the FeB pair has become a standard test for iron contamination in Si within the PV industry. However, care must be exercised as at high excitation densities the FeB pair is a more effective recombination center and so the excitation density affects the lifetime change associated with the dissociation and hence the calibration.

Figure 3.15 taken from Macdonald et al. [35] shows the effect on the change of lifetime resulting from dissociation of the FeB pairs. Inserted into the diagram are arrows showing typical excess carrier densities used with various lifetime measurement techniques: Surface Photo Voltage, PL imaging, Quasi Steady State Photo Conductivity and Microwave Photo Conductive Decay. The latter two operate at higher excitation densities than the cross-over point while SPV and PL use lower excitation densities. At very high excitation densities Auger recombination starts to have an important effect on the lifetime invalidating the measurement.

Normally the dissociation of the FeB pairs is effected by shining light onto the sample at a high excitation density. It is evident that some dissociation will also occur during the measurement of lifetime while the Fe is paired with the B and care needs to be exercised if accurate results are to be obtained particularly using

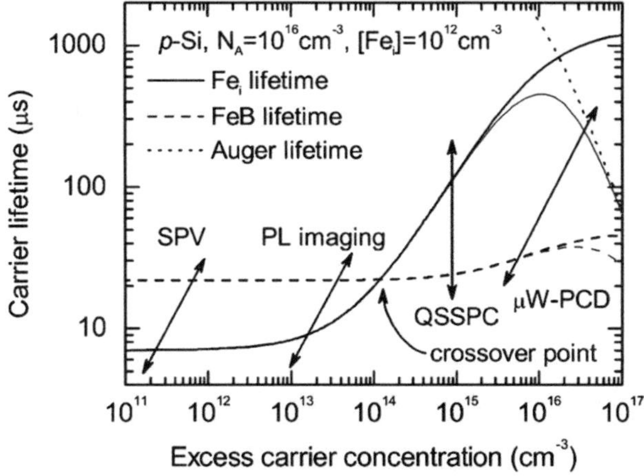

Fig. 3.15 Effect of the dissociation of FeB pairs by illumination at various excess carrier densities. It can be seen that at high excitation densities the lifetime is increased by the dissociation (FeB is a more powerful recombination center) while at low excitation densities Fe_i is the more powerful center (Reproduced with permission from Macdonald et al. [35])

the techniques which operate above the cross-over point [36]. The actual cross-over point depends on temperature and the boron concentration but in commercial lifetime machines such an iron test procedure is often built into the operating system so ensuring that appropriate test conditions and interpretation are always used.

3.5 Optical Methods for Defect Evaluation

3.5.1 Optical Absorption

Optical transmission or absorption measurements are widely used in the semiconductor industry for a number of applications. The physical quantity that is measured during optical absorption measurements is the absorption coefficient, $\alpha(\nu)$. From measurements of the absorption coefficient as a function of the photon energy the semiconductor band gap can be determined and information about electronic and vibrational properties of some lattice defects can be obtained.

The optical absorption coefficient is usually determined by relating the intensities of the transmitted and incident light. In the case of the absence of internal interference (due to multiple reflections) and equal front and back reflection coefficients for a measured sample the ratio of the intensities of the transmitted to incident light can be expressed as

$$T = \frac{(1-R)^2 e^{-\alpha d}}{1 - R^2 e^{-2\alpha d}} \tag{3.22}$$

where d is the sample thickness and R is the reflection coefficient given by

$$R = \frac{(n-1)^2 + k^2}{(n+1)^2 + k^2}, \tag{3.23}$$

where n and k are the refractive index and extinction coefficient, respectively. The R value can be determined from that part of the transmittance spectrum where $\alpha \approx 0$ and then be used for the calculation of the absorption coefficient from the T spectrum on the basis of Eq. (3.22).

The optical absorption measurements of defects in silicon crystals are usually undertaken using Fourier Transform Infrared (FTIR) spectroscopy. In FTIR spectrometers the basic optical component is the Michelson interferometer. Usually the interferograms with and without the sample inserted into a spectrometer are recorded and then converted into single-beam intensity spectra by a Fourier transform. Subsequently, a spectrum of absorbance versus light frequency is calculated as:

$$A = \ln\left(\frac{I_B}{I_S}\right), \tag{3.24}$$

where I_B and I_S are intensities of the single-beam spectra recorded without and with the sample inserted, respectively. The spectral dependence of the absorption coefficient, $\alpha(\nu)$, is obtained by normalization of the calculated absorbance values to the sample thickness. In studies of impurity- and defect-related optical absorption, an absorption spectrum due to vibrations of the host atoms (recorded with a high-purity defect-free reference sample) is usually subtracted from a spectrum for a sample with defects. So, only the absorption features related to defects occur then in the resulting difference spectrum.

3.5.1.1 Electronic Transition Related IR Absorption

Electronic transition related infrared absorption spectroscopy is based on the optical excitation of electrons or holes from ground states of the donor or acceptor defects into excited states. This type of spectroscopy has been proven to be very useful for identification and characterization of shallow level defects in silicon, electronic properties of which can be described in the frame of the effective mass theory (EMT), early developed by Kohn [37, 38] and later reviewed by Stoneham [39] and Ramdas and Rodriguez [40]. Let us consider a simplified case of a singly ionized group-V or doubly ionized group-VI atom replacing a host atom from a group-IV crystal with an isotropic conduction band minimum and permittivity ε. Four valence electrons of the atom may be assumed to form sp^3 hybrid orbitals with four host neighboring atoms. The excess positive charge Z at the impurity generates a screened Coulomb potential given by

$$V_i = -\frac{Z}{\varepsilon r} \tag{3.25}$$

This equation is not valid in the neighborhood of the impurity atom (often referred as a central cell region) but is a good approximation for r exceeding the lattice constant. An unpaired electron in the vicinity of the impurity atom will move as a species with the effective mass m_e* in the hydrogen like potential V_i, and similarly to the hydrogen atom, a series of bound states at E_{bn} below the conduction band minimum will appear with:

$$E_{bn} = \frac{Z^2 m_e^*}{2\varepsilon^2 n^2} \tag{3.26}$$

where $n = 1, 2\ldots$ For a singly ionized group-VI impurity in the ground state, partial screening occurs and it can be assumed $Z \approx 1$ when considering the second weakly bound electron.

To observe electronic transition related lines in absorption spectra the measurements should be carried out at low temperatures in order to populate the ground state of defects with the respective charge carriers. Upon illumination with photons of energy smaller than or approximately equal to the binding energy of a charge

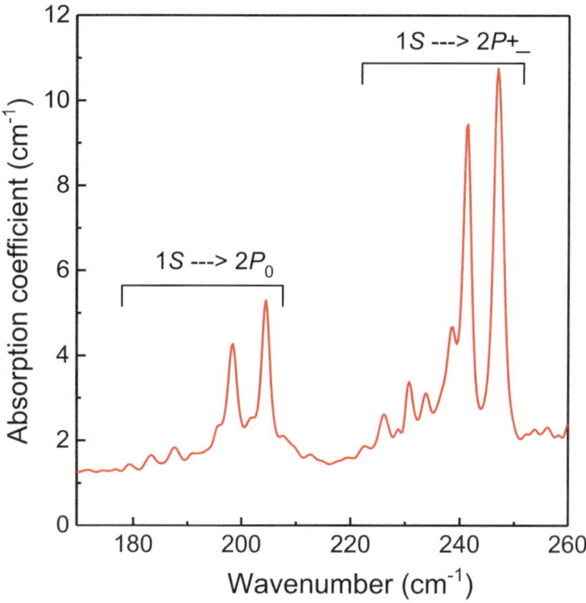

Fig. 3.16 Absorption spectra at 10 K for a hydrogenated and heat-treated (for 10 h at 470 °C) Czochralski-grown Si sample

carrier in the ground state, two optical absorption processes can occur: (i) electrons (holes) can be excited from the ground state to the conduction (valence) band giving rise to a broad absorption continuum; (ii) electrons (holes) can be excited from the ground state to one of several excited states resulting in sharp absorption lines characteristic of the shallow-level impurities. As an example, Fig. 3.16 shows an infrared absorption spectrum with sharp absorption lines related to ground-to-excited state electronic transitions due to hydrogen-related shallow thermal donors in silicon [41, 42].

It should be pointed out that the binding energies of a charge carrier in the ground state of the shallow level defects deviate from the value predicted by the effective mass theory and are dependent on the chemical nature of the defect. These deviations result in some shifts of the electronic transition related absorption lines, which are different for different centers. Positions of infrared absorption lines related to electronic transitions due to many shallow level defects in silicon have been identified and can be found in Ref. [43]. Further, as each impurity has unique absorption peaks and in many cases the calibration coefficients linking intensities of the absorption peaks with concentration of the corresponding defects have been found [43], it is possible to identify the impurities in the Si crystals and, furthermore, to determine their concentration. With the use of FT-IR techniques high sensitivity is obtained and the detection limits are very low. Concentrations of shallow dopant impurities as low as 5×10^{11} cm^{-3} can be measured in silicon [43].

3.5.1.2 Infrared Vibrational Spectroscopy

The theory of lattice atoms vibrations in a solid is well developed. The phonon dispersion curve depends on the masses of the host atoms and the effective spring constants describing their interactions. The presence of defects in a crystalline host destroys the translational symmetry of the lattice and modifies the storage of vibrational energy in two main ways: by perturbing the vibrational density of states, where any new mode strongly overlaps with the existing host modes, or by introducing modes with such frequencies that lie in spectral regions which are forbidden in the perfect crystal. In the first case the mode is called resonant, whereas in the second case the mode is said to be localised. A light impurity introduced into the crystal normally oscillates with frequencies much higher than the phonon frequencies. Coupling of these oscillations with the lattice phonons is weak, and the modes of the oscillations are denoted as local vibrational modes (LVM's). The principle of LVM absorption spectroscopy is to excite the oscillating defect from one vibrational state to another by absorption of infrared light of energy hν. An LVM is seen as a sharp line in the absorption spectrum at the corresponding frequency. The integrated absorption (IA) of an IR absorption band is proportional to the concentration of oscillating defects. Dawber and Elliot [44] derived the commonly used expression for the IA of a non-degenerate mode of a defect with effective mass m as:

$$IA = \int \alpha(v)dv = \frac{\pi \eta^2 N}{mnc^2} \tag{3.27}$$

where N is the concentration of absorbing centers and c is the velocity of light. The quantity $\eta = |d\mu/dQ|$ has unit of electric charge, and is normally referred to as the effective charge of the mode Q. It should be noted that η is not related to the electrical charge of the defect. In fact, neutral defects with values of $\eta = 2$ or even higher have been reported [45].

The chemical nature of a defect can sometimes be determined from the LVM spectroscopy with the use of isotopic substitutions, which shift the value of the vibrational frequency. The technique is also used in combination with uniaxial stress for symmetry determination and defect alignment experiments. Some of these applications will be briefly discussed below but more detailed sources and further information can be found in works by Newman [45], Stavola [46], McCluskey [47] and Pajot [48].

Identification of the chemical composition of defects in semiconductors is one of the primary goals in spectroscopic studies. IR vibrational spectroscopy, can provide some clues regarding the atoms, which are incorporated into a defect, from band shifts that occur when samples containing different isotopes are investigated. It is known from a ball and spring model that a diatomic molecule vibrates with a frequency $\omega \propto \sqrt{1/\mu}$,, where μ is the reduced mass of the atomic pair. This empirical diatomic model was applied for estimating isotopic shifts of impurities in crystals [45]. The LVM frequency in the crystal can be given by

$$\omega_L = \sqrt{f\left(\frac{1}{\alpha M} + \frac{1}{m}\right)} \tag{3.28}$$

where $1 \leq \alpha \leq 4$ is a parameter that takes into account the coupling of the impurity with the host atoms. For example, the interstitial oxygen in silicon sits at the bond center site, where ^{16}O and ^{18}O isotopes produce absorption bands at 1136 and 1084 cm^{-1} at liquid He temperature, respectively. The ratio of the measured frequencies $^{16}\omega/^{18}\omega = 1.048$ can be compared with the value calculated by using Eq. (3.28), where $^{16}\omega/^{18}\omega = 1.046$ is obtained with $\alpha = 2$ to account for bonding with two Si neighbours.

Another important phenomenon which can produce a shift of an LVM is associated with charge state effects. When a particular defect is electrically active, the capture or emission of carriers can change the defect local environment dramatically. The bonding properties (which can be interpreted as the spring force constants), are then perturbed, and the vibrational mode frequency will differ from that of the uncharged defect. The vacancy-oxygen (VO) complex in Si is an obvious example, where the absorption band associated to the antisymmetric vibrational mode of the Si-O-Si unit is shifted by ~50 cm^{-1} when comparing the negative charge state with the neutral defect [49]. Figure 3.17 illustrates the isotope and charge state induced shifts of the local vibrational mode lines associated with the VO defect in Si.

Positions of infrared absorption lines related to local vibrational modes of many defects associated with light impurities (oxygen, carbon, nitrogen, and hydrogen) in crystalline silicon have been identified and can be found in Ref. [48]. For a number of the LVM lines the calibration coefficients linking their integrated intensities or absorption coefficients in the peak maxima with concentrations of the corresponding defects have been determined [48]. The LVM spectroscopy is widely used in the silicon industry for detection and measurements of concentrations of some defects, which are electrically neutral and cannot be studied by electrical characterization methods. The most prominent examples are interstitial oxygen and substitutional carbon impurities [48, 50–52]. Some details of the procedures used for the determination of oxygen and carbon concentrations in Si crystals from IR absorption measurements can be found in Refs. [48, 53, 54].

3.5.2 Photoluminescence

Photoluminescence is one of the key tools used in the assessment of III-V semiconductors. Until recently its application to silicon research and technology has been more limited. In the silicon industry it has been used to quantify low

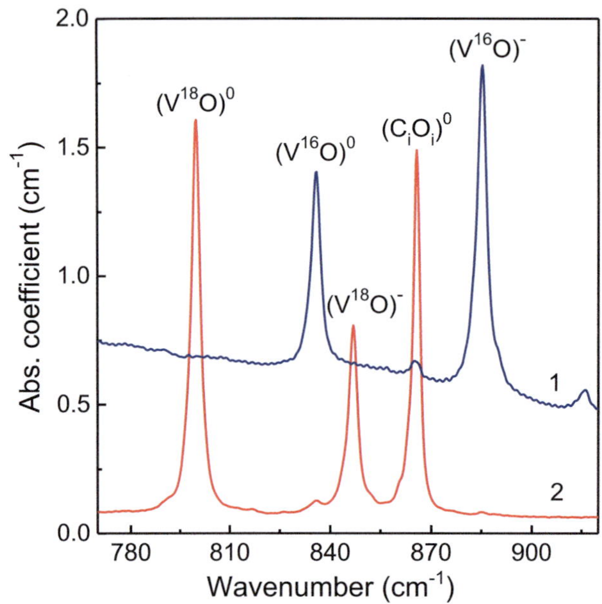

Fig. 3.17 Absorption spectra at 10 K for electron-irradiated ($F = 6 \times 10^{17}$ cm^{-2}) highly doped Si crystals with (1) natural abundance of oxygen isotopes ($[^{16}O] = 9 \times 10^{17}$ cm^{-3}) and (2) ^{18}O enrichment ($[^{16}O] = 5 \times 10^{16}$ cm^{-3}, $[^{18}O] = 1.5 \times 10^{18}$ cm^{-3})

concentrations of dopant in feedstock for silicon ingot growth using the technique invented by Tajima in 1978 [55] and subsequently adopted as a standard method worldwide [56, 57]. In the last decade photoluminescence in general has become a crucially important tool in the assessment of solar silicon. Several commercial instruments are available to assess material and the impact of defects using photo- or electro-luminescence techniques. In this section we outline some of the physical concepts of luminescence and then look in more detail at the major applications.

Photoluminescence equipment consists basically of an above band gap light source (typically a gas or solid state laser), a sample (which is often cooled to cryogenic temperatures), and a spectrometer. The wavelength of the emitted light depends on the recombination process and can involve impurities and defects. Figure 3.18 shows some of the processes which can give rise to luminescence in order of decreasing photon energy from left to right. At low temperatures many of the features are associated with excitons. An exciton is a bound state of an electron and a hole which are attracted to each other by electrostatic (Coulombic) forces. It is electrically neutral and has formation energy E_{FX} which in silicon is 14.7 meV. At cryogenic temperatures (where $kT < E_{FX}$) excitonic features dominate the photoluminescence spectrum.

The first process in Fig. 3.18 shows recombination resulting from a free exciton (FE). In silicon due to its indirect gap this luminescence transition must involve

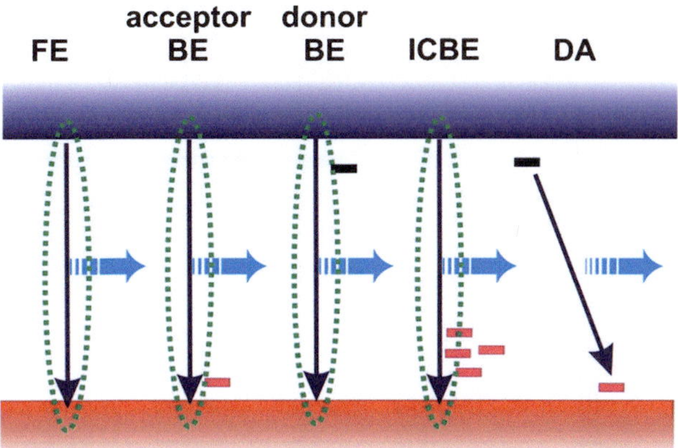

Fig. 3.18 Schematic of some low temperature photoluminescent process in silicon related to free and bound excitons and to donor acceptor pairs. See text for details

a phonon and is very weak compared to direct gap materials. The free exciton luminescence spectrum has a Maxwell-Boltzmann shape with linewidth δ that exhibits thermal broadening to the extent that $\delta \approx 1.8$ kT The emitted energy of luminescence due to free excitons is given by:

$$h\nu_{\mathrm{FE}} = E_G - E_{FX} - \hbar\omega \tag{3.29}$$

Where E_G is the band gap, E_{FX} the formation energy of the exciton and $\hbar\omega$ the energy of the momentum conserving phonon associated with the transition. In Si the dominant phonon replica is the transverse optical (TO) which is displaced from the no phonon line by 63 meV, but the longitudinal optical (LO) and transverse acoustic (TA) phonon also produce detectable excitonic luminescence lines.

The second and third processes in Fig. 3.18 result from the exciton being bound to a neutral impurity. These complexes are referred to as bound excitons (BE) as they do not have the freedom of movement associated with free excitons. This has an important impact on the linewidth of the luminescent feature which is much narrower than the free exciton. The energy is given by:

$$h_{\mathrm{FE}} = E_G - E_{FX} - E_{BX} - \hbar\omega \tag{3.30}$$

The energy of the emitted photon is reduced compared to the free exciton case by the binding energy to the impurity E_{BX}. In general, E_{BX} is approximately a tenth of the ionization energy of the donor or acceptor. This important relationship was first reported by Haynes [58].

Figure 3.19 is taken from Haynes original paper [58]. Part (a) in Fig. 3.19 shows PL spectra taken on undoped and arsenic doped silicon . It shows the TO phonon replica of the free exciton line and the narrow bound exciton lines. The bound exciton lines are much sharper than the free exciton line because the excitation

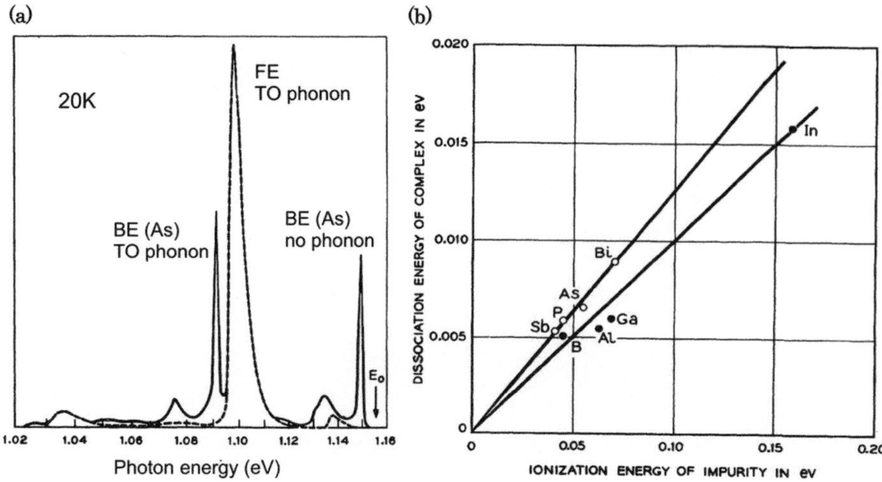

Fig. 3.19 (**a**) Photoluminescence spectra of undoped silicon (*dotted line*) and silicon doped with 8×10^{16} As (*solid line*). The bandgap energy minus the exciton binding energy is marked as E_0, the free exciton peak is displaced from this by 63 meV (the TO phonon energy). The sharp bound exciton lines are displaced in energy by 6.5 meV from the FE TO line in the case of the phonon replica and the E_0 mark in the case on the no phonon emission. (**b**) The separation of the BE and FE lines interpreted as the dissociation energy of the exciton – impurity complex for dopants in silicon (Taken from Haynes [58] with permission)

is localized on an impurity, there is no net translational kinetic energy as in the free exciton case. Part (b) in Fig. 3.19 shows the relationship between the impurity ionisation energy and the exciton binding energy.

Because the energies of the bound-exciton luminescence lines are impurity-specific with separations greater than the linewidths these luminescence lines provide evidence of the presence and identity of the donor and acceptor species even at very low concentrations. In addition, both donors and acceptors can be detected simultaneously.

The fourth process in Fig. 3.18 is only seen at higher doping levels ($>2 \times 10^{16}$ cm^{-3}). Here the exciton is bound to a cluster of donors or acceptors and so the term E_{BX} does not have a unique value and the line is broadened. Such features have been observed by several groups and are referred to generically as impurity associated localized carriers (IALC) or more specifically as impurity-cluster bound excitons (ICBE) [59].

The fifth process is of quite different origin and is due to recombination between donors and acceptors it is only seen where high concentrations of both donors and acceptors are present. This is the case in some types of solar silicon feedstock. In this case a sequence of lines is seen in the spectrum due to different lattice spacing between the donor and acceptor. The energy of the photoluminescence lines are given by:

$$h\nu_{\mathrm{DA}} = E_G - (E_D + E_A) + \frac{e^2}{4\pi\varepsilon_0\varepsilon_r r}\left\{1 - \left(\frac{\alpha}{r}\right)^5\right\} - \hbar\omega \tag{3.31}$$

Where E_D and E_A are the donor and acceptor ionization energies, $\varepsilon_0\varepsilon_r$ is the low-frequency dielectric constant, and r is the distance between the donor and acceptor. The third term is due to Coulomb and van der Waals interactions and α is the effective van der Waals constant for the specific DA pair species. The end result is a complex series of lines particularly so due to the phonon replica terms [59].

The intensity of the various features varies with the concentration of impurities but also with measurement temperature and excitation density. However, by comparing the ratio of intensities of the bound and free exciton emission under standardised conditions, photoluminescence can provide a quantitative measure of the donor and acceptor concentrations in silicon. Figure 3.20 shows such a calibration for doping in the range 10^{11}–10^{17} cm^{-3} taken from works by Tajima [55]. At low concentrations the 4.2 K measurement enables the magnitude of the free exciton and bound exciton lines to be measured. However at concentrations >10^{15} cm^{-3} this becomes increasingly difficult as the bound exciton luminescence increases at the expense of the free exciton. If the temperature is increased to 20 K the population of bound excitons decreases due to dissociation of the exciton from the impurity (the binding energy of the exciton to the neutral impurity is less than the binding energy between the hole and electron).

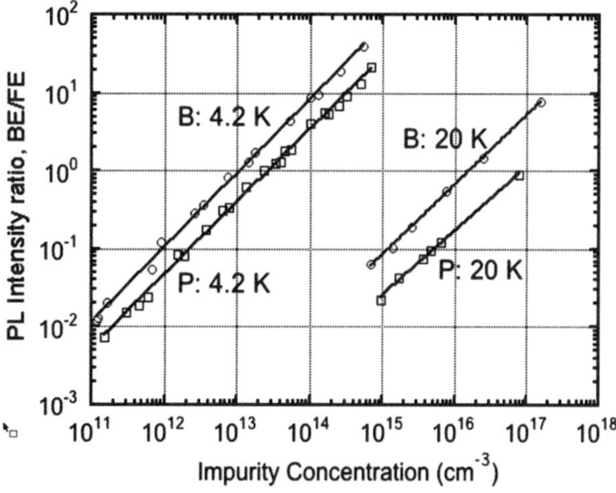

Fig. 3.20 The dependence of the intensity ratio of the bound exciton to the free exciton as a function of doping for the cases of boron and for phosphorus. The lower concentrations are measured at 4.2 K while the higher concentrations must be measured at higher temperature (20 K) in order to enhance the free exciton emission (Reproduced with permission from [55])

Low temperature photoluminescence has played an important role in detecting and analysing defects deep in the silicon band gap particularly transition metals. A very comprehensive review was undertaken by Davies [60] and more recently deeper understanding of the behaviour of some metals particularly copper has been achieved by isotope studies [61].

Structural defects, and in particular dislocations give rise to features in the photoluminescence spectrum. These produce very characteristic D lines (D1–D6) which have been shown to be associated with particular types of dislocation [62]. Recent work on dislocation-related luminescence both at low temperature and room temperature has proved to be particularly valuable in the analysis of solar silicon. The ability to see dislocation related luminescence at room temperature has enabled mapping of extended defects to be undertaken using photoluminescence [63].

Room temperature photoluminescence is also widely used as a technique to map the carrier lifetime of silicon wafers and ingots, an application which has been discussed in Sect. 3.4.5 of this chapter.

3.5.3 Raman Spectroscopy

Raman spectroscopy is used primarily for measuring localised strain in the semiconductor field although it has many other capabilities related to crystallographic orientation, doping and composition of both bulk and thin film materials and devices. These capabilities related to silicon have been reviewed in detail recently [64, 65].

The technique consists of analysing scattered light from a semiconductor using a monochromatic source. The majority of the scattered light is of the same wavelength as the source (Raleigh scattering) but some light interacts with the lattice phonons to produce light of lower energy (Stokes-shifted) or of higher energy (anti-Stokes-shifted). The interaction of interest is with optical phonons which is the Raman effect but shifts can also be observed due to acoustic phonons (Brillouin scattering). The phonon energies are quite small, the energy of an optical phonon in silicon is $\sim$64 meV and acoustic phonons much less. Inelastic scattering is experienced by only a very small proportion of incident photons (somewhere between 1 in 10^6 and 1 in 10^{11}). As a result the Raman signal is very weak and very close to the wavelength of the incident light.

This places considerable demands on the instrumentation required for these measurements requiring narrow linewidth lasers with stable modes, double or triple monochromators and state of the art filters. In the last decade tremendous progress has been made and today several very high performance instruments are available. The wavelength of the laser can be changed to obtain different penetration depths and the laser can be focused using microscope objectives giving a spatial resolution $<1\ \mu$m with a 633 nm laser.

Without stress the Raman peaks due to the longitudinal and two transverse optical phonons of Si are degenerate resulting in a single Raman shift of 520 cm^{-1}. With stress the degeneracy is lifted. Tensile stress reduces the Raman shift and with compressive stress it is greater. The precise values depend on crystal orientation but as a guide a biaxial stress of ~250 MPa in the (100) plane will give a change in shift of 1 cm^{-1}.

The intensity of the Raman signal is dependent on the crystal orientation. This is very valuable in analysing multi-crystalline solar silicon where a map of intensity reveals grain boundaries and if the orientation is derived enables more precise stress analysis to be performed [66].

High doping produces measurable effects in the Raman line-shape due to Fano resonances. Quite precise quantitative measures of doping at high levels ($>10^{18}$ cm^{-3}) can be obtained by fitting the line-shape [67]. Figure 3.21, taken from Ref. [64], shows Raman images recorded using the techniques described above.

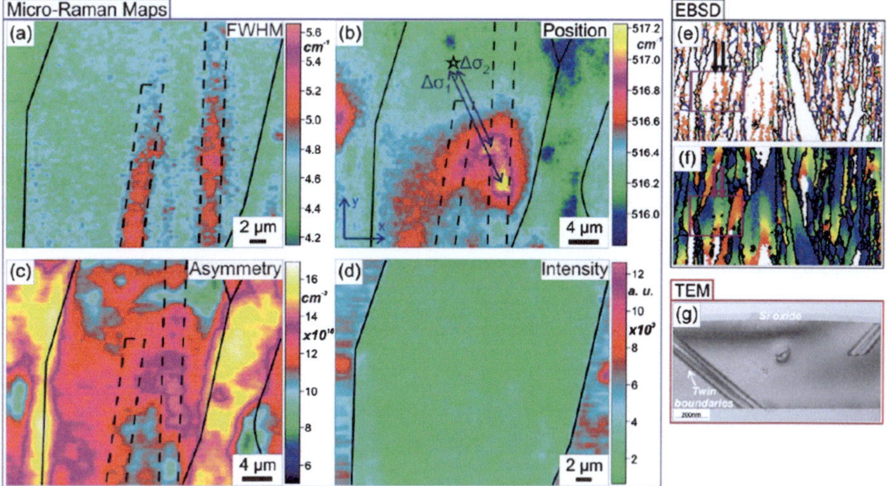

Fig. 3.21 Micro-Raman maps of a laser crystallized silicon seed layer of a thin film solar cell on glass (110 nm thick, nominal boron doping of 2.1×10^{19} cm^{-3}) obtained from the fitting of the first-order Raman spectra of silicon: (**a**) peak FWHM – defect density map, (**b**) peak position – internal stress map with two lateral stress gradients $\Delta\sigma 1 = 227$, $\Delta\sigma 2 = 197 \pm 12$ MPa (von Mises stresses), (**c**) peak asymmetry – doping map, (**d**) peak intensity – grain orientation map. EBSD maps: (**e**) grain boundary map including high angle GBs (*black lines*), low angle GBs (*orange lines*), Σ3 GBs (*blue lines*), Σ9 GBs (*green lines*), (**f**) Inter-granular mis-orientation gradient map. The two *vertical arrows* indicate low angle GBs corresponding to the areas delineating by *dashed lines* in (**a**, **b** and **c**). (**g**) TEM cross-section image (Reproduced with permission from Sarau et al. [64])

3.6 Characterization of Defects in Electronic Grade Silicon

3.6.1 General Requirements in Relation to Device Requirements

The term "electronic grade silicon" is generally used to identify silicon employed in the microelectronics industry, specifically excluding solar photo-voltaic applications. It includes silicon for extremely scaled integrated circuits with devices of nano-metric dimensions through to large power devices with dimensions of several centimeters. The common factor is a requirement for very low concentrations of defects including structural defects and unwanted impurities. As discussed previously in this chapter, in high quality electronic grade silicon material the concentration of unwanted impurities is extremely low, e.g. $<10^{10}$ cm^{-3}. Indeed they are often undetectable by easily available techniques. This is true for material grown by Czochralski techniques and the Float Zone material used for large power devices which has an extremely uniform distribution of intentionally added dopants. In addition the material has a near perfect crystal lattice with each atom aligned in a cubic structure. The control of intrinsic defects during Czochralski growth is described in Chap. 4 and more general aspects of defect engineering in Chap. 9 of this book.

The end result is that the high quality electronic grade material delivered by manufacturers has a very low concentration of defects. However during processing some defects are generated by the displacement of atoms caused by ion implantation or excessive stress. In addition impurities can be introduced as contamination at some technological stages. It is these processing induced defects which the techniques described in this chapter are most valuable in detecting and analyzing.

3.6.2 Contamination During Silicon Processing

The fabrication of any silicon electronic devices and solar cells inevitably has some processing steps which are carried out at relatively high temperatures in the region around 1000 °C. These include epitaxial growth, oxidation of the surface, diffusion of impurities, ion implantation followed by annealing for activation of the implanted atoms, etc. The cleanliness of the wafer before it is heated and also the heating system is of paramount importance for these technological steps. Careless handling of the processed wafers results in their contamination with metal impurities, and, further, in degradation of characteristics of manufactured devices. A few examples of such contamination can be seen in Fig. 3.22, which shows lifetime maps of the wafers with the defect areas originating from unclean processing.

Electronic signatures and optical fingerprints of the main (Fe, Cu, Ni, Co) and also less common contaminants in silicon device technology have been identified in extensive studies with the use of DLTS and photoluminescence and can be found in

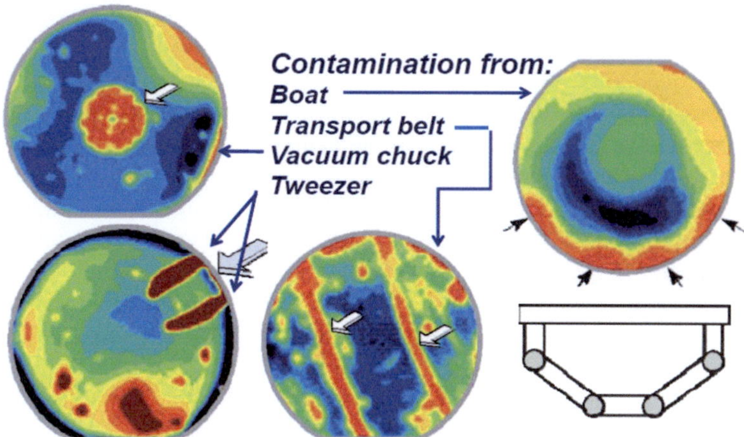

Fig. 3.22 Lifetime maps of some heat-treated silicon wafers with defects areas, which originate from careless wafer handling and unclean equipment (Courtesy of Semilab)

review papers [68, 69] and a comprehensive book [70] on the problem. Some recipes for the suppression of processing induced defects were proposed and unresolved problems discussed in Ref. [70].

3.6.3 Ion Implantation and Radiation Damage

Defects induced in crystalline silicon by electron irradiation and ion implantation have been intensively studied starting from 1950s and are still the object of research. The main reason for this is that both ion implantation and electron irradiation have been intensively applied in silicon technology. In particular, ion implantation is the most widely used technique for dopant introduction into many modern silicon electronic devices.

Much of our understanding of ion implantation defects in silicon and their annealing kinetics is based on studies of electron irradiated silicon. A high energy electron displaces an atom in the semiconductor matrix to produce a vacancy and an interstitial as a Frenkel pair. It was found that both the vacancy (V) and the self-interstitial (I_{Si}) are sufficiently mobile in silicon at room temperature to diffuse in the lattice and react with themselves and with other species so that additional defect reactions involving the formation of complexes also occur. The most probable reactions for the vacancy and self-interstitial in Si have been identified and structure, electronic properties and thermal stability of the resulting complexes have been determined [71]. Particularly, it has been found that in Czochralski-grown silicon interstitial oxygen atoms are very effective sinks for mobile vacancies. The vacancy-oxygen complex was first identified in an EPR study of electron-irradiated Cz-grown

Si crystals [72], and its DLTS electronic signatures and LVM-related absorption lines were then determined [73, 74]. For the silicon self-interstitial the most effective sinks were found to be substitutional carbon and boron atoms and the reactions to occur through the so-called Watkins displacement mechanism, which consists of the displacement of C_s and B_s atoms into interstitial positions by I_{si} [75]. Electronic signatures and LVMs of the C_i and B_i centres have been identified [76]. It should be noted that both the C_i and B_i defects are mobile at temperatures slightly above 300 K and can interact with other defects in the Si lattice (particularly, with oxygen), so creating complexes of the next generation (e.g., C_iO_i, B_iO_i, etc.).

However there are very substantial differences between the point defects introduced by electron damage and the defects commonly seen in ion implanted silicon. As the mass and energy of the implanted species increases, a more complex population of defects is created. There is clear evidence that this is due to clusters of interstitials and/or vacancies depending on the depth below the surface relative to the implant range. Many of the vacancies and interstitials recombine to reconstitute the lattice, others react with each other to form clusters of vacancies or clusters of interstitials. Such clusters result in relatively stable entities which may dissociate during the annealing process used to activate the implanted dopants. The released point defects can result in substantial changes in the diffusivity of the dopants. By far the best known example is the enhancement of boron diffusivity in silicon by excess self-interstitials in which the diffusivity is increased by several orders of magnitude.

The generation of intrinsic defects is very non-uniform throughout the ion path, the majority of defects being created near the end of range. Redistribution of the defects at the implant temperature results in a vacancy rich region near the surface and an interstitial rich region just beyond the concentration peak of the implanted species. On annealing, particularly under oxidising conditions where the surface cannot act as a sink for the interstitials, extrinsic stacking faults with their bounding Frank partial dislocation loops can be formed at the end of the implant range. These accommodate the excess self-interstitials locally as an extra plane of atoms. Some of the defects react with impurities (dopants, carbon, oxygen and hydrogen) during the anneal process. However the majority of these defects can be removed from the active region of the device provided the device design will withstand a sufficient thermal budget.

3.7 Characterization of Defects in Solar Grade Silicon

3.7.1 Solar Grade Silicon and Its Defects

Today's PV market is dominated by cells made from bulk crystalline or multi-crystalline silicon, which account for $\sim$90 % of production. Silicon's abundance, chemical stability, density, band gap, and non-toxic nature mean that it is certain to play a leading role in at least the medium term ($>$10 years).

The last decade has witnessed significant progress in silicon photovoltaics. Among the many factors involved in this are innovative cell design, the use of low cost materials and advances in power management electronics. In particular, the cost of silicon in the cell, which used to be ~50 % of the cell value, has been reduced by using thinner slices and moving away from electronic grade single crystal material to less pure "solar grade" silicon, which is mainly cast in multi-crystalline blocks. Although cells produced from such material have lower conversion efficiency than those from electronic grade silicon, a cost reduction per kWh delivered is achieved because the reduction in efficiency is more than compensated for by the lower price of the silicon.

The term "solar grade silicon" is not uniquely defined. It has been used to describe block cast multi-crystalline material from electronic grade feedstock, Cz silicon grown from silicon scrap feedstock, or silicon grown from feedstock that has been produced by refining processes which are less energy consuming (and less effective) than the Siemens process used for the production of electronic grade silicon feedstock. These include fluidized bed methods and in the extreme case improved metallurgical (IMG) silicon. These materials contain higher concentrations of metals than electronic grade material and in some cases higher carbon content. This can originate from upgraded metallurgical extraction methods or from casting highly refined silicon in graphite crucibles. Such material can contain concentrations of carbon in solution around the equilibrium solubility at the melting point (3.5×10^{17} cm^{-3}) and SiC particles.

In cast, multi-crystalline silicon the presence of grain boundaries and dislocations makes the material inhomogeneous in terms of its electronic properties. An example of this inhomogeneity can be seen in Fig. 3.23 which shows a very strong variation in lifetime map of a multi-crystalline p-type silicon wafer. Comparison with a map of the grains in this slice shows that the excess recombination is not associated with the grain boundaries but rather with regions within specific grains, which we associate with decorated dislocation clusters. A lifetime map of present day state of the art electronic grade single crystal material (Cz or FZ) shows a variation of minority carrier lifetime of ~1 % with edge exclusion.

The combination of the structural defects, metallic impurities and precipitates of various sizes make defect reactions extremely complex and the material very difficult to assess by conventional techniques. The pragmatic approach of the solar industry is to measure average properties assigning a single lifetime to the whole slice (~5 inch square). In many cases, this works quite well at least in terms of deciding if material is worth processing into cells but such averaging is not adequate in trying to understand the localized recombination paths and improving the material.

Originally, it was thought that grain boundaries acted as important recombination centers in mc-Si but it now seems that this is not often the case. In recent material, the grains are very much larger than the diffusion length in equivalent single crystal

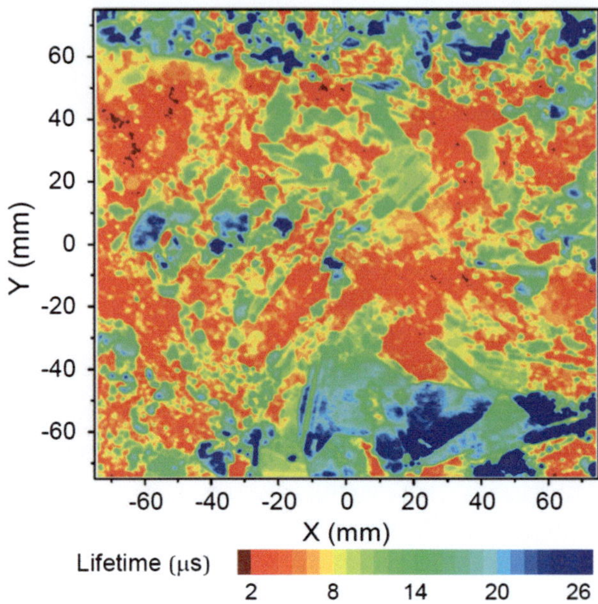

Fig. 3.23 Map of the minority carrier lifetime of a multi-crystalline p-type silicon wafer. Before measurements the wafer was etched in a KOH solution. The map was taken using a Semilab WT-2000 PVN in the μPCD mode with a 0.5 mm diameter 904 nm laser probe. The wafer surface was passivated with an iodine-ethanol solution

material. In addition, the grain boundaries are mostly parallel to the direction of carrier flow in the cell. These factors result in recombination at grain boundaries being a relatively small perturbation in terms of cell efficiency or minority carrier lifetime whereas decorated dislocations and dislocation clusters are much more significant as will be discussed later.

In the following we will consider solar silicon in two stages. Firstly, experimental results on single crystal materials containing metallic impurities will be reviewed. Passivation of defects with hydrogen will also be briefly touched and some results on the defects responsible for the light-induced degradation of solar cell produced on silicon material doped with boron and oxygen will be discussed.

We then consider multi-crystalline solar silicon with similar contamination and explore the role of grain boundaries, dislocations and precipitates in the minority carrier recombination processes. This necessitates localized measurements of specific grains and grain boundary regions. In combination with hydrogen passivation this provides us with a basis for understanding some of the efficiency limitations, which occur in this very complex material system.

3.7.2 *Defects in Single Crystal Silicon Material for Photovoltaic Applications*

3.7.2.1 Metallic Impurities in Single Crystal Silicon for Photovoltaic Applications

In general, transition metals are common contaminates in lower grade silicon feedstock. The majority of the metals act as effective recombination centers if present in the finished solar cell on specific sites. So, metallic impurities are a serious problem for silicon grown from impure feedstock as a solar material [77].

There is a vast amount of literature on metals in silicon with an emphasis on the transition metals, some of these publications have been focused on photovoltaic problems. In a comprehensive early experiment, known as the Westinghouse study [78], contaminants were added to a Czochralski melt and small ingots were pulled of both n and p-type material. Chemical concentrations of impurities in slices cut from the ingots were measured by means of neutron activation analysis and spark source mass spectrometry. Concentrations of electrically active defects were measured in some slices using DLTS. Solar cells were fabricated (both n^+p and p^+n) with and without an antireflection layer and the efficiency measured. In all approximately 200 ingots were grown and processed into solar cells. Although there was no intentional hydrogen passivation the contacts were sintered at 300–550 °C in hydrogen.

The impact of various impurities is shown in Fig. 3.24 in terms of the effect of the chemical concentration of metals in the slices on the solar cell efficiency for the case of a p-type base. It is important to emphasize that these results convolute many factors apart from the Shockley-Read-Hall (SRH) properties of the defect state associated with the metal. Not all the metal will be electrically active in the slice, the n^+ layer diffusion will tend to getter metals with high diffusivity and growth of precipitates may occur during cell processing. The details of the observed effects of different impurities on solar cell efficiency were discussed in the original study [78] and further works [79, 80] and are reasonably well understood.

3.7.2.2 Passivation of Defects with Hydrogen in Silicon Photovoltaics

Passivation of recombination (both surface and via defects in the bulk) is a crucial part of silicon solar cell manufacture. It is used in conjunction with gettering to reduce the recombination paths competing with the extraction of carriers for power generation. In most single crystal and mc-Si cells gettering is done as part of the n^+ front face emitter diffusion and passivation is done by the deposition of hydrogenated silicon nitride on the n^+ emitter of a cell. Almost all cells are made with a p-type base but the hydrogenation is similar for n based cells. The nitride is multipurpose acting as the antireflection layer as well as a source of hydrogen for passivation. Silicon nitride deposited by plasma assisted CVD from silane and ammonia at ~350 °C contains about 25 at% hydrogen [81]. On annealing, hydrogen

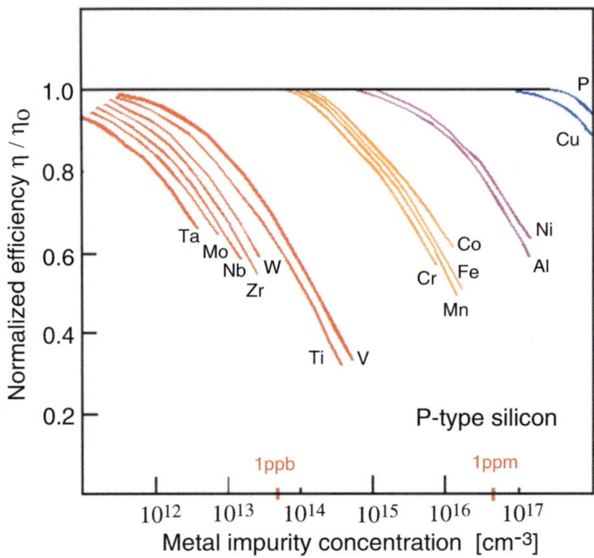

Fig. 3.24 The effect of metallic contaminants on solar cell efficiency. The impurity concentration is that in the slice prior to cell processing (The data are taken from the Westinghouse study [78] and re-plotted by Pizzini [79])

is freed and diffuses into the Si where it passivates some recombination centres and increases the minority carrier lifetime considerably.

The classic role of hydrogen as a passivator is to saturate dangling bonds at metal-oxide-semiconductor (MOS) interfaces. Actually the great success of silicon MOS technology has been achieved due to excellent quality of the Si-SiO_2 interfaces passivated with hydrogen. The hydrogen bond to silicon at the surface is not very strong (the binding energy is about 2.5 eV) but is enough for long term retention at device operating temperatures although recombination events can perturb this [82]. In the silicon bulk the isolated hydrogen atom has a negative U character; it can exist in positive and negative charge states but the neutral state is unstable [83]. Hence it will be attracted to donor like states in n-type and acceptor like states in p-type. The bonding energy of H to boron is 1.28 eV and to phosphorus 1.20 eV. Transition metals can bond multiple hydrogen atoms. An example which has been investigated in detail is gold which has electrically active states for the case of 1, 2, and 4 hydrogen atoms. AuH, AuH_2 have been observed with the use of Laplace DLTS [84], AuH_3 is inactive (passivated) while AuH_4 is calculated to have electrical activity but has not been observed. The bonding of the hydrogen is weak and the dissociation processes can be observed with LDLTS between 200 °C and 250 °C as hydrogen is lost [84]. Data on hydrogenation of the technologically important 3d transition metals is less detailed but seems to follow a similar pattern, i.e. passivation by trapping of single or multiple hydrogen atoms which are released at ~200 °C [19, 85]. Hydrogen is quite mobile at these temperatures and so the passivation is fragile unless there is a substantial reservoir of hydrogen available to recombine with the defects on cooling. In silicon solar cells PCVD SiN provides such a reservoir under appropriate thermal conditions but not sufficient to passivate all defects throughout the entire thermal history of the cell. Recent work on hydrogen passivation seems to indicate that stable passivation is possible by manipulating the charge state of hydrogen and so reducing recombination due to states which have not been removed through other routes [86, 87].

3.7.2.3 Light-Induced Degradation of Si Solar Cells Doped with Boron and Oxygen

Perhaps the complex which has gained most attention in recent years is a defect which limits the efficiency of solar cells produced from Cz-grown p-type silicon and is associated with a complex incorporating boron and oxygen atoms. The efficiency of solar cells fabricated from boron-doped Cz-Si was found to degrade when excess electrons are created by illumination [88, 89]. From the carrier lifetime spectroscopy an energy level at about 0.41 eV from the conduction band and with $\sigma_n/\sigma_p = 9.3$ was assigned to the defect responsible for the degradation [89, 90].

Further, it was suggested that the defect could be B_sO_{2i}. The basis for the structural assignment was a dependence of the concentration of the assumed recombination centre (derived from the lifetime degradation) on $[O_i]^2$ and $[B_s]$ [89]. The formation mechanism was suggested to be the migration of the oxygen dimer to B_s via an enhanced diffusion mechanism (the dimer is normally immobile at room temperature) [91]. This enhanced diffusion was proposed as being due to a Bourgoin-Corbett mechanism under minority carrier injection [92, 93]. This occurs in other defect systems but necessitates in this case a change in the charge state of the dimer upon capture of a minority carrier, i.e. two charge states of the dimer must exist.

Two important studies have subsequently shown that the defect in question is unlikely to be B_sO_{2i}. Firstly it was observed that in compensated material the degradation depended on the carrier concentration rather than the boron concentration [94]. Secondly a study of the optical absorption spectra of the oxygen dimer showed that only one charge state exists making the Bourgoin-Corbett mechanism impossible in the case of the oxygen dimer [95].

An alternative model has been proposed by Voronkov and Falster in which the defect is B_iO_{2i} formed during ingot growth [96]. The defect is then activated to the recombination active configuration from the latent form with weak recombination activity by the injection of minority carriers. It appeared, however, that the "B_iO_{2i}" model could not explain a number of experimental findings including evidence that the defect density is not fixed during cooling [97] and quantitative disagreement with observations in boron compensated n-type material [98].

A crucially important dilemma is that despite many attempts no SRH defect has been observed by DLTS or related techniques which fits the observed lifetime degradation behavior. It is evident that despite an immense amount of data and efforts by many groups throughout the world we do not as yet have a fundamental understanding of a process which was first observed more than 40 years ago and is now of very considerable commercial importance.

3.7.3 Defects in Multicrystalline Silicon

The situation with defect characterization in multicrystalline silicon is much more complex than in single crystal material. The mc-Si ingots contain grain boundaries with various degrees of mis-orientation between the grains. The multicrystalline material also contains dislocations and dislocation clusters within the grains. Further, mc-Si often has metallic contamination which originates either from the crucible material used or from the use of less pure feedstock.

Neutron activation analysis (NAA) and inductively coupled plasma mass spectrometry (ICP-MS) have been used by several researchers to determine the level of contamination with transition metals in mc-Si. Istratov et al. [99] found from an NAA study that the dominant metal impurities were Fe (6×10^{14} cm^{-3} to 1.5×10^{16} cm^{-3}), Ni (up to 1.8×10^{15} cm^{-3}), Co (1.7×10^{12} cm^{-3} to 9.7×10^{13} cm^{-3}), Mo (6.4×10^{12} cm^{-3} to 4.6×10^{13} cm^{-3}), and Cr (1.7×10^{12} cm^{-3} to 1.8×10^{15} cm^{-3}). Shabani et al. [100] performed ICP-MS analysis of metallic impurities in many samples of cast multicrystalline solar silicon from several vendors. They found that the major metal impurities in cast solar grade m-Si are Al, Fe, Ni, and Cu with concentrations in the range from 10^{12} to 10^{15} cm^{-3}. Impurities with the medium levels of contamination from 10^{10} to 10^{13} cm^{-3} were Ti, Cr, Mn, and Co.

Macdonald et al. [101] investigated distribution of metals in different regions of cast ingots. The highest concentrations of Fe, Co, and Cu were found near the top of the ingots. These is because metals accumulate in the melt during solidification because of the high segregation coefficient between molten and solid silicon. This produces high concentrations of the metals near the top of the solidifying ingot. Further, the concentrations were found to be high near the bottom of the ingots due to diffusion from the crucible after crystallization has occurred. In general concentrations similar to those values mentioned above were found. Zn, Cr and Ni were found to be more uniformly distributed in the ingots (2×10^{12} to 3×10^{13} cm^{-3}).

An investigation of the effects of specific metals on the electrical behaviour of mc-Si was undertaken by Pizzini et al. [102]. In this study Cz ingots were grown from intentionally contaminated feedstocks. Necking the ingots after seeding was omitted, therefore multi-crystalline growth occurred. The overall conclusion of the work was that the presence of grain boundaries and dislocations did not enhance the detrimental effect of the metallic impurities beyond that expected for similar densities of the metallic impurities in single crystal silicon.

Recently a study of the effects of metallic impurities on performance of solar cells produced from cast mc-Si was undertaken under the EU project "CrystalClear"

[103]. Materials for the work were produced by directional solidification (Bridgeman technique) with ingots of 250 mm diameter and 110 mm height. High purity feedstock and furnace components were used so without added contaminants the material had a lifetime of 60 μs. Solar cells with a p-type base of area 156 cm^2 were made using a classic industrial process with silicon nitride passivating/anti reflection layer, 185 μm thick with aluminium back surface field. The uncontaminated efficiency was 15.5 %. The effects of Fe, Cr, Ni, Ti and Cu on the efficiency were investigated and the Fe study has been published in considerable detail [104].

A known amount of the impurity was added to the melt and the effect on the cell efficiency determined. In the iron case the electrically active iron content was measured in the slice. In terms of the concentration of impurity added to the melt it was found that 8 ppm wt of Cr, 11 ppm wt of Fe, 0.1 ppm wt of Ti, 13 ppm wt of Ni and 8 ppm wt of Cu are equivalents in terms of their effect on solar-cell performance giving a degradation of about 2 %. Taking into account the segregation coefficients the rank order is nearly the same as that obtained for the single crystalline silicon [78] and the absolute concentration added to the melt for a similar degradation is of the same order. This is rather surprising when we consider that the metal distribution would be expected to be non-uniform across the slice and that far fewer atoms might be expected to be in solution due to accumulation at grain boundaries and dislocations.

On the other hand maybe this is offset by localised recombination. Sopori et al. have studied such localisation in cast mc-Si [105] and find that regions with clusters of dislocations exhibit very high recombination. In cases where the same number of dislocations are spread evenly across the cell, recombination is much reduced. This was attributed to gettering and hydrogenation being much more effective in the case of distributed dislocations. This is consistent with our work on decoration of dislocations.

Our observation is that grain boundaries are not effective recombination centres after hydrogenation even in the presence of metals and presumably after decoration. As discussed previously in the case of large grains it is unlikely that grain boundaries will have much effect on the carrier lifetime simply because the grain size is very much larger than the diffusion length. In addition in the solar cell the minority carrier flux is parallel to the grain boundaries so will not have much effect on the photocurrent.

We have undertaken DLTS studies on solar grade mc-Si with grains in the size range 1–15 mm. The technique we have used is shown in Fig. 3.25 where semi-transparent Schottky diodes have been deposited on mc-Si. It is possible to see the grain boundaries in the optical micrographs or by mapping the magnitude of the Raman scattering. The lifetime distribution around the diodes can also be seen.

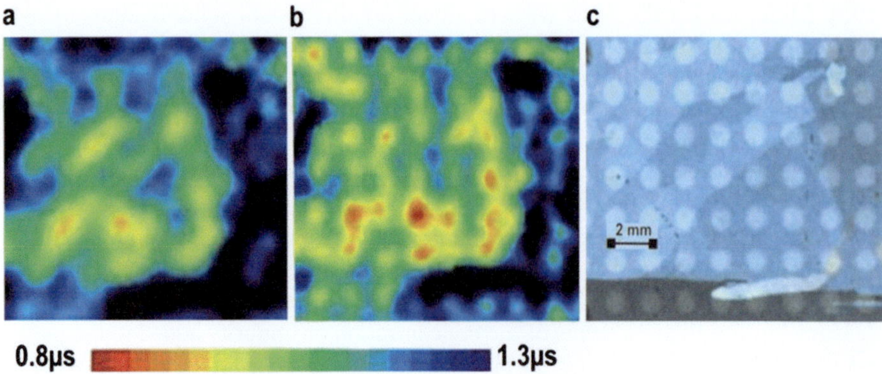

Fig. 3.25 The technique for combining DLTS measurements with minority carrier lifetime maps on a multi-crystalline p-type silicon wafer. Shown are: (**a**) μPCD lifetime map of the original slice, (**b**) μPCD lifetime map of the same region with semitransparent Schottky diodes deposited, (**c**) an optical micrograph of the same slice with the diodes

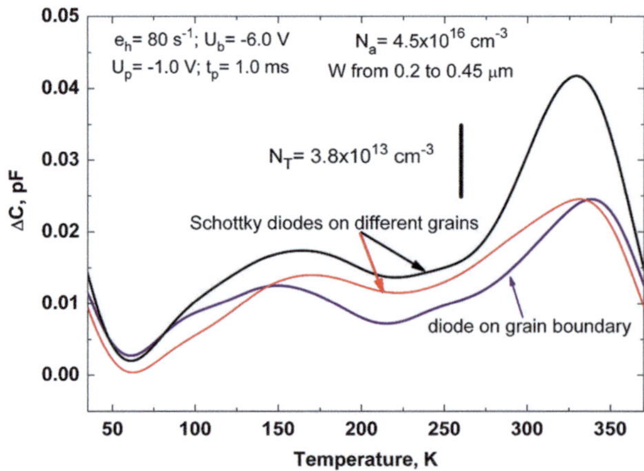

Fig. 3.26 DLTS spectra of three regions of a mc-Si wafer using samples of the type shown in Fig. 3.25

In some slices the DLTS signatures from diodes on grain boundaries are quite different to those relating to defects in the centre of the grains. In particular negative DLTS peaks indicating emission of minority carriers are often seen at the grain boundaries [106]. However, in many cases the DLTS spectra from different regions of the same slice show similarities. Figure 3.26 shows the case of a slice containing a broad distribution of dislocation clusters. The peak on the right at around 330 K was greatly reduced by hydrogenation and showed the logarithmic filling behaviour of an extended defect. The family of peaks in the range 100–150 K was essentially unaffected by hydrogenation and had point defect like carrier capture. These defects have all the characteristics of powerful recombination centres.

In general metallic precipitates show very limited recombination activity compared to the uniform distribution of the same number of atoms and hence gettering to extended defects even in the active region of a solar cell is generally regarded as beneficial provided that the precipitate does not provide a significant leakage path. These issues are complex and are revisited in Chap. 9 of this book.

References

1. Shockley, W., Read, W.T.: Phys. Rev. **87**, 835 (1952)
2. Hall, R.N.: Phys. Rev. **87**, 387 (1952)
3. Schroder, D.K.: Semiconductor Material and Device Characterization, 3rd edn. Wiley, Hoboken (2005)
4. Dziewior, J., Schmid, W.: Appl. Phys. Lett. **31**, 346 (1977)
5. Richter, A., Glunz, S.W., Werner, F., Schmidt, J., Cuevas, A.: Phys. Rev. B. **86**, 165202 (2012)
6. Hangleiter, A.: Phys. Rev. B. **35**, 9149 (1987)
7. Ambridge, T., Faktor, M.M.: J. Appl. Electrochem. **5**, 319 (1975)
8. Sah, C.T., Forbes, L., Rosier, L.C., Tasch, A.F.: Solid State Electron. **13**, 759 (1970)
9. Lang, D.J.: J. Appl. Phys. **45**, 3023 (1974)
10. Istratov, A.A.: J. Appl. Phys. **82**, 2965 (1997)
11. Peaker, A.R., Markevich, V.P., Hawkins, I.D., Hamilton, B., Bonde Nielsen, K., Gościński, K.: Phys. B. Condens. Matter. **407**, 3026 (2012)
12. Dobaczewski, L., Kaczor, P., Hawkins, I.D., Peaker, A.R.: J. Appl. Phys. **76**, 194 (1994)
13. Dobaczewski, L., Peaker, A.R., Bonde Nielsen, K.: J. Appl. Phys. **96**, 4689 (2004)
14. Hamilton, B., Peaker, A.R., Wight, D.R.: J. Appl. Phys. **50**, 6373 (1979)
15. Brunwin, R., Hamilton, B., Jordan, P., Peaker, A.R.: Electron. Lett. **15**, 349 (1979)
16. Lemke, H.: Phys. Status Solidi A. **64**, 215 (1981)
17. Kimerling, L.C., Benton, J.L.: Phys. B. **116**, 297 (1983)
18. Istratov, A.A., Hieslmair, H., Weber, E.R.: Appl. Phys. A. **69**, 13 (1999)
19. Leonard, S., Markevich, V.P., Peaker, A.R., Hamilton, B., Murphy, J.D.: Appl. Phys. Lett. **107**, 032103 (2015)
20. Zoth, G., Bergholz, W.: J. Appl. Phys. **67**, 6764 (1990)
21. Kang, J.S., Schroder, D.K.: Phys. Status Solidi (a) **89**, 13 (1985)
22. Stevenson, D.T., Keyes, R.J.: J. Appl. Phys. **26**, 190 (1955)
23. Kunst, M., Beck, G.: J. Appl. Phys. **60**, 3558 (1986)
24. Horanyi, T.S., Pavelka, T., Tutto, P.: Appl. Surf. Sci. **63**, 306 (1993)
25. Batra, N., Kumar, S., Sharma, M., Srivastava, S.K., Sharma, P., Singh, P.K.: Sol. Energy Mater. Sol. Cells **100**, 43 (2011)
26. Chabbra, B., Bowden, S., Opila, R.L., Honsberg, C.B.: Appl. Phys. Lett. **96**, 063502 (2010)
27. SEMI MF1535-0707. Test method for carrier recombination lifetime in silicon wafers by noncontact measurement of photoconductivity decay by microwave reflectance, (2007) from http://www.semi.org
28. Johnson, E.O., Johnson, L.G.: J. Appl. Phys. **28**, 1349 (1957)
29. Sinton, R.A., Cuevas, A.: Appl. Phys. Lett. **69**, 2510 (1996)
30. Macdonald, D., Sinton, R.A., Cuevas, A.: J. Appl. Phys. **89**, 2772 (2001)
31. McIntosh, K.R., Paudyal, B.B., Macdonald, D.: J. Appl. Phys. **104**, 084503 (2008)
32. Isenberg, J., Riepe, S., Glunz, S.W., Warta, W.: J. Appl. Phys. **93**, 4268 (2003)
33. Schubert, M.C., Isenberg, J., Warta, W.: J. Appl. Phys. **94**, 4139 (2003)
34. Giesecke, J.A., Schubert, M.C., Walter, D., Warta, W.: Appl. Phys. Lett. **97**, 092109 (2010)
35. Macdonald, D., Tan, J., Trupke, T.: J. Appl. Phys. **103**, 073710 (2008)
36. Macdonald, D., Deenapanray, P.N.K., Diez, S.: J. Appl. Phys. **96**, 3687 (2004)

37. Kohn, W.: Phys. Rev. **98**, 915 (1955)
38. Kohn, W.: Solid State Phys. **5**, 257 (1957)
39. Stoneham, A.M.: Theory of Defects in Solis. Oxford University Press, London (1975)
40. Ramdas, A.K., Rodriguez, S.: Rep. Prog. Phys. **44**, 1297 (1981)
41. Markevich, V.P., Suezawa, M., Sumino, K., Murin, L.I.: J. Appl. Phys. **76**, 7347 (1994)
42. Newman, R.C., Ashwin, R.J., Pritchard, R.E., Tucker, J.H.: Phys. Status Solidi (b) **210**, 519 (1998)
43. Pajot, B.: Optical Absorption of Impurities and Defects in Semiconducting Crystals: I. Hydrogen-Like Centers. Springer Series in Solid State Sciences. Springer, Berlin/Heidelberg (2010)
44. Dawber, P.G., Elliot, R.J.: Proc. Phys. Soc. **81**, 453 (1963)
45. Newman, R.C.: Infrared Studies of Crystal Defects. Taylor & Francis Ltd., London (1973)
46. Stavola, M.: Identification of defects in semiconductors. In: Stavola, M. (ed.) Semiconductors and Semimetals, vol. 51B, p. 153. Academic, San Diego (1999)
47. McCluskey, M.D.: J. Appl. Phys. **87**, 3593 (2000)
48. Pajot, B.: Optical Absorption of Impurities and Defects in Semiconducting Crystals: Electronic Absorption of Deep Centres and Vibrational Spectra. Springer Series in Solid State Sciences. Springer, Berlin/Heidelberg (2013)
49. Lindström, J.L., Murin, L.I., Markevich, V.P., Hallberg, T., Svensson, B.G.: Phys. B. **273–274**, 291 (1999)
50. Baghdadi, A., Bullis, W.M., Croarkin, M.C., Li, Y.-Z., Scace, R.I., Series, R.W., Stallhofer, P., Watanabe, M.: J. Electrochem. Soc. **136**, 2015 (1989)
51. Bullis, W.M.: Oxygen in silicon. In: Shimura, F. (ed.) Semiconductors and Semimetals, vol. 42, p. 95. Academic, San Diego (1994)
52. Regolini, J.L., Stoquert, J.P., Ganter, C., Siffert, P.: J. Electrochem. Soc. **133**, 2165 (1986)
53. ASTM Standard F 1391–1393, Standard Test Method for Substitutional Atomic Carbon Content of Silicon by Infrared Absorption. 1996 Annual Book of ASTM Standards. American Society for Testing and Materials, West Conshohocken (1996)
54. ASTM Standard F 1188–1193a, Standard Test Method for Interstitial Atomic Oxygen Content of Silicon by Infrared Absorption. 1996 Annual Book of ASTM Standards. American Society for Testing and Materials, West Conshohocken (1996)
55. Tajima, M.: Appl. Phys. Lett. **32**, 719 (1978) and Awai, T., Tajima, M., Ogura, A.: Phys. Status Solidi C **8**, 792 (2011)
56. Japanese Industrial Standard, JIS H 0615 (1996)
57. Semiconductor Equipment and Materials International, SEMI MF1389–0704 (2004)
58. Haynes, J.R.: Phys. Rev. Lett. **4**, 361 (1960)
59. Tajima, M., Iwai, T., Toyota, H., Binetti, S., Macdonald, D.: J. Appl. Phys. **110**, 043506 (2011)
60. Davies, G.: Phys. Rep. **176**, 83 (1989)
61. Steger, M., Yang, A., Sekiguchi, T., Saeedi, K., Thewalt, M.L.W., Henry, M.O., Johnston, K., Riemann, H., Abrosimov, N.V., Churbanov, M.F., Gusev, A.V., Kaliteevskii, A.K., Godisov, O.N., Becker, P., Pohl, H.-J.: J. Appl. Phys. **110**, 081301 (2011)
62. Shreter, Y.G., Rebane, Y.T., Peaker, A.R.: Phys. Status Solidi (a) **138**, 681 (1993)
63. Tajima, M.: IEEE J. Photovolt. **4**, 1452 (2014). Tajima, M., Iwata, Y., Okayama, F., Toyota, H., Onodera, H., Sekiguchi, T.: J. Appl. Phys. **111**, 113523 (2012)
64. Sarau, G., Bochmann, A., Lewandowska, R., Christiansen, S.: From micro– to macro–Raman Spectroscopy: solar silicon for a case study, chapter 8. In: Dr. Muhammad Akhyar Farrukh (ed.) Advanced Aspects of Spectroscopy. ISBN: 978-953-51-0715-6, InTech (open access) (2012). doi:10.5772/48160. http://www.intechopen.com/books/advanced-aspects-of-spectroscopy/from-micro-to-macro-raman-spectroscopy-solar-silicon-for-a-case-study
65. Sarau, G., Christiansen, S., Lewandowska, R., Roussel, B.: Proceedings 35th IEEE PVSC: 001770–001775 (2010)
66. Becker, M., Scheel, H., Christiansen, S., Strunk, H.P.: J. Appl. Phys. **101**, 063531 (2007)

67. Kunz, T., Hessmann, M.T., Seren, S., Meidel, B., Terheiden, B., Brabec, C.J.: J. Appl. Phys. **113**, 023514 (2013)
68. Istratov, A.A., Weber, E.R.: Appl. Phys. **A66**, 123 (1998)
69. Kitagawa, H.: Solid State Phenom. **71**, 51 (2000)
70. Graff, K.: Metal Impurities in Silicon-Device Fabrication. Springer Series in Materials Science. Springer, Berlin/Heidelberg (2000)
71. Watkins, G.D.: Vacancies and interstitials and their interaction with impurities in c-Si. In: Hull, R. (ed.) Crystalline Silicon. EMIS Data Reviews Series, vol. 20, p. 643. INSPEC, London (1999)
72. Watkins, G.D., Corbett, J.W.: Phys. Rev. **121**, 1001 (1961)
73. Kimerling, L.C.: Defect states in electron bombarded silicon: Capacitance transient analysis. In: Urli, N.B., Corbett, J.W. (eds.) Radiation Effects in Semiconductors 1976, p. 221. Institute of Physics, Bristol (1977)
74. Newman, R.C.: Rep. Prog. Phys. **45**, 1163 (1982)
75. Watkins, G.D.: Materials Science and Technology, vol. 4. VCH, Weinheim (1991). ch. 3
76. Kimerling, L.C., Asom, M.T., Benton, J.L., Drevinsky, P.J., Caefer, C.E.: Mater. Sci. Forum **38–41**, 141 (1989)
77. Libal, J., Novaglia, S., Acciari, M., Binetti, S., Petres, R., Arumughan, J., Kopecek, R., Prokopenko, A.: J. Appl. Phys. **104**, 104507 (2008)
78. Davies Jr, J.R., Rohatgi, A., Hopkins, R.H., Blais, P.D., Rai-Choudhury, P., McCormick, J.R., Mollenkopf, H.C.: IEEE Trans. Electron. Dev. **ED-27**, 677 (1980); Rohatgi, A., Rai-Choudhury, P.: Solar Cells **17**, 119 (1986)
79. Pizzini, S., Acciarri, M., Binetti, S.: Phys. State Solid (a) **202**, 2928 (2005)
80. Peaker, A.R., Markevich, V.P., Hamilton, B., Parada, G., Dudas, A., Pap, A., Don, E., Lim, B., Schmidt, J., Yu, L., Yoon, Y., Rozgonyi, G.: Phys. Status Solidi A. **209**, 1884 (2012)
81. Lanford, W.A., Rand, M.J.: J. Appl. Phys. **49**, 2473 (1978)
82. van de Walle, C.G.: J. Vac. Sci. Tech. **A16**, 1767 (1998)
83. Peaker, A.R., Markevich, V.P., Dobaczewski, L.: Hydrogen-related defects in silicon, germanium and silicon-germanium alloys. In: Fleetwood, D.M., Pantelides, S.T., Schrimpf, R.D. (eds.) Defects in Microelectronic Materials and Devices, p. 27. CRC Press, Boca Raton (2009)
84. Deixler, P., Terry, J., Hawkins, I.D., Evans-Freeman, J.H., Peaker, A.R., Rubaldo, A.L., Maude, D.K., Portal, J.-C., Bonde Nielsen, K., Nylandsted Larsen, A., Mesli, A.: Appl. Phys. Lett. **73**, 3126 (1998)
85. Leonard, S., Markevich, V.P., Peaker, A.R., Hamilton, B.: Appl. Phys. Lett. **103**, 132103 (2013)
86. Hallam, B.J., Hamer, P.G., Wenham, S.R., Abbott, M.D., Sugianto, A., Wenham, A.M., Chan, C.E., Xu, G., Kraiem, J., Degoulange, J., Einhaus, R.: IEEE J. Photovolt. **4**, 88 (2014)
87. Liu, A., Sun, C., Macdonald, D.: J. Appl. Phys. **116**, 194902 (2014)
88. Fisher, H., Pschunder, W.: Proceedings of the 10th IEEE Photovolataic Specialist Conference, p. 404. IEEE, New York (1973)
89. Bothe, K., Schmidt, J.: J. Appl. Phys. **99**, 013701 (2006)
90. Rein, S., Glunz, S.W.: Appl. Phys. Lett. **82**, 1054 (2003)
91. Schmidt, J., Bothe, K.: Phys. Rev. B. **69**, 024107 (2004)
92. Adey, J., Jones, R., Palmer, D.W., Briddon, P.R., Öberg, S.: Phys. Rev. Lett. **93**, 055504 (2004)
93. Du, M.-H., Branz, H.M., Crandall, R.S., Zhang, S.B.: Phys. Rev. Lett. **97**, 256602 (2006)
94. Macdonald, D., Rougieux, F., Cuevas, A., Lim, B., Schmidt, J., Di Sabatino, M., Geerligs, L.J.: J. Appl. Phys. **105**, 093704 (2009)
95. Murin, L.I., Tolkacheva, E.A., Markevich, V.P., Peaker, A.R., Hamilton, B., Monakhov, E., Svensson, B.G., Lindström, J.L., Santos, P., Coutinho, J., Carvalho, A.: Appl. Phys. Lett. **98**, 182101 (2011)
96. Voronkov, V.V., Falster, R.: J. Appl. Phys. **107**, 053509 (2010)

97. Rougieux, F.E., Lim, B., Schmidt, J., Forster, M., Macdonald, D., Cuevas, A.: J. Appl. Phys. **110**, 063708 (2011)
98. Macdonald, D., Liu, A., Cuevas, A., Lim, B., Schmidt, J.: Phys. Status Solidi A. **208**, 559 (2011)
99. Istratov, A.A., Buonassisi, T., McDonald, R.J., Smith, A.R., Schindler, R., Rand, J.A., Kalejs, J.P., Weber, E.R.: J. Appl. Phys. **94**, 6552 (2003)
100. Shabani, M.B., Yamashita, T., Morita, E.: ECS Trans. **16**, 179 (2008)
101. Macdonald, D., Cuevas, A., Kinomura, A., Nakano, Y., Geerligs, L.J.: J. Appl. Phys. **97**, 033523 (2005)
102. Pizzini, S., Bigoni, L., Beghi, M.: J. Electrochem. Soc. **133**, 2363 (1986)
103. Coletti, G., Bronsveld, P.C.P., Hahn, G., Warta, W., Macdonald, D., Ceccaroli, B., Wambach, K., Le Quang, N., Fernandez, J.M.: Adv. Funct. Mater. **21**, 879 (2011)
104. Coletti, G., Kvande, R., Mihailetchi, V.D., Geerligs, L.J., Arnberg, L., Øvrelid, E.J.: J. Appl. Phys. **104**, 104913 (2008)
105. Sopori, B., Rupnowski, P., Shet, S., Budhraja, V., Call, N., Johnston, S., Seacrist, M., Shi, G., Chen, J., Deshpande, A.: Proceedings of the 35th Photovoltaic Specialists Conference, p. 2233. IEEE, Honolulu (2010)
106. Chen, J., Cornagliotti, E., Hieckmann, E., Behrendt, S., Weber, J., Simoen, E., Poortmans, J.: ECS Trans. **33**, 71 (2011)

Chapter 4
Control of Intrinsic Point Defects in Single-Crystal Si and Ge Growth from a Melt

Jan Vanhellemont, Kozo Nakamura, Eiji Kamiyama, and Koji Sueoka

Abstract The so called Voronkov criterion defines a critical value Γ_{crit} of the ratio $\Gamma = v/G$ of the pulling rate v over the thermal gradient G at the crystal-melt/interface of a growing crystal. For $\Gamma > \Gamma_{crit}$, the crystal is vacancy-rich and can contain large vacancy clusters that are detrimental for gate oxide performance and for thin film epitaxial growth. For $\Gamma < \Gamma_{crit}$, the crystal is self-interstitial-rich and in the worst case will contain dislocation clusters. For $\Gamma \approx \Gamma_{crit}$, the crystal is free of grown-in intrinsic point defect clusters and optimal for device processing. Analytical expressions have been derived describing Γ_{crit} as function of intrinsic point defect parameters. The impact of thermal stress σ_{th} at the crystal-melt interface and of crystal doping on Γ_{crit} will be clarified. As σ_{th} increases with increasing crystal diameter, controlling G and v will become a real challenge for the development of future 450 mm diameter, defect free Si crystals. The possible application of the Voronkov criterion for Ge single-crystal growth from a melt will also briefly be discussed. Besides the impact of stress on intrinsic point defect formation energies and diffusivities, DFT calculations also suggest that near the crystal-melt interface, assumed to be stress free, the formation energy of the intrinsic point defects is lower than in the bulk of the crystal. This leads to thermal equilibrium concentrations of intrinsic point defects at the crystal-melt interface that are considerably different from those in the bulk which should be taken into account when applying the Voronkov criterion and also for intrinsic defect engineering in general. The Voronkov criterion was established for a flat interface. During crystal growth the crystal-melt interface is however curved which will have a significant impact on the diffusion of the intrinsic point defects. The impact of this curvature is discussed in detail both theoretically and experimentally.

J. Vanhellemont (✉)
Department of Solid State Sciences, Ghent University, Krijgslaan 281-S1, Gent B-9000, Belgium
e-mail: jan.vanhellemont@ugent.be

K. Nakamura • E. Kamiyama • K. Sueoka
Department of Communication Engineering, Okayama Prefectural University, 111 Kuboki, Soja, Okayama 719-1197, Japan
e-mail: kozo_nakamura@nifty.ne.jp; ejkamiyama@aol.com; sueoka@c.oka-pu.ac.jp

Y. Yoshida, G. Langouche (eds.), *Defects and Impurities in Silicon Materials*,
Lecture Notes in Physics 916, DOI 10.1007/978-4-431-55800-2_4

Keywords Silicon • Germanium • Single crystal pulling • Intrinsic point defect • Grown-in defect • Doping effect • Stress effect • Fermi-level effect

4.1 Introduction: A Very Brief History of Si and Ge Crystal Pulling

The first Si single-crystals were pulled in the early 1950s and had a diameter of about 10 mm and a weight of the order of 100 g [91]. Driven mainly by the expected cost reduction of fabricated devices, since that time, the silicon wafer diameter and thus also the diameter of Si crystals grown by the Czochralski (CZ) technique increased in steps with standard wafer diameters of 50, 75, 100, 125, 150, 200 and then to reach in one step the 300 mm by the end of the 1990s with crystal weights of about 300 kg as shown in Figs. 4.1 and 4.2.

By the end of the 1990s, The Japanese government also initiated the Super Silicon (SSi) project which was aiming at a feasibility study of 400 mm crystal growth [58]. The obtained results indicated that it was possible to grow vacancy-rich and interstitial-rich 400 mm crystals and that therefore it should also be possible to grow so called perfect Si 400 mm crystals, not containing large intrinsic point defect clusters and therefore suitable for advanced device mass production. At this

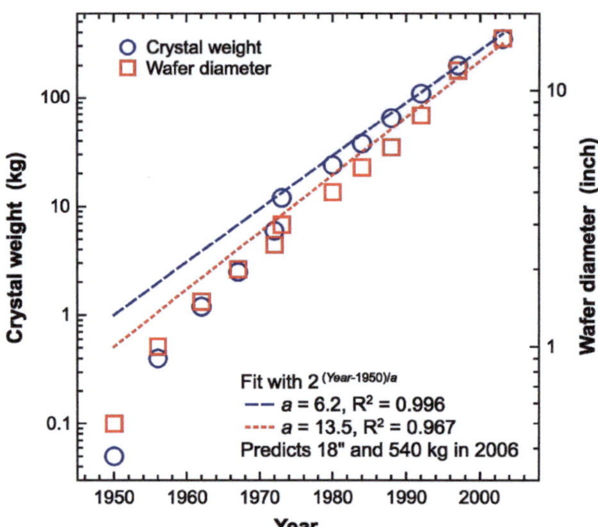

Fig. 4.1 Evolution of CZ Si wafer diameter and crystal weight with time [91]. The weight data are typical values. The data points correspond with the year when the diameter was for the first time produced in larger numbers. Since 1970, the crystal diameter doubles approximately every 13 years while the average weight of the crystal doubles every 6 years

Fig. 4.2 *Left:* State of the art 300 mm diameter CZ Si ingot. Courtesy Siltronic AG. Ingots of this diameter weigh several hundred kilos. *Right:* First 450 mm diameter, dislocation free CZ Si ingot grown by SunEdison (Reprinted from [33], Copyright 2011, with permission from Elsevier. Courtesy SunEdison)

moment the pulling processes for the move to 450 mm are in full preparation [33, 86] but there is a clear delay of several years compared to what could be expected when extrapolating the data in Fig. 4.1. There are several reasons for this delay such as the development cost which increases very rapidly with increasing crystal diameter and also the uncertainty on the number of potential customers decreasing rapidly with increasing wafer diameter, the availability of suitable crystal pullers and 450 mm wafer characterization tools, and last but not least, the fact that one is now approaching the limit of what is possible based on the Si material properties as will become clear further in the chapter.

The main driver towards ever larger crystal diameters both for CZ and FZ Si crystal production has been reduction of cost per unit of wafer surface. While the CZ technique is coming close to what can be achieved with that respect, FZ crystal pulling might still be improved considerably as about 50 % of the cost of the crystal is due to the expensive polycrystalline Si feed rods. A new FZ-like technique was proposed using continuous feed of inexpensive Si granules instead of a feed rod [80].

For Ge substrates, the most used wafer diameter is 100 mm, although also 200 mm wafers are commercially available and 300 mm dislocation-free crystals

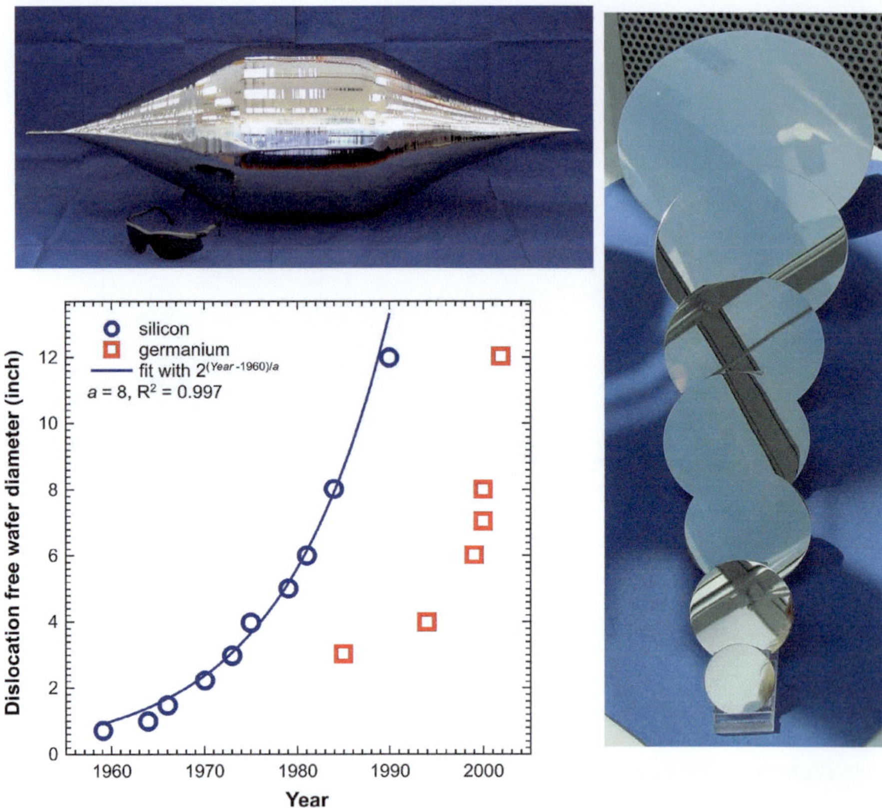

Fig. 4.3 *Top left:* One of the first 300 mm diameter, dislocation free Ge crystals, pulled in 2004. Courtesy Umicore Electro-Optic Materials (EOM). *Bottom left:* Very rapid development of dislocation-free 200 and 300 mm Ge wafers at Umicore EOM [47] compared with the wafer diameter development for CZ Si. *Right:* Electronic grade CZ Ge wafers with diameters ranging from 50 to 300 mm (Reprinted from [13], Copyright 2007, with permission from Elsevier. Courtesy Umicore EOM)

and wafers have been demonstrated as illustrated in Fig. 4.3 [13, 47]. In view of the limited Ge reserves available, application of Ge in mass-produced electronic components will always be under the form of thin Ge films mostly on large diameter Si substrates. One of the possibilities is to use the so called "smart cut" process to produce Ge On Insulator (GOI) wafers consisting of a thin mono-crystalline Ge film separated from a standard Si substrate by a thin silicon oxide layer [61]. For that purpose a high quality Ge "mother wafer" is needed with the same diameter as the Si substrate. This necessitated the development of 200 and 300 mm Ge crystal pulling and wafering processes.

4.2 Grown-in Defects in Single-Crystal Silicon Grown from a Melt

During the long history of single-crystal Si and Ge growth from a melt, grown-in defects have been a major concern. Initially the main problem was dislocation generation due to the thermal shock when dipping the crystal seed in the melt but this was soon solved by using the technique of Dash necking [9] leading to a complete removal of dislocations as illustrated in Fig. 4.4. After that, for a long time it was believed that grown-in defects were no longer an issue, all crystals were pulled fast enough to avoid dislocation clusters formed by self-interstitial "precipitation" during cooling of the pulled crystal. By the end of the 1980s, the device makers however noticed the presence of so called Crystal Originated Particles (COP's) that were observed by wafer surface inspections tools based on visible light scattering after standard RCA wafer cleaning. The number of COP's increased when repeating or extending the RCA cleaning hence it was concluded that the origin had to be the substrate. Closer investigation soon revealed that the so called particles where in fact crystallographic pits on the polished wafer surface [35] that formed by vacancy clustering during crystal cooling. As the COP's were shown to lead to gate oxide degradation [36] it was quite important to improve the crystal pulling processes so that COP free wafers could be obtained. All commercial crystals in those days were

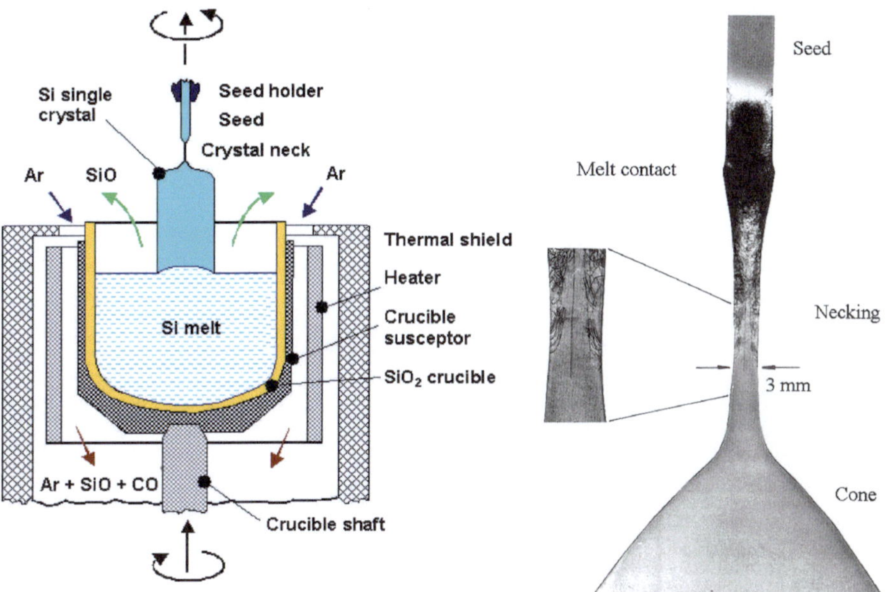

Fig. 4.4 *Left:* Schematic representation of the Czochralski pulling process [19]. *Right:* X-ray topographs showing dislocation removal by dash necking. Dislocations are formed due to the thermal shock when the seed touches the melt and glide out in the neck area [19]

vacancy-rich. Todays state of the art crystal pulling is aimed at COP and dislocation cluster free wafers.

4.2.1 State of the Art CZ and FZ Single-Crystal Pulling

Today the state of the art commercial single-crystal diameters are 300 mm for Czochralski pulled Si and Ge and 200 mm for Floating Zone (FZ) pulled Si (Figs. 4.2 and 4.5). As yet, no FZ pulling process for Ge is available. This is in

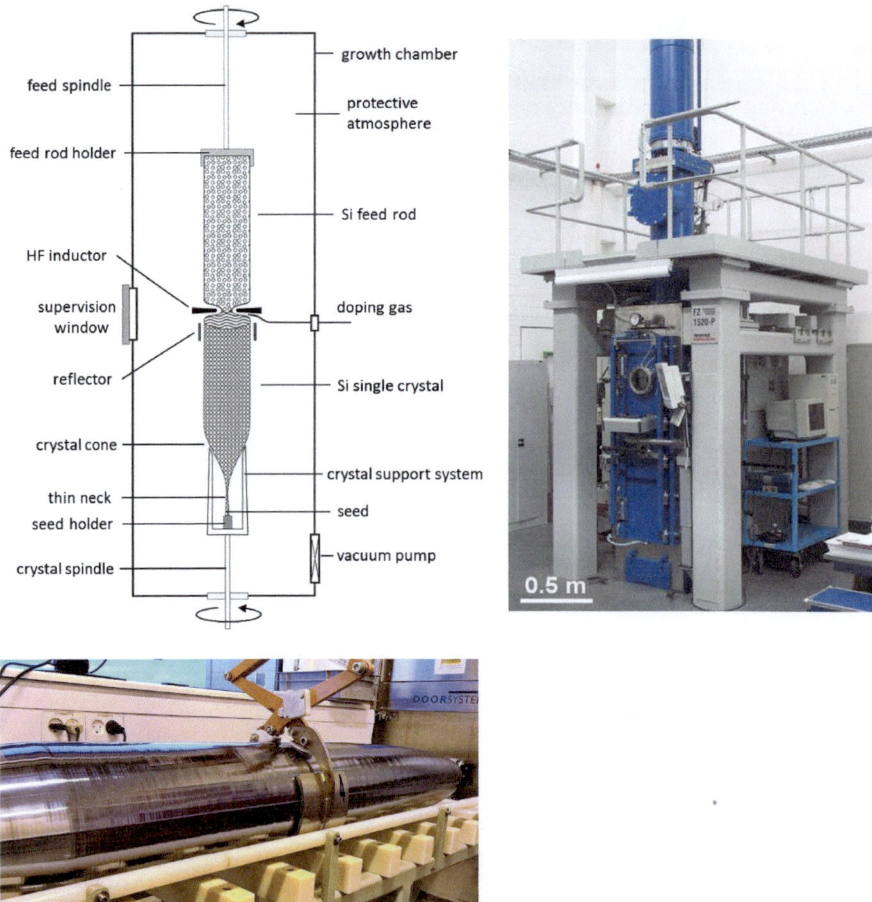

Fig. 4.5 *Top left:* Scheme of the FZ process for large-diameter Si single-crystals. *Right:* Puller for large diameter FZ crystal pulling at IKZ in Berlin [34] (Courtesy Robert Menzel). *Bottom left:* 8 in. FZ crystal, just taken out of the puller (Courtesy Topsil Semiconductor Materials)

part due to the larger weight of Ge, making it difficult to confine the molten zone during the FZ process. Recent attempts seem however promising and showed the feasibility to grow 35 mm diameter crystals [87]. The use of a magnetic field to contain the melt will probably be needed in order to make larger diameters FZ Ge crystal growth possible.

450 mm diameter CZ crystal pulling processes are being developed for Si (Fig. 4.2), requiring a more profound understanding of the various process parameters influencing intrinsic point defect behavior in order to be able to pull grown-in defect-free crystals using a commercially viable pulling process.

4.2.2 Experimental Observations on Grown-in Defects

4.2.2.1 Detection and Characterization of Grown-in Defects

Axial and radial distributions of grown-in defects in as-grown crystals can be observed using Cu decoration and X-ray topography as was already done in the early days of Si crystal growth from a melt using the Floating Zone or Czochralski pulling technique as illustrated in Fig. 4.6 [2, 4].

Grown-in defects can also be observed after using dedicated defect etching techniques. So called flow pattern defects (FPD's) and Secco etch pits (SEP's) are observed on wafer surfaces after immersion of the wafer piece with the polished surface in vertical position in unstirred Secco etchant for a prolonged etching time of the order of half an hour [88]. FPD's correspond with so called D-defects which are vacancy cluster related while SEP's are observed in regions that contain a lot of A-defects which are related to self-interstitial clusters. Although developed for moderately or low doped silicon substrates, a modified etch recipe can also be

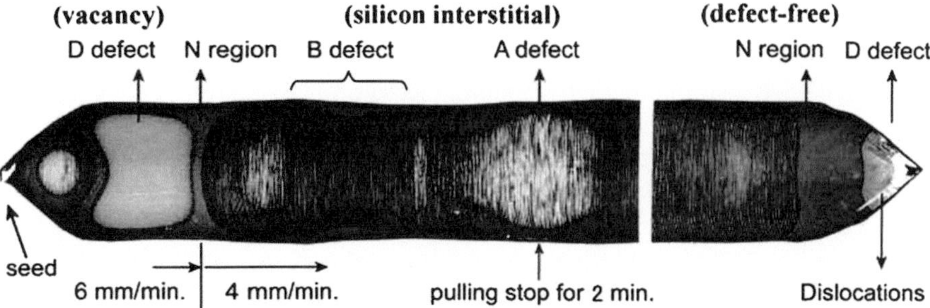

Fig. 4.6 X-ray topograph after Cu decoration of a 20 mm diameter FZ silicon crystal from seed to tail. During crystal growth, the pulling speed was varied while keeping the diameter constant resulting in vacancy type defect (D) and interstitial type defect (A and B) containing crystal parts as well as regions where no grown-in defects are observed (N region) (Reprinted from [4], Copyright 2011, with permission from Elsevier)

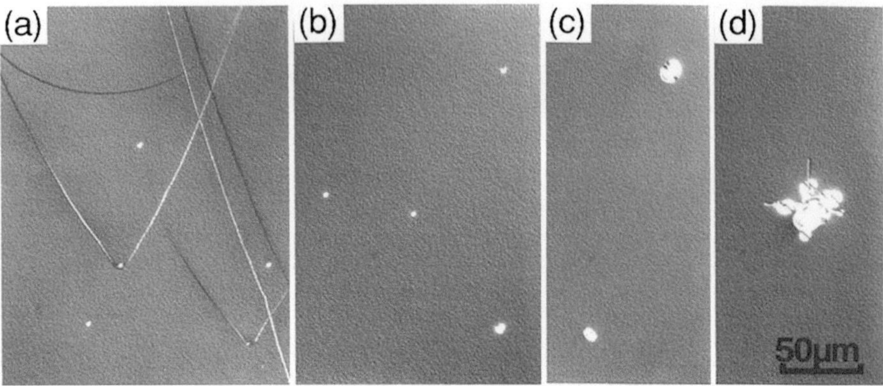

Fig. 4.7 Optical micro-graphs after FPD etching. Pulling rates of the 150 mm Si crystal: 1.1 mm/min **(a)** leading to a vacancy-rich crystal with D-defects that are revealed by FPD's; 0.7 mm/min, close to OSF-ring **(b)**; outside OSF-ring **(c)** which is an interstitial-rich part of the crystal containing A-defects that are revealed as SEP's; 0.4 mm/min **(d)** leading to an interstitial-rich crystal containing a low density of dislocation clusters revealed as so called large pit defects [48] (Copyright 1993 The Japan Society of Applied Physics. Courtesy SUMCO)

applied for low resistivity wafers [89]. Examples of delineated A and D defects are shown in Fig. 4.7.

4.2.2.2 Vacancy Type Defects

Vacancy type grown-in defects in CZ Si or Ge are observed on polished wafer surfaces as so called Crystal Originated Particles (COP's) by wafer surface inspection tools based on the detection of scattered laser light by surface irregularities [68, 69, 74]. These advanced wafer surface inspection tools not only allow to detect and differentiate crystal defects from particles bu also to determine the coordinates of these defects on the wafer surface. This makes it possible to investigate the nature of these defects using tools like scanning electron microscopy (SEM) or atomic force microscopy (AFM) after transfer of the defect coordinates to the sample/wafer holder of these instruments. This approach showed that the crystal defects observed on wafers prepared from vacancy-rich crystals are in reality crystallographic pits bounded by {111} planes that intersect the (001) wafer surface as a square with <110> edges. Often double pits are observed as show by the AFM images in Fig. 4.8. COP sizes in the range between 50 and 150 nm are commonly observed on Si wafers. The pit morphology and size are in good agreement with the octahedral voids that are observed by TEM (Figs. 4.9 and 4.10).

A similar approach but this time using a light scattering tool with an infra-red light laser beam, allows to detect and determine the 3D coordinates inside the silicon wafer of the lattice defects leading to COP formation on the polished wafer surfaces. Transferring the coordinates to focused ion beam specimen preparation tools for

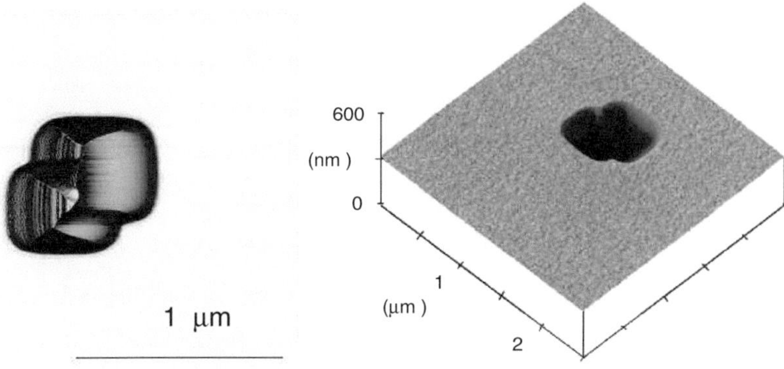

Fig. 4.8 AFM images of a typical double COP after 4h SC1 delineation. *Left:* top view. *Right:* 3D-view [68]

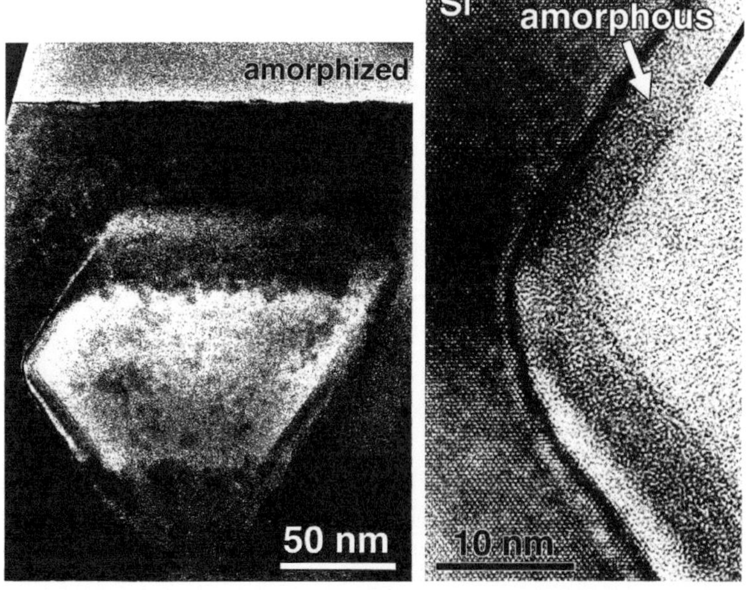

Fig. 4.9 *Left:* Cross-section TEM image of a void in CZ Si. *Right:* HREM image of the void/silicon matrix interface. A thin ($\approx$10 nm) silicon oxide layer is formed during crystal cooling [8] (Copyright 1997 The Japan Society of Applied Physics)

transmission electron microscope (TEM) investigation, allows to study the nature of the defects in the as-grown crystal. An example of the observation of grown-in defects by infra-red light scattering tomography is shown in Fig. 4.11.

Common techniques for non-destructive bulk inspection of grown-in or processing induced point defect clusters such as cross-section infra Red Light Scattering

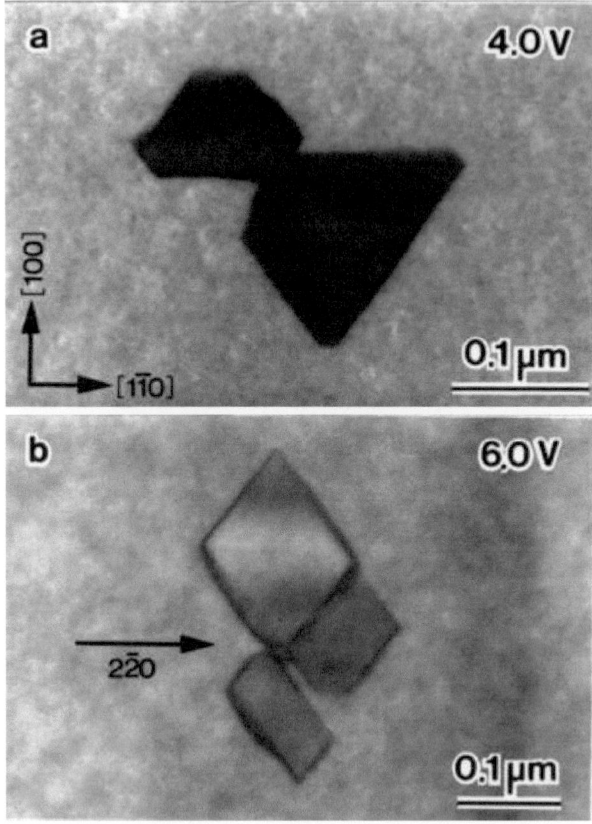

Fig. 4.10 TEM micrographs of grown-in voids in CZ Si grown at a pulling rate of 1.0 mm/min. *Top:* taken from the [110] pole. *Bottom:* taken with a [2$\bar{2}$0] Bragg condition [25] (Copyright 1998 The Japan Society of Applied Physics)

Tomography (IR-LST) [28, 69], plan view inspection with a Scanning Infra Red Microscope (SIRM) which based on detecting backscattered light [76, 89, 90] or differential interference contrast microscopy [62], are based on scattering or interference of near infra red light in the 1–1.3 μm wavelength range for which Si is transparent.

When the size d of the light scattering defect is much smaller than the wavelength of the probing beam, Rayleigh scattering occurs and the Rayleigh scattering cross-section σ_s is given by [74]

$$\sigma_s = \frac{2\pi^5}{3}\frac{d^6}{\lambda^4}\left(\frac{n^2-1}{n^2+2}\right)^2, \tag{4.1}$$

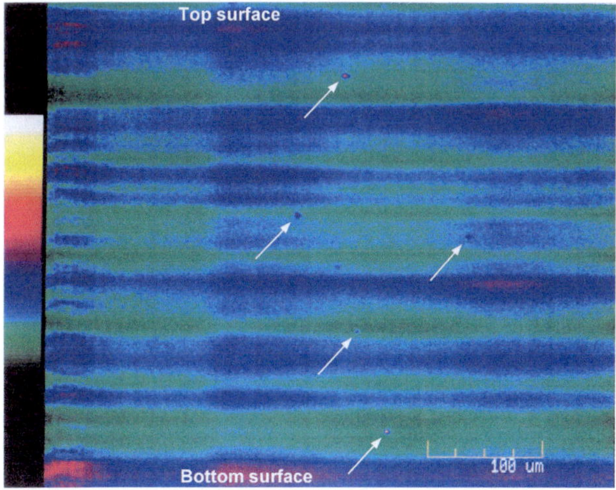

Fig. 4.11 IR-LST cross-section image of a cleaved wafer prepared from a vacancy-rich Si crystal revealing the presence of a large number of grown-in vacancy clusters indicated by *arrows*. The false *color scale* at the left corresponds with the scattering intensity and thus defect size, increasing from *green* to *white* (Courtesy Gudrun Kissinger)

with λ the wavelength of the probing light, n the refractive index of the particle material and d its diameter. From (4.1), and taking into account the difference in refractive index of the Si matrix and the defect phase, e.g. vacuum or SiO_2, a lower size detection limit of about 20 nm (diameter) can be estimated for voids while that for oxide precipitates is only about 4 % larger.

When the inclusion is of the same order of magnitude or even larger than the wavelength of the probing light, assuming Mie scattering is a better approximation to describe the scattering process [74].

The advantage of working in backscattering like with SIRM is that measurements of grown-in void distributions [89] or oxide precipitate distributions after thermal treatments [90] are also possible on low resistivity material as illustrated in Fig. 4.12. The limited penetration depth should thereby be taking into account and corrections should be made for the reduced scattering intensity due to the increased light absorption in the substrate.

A typical void in as-grown CZ Si is e.g. shown in Fig. 4.9. The about 150 nm large void is a cluster of about 3×10^6 vacancies and the typical void density is of the order of $5 \times 10^6\,cm^{-3}$. This corresponds with a total vacancy concentration of the order of $10^{13}\,cm^{-3}$, which is about 1 % of the thermal equilibrium vacancy concentration at melting temperature.

COP's are also observed on polished CZ Ge wafers (Fig. 4.13, right). They are typically one order of magnitude larger than those in Si. At the same time, the void density in CZ Ge is three orders of magnitude lower, suggesting a V thermal equilibrium concentration at melting temperature of the same order of magnitude as in Si.

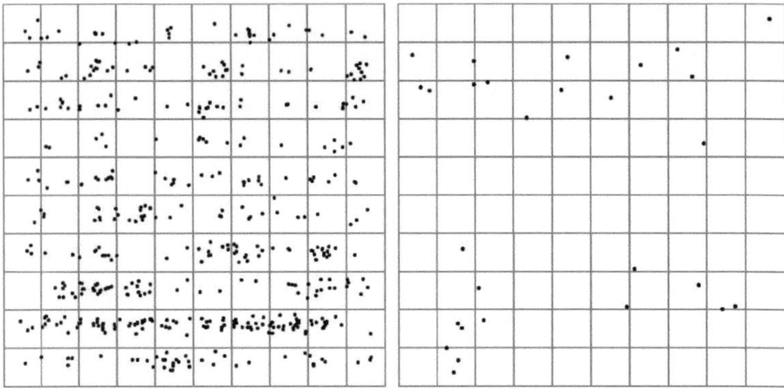

Fig. 4.12 Typical void maps obtained with SIRM in a 1.81.8 mm^2 area in the center of two 100 mm diameter wafers. *Left:* 10^{20} Ge cm^{-3} doped crystal. *Right:* Same type of crystal but co-doped with about 3×10^{19} B cm^{-3} (3–5 mΩcm), after measuring a matrix of 10×10 adjacent areas of $180 \times 180\,\mu m^2$ (Reproduced with permission from [89]. Copyright 2011, The Electrochemical Society)

In fast pulled crystals that are very vacancy-rich, the total concentration of vacancies incorporated in voids is nearly independent of the crystal cooling rate at 1120 °C as illustrated in Fig. 4.14 [39].

4.2.2.3 Interstitial Type Defects

Interstitial type defects can be observed after Secco etching or after dislocation etching. In slow pulled crystals large etch pits are observed, corresponding with large dislocation clusters [48]. Such dislocations are very detrimental as they will propagate and multiply in epitaxial layers that are grown on such substrates [49]. A few typical examples are given in Fig. 4.15. The TEM image on the top shows a plan view of a large dislocation cluster in a slow pulled crystal [48]. The bottom micrographs show cross-section TEM images of dislocations in a 3 μm thick epitaxial layer grown on a polished wafer prepared from a slow pulled 200 mm diameter, low resistivity, B doped Si crystal. From the dislocation cluster size in the substrate and taking into account the Burgers vectors, it can be estimated that one grown-in defect contains about 10^{10} self-interstitials [49]. Taking into account that about 4×10^3 cm^{-3} grown-in defects are observed by Secco etching, the total number of self-interstitials in the grown-in defects is about 4×10^{13} cm^{-3} , which is again close to 1 % of the thermal equilibrium self-interstitial concentration at melting temperature taking into account the relatively large uncertainty on the number of self-interstitials in the grown-in dislocation.

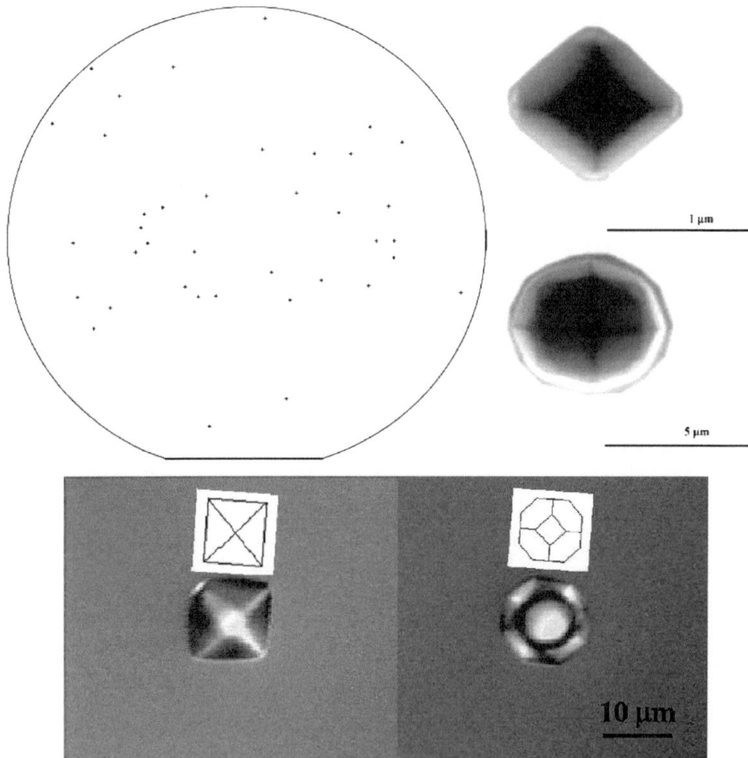

Fig. 4.13 *Left:* Wafer image obtained with a SURFSCAN surface inspection tool revealing the presence of a low density of large LPD's on a polished germanium wafer (Reprinted from [71], Copyright 2007, with permission from Elsevier). *Right:* SEM micro-graphs showing COP's corresponding with LPD's on the same wafer (Reprinted from [74], Copyright 2008, with kind permission from Springer Science and Business Media). *Bottom:* In extreme cases the COP's can be observed by the naked as strong light scatterers. Optical microscopy reveals a similar morphology as for the small COP's observed with SEM [22]. The COP's are formed by the intersection with the polished Ge wafer surface of large octahedral or truncated octahedral voids in the as-grown crystal

4.2.2.4 Transient Defect Phenomena

Abe and Takahashi [4] investigated the formation of grown-in defects in crystals grown with rapidly varying pulling rates. 50, 100 and 200 mm diameter crystals were grown in the same hot zone with a constant pulling rate of 1.0 mm/min until a steady state grown-in defect distribution was reached. Subsequently, the pulling rate was rapidly lowered to 0.3 mm/min, kept constant at that value for 30 min and then rapidly increased again to 1.0 mm/min.

Carrier lifetime maps of transversal sections of these crystals are shown in the top figure of Fig. 4.16 revealing the grown-in defect distributions. The OSF ring region is represented by the red area of low lifetime surrounded by narrow yellowish

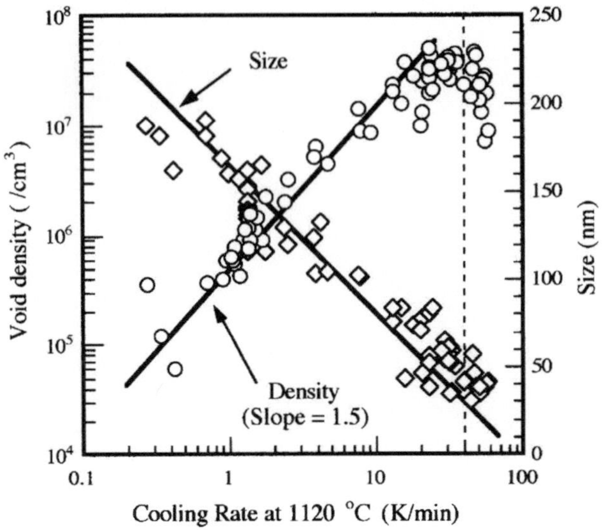

Fig. 4.14 Relationship between the crystal cooling rate at 1120 °C and the density and size of voids. The v/G value for the investigated crystals is more than double the critical value, yielding crystals that contain close to the maximum vacancy-concentration (Reprinted from [39], Copyright 2002, with permission from Elsevier)

regions of medium lifetime. Lifetimes are somewhat higher in the interstitial-rich region in the center, where interstitial clusters are present. Vacancy clusters are outside the OSF ring area where the lifetime is highest. The images also show that the interstitial-rich area extends well into the parts of the crystal where the original pulling rate of 1 mm/min was used and also that this extension is asymmetric.

Dornberger et al. [16] reproduced these experimentally observed features by simulation of the intrinsic point defect behavior (bottom figure of Fig. 4.16. They showed that the observations could be explained by an enhanced in-diffusion of self interstitials from the interface due to the lowered pulling rate. Resuming the original pulling rate generates a vacancy-rich crystal with the observed embedded ellipsoidal interstitial-rich areas in the 50 and 100 mm diameter crystals.

In another experiment, Abe and Takahashi [4] investigated the extreme case of pulling speed change by detaching the crystal from the melt. The results are shown in Fig. 4.17 revealing that even for a very slow pulled crystal that is thus very interstitial-rich after growth, a thin layer in which Anomalous Oxygen Precipitation (AOP) occurs can be observed close to the crystal-melt interface before detaching. AOP is an indication of a vacancy-rich crystal as the presence of a supersaturation of vacancies enhances oxide precipitate nucleation and growth. These detaching experiments clearly show that in all Si crystals pulled from a melt, the part of the crystal close to the melt is vacancy-rich with the axial extent of this vacancy-rich layer determined by the pulling rate and temperature gradient for the given hotzone.

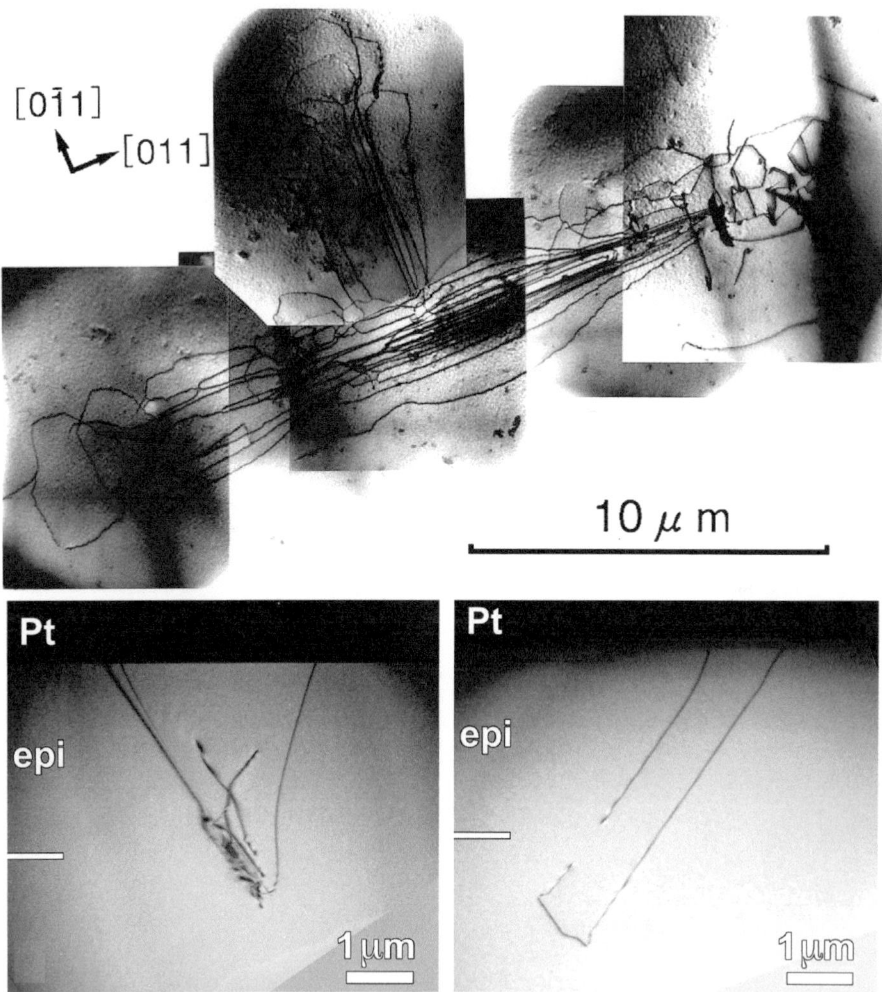

Fig. 4.15 *Top:* (100) plan view TEM image showing large dislocation clusters in a 150 mm Si crystal grown at 0.4 mm/min [48] (Copyright 1993 The Japan Society of Applied Physics. Courtesy SUMCO). *Bottom:* Cross-section TEM images of dislocations in a 3 μm thick epitaxial Si layer grown at 1100 °C on a wafer from a slow pulled 200 mm p^+ Si crystal. The dislocations nucleate at grown-in self-interstitial defects in the Si substrate [49] (Courtesy Hugo Bender, IMEC)

4.2.3 Simulation of Intrinsic Point Defect Cluster Formation During Crystal Pulling

Commercial simulators are available nowadays that allow a detailed simulation of the complete crystal growth process including temperature and thermal stress distributions, gas and melt flows and intrinsic point defect incorporation, distribution,

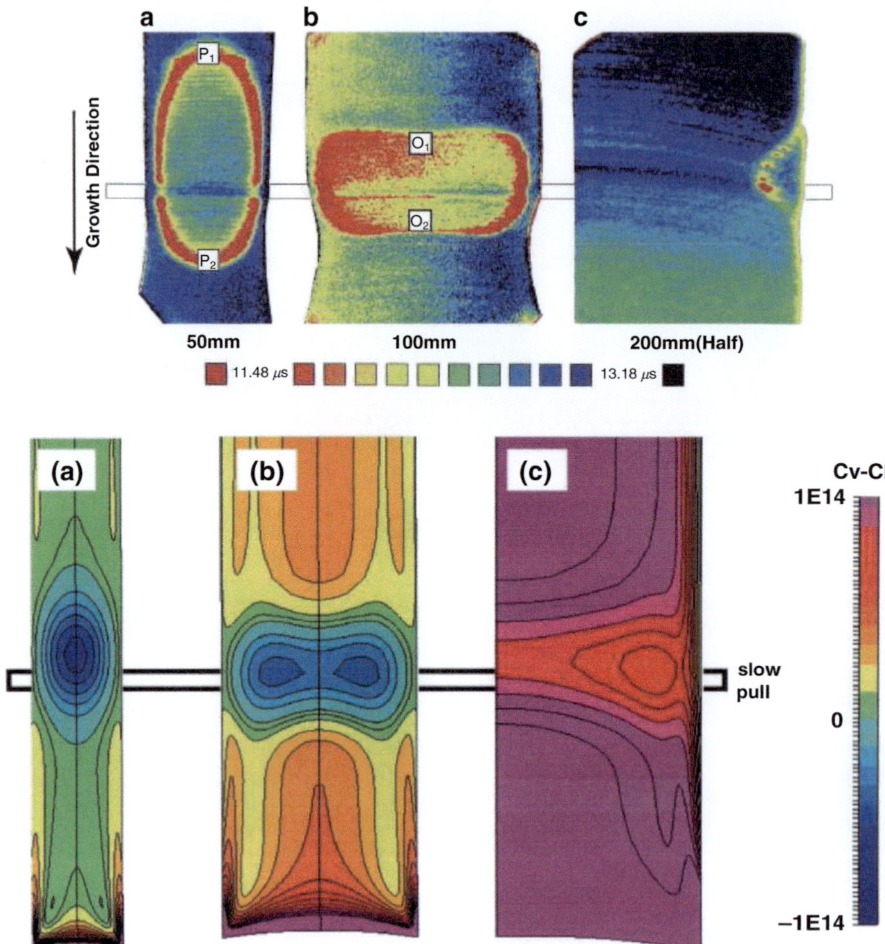

Fig. 4.16 *Top:* Minority carrier lifetime maps for 50 (**a**), 100 (**b**) and 200 mm (**c**) CZ silicon crystals grown with a pull rate of 1.0 mm/min and a temporary decrease to 0.3 mm/min (Reprinted from [4], Copyright 2011, with permission from Elsevier). *Bottom:* Simulated ($C_V - C_I$) distributions for the same crystal diameters and pulling conditions (Reprinted from [16], Copyright 2001, with permission from Elsevier)

recombination and clustering. This is illustrated in Fig. 4.18 (top) showing the result of a simulation of the formation of voids in Si and Ge by homogeneous nucleation of vacancy clusters which is driven by the increasing vacancy supersaturation during crystal cooling [22, 73]. The growth of the vacancy clusters is assumed to be vacancy diffusion controlled. The larger void/COP sizes in germanium compared to silicon are due to the much higher diffusivity of vacancies in germanium. Figure 4.18 (bottom) illustrates that good agreement can be obtained between measurement and simulation.

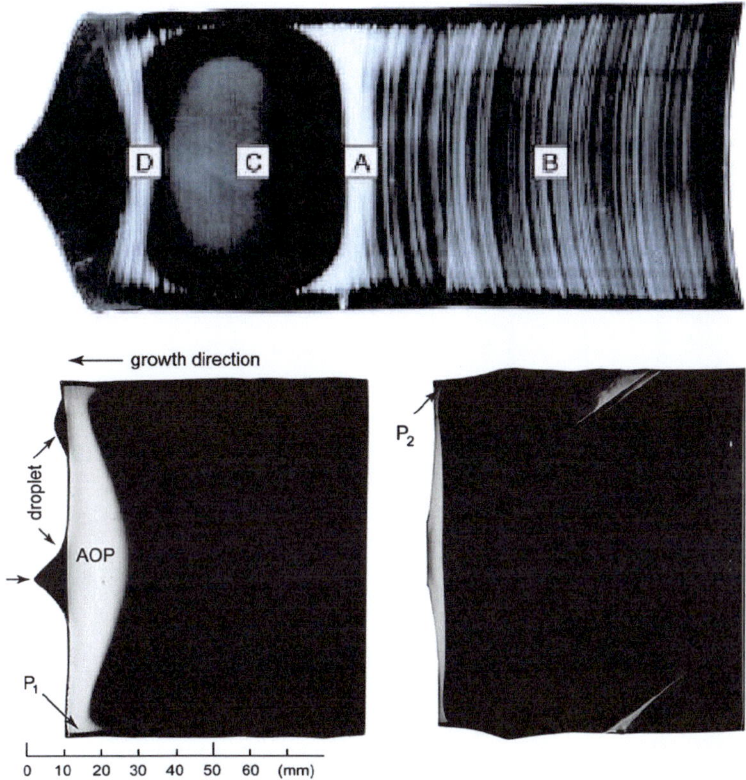

Fig. 4.17 *Top:* AOP in a fast pulled detached 70 mm CZ crystal (1.5 mm/min, vacancy-rich). *Bottom left:* AOP in a medium speed pulled detached 105 mm CZ crystal (0.47 mm/min, interstitial-rich). *Bottom right:* AOP in a slow pulled detached 110 mm CZ crystal (0.15 mm/min, very interstitial-rich) (Reprinted from [4], Copyright 2011, with permission from Elsevier)

Despite this good agreement between simulation and measurement for the various models reported in literature, it should be mentioned at this stage that the point defect properties that are extracted from a best fit between experimental data and simulation, can vary widely depending on the model that is used to simulate the crystal pulling process and the point defect clustering.

A good illustration of this uncertainty on the properties of both intrinsic point defects is the value of the recombination coefficient k_{IV} between vacancies and self-interstitials. In processes such as crystal pulling, plastic deformation and ion implantation, both types of intrinsic point defects are generated and/or coexist. An important parameter controlling the concentration of both point defects is k_{IV} which is given by the well-known expression

$$k_{IV} = 4\pi a_c(D_V + D_I)\exp[-\Delta G_{rec}/kT], \tag{4.2}$$

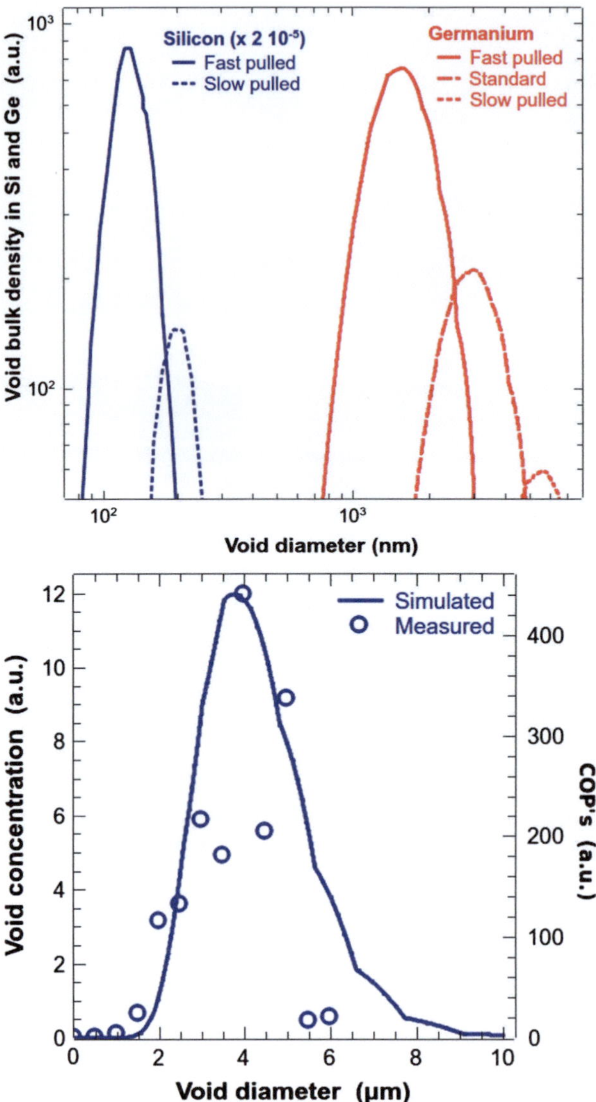

Fig. 4.18 *Top:* Simulated void size/density distributions in Si and Ge, illustrating the impact of crystal pulling speed. The simulations also predict the observed one order of magnitude larger void size in Ge which is associated with a three orders of magnitude lower volume density [73, 75] (Reprinted from [73], Copyright 2008, with permission from Cambridge University Press). *Bottom:* Comparison between measured COP distribution on a polished Ge wafer and the simulated distribution of voids in the bulk [22]. The COP distribution was measured using a confocal review station

with a_c the capture radius of the order of 1 nm and ΔG_{rec} the free energy barrier against recombination.

Assuming an Arrhenius-type behavior, k_{IV} can also be written as [32]

$$k_{IV} = k_m \exp[-(E_{rec}/kT_m)(T_m/T - 1)], \tag{4.3}$$

with k_m a constant and E_{rec} the activation energy for recombination.

Applying (4.2) and (4.3) for data obtained from grown-in defect distributions in floating zone and Czochralski grown Si crystals [18, 30], yields the k_{IV} dependence on the temperature normalized to the melt temperature as shown in Fig. 4.19. It is clear that a large uncertainty (still several orders of magnitude near the melting temperature) exists with respect to the value of k_{IV} and that it strongly depends on the model that is used to simulate the crystal growth process. For Ge, data for the self-interstitial are to a large extent still lacking making it very difficult to estimate the recombination factor.

To some extent the crystal pulling process and defect simulators should therefore be considered as sophisticated extra- or interpolation tools. Only a further refinement of the physical models used in the simulators will allow to extract more reliable intrinsic point defect properties from the fit to the experiments.

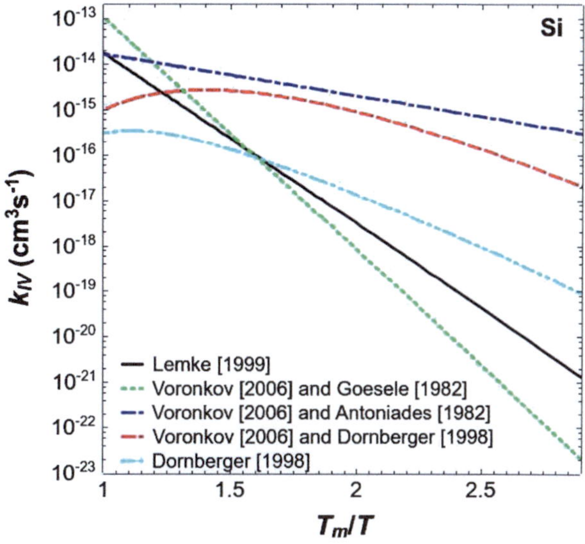

Fig. 4.19 k_{IV} in Si as a function of the temperature normalized with respect to the melting temperature [72, 73] (Reprinted from [73], Copyright 2008, with permission from Cambridge University Press)

4.2.4 The Voronkov Criterion for Defect-Free Crystal Growth

Already in 1982, Voronkov presented a model describing intrinsic point defect behavior during the growth of single-crystal silicon from a silicon melt and derived a criterion to predict if the crystal was vacancy- or self-interstitial-rich [82]. The "Voronkov criterion" was further refined in the following decades [84] and is based on the experimental observation that the intrinsic point defect balance near the crystal-melt interface is determined by Γ, the ratio of pulling speed v over temperature gradient G at the crystal-melt interface. When this ratio is larger than a critical value Γ_{crit}, the crystal is vacancy-rich; when the ratio is below the critical value, the crystal is self-interstitial-rich. Published values for Γ_{crit} range between 1.3 and $2.2\times10^{-3}\,cm^2\,min^{-1}\,K^{-1}$ and depend among others on the simulator that is used to calculate the thermal gradient and also on the doping and resistivity of the crystal [14, 15, 23, 31, 81, 83]. The Voronkov model has been widely accepted and is used in crystal pulling simulators, allowing to predict intrinsic point defect clustering in single-crystal silicon grown by the Czochralski or floating zone techniques. The growth of so called perfect silicon crystals up to 300 mm diameter, without observable intrinsic point defect clusters is largely based on the application of the Voronkov criterion.

Despite the success in applying the Voronkov criterion to improve crystal quality there are also a number of remarks on its validity. Abe claimed already 30 years ago that the axial thermal gradient G near the crystal-melt interface in growing FZ and CZ silicon crystals decreases with increasing pulling speed instead of increasing, as mostly assumed [1, 3, 5]. The FEMAG simulator predicts that G increases with decreasing pulling rate [81], while other simulators give opposite results [60]. This also suggests that the pulling speed and the thermal gradient might not be independent parameters in the Voronkov criterion.

Another issue with the Voronkov criterion is that the influence of the stress introduced by the thermal gradient at the crystal-melt interface on the intrinsic point defect formation enthalpy is not explicitly taken into account. As the thermal stress increases with crystal diameter, this might become an important issue in the development of 450 mm diameter Si crystals.

Application of the criterion for the development of even larger diameter crystals and for heavily doped crystals needs thus a better understanding of the various material and crystal pulling process parameters that influence the intrinsic point defect properties and thus also the value of Γ_{crit}. A side effect would also be that more accurate intrinsic point defect parameter could be extracted from grown-in defect studies in dedicated as-grown crystals, even using smaller diameters thus making the experiments less expensive. This paper discusses the present state of understanding of the impact of these parameters and also further work that is needed to improve the Voronkov criterion for ultra large diameter crystals.

4.2.4.1 Intrinsic Point Defect Diffusion and Recombination

The diffusion and recombination reaction of point defects during the crystal growth process can be described by Eqs. (4.4), (4.5), (4.6) and (4.7) [29, 40, 43, 51, 52].

$$J_V = -D_V \nabla C_V - \frac{D_V C_V Q_V^*}{k_B T} \nabla T \tag{4.4}$$

$$J_I = -D_I \nabla C_I - \frac{D_I C_I Q_I^*}{k_B T} \nabla T \tag{4.5}$$

$$\frac{\partial C_V}{\partial t} = -\nabla J_V - v\frac{dC_V}{dz} - k_{IV}\left(C_V C_I - C_V^{eq} C_I^{eq}\right) \tag{4.6}$$

$$\frac{\partial C_I}{\partial t} = -\nabla J_I - v\frac{dC_I}{dz} - k_{IV}\left(C_V C_I - C_V^{eq} C_I^{eq}\right) \tag{4.7}$$

where J is the diffusion flux, C is the actual concentration of the point defect and the subscripts V and I denote the vacancy and self-interstitial, respectively. T is absolute temperature and k_B is the Boltzmann constant. The second term in the right part of Eqs. (4.4) and (4.5) expresses the effect of the temperature gradient on the diffusion. Q^* is the reduced heat of transport. The second term in the right part of Eqs. (4.6) and (4.7) denotes the effect of the crystal pulling on the concentrations with v the pulling rate of the crystal. The third term in the right part of Eqs. (4.6) and (4.7) describes the recombination reaction between V and I with k_{IV} the reaction constant given by (4.2).

C^{eq} and D depend exponentially on temperature

$$C_{I,V}^{eq}(T) = C_{I,V}^{0} \exp(-\frac{E_{I,V}^{f}}{k_B T}) \tag{4.8}$$

$$D_{I,V}(T) = D_{I,V}^{0} \exp(-\frac{E_{I,V}^{m}}{k_B T}), \tag{4.9}$$

with $E_{I,V}^{f}$ and $E_{I,V}^{m}$ the intrinsic point defect formation and migration energies, respectively. $C_{I,V}^{0}$ and $D_{I,V}^{0}$ are the corresponding pre-factors.

Table 4.1 lists recent parameters of the intrinsic point defects in silicon crystal derived from the analysis of grown-in defect patterns [29, 43, 46, 51]. These values reported in literature fall within a relatively narrow range. Figure 4.20 shows the corresponding $C_{I,V}^{eq}$ and $D_{I,V}$ as function of the temperature. In the same figure also the transport capacity $C_{I,V}^{eq} D_{I,V}$ is shown, revealing a cross-over between 800 and 950 °C. At higher temperature the transport capacity of the interstitial is larger than that of the vacancy and vice versa below that temperature. It is this cross-over that is the reason that the Voronkov criterion "works". The transport capacities calculated with the Nakamura et al. [43] and the Nishimo et al. [46] data nearly coincide.

Figure 4.21a, b show the axial distributions of V and I concentration for the case of low v/G condition and high v/G condition, respectively. The parameters of point

Table 4.1 Intrinsic point defect parameters in silicon derived from grown-in defect distributions

	[43]	[51]	[29]	[46]
$C_V^{eq}(T_m)$ ($\times 10^{14}$ cm^{-3})	6.49	8.73	8.40	4.58
$C_I^{eq}(T_m)$ ($\times 10^{14}$ cm^{-3})	4.84	7.1	6.89	3.17
$D_V(T_m)$ ($\times 10^{-5}$ cm^2 s^{-1})	4.45	3.4	4.0	6.31
$D_I(T_m)$ ($\times 10^{-4}$ cm^2 s^{-1})	5.0	3.76	4.0	7.66
E_V^f (eV)	3.94	3.70	4.00	3.84
E_I^f (eV)	4.05	4.00	4.00	3.77
E_V^m (eV)	0.3	0.457	0.4	0.4
E_I^m (eV)	0.9	0.937	0.9	1.18
Q_V^* (eV)	0	0	0	0
Q_I^* (eV)	1.01	0	0	0
$C_V^{eq}(T_m) - C_I^{eq}(T_m)$ ($\times 10^{14}$ cm^{-3})	1.7	1.6	1.51	1.41

defects that were used are the values of Nakamura et al. [43] in Table 4.1. We can see that self-interstitial becomes the dominant species at the steady state away from the crystal-melt interface in the case of small v/G condition and that the vacancy becomes the dominant species in the case of large v/G condition.

Figure 4.22 shows the relationship between v/G and point defect concentrations at the steady state. We can see that there is a critical v/G at which the dominant species changes. Both concentrations of vacancy and self-interstitial become very low at $[v/G]_{crit}$. Therefore, the grown-in defect-free crystals can be grown under the condition of $[v/G]_{crit}$.

Voronkov and Falster have derived Eq. (4.10) for $[v/G]_{crit}$ at which J_V and J_I from the crystal-melt interface to the crystal become equal [84].

$$\left(\frac{v}{G}\right)_{crit} = \frac{D_I(T_m)C_I^{eq}(T_m)\left(E + Q_I^*\right) - D_V(T_m)C_V^{eq}(T_m)\left(E + Q_V^*\right)}{\left[C_V^{eq}(T_m) - C_I^{eq}(T_m)\right] k_B(T_m)^2} \tag{4.10}$$

with

$$E = \frac{E_V^f + E_I^f}{2}.$$

T_m is the melting temperature, C^{eq} the equilibrium concentration and E^f the formation energy of the intrinsic point defect. Q^* is the reduced heat of transfer which is not very well known with values in literature ranging between 0 and $-E^f$ [38] in agreement with the expression derived for Q^* by Tan [59]

$$Q^* = E^m - E^f + k_B \ln \frac{C}{C^{eq}}. \tag{4.11}$$

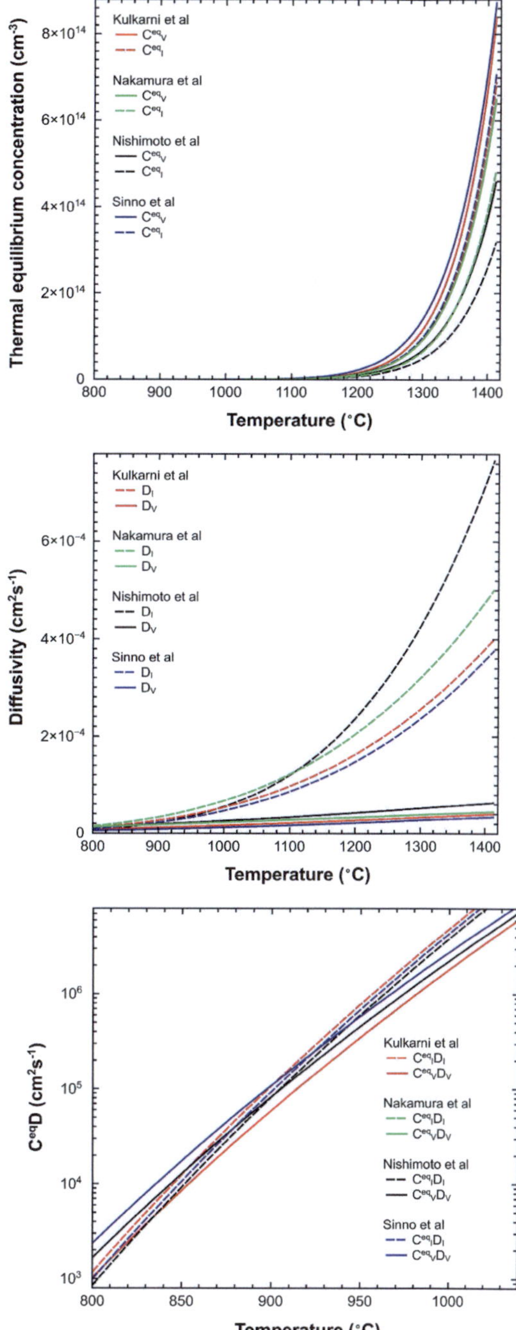

Fig. 4.20 Thermal equilibrium concentrations (*top*) and diffusivities (*middle*) of the vacancy and the self-interstitial corresponding with the point defect parameters of Table 4.1 [29, 43, 46, 51]. *Bottom:* Transport capacity $C^{eq}D$ as function of temperature. At high temperatures the transport capacity of the self-interstitial is larger, at low temperatures that of the vacancy

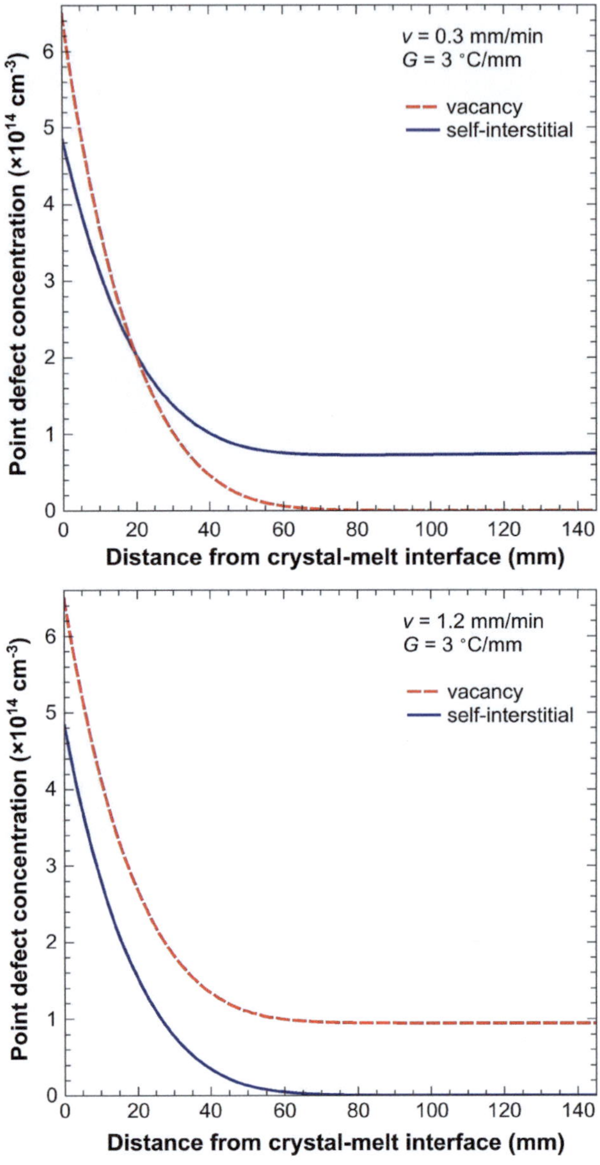

Fig. 4.21 Axial distribution of point defect concentrations during crystal growth. *Top:* Self-interstitial dominant condition (small v/G). *Bottom:* Vacancy dominant condition (large v/G) [37]

Neglecting the possible impact of Q^*, a more simple expression can be derived for the Voronkov criterion [66]

$$\left(\frac{v}{G}\right)_{crit} = \frac{D_I(T_m)C_I^{eq}(T_m)E_I^f - D_V(T_m)C_V^{eq}(T_m)E_V^f}{\left[C_V^{eq}(T_m) - C_I^{eq}(T_m)\right]k_B(T_m)^2} \tag{4.12}$$

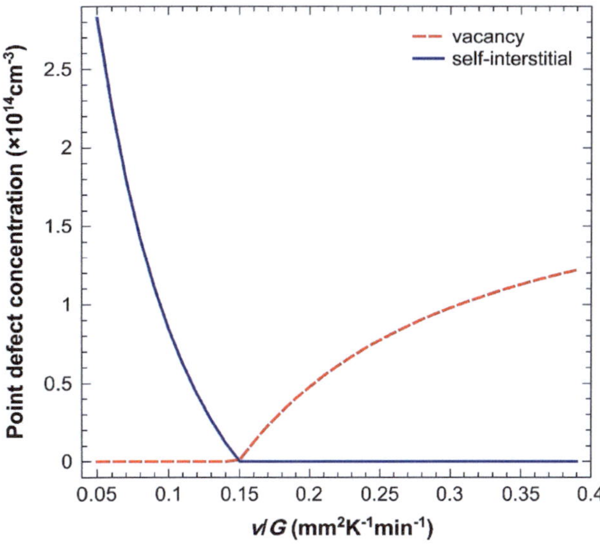

Fig. 4.22 Relationship between v/G and point defect concentrations at the steady state in the sufficiently distance away from the crystal-melt interface [37]

$[v/G]_{crit}$ is only determined by the physical parameters of the point defects as shown in the right hand side of Eqs. (4.10) and (4.12). Grown-in-defect-free crystals basically can be grown by controlling v/G near the critical value at all radial positions. Ideally this means that G is constant in the radial direction and v can then be calculated from Eqs. (4.10) or (4.12), using the appropriate intrinsic point defect parameters.

An example of result of the calculation of the relative point defect concentration difference $(C_V - C_I)/(C_V^{eq} - C_I^{eq})$ using Eq. (4.12) is shown in Fig. 4.23 illustrating that for all crystals, even the most interstitial-rich ones, there is a layer close to the crystal-melt interface that is vacancy-rich [66]. The thermal gradients of 170, 85 and 20 K/cm correspond roughly with crystal diameters of 50, 100 and 300 mm, respectively. The simulation results suggest that for interstitial-rich crystals pulled close to $[v/G]_{crit}$, the vacancy-rich layer can easily extend over a distance of the order of the crystal radius from the crystal-melt interface.

Equations (4.10) and (4.12) are, however, derived by using a one dimensional diffusion model. That is a good approximation near the crystal center where the crystal-melt interface is quasi flat. At other radial positions, the crystal-melt interface is however curved and due to that, the intrinsic point defect diffusion from the crystal-melt interface has a radial component. This radial component is determined not only by the interface curvature but also by radial out diffusion towards the free surface the crystal. The effect of the crystal-melt interface shape on $[v/G]_{crit}$ is discussed in next section based on the use of Eq. (4.10).

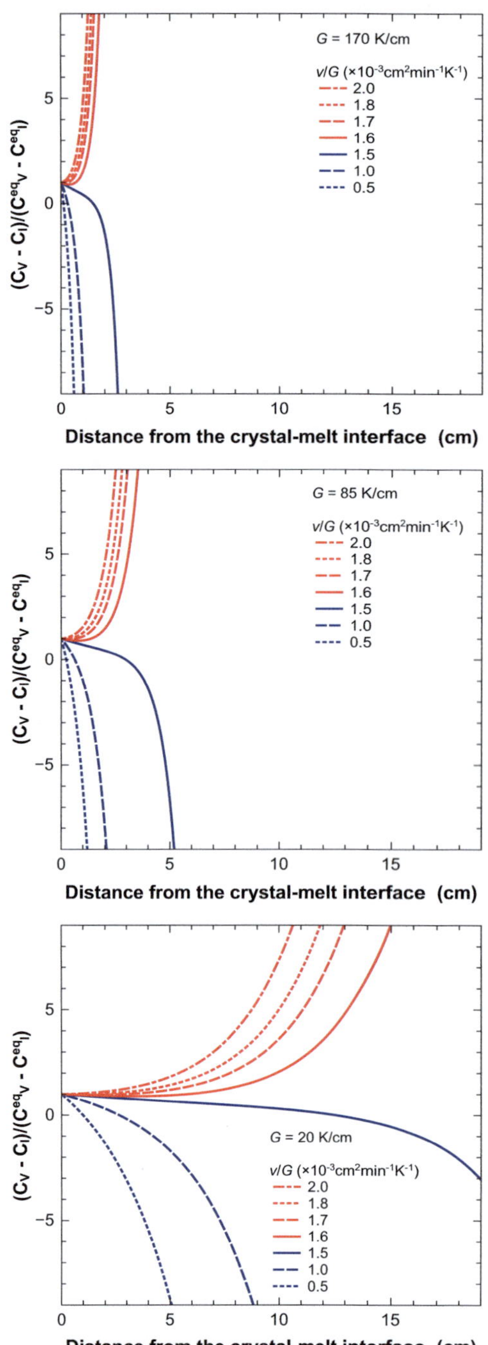

Fig. 4.23 $(C_V - C_I)/(C_V^{eq} - C_I^{eq})$ as function of the distance from the crystal-melt interface with v/G as parameter without taking stress effects into account for three different thermal gradients G [66]. The figures illustrate that the crystal remains free from point defect clusters over a length which depends on v/G and on G (and thus also on the crystal radius). Further more, for all diameters and pulling conditions, the crystal is vacancy-rich, corresponding with $(C_V - C_I)/(C_V^{eq} - C_I^{eq}) > 0$, close to the crystal-melt interface. This explains the observations shown in Fig. 4.17

Besides the curved crystal-melt interface, also material or process parameters affecting the intrinsic point defect thermodynamic parameters, will change the value of $(v/G)_{crit}$. The impact of thermal stress and of dopants will be discussed in the final part of the chapter.

4.3 Impact of the Crystal-Melt Interface Shape on *v/G*

4.3.1 Change of the Critical v/G by the Interface Shape

$[v/G]_{crit}$ changes by the shape of the crystal-melt interface [40, 43] as illustrated in Fig. 4.24, showing the relationship between $[v/G]_{crit}$ and the inverse radius of curvature of the interface at the crystal center of 200 mm diameter crystals with different interstitial oxygen concentration. In first order approximation, $[v/G]_{crit}$ is inversely proportional to the radius of curvature of the interface. This phenomenon has been explained by simulating the intrinsic point defect behavior in a growing crystal [40].

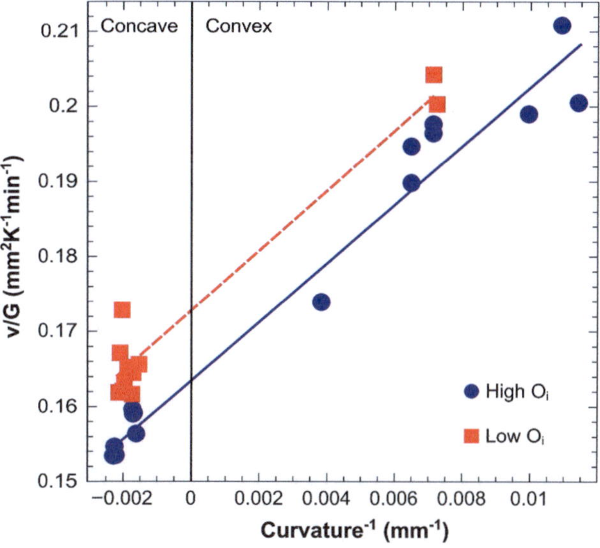

Fig. 4.24 Relationship between $(v/G)_{crit}$ at the crystal center and the radius of curvature of the crystal-melt interface for two interstitial oxygen concentrations (high: $> 13 \times 10^{17}$ O_i/cm^3; low: $< 13 \times 10^{17}$ O_i/cm^3, old ASTM) [40, 43]

4.3.2 Temperature Gradient Distribution Dependence on Interface Shape

Numerical simulation shows that the radial distribution of G changes with the shape of the crystal-melt interface. The boundary conditions that determine the temperature distributions in the crystal are the temperature distribution at the crystal periphery and at the crystal-melt interface. Because the temperature at the crystal-melt interface is the melting temperature, the shape of crystal-melt interface and the temperature distribution at the crystal periphery determine the temperature distribution in the crystal. For a series of crystal-melt interface shapes, the G distribution at the interface was calculated using the same temperature distribution at the crystal periphery. Results are shown in Fig. 4.25 illustrating that G is quite sensitive to the interface shape and thus also v/G.

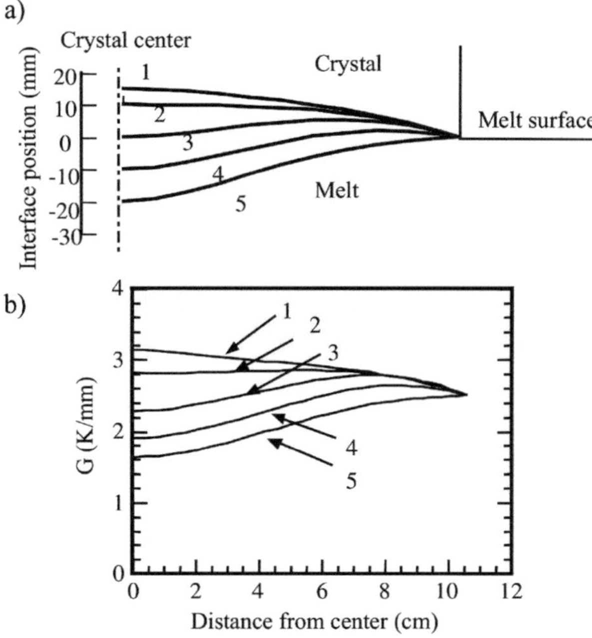

Fig. 4.25 Dependence of the axial thermal gradient G on the interface shape *top:* crystal-melt interface shapes that were studied. *Bottom:* Calculated radial G profiles. The temperature distribution of the crystal outer surface is assumed to be T (in K) $= 1685 - 2.5 \times z$, with z (in mm), the distance along the crystal axis direction [37]

4.3.3 Change of the Interface Shape by the Pulling Condition

The interface shape is thus an important factor for the control of intrinsic point defect incorporation as shown in the previous paragraph. The interface shape itself is determined by the balance of heat fluxes in the crystal and the melt and by the latent heat at the crystal-melt interface. Figure 4.26 shows the relationship between the interface height at the crystal center and the average v/G ($[v/G]_{ave}$) for a 200 mm diameter silicon crystal using horizontal magnetic fields with different strengths [50].

G_{ave} is the average axial temperature gradient defined in Fig. 4.26 and represents the heat extraction capacity of the CZ furnace and hot zone. As the heat flux in the melt changes with the strength of the horizontal magnetic field, the dependence of the interface height on v/G_{ave} also changes with the horizontal magnetic field.

It was found that v/G can be used to predict the height h of the crystal-melt interface, according to the semi-empirical linear equation [50]

$$h = \frac{a\alpha H}{\lambda_S}\frac{v}{G} + a(\alpha - 1) + b. \tag{4.13}$$

Hereby, α is a parameter that depends on the crystal diameter, H the latent heat and λ_S the thermal conductivity of the Si crystal. a and b are fitting parameters which can be obtained statistically for a given crystal diameter and magnetic field by analyzing many sets of actual interface heights and calculating the corresponding

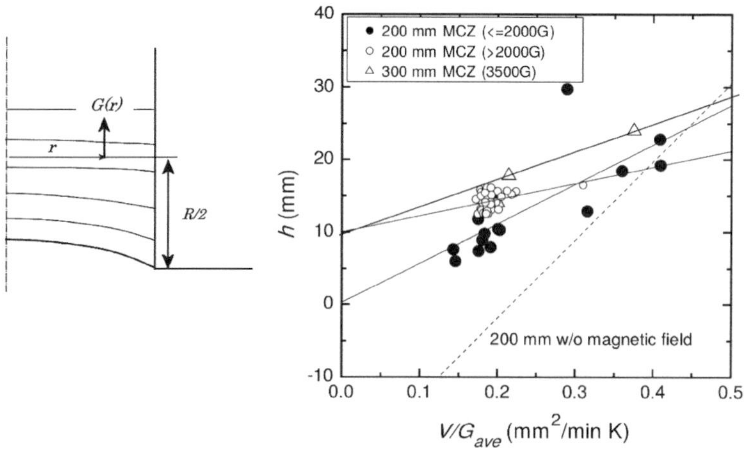

Fig. 4.26 *Left:* Schematic representation used for the definition of the average temperature gradient $G_{ave} = [2\pi \int_0^R G(r)]/[\pi r^2]$. *Right:* The relationship between the experimental interface height h and $[v/G]_{ave}$ for 150, 200 and 300 mm Si crystals grown with magnetic fields of different strengths. The *dashed line* is for a 200 mm crystal grown without magnetic field (Reprinted from [50], Copyright 2004, with permission from Elsevier)

temperature distributions for various types of hot-zones under different process conditions.

in addition, it is possible to estimate the maximum crystal growth rate v_{max} for a given hot zone from [50]

$$v_{max} = \frac{\alpha \lambda G}{(1-\alpha)H}. \tag{4.14}$$

4.4 Simulation of Intrinsic Point Defects During Crystal Growth

The optimum condition for pulling a grown-in defect-free crystal was determined based on intrinsic point defect simulations [40]. First, the results of the point defect simulation were compared with the experimental observations on grown-in intrinsic point defect cluster distributions. The temperature distribution was estimated by a global heat transfer analysis method that took the melt convection into account and fitted the shapes of the crystal-melt interface in the calculations to the actual shapes using a turbulence parameter [40, 43]. The FEMAG simulation code was used for the calculation of the temperature field with the Mixing-Length (M-L) model to take into account convection [17]. The shape of the crystal-melt interface was also evaluated experimentally by investigating the growth striations using X-ray topography.

Figure 4.27 shows how changes in the growth rate influence the distribution of the defect regions for an almost uniform G in the radial direction. Figure 4.27a shows the change in the growth rate and Fig. 4.27b shows a cross section X-ray topograph of a crystal after thermal annealing. When the growth rate is decreased, the OSF ring closes and self-interstitial type dislocation clusters appear. Regions with and without oxygen precipitates can be seen between the OSF ring and the so called B band. These regions are called P_V (vacancy rich grown-in-defect-free region) and P_I (self-interstitial rich grown-in-defect-free region), respectively. The boundary between the two regions is the actual transition between the V-rich and I-rich parts [85].

Figure 4.27c shows the contour lines of $\Delta C = C_V - C_I$ at 1000 °C using the intrinsic point defect parameters of Nakamura [43] in Table 4.1. When the contour lines of Fig. 4.27c and the defect distribution of Fig. 4.27b are compared, the shapes of the contour line of ΔC agree well with the boundary lines for the defect regions. This shows that ΔC can be used to define the domain in which each type of defect is generated.

Figure 4.28b shows an X-ray topograph made using specimens cut along the axis of another crystal grown with a changing pulling rate as shown in Fig. 4.28a, and subjected to a two-step thermal anneal to precipitate the interstitial oxygen. The position of the OSF ring is again clearly changed by the pulling rate. Usually, the difference between the concentration of vacancies and self-interstitials has

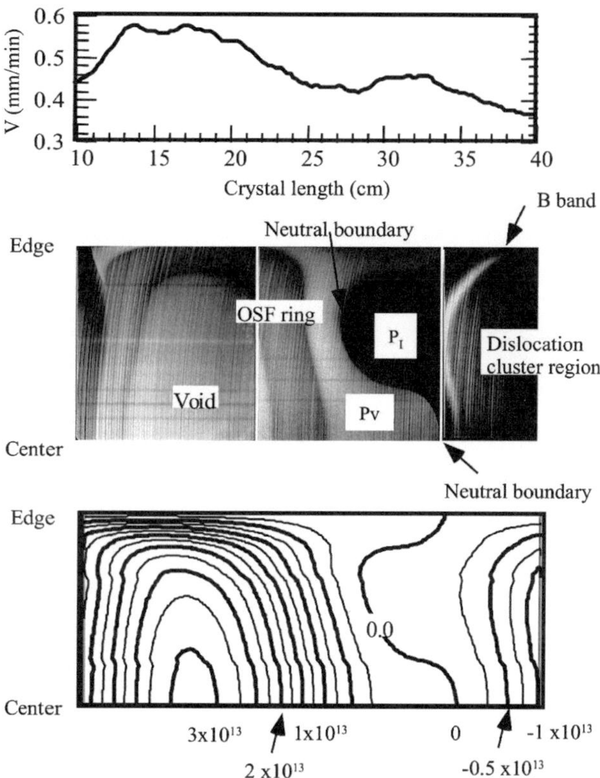

Fig. 4.27 Comparison between the defect pattern and the calculated *contour line* of the intrinsic point defect concentration difference. *Top:* Pulling rate over the crystal length. *Middle:* Cross-section X-ray topograph after 2-step annealing. *Bottom:* Calculated contour lines of $\Delta C = C_V - C_I$ at 1000 °C. P_V and P_I are the vacancy-rich and interstitial-rich parts of the crystal, respectively, without grown-in intrinsic point defect clusters. For the calculation only point defect diffusion and pair annihilation are considered as described by Eqs. (4.4), (4.5), (4.6) and (4.7) (Reproduced with permission from [42]. Copyright 2002, The Electrochemical Society)

been used as the criterion for dominant species of point defects [87]. Because the experimental observation shown in Fig. 4.28b is revealing differences in interstitial oxygen precipitation, one needs to take into account the effect of intrinsic point defects on oxygen precipitation for the further interpretation of the observations. This effect can be expressed by the critical oxide precipitate radius r_c [64, 67]

$$r_c \approx \frac{2\sigma}{\frac{xk_BT}{\Omega_{SiO_x}} ln \frac{C_{O_i}}{C^*_{O_i}} \left(\frac{C_V}{C^*_V}\right)^{\beta} \left(\frac{C^*_I}{C_I}\right)^{\gamma}}, \tag{4.15}$$

with σ the interface energy between SiO_x and Si and Ω_{SiO_x} the molecular volume of the oxide phase that is formed, e.g. amorphous SiO_2 or SiO. k_B is the

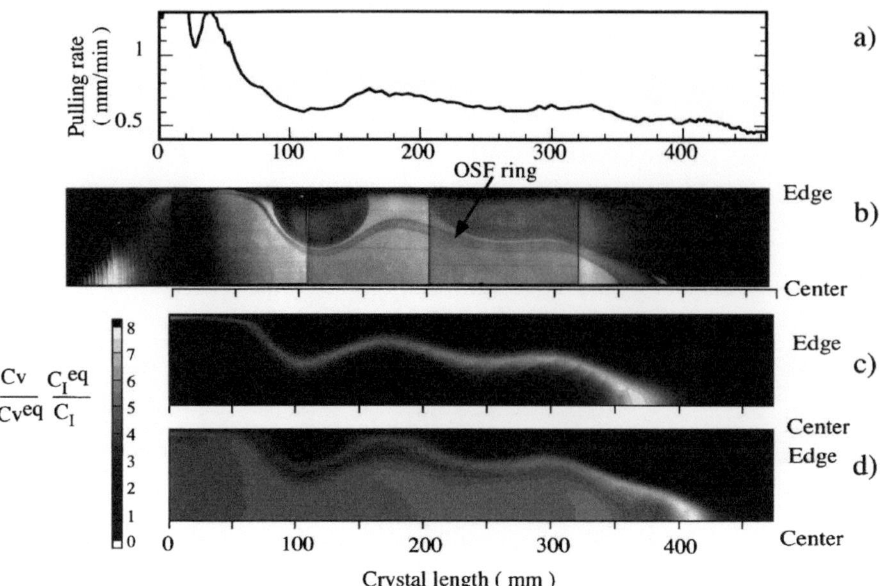

Fig. 4.28 (**a**) Change of pulling rate. (**b**) X-ray topograph made using specimens cut along the axis of crystals grown with the change of pulling rate shown in (**a**) and subjected to the two-step anneal. (**c**) and (**d**) Calculated maps of the ratio of the supersaturation of vacancies over the supersaturation of self-interstitials when each crystal position passes the position at 950 and 700 °C, respectively (Reprinted from [38], Copyright 2000, with permission from Elsevier). Besides point defect diffusion and pair annihilation as described by Eqs. (4.4), (4.5), (4.6) and (4.7), also the formation and growth of vacancy clusters is taken into account in this calculation. The model includes the absorption of intrinsic point defects by voids

Boltzmann constant. C_V, C_V^*, C_I, C_I^* and C_{O_i}, $C_{O_i}^*$ are the actual and the thermal equilibrium concentrations of vacancies V, self-interstitials I and interstitial oxygen O_i, respectively. β and γ are the number of vacancies and interstitials, respectively, that are absorbed per precipitating oxygen atom.

This equation describes the relationship between the critical radius of oxide precipitates and the point defect concentrations when the strain is assumed to be entirely released by the intrinsic point defect emission or absorption. It describes how the increase of the ratio of supersaturation of $\left(C_V/C_V^{eq}\right)\left(C_I^{eq}/C_I\right)$ decreases the critical radius of oxide precipitates and thus enhances oxygen precipitation. Therefore, in the following, the ratio $\left(C_V/C_V^{eq}\right)\left(C_I^{eq}/C_I\right)$ is used as the criterion instead of $(C_V - C_I)$ [53] or $\left(C_V - C_V^{eq}\right) - \left(C_I - C_I^{eq}\right)$ [14].

Figure 4.28c shows the calculated map of $\left(C_V/C_V^{eq}\right)\left(C_I^{eq}/C_I\right)$ when each crystal position passes the position at 950 °C. As shown, the peak position of $\left(C_V/C_V^{eq}\right)\left(C_I^{eq}/C_I\right)$ corresponds with the position of the OSF ring. Because the oxide precipitates leading to the formation of the stacking faults in the OSF ring are formed in the temperature range between 900 and 1000 °C [24] during the crystal pulling process, formation of the OSF ring is assumed to be caused by the strong supersaturation of vacancies at 950 °C.

The peak of $\left(C_V/C_V^{eq}\right)\left(C_I^{eq}/C_I\right)$ forms via the following process. The vacancy concentration at the center of the crystal is always higher than at the periphery side due to the dependence of G on the radial position. Therefore, the vacancy clusters in the central part begin to form earlier, and the vacancy concentration decreases with the absorption of vacancies by the vacancy clusters. Due to that the vacancy concentration becomes lower in the center than at the periphery where the vacancy clusters did not yet form. As a result, the peak of vacancy supersaturation forms just after the void formation in the central part of the crystal. Calculations allowed to reproduce OSF ring formation based on the the proposal of Harada et al. [21] that large grown-in oxide precipitates form the nuclei for the stacking faults leading to the OSF ring.

Figure 4.28d shows the calculated map of $\left(C_V/C_V^{eq}\right)\left(C_I^{eq}/C_I\right)$ when each crystal position passes the position at 700 °C. As shown, the distribution of $\left(C_V/C_V^{eq}\right)\left(C_I^{eq}/C_I\right)$ at 700 °C corresponds well with the distribution of oxygen precipitates formed by the two step annealing. Assuming that the nuclei that grow to full precipitates by annealing below 1000 °C are formed in the temperature range between 600 and 750 °C during the crystal growth process, the distribution of oxygen precipitation corresponds well with the intrinsic point defect supersaturation ratio at 700 °C.

Although it has been assumed by most researchers that the precipitated phase is SiO_2 and most precipitation simulators are based on that assumption, there is more and more evidence that the precipitated phase is SiO_x with x close to 1 [10, 45, 54, 65, 70] which will have a significant influence on the critical precipitate size given by (4.15) and thus on the BMD density which might change the results shown in Fig. 4.28c, d unless also the intrinsic point defect parameters are adapted.

From these figures one can define different zones in the crystal according to the value of $\Delta C = C_V - C_I$:

Vacancy-rich

- Void region: $>1.5 \times 10^{13}\,\text{cm}^{-3}$
- OSF region: 1.5×10^{13}–$0.5 \times 10^{13}\,\text{cm}^{-3}$
- Defect-free region (PV): 0–$0.5 \times 10^{13}\,\text{cm}^{-3}$

Interstitial-rich

- Defect-free region (PI): 0 to $-0.25 \times 10^{13}\,\text{cm}^{-3}$
- I type defect region: $< -0.25 \times 10^{13}\,\text{cm}^{-3}$

Since this result is obtained by fitting only one growth condition, one needs to verify whether or not this criterion of ΔC calculated using this parameter set also holds for other growth conditions. To verify this, the growth rates at which the defect types changed were investigated using 200 mm diameter crystals grown under 14 different growth conditions. V_{osf}, the pulling rate at which the OSF ring closes, $V_{neutral}$, the pulling rate at which the intrinsic point defects become neutral and do not cluster, and V_{id}, the pulling rate at which the self-interstitial type defects appear, were investigated in particular. These pulling rates were investigated at the center of the crystal and at 90 mm from the center. Calculations were performed using the temperature distributions calculated by a global heat transfer analysis method that took into account melt convection and the actual shapes of the crystal-melt interface. In Fig. 4.29a, b, the horizontal axis shows the values of V_{osf}, $V_{neutral}$, and

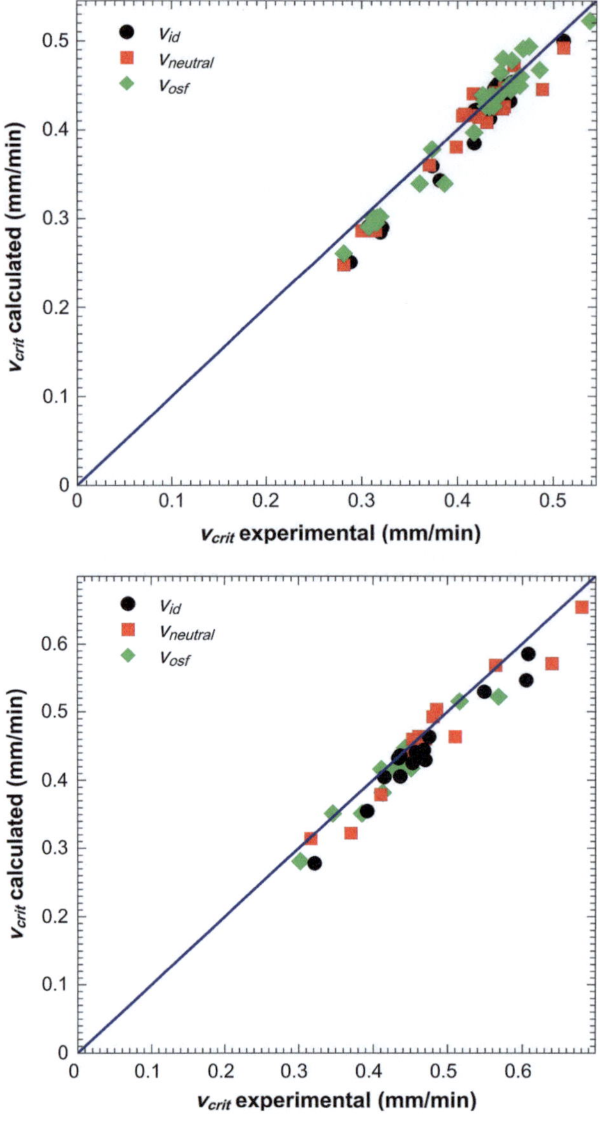

Fig. 4.29 Comparison between experimental and predicted values of V_{osf}, $V_{neutral}$, and V_{id} [42]. *Top:* Crystal center. *Bottom:* 90 mm from the crystal center

V_{id} obtained experimentally, while the vertical axis shows the values predicted by calculation. The criterion of ΔC calculated using the Nakamura et al. parameter set corresponds well to the experimental result.

4.5 Optimum Condition for a Grown-in Defect Free Intrinsic Crystal

Using the intrinsic point defect parameters of Nakamura et al. [43] in Table 4.1, the growth condition whereby the concentrations of intrinsic point defects are such that all positions in the radial direction are free from grown-in defects, was examined. In other words, the condition at which ΔC falls in the range between -0.25×10^{13} and $0.5 \times 10^{13}\,\mathrm{cm}^{-3}$ at all positions in the radial direction, was examined. This state is obtained for the crystal length between 26 and 34 cm as shown in Fig. 4.27. The temperature distribution at the crystal periphery and the shape of crystal-melt interface determine the temperature gradient (G) inside the crystal. Therefore, the growth conditions that determine the distribution of the intrinsic point defect distribution are: (1) the temperature distribution at the crystal periphery, (2) the shape of the crystal-melt interface, and (3) the growth rate. Using typical data of 200 mm-diameter crystals, a series of crystal-melt interface shapes (Fig. 4.30 was established (a)) as well as axial-temperature distributions at the crystal periphery (Fig. 4.30b). Subsequently the relationship between v and the radial distribution of ΔC was calculated using combinations of both sets of data.

Figure 4.31 illustrates the change of the formation boundary of the self-interstitial type defects ($\Delta C = -0.25 \times 10^{13}\,\mathrm{cm}^{-3}$) and OSF ($\Delta C = 0.5 - 1.5 \times 10^{13}\,\mathrm{cm}^{-3}$) and of the voids ($\Delta C > 1.5 \times 10^{13}\,\mathrm{cm}^{-3}$) by the crystal growth rate. The temperature gradients of the crystal peripheries (G_{outer}) in Fig. 4.31 are 3 K/mm near the melting point and the height differences of the center and periphery of the crystal-melt interface ($H = H_{center} - H_{outer}$) are -15 mm (left figure) and $+15$ mm (right figure), respectively. In this way a growth rate was obtained at which all radial positions become defect-free (right figure) and wafers without grown-in defects can be produced from a crystal grown within that growth rate range. In the right figure, the tolerance (Δv) of the growth rate in order to obtain a grown-in-defect-free crystal is only 0.02 mm/min. Such growth rate window could not be obtained for the left figure and one of the defect types will always be found at some position in the radial direction.

The tolerance (Δv) of the growth rate to obtain a grown-in-defect-free crystal with all combinations of G_{outer} and H was examined and the results are summarized in Fig. 4.30. Figure 4.32 shows the contour lines of Δv for G_{outer} and H. As the figure shows, a grown-in-defect-free crystal (positive Δv) can be found in a specific region for G_{outer} and H. The explanation for this is relatively straightforward. According to heat transfer analysis, an increase of G_{outer} increases the ratio of G_{outer}/G_{center} with G_{center} the temperature gradient at the crystal center, hence the

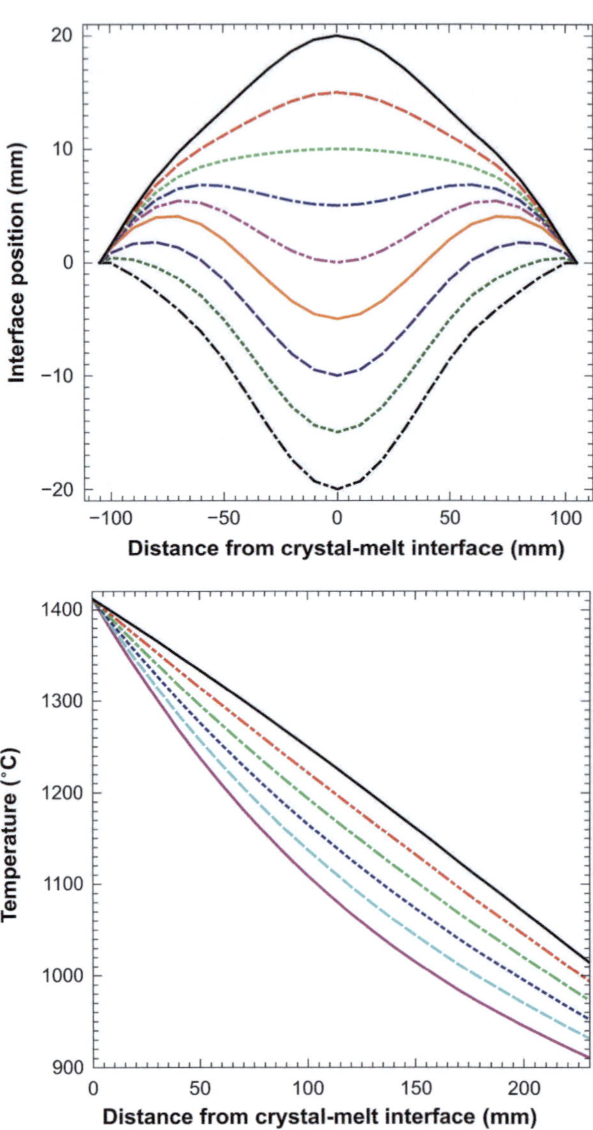

Fig. 4.30 Series of the shapes of the crystal-melt interface (*top*) and axial-temperature distributions at the crystal periphery used in the calculations (*bottom*) [42]. All values are derived from typical data for 200 mm diameter crystals. G at the crystal periphery (G_{outer}) is the temperature gradient from melting point to 1350 °C. G_{outer} varies between 1.5 and 4.5 K/mm. The height at the crystal center (H) is −20 to +20 mm

distribution of G becomes non-uniform in the radial direction. The increase of H, meanwhile, increases G_{center}, hence the distribution of G in the radial direction is made uniform by the increase of H under a condition with a higher G_{outer}. To obtain a near neutral state of point defects in all positions in the radial direction, the distribution of G (in other words, v/G) in the radial direction must be uniform. Therefore, G_{outer} in the region free of grown in defects is dependent on H, as shown in Fig. 4.32.

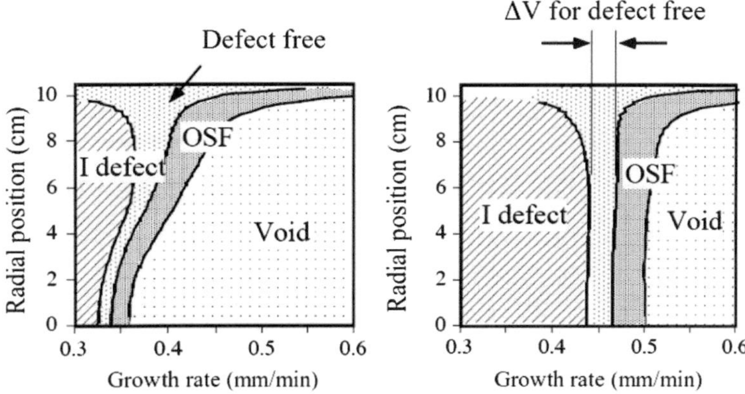

Fig. 4.31 Change in the formation boundary of the self-interstitial type defect ($\Delta C = -0.25 \times 10^{13}\,\text{cm}^{-3}$) and OSF ($\Delta C = 0.5 \times 10^{13}$–$1.5 \times 10^{13}\,\text{cm}^{-3}$) by the growth rates. $G_{outer} = 3\,\text{K/mm}$, *left:* $H = -15\,\text{mm}$, **right:** $H = +15\,\text{mm}$ (Reproduced with permission from [42]. Copyright 2002, The Electrochemical Society)

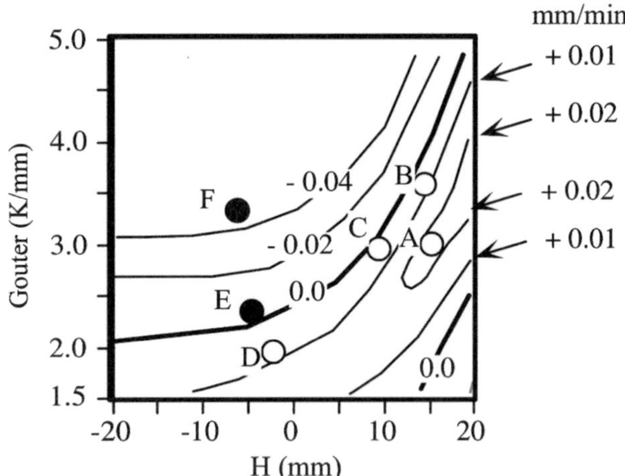

Fig. 4.32 Contour line of Δv (mm/min) for G_{outer} and H. Δv is the tolerance of the growth rate for the grown-in-defect-free state. Negative Δv indicates that the defect-free state does not exist (Reproduced with permission from [42]. Copyright 2002, The Electrochemical Society)

Table 4.2 Δv for the conditions of the plots shown in Fig. 4.32

	G_{outer} (K/mm)	H (mm)	Δv (mm/min)
A	3.3	15.0	0.020
B	3.5	14.0	0.009
C	3.0	9.3	0.002
D	1.9	−2.5	0.005
E	2.2	−5.0	−0.002
F	3.3	−6.0	−0.040

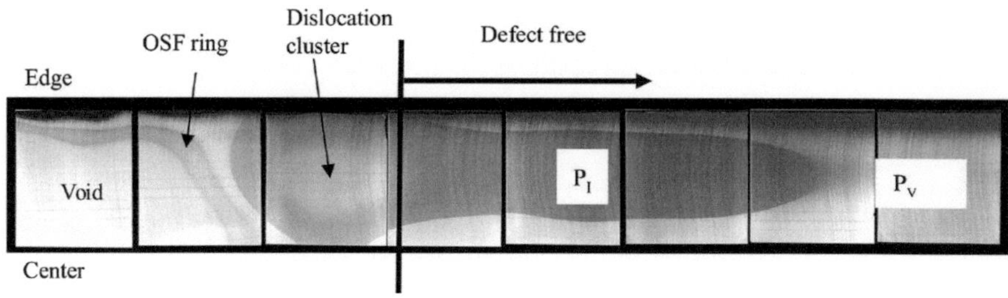

Fig. 4.33 Cross-sectional X-ray topographs of a grown-in-defect-free 200 mm diameter Si crystal after the 2-step annealing [37]. The interstitial oxygen concentration is about 1.4×10^{18} (atoms/cm^3, old ASTM). P_V: Vacancy rich grown-in-defect-free crystal, P_I: Interstitial rich grown-in-defect-free crystal

To verify Fig. 4.32, experiments were performed similar to those shown in Fig. 4.27 and evaluated Δv for each growth condition. Table 4.2 shows Δv for the conditions of the plots shown in Fig. 4.32. The calculation results correspond well to the experimental results. Figure 4.33 shows the example of a grown-in-defect-free crystal grown under the condition D in Fig. 4.32.

From Fig. 4.32, the maximum of tolerance (Δv) of the growth rate for the grown-in-defect-free state is around 0.02 mm/min. Thus, the following three problems must be solved in order to be able to obtain crystals free of grown-in defects requiring a tight control of the crystal pulling conditions to have a high yield production:

- High-precision control of the growth rate, ± 1–$3\,\%$.
- Reproducibility of G, ± 1–$3\,\%$.
- The control of the growth rate depends on the crystal length (G_{outer} and the shape of the crystal-melt interface change in crystals of different lengths).

4.6 Impact of Stress on *v*/*G*

Recently, Vanhellemont [66] predicted that the effect of thermal stress on the equilibrium concentrations of point defects significantly changes $[v/G]_{crit}$. Nakamura et al. [44] were the first to report experimental results for 300 mm diameter crystals clearly showing that compressive thermal stress increases the incorporated vacancy concentration and reduces $[v/G]_{crit}$. Using the same experimental data it was shown that one should take into account that previously determined intrinsic point defect parameters that were used are "contaminated" by not taking into account the effect of stress to explaining the experimental observation on grown-in intrinsic point defect clusters. This "built-in" stress effect can be taken into account to some extent by fitting the experimental data using a stress dependent Voronkov criterion yielding a consistent estimate of the pre-factors of the exponential equations describing the diffusivity and thermal equilibrium concentration of the intrinsic point defects [78]. Subsequently it was shown that when assuming plane stress, the experimental data of Nakamura et al. can be reproduced accurately using the ab initio calculated impact of plane stress on the intrinsic point defect formation energies [56]. It is interesting to remark also that an arbitrary linear fit lies very close the best fit using ab initio formation energies obtained for plane stress as is illustrated in Fig. 4.34. The linear fit using all data points yields [79]

$$[v/G]_{crit} = (0.1720 \pm 0.0007) + (0.00146 \pm 0.00007)\sigma_{ave}, \tag{4.16}$$

with $[v/G]_{crit}$ in mm^2 K^{-1} min^{-1} and the average plane stress σ_{ave} in MPa. The goodness of fit R^2 is 0.844.

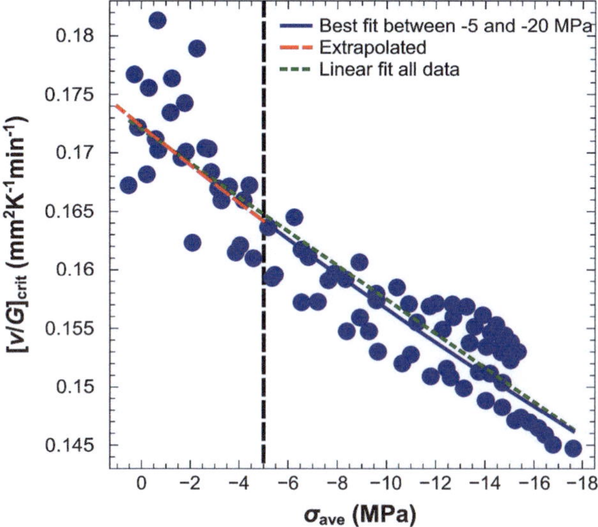

Fig. 4.34 Γ_{crit} as function of planar stress at the crystal-melt interface fitted to experimental data for 300 mm diameter CZ silicon crystals, using the data and point defect parameters of Nakamura et al. [43]. The fit was performed for average planar stresses between −5 and 20 MPa, and suggests that the intrinsic point defect data of Table 4.1 were obtained for an average stress level of −5.1 MPa. A linear fit using all data points, lies very close [79]

When the crystal diameter is 300 or 450 mm and the shape of crystal-melt interface is concave (center position is higher than that of periphery), the thermal stress is thus high enough to affect significantly $[v/G]_{crit}$. In this case, the thermal stress at the center of crystal-melt interface is compressive and that at the periphery is much lower and close to zero. Therefore, a homogeneous distribution of G to the radial direction is not appropriate any more to grow a grown-in-defect-free crystal and it is necessary to modify the G distribution for the compensation of the mentioned stress effect.

The described impact of the thermal stress on $[v/G]_{crit}$ will thus become a critical issue in the development of 450 mm diameter Si crystal pulling processes and will also pose a serious challenge for surface inspection tools based on light scattering as small voids can be formed inside nominal "COP-free" regions due to compressive stress variations [27].

4.7 Impact of Doping: Dopant Induced Stress and Trapping

4.7.1 Reported Experimental Observations

The impact of doping on intrinsic point defect cluster formation during single-crystal silicon growth was already realized in the early days and possible mechanisms to explain this impact were already identified, i.e. change of Fermi level, dopant atom size related stress, dopant atoms acting as traps for intrinsic point defects and complex formation of intrinsic point defects with dopant atoms [6, 11, 12]. Transmission electron microscopy investigations revealed e.g. small self-interstitial type dislocation clusters in 3×10^{18} B cm^{-3} doped CZ samples while in 1.2×10^{18} Sb cm^{-3} doped samples only vacancy type inclusions were observed, most probably the first direct observation of voids formed by vacancy clustering during crystal growth [11]. Main conclusions of that work were that for CZ-grown crystals, the formation of self-interstitial clusters was suppressed by doping with donors (Sb, P and As) while doping with acceptors (B, Ga) suppresses formation of vacancy clusters whereby the dopant concentrations should at least 10^{17} cm^{-3} to have a noticeable effect [12]. In addition to the effect of electrically active dopants, As shown in Figs. 4.35 and 4.36, Abe et al. [2, 6] also studied the effect of various dopants on grown-in defects in FZ crystals. A strong suppression of vacancy type point defect clusters was e.g. observed in case of N and C doping while O doping led to an increase of the vacancy cluster formation.

4.7.1.1 Impact of Dopants

The impact of B doping of COP distribution was studied and results are shown in Figs. 4.37 and 4.38. Heavy B doping makes the crystal more interstitial rich than a

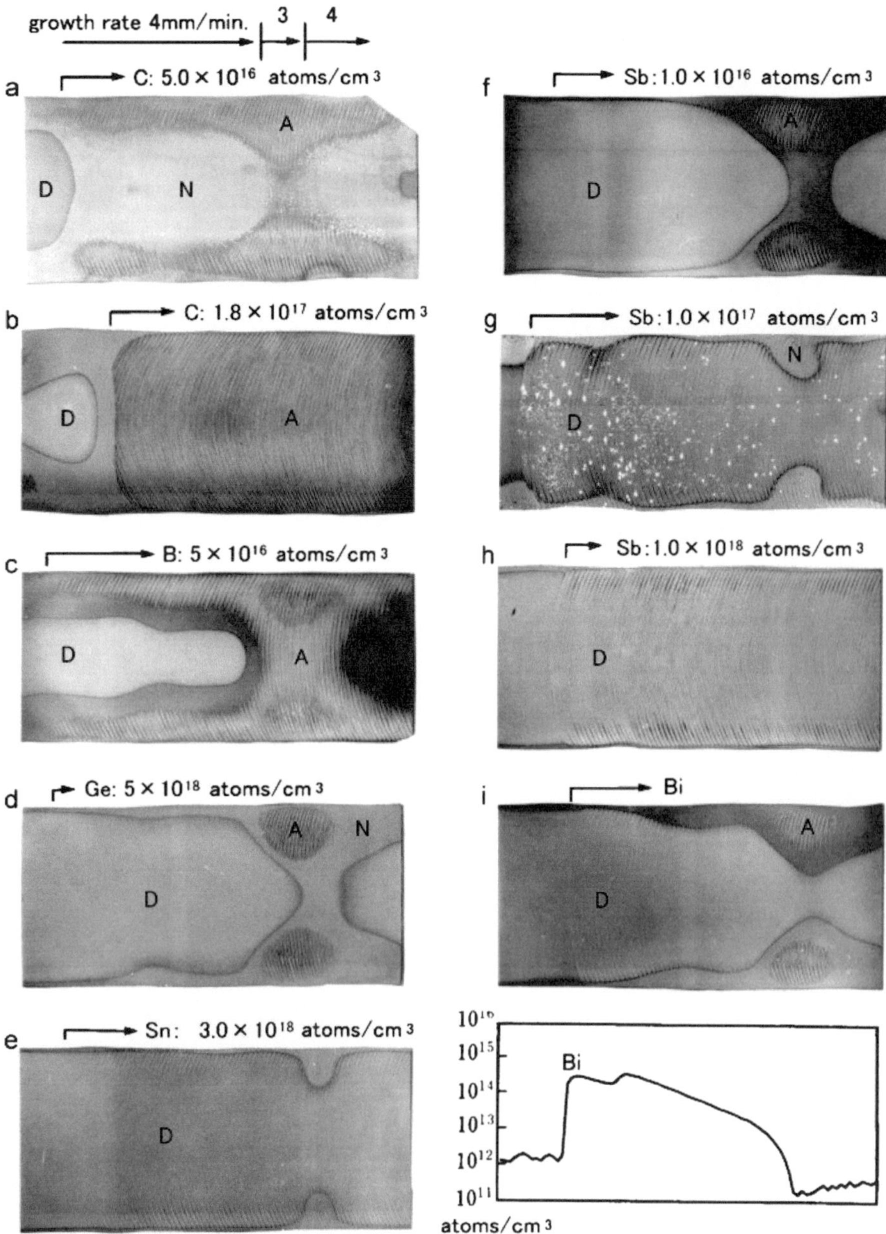

Fig. 4.35 X-ray topographs of longitudinally cut wafers after Cu decoration of 42 mm FZ crystals doped with different impurities (Reprinted from [2], Copyright 2011, with permission from Elsevier). Two different C concentrations (**a**) and (**b**), B (**c**), Ge (**d**), Sn (**e**), three different Sb concentrations (**f**), (**g**) and (**h**), and Bi with concentration profile (**i**)

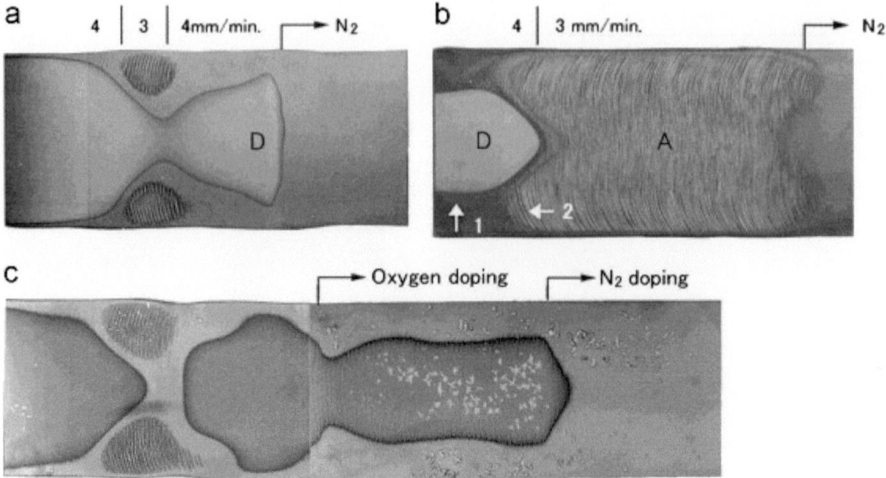

Fig. 4.36 X-ray topographs of FZ specimens after Cu decoration (Reprinted from [2], Copyright 2011, with permission from Elsevier). *Top left:* Generation of D-defects and *top right:* A-defects is suppressed by N doping. *Bottom:* The effect of O and N doping on D-defect formation

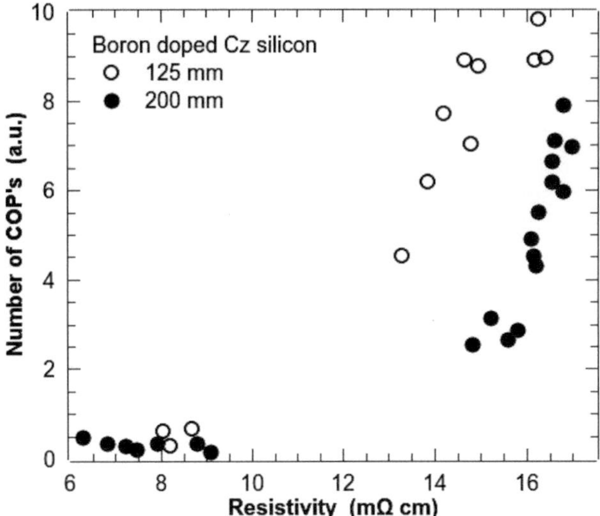

Fig. 4.37 Number of COP's after 4h SC1 in 125 and 200 mm boron doped CZ Si crystals [68]

moderately one for the same v/G. This leads to a decrease of the COP density with ultimately the disappearance of COP's. As illustrated in Fig. 4.38 the OSF ring also moves inwards with increasing B concentration confirming that the crystal gradually changes from vacancy-rich to interstitial-rich.

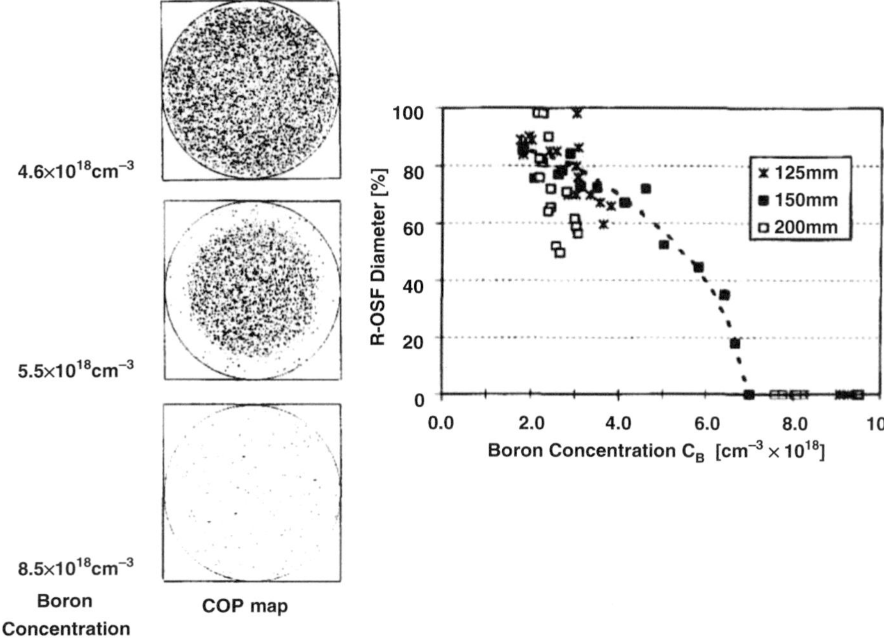

Fig. 4.38 *Top:* Radial COP distribution on wafers prepared from 150 mm diameter Si crystals pulled with similar v/G but with different boron concentration C_B. *Bottom:* OSF-ring diameter as function of C_B for 125, 150 and 200 mm crystals (Reprinted from [15], Copyright 1997, with permission from Elsevier)

A similar effect on COP's is observed when doping with nitrogen as illustrated in Fig. 4.39. The bottom figure shows results obtained on oxygen doped FZ Si illustrating that the presence of interstitial oxygen is needed in order to form COP's that are large enough to be observed. Both results are in agreement with the X-ray topography observations on nitrogen doped FZ as shown in Fig. 4.36.

Dornberger et al. [15] reported a linear dependence of Γ_{crit} on active boron concentration based on the study of the dependence of the stacking fault ring position on the level of boron doping. Similar results were obtained by Valek et al. [63]. Nakamura et al. [43] published results of an extensive study of the effect of seven impurities, i.e. B, C, O, N, Sb, P and As, on grown-in defects in CZ-grown silicon crystals, discussing also the different mechanisms mentioned above. They concluded that doping with high concentrations of acceptors or donors (e.g. about $5 \times 10^{18}\,\text{cm}^{-3}$ B and about $5 \times 10^{19}\,\text{cm}^{-3}$ As and P) leads to a significant change of the intrinsic point defect equilibrium concentrations and thus also to significant changes of Γ_{crit} whereby acceptor doping makes the crystal more self-interstitial-rich, while donor doping leads to a more vacancy-rich crystal. At the same time, the impurities that enhance the incorporation of one type of intrinsic point defect also suppress partially the clustering of these point defects into grown-in defects

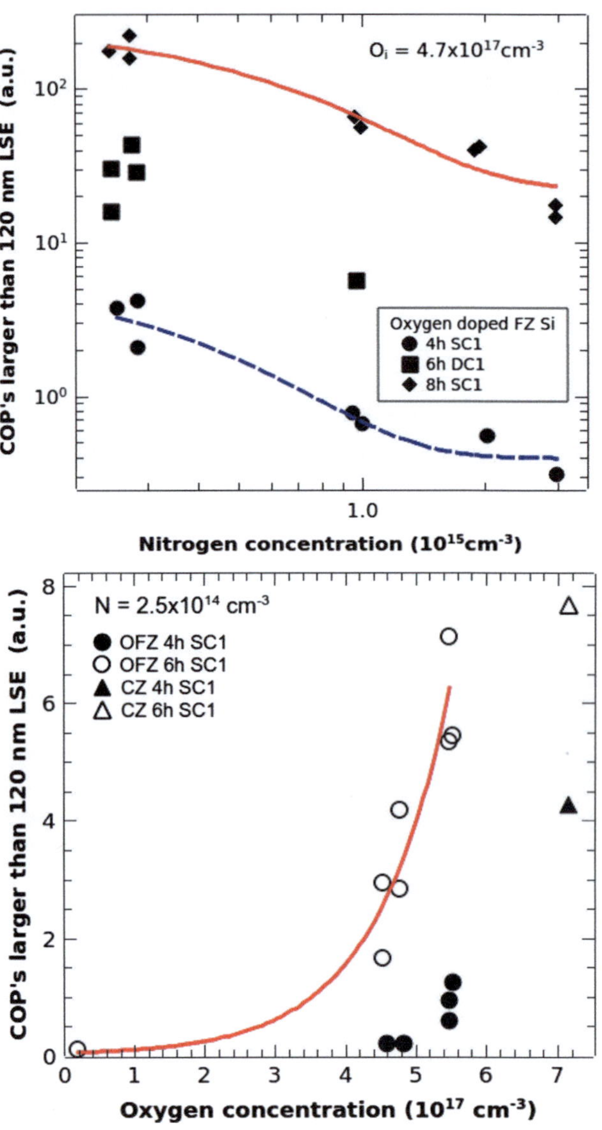

Fig. 4.39 The number of COP's larger than 120 nm LSE, as function of the nitrogen (*top*) and oxygen (*bottom*) concentration in 125 mm diameter, oxygen doped FZ and CZ crystals [20, 68]. The lines are empirical fits with an exponential function to guide the eye

when the impurities strongly bind with the intrinsic point defects. This is the case for N, P and As which are dopants that enhance vacancy incorporation as evidenced by the shift of the OSF ring position but at the same time suppress void formation [41]. This is illustrated for P and As doping in Fig. 4.40. B and C doping, on the other hand enhance self-interstitial incorporation [15] but suppress the formation of dislocation clusters [7].

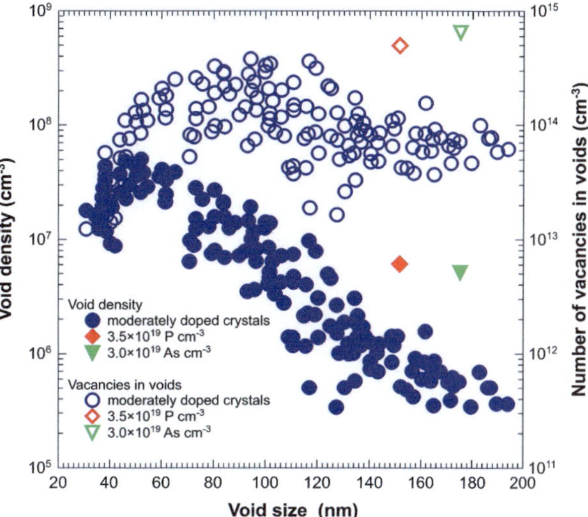

Fig. 4.40 Void density vs. void size for moderately doped CZ Si crystals and for a heavily P doped and a heavily B doped CZ Si crystal [43]. An estimate of the total vacancy concentration incorporated in the voids, assumed to be octahedral, is also shown

4.7.2 Ab Initio Calculation of Dopant Impact on Uncharged Intrinsic Point Defects

4.7.2.1 Calculation Details

The formation energies of uncharged *V* and *I* at all sites within a sphere with 0.6 nm radius around the dopant atom for *V* and with 0.5 nm radius for *I* are calculated by DFT. Substitutional p-type (B and Ga), neutral (C, Ge, and Sn) and n-type (P, As, Sb, and Bi) dopants were considered.

The formation energy of *V* within a sphere with 0.6 nm radius around the dopant atom is calculated as follows. The cell size of a perfect 216-atom supercell after its geometry is optimized, is 1.6392 nm. A dopant atom is introduced at the center of perfect 216-atom supercells and a vacancy is placed at the 1st to 5th neighbors from the dopant atom. It turns out that there are 46 possible sites for *V* within the 0.6 nm radius sphere around the dopant atom. The formation energy of *V* at each site is calculated by fully relaxing the ionic coordinates. The number of sites at 1st to 5th neighbors from the dopant atom are 4 (1st), 12 (2nd), 12 (3rd), 6 (4th) and 12 (5th), respectively.

The formation energy of *I* within a sphere with 0.5 nm radius around the dopant atom is calculated as follows. A self-interstitial *I* is placed at all interstitial sites around the dopant atom. Hereby *I* at the tetrahedral (T)-, hexagonal (H)-, [110] dumbbell (D)-, [100] D-, and [114] D-sites is considered as shown in Fig. 4.41. The formation energy for each site is calculated by fully relaxing the ionic coordinates. Further details on the calculation procedures can be found in [55].

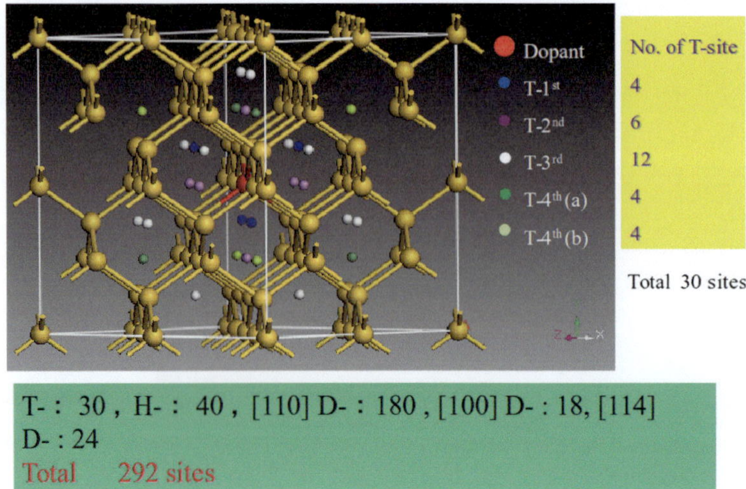

Fig. 4.41 Possible sites of a vacancy within a sphere with 0.6 nm radius around dopant atom (*red atom*) in a 64-atom supercell [55]

4.7.2.2 Intrinsic Point Defect Formation Energy

As an illustration, Fig. 4.42 shows the calculated vacancy formation energy as function distance from common neutral and n-type dopant atoms. The dotted lines from the 1st to the 5th position in the figure indicate the distance from the dopant before the cell size and ionic coordinates are relaxed. It is clear that $E^f_{V,dope}$ at the 1st site differs for the different dopants. The formation energy of vacancies with larger dopants is smaller than with smaller dopants. Since the electrical state is almost the same for the same types of dopants, this result is mainly due to the difference in local strain. Furthermore, $E^f_{V,dope}$ at and far from the 2nd sites are close for neutral dopants without changing the electrical state, and close to that in undoped Si. This indicates that local strain effects are only important at the 1st site from the dopant atom. The type and magnitude of local strain differ for the n-type dopants P, As, Sb, and Bi. However, starting from the 2nd site, $E^f_{V,dope}$ is nearly the same for all n-type dopants and about 0.3–0.4 eV lower than that in perfect Si. This illustrates that not local strain but the electrical state around n-type dopants mainly determines the V formation energy.

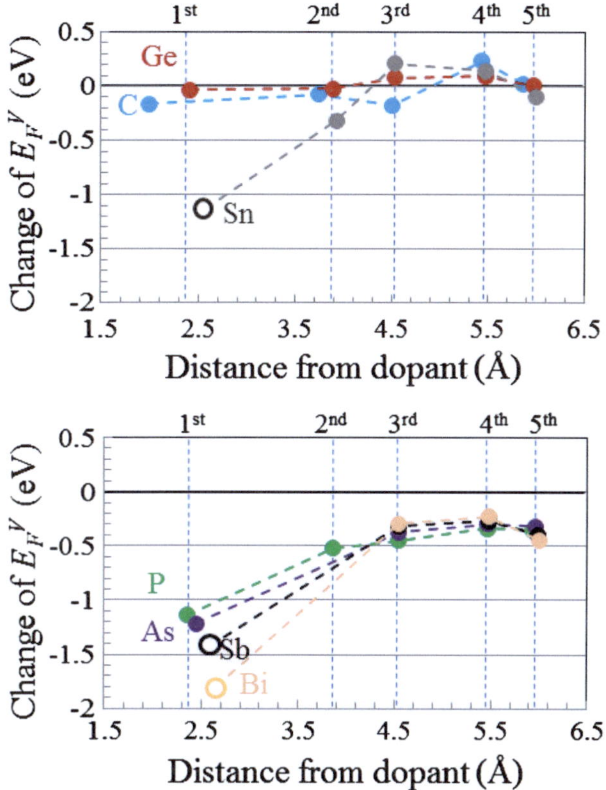

Fig. 4.42 *Top:* Dependence of the change of vacancy formation energy from that for Jahn-Teller distortion in a perfect Si crystal on the distance from neutral and *bottom:* n-type dopant atoms. *Open circles* for Sn, Sb, and Bi indicate split vacancies. *Dotted lines* from 1st to 5th indicate the distance from the dopant before cell size and ionic coordinates are relaxed (Reprinted with permission from [55]. Copyright 2013, AIP Publishing LLC)

More results for $E^f_{V,dope}$ in case of p-type dopants and also similar results for the formation energy $E^f_{I,dope}$ of the self-interstitial in case of neutral-, n- and p-type dopants, can be found elsewhere [55].

The DFT calculations also allow to calculate $C_V - C_I$ as function of dopant concentration and type and thus give an indication when a crystal changes from vacancy-rich to self-interstitial-rich as illustrated in Fig. 4.43 for common dopants in Si [55]. Excellent agreement between calculation and experiment is obtained.

- Top graph of Fig. 4.43:
  - Sb > 10^{17} cm^{-3}. D-defects increase;
  - Sb = 1×10^{18} cm^{-3}, Sn = 3×10^{18} cm^{-3}. Similar impact on D-defects increase;
  - Bi = 10^{15} cm^{-3}. No impact on D-defects [2].

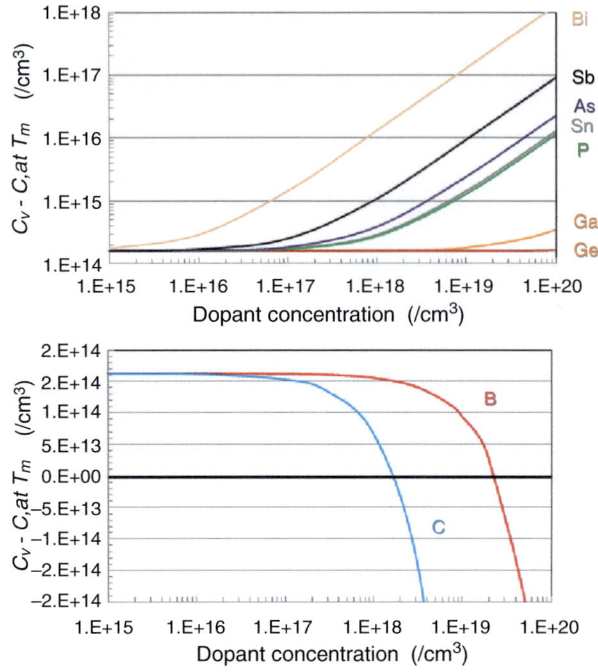

Fig. 4.43 Calculated dependence of $C_V - C_I$ at melting temperature of Si on dopant concentration and type. The parameters of Table 4.1 were used for the calculation. Excellent agreement between calculation and experiment is obtained (Reprinted with permission from [55]. Copyright 2013, AIP Publishing LLC)

  - As $> 2 \times 10^{18}\,\text{cm}^{-3}$. Voids increase [57].
  - Ge $= 10^{20}\,\text{cm}^{-3}$. No impact on voids [76].

- Bottom figure of Fig. 4.43:
  - B $> 5 \times 10^{18}\,\text{cm}^{-3}$. OSF-ring shrinks [15].
  - B $= 2 \times 10^{19}\,\text{cm}^{-3}$. I-rich crystal;
  - C $= 6 \times 10^{16}\,\text{cm}^{-3}$. V decreases [43].

One could summarize the results of dopant effects as follows: Self-interstitials for p-type dopants are rather stable at T-sites, while self-interstitials for neutral and n-type dopants are rather stable at D-sites. Furthermore, $E^f_{I,dope}$ differs for the types of dopants as follows:

- In case of p-type dopants, $E^f_{I,dope}$ at T-sites up to ≈0.6 nm is reduced by about 0.5–1.3 eV compared to that in perfect Si. No remarkable differences in $E^f_{I,dope}$ are obtained for B and Ga atoms. These results are due to the Coulomb (long-range) interaction between acceptor and positively charged I at the T-site.

- In case of neutral dopants, $E^f_{I,dope}$ at the D-sites up to $\approx$0.3 nm from C atom is reduced by about $\approx$0.7–1.3 eV compared to that in perfect Si while Ge and Sn atoms have no impact on $E^f_{I,dope}$. These results are due to the larger local tensile strain introduced by the C atom, which reduces the formation energy of the neutral I at the D-site. In case of n-type dopants, $E^f_{I,dope}$ at the D-sites up to $\approx$0.3 nm is reduced by about 0.5 eV compared to that in perfect Si. P, which gives local tensile strain, shows the largest impact on the neutral I at D-site among the n-type dopants.

Impact of doping on Γ_{crit}. For simplicity it was assumed in case of heavy doping that $C(T_m) = C^{eq}(T_m) = C^{eq,tot}(T_m)$, that E^f_I and E^f_V are the intrinsic values which is a reasonable assumption close to melting temperature and for not too high doping, and that D_I and D_V are not affected by doping.

Figure 4.44 shows the calculated dependence on dopant concentration of Γ_{crit} normalized with respect to the intrinsic value Γ^0_{crit}, assuming zero stress. The circles in the figures for B and C doping are the experimental results obtained by Nakamura et al. [43]. The calculated results for heavy B doping agree well with the experimental results. Although there is only one experimental plot for C doping, it is also close to the calculated line. To the best of our knowledge, no experimental results for the impact of n-type dopants on critical $(v/G)_{crit}$ have yet been reported in literature.

Summarizing the main DFT results: a model was proposed explaining quantitatively the intrinsic point defect behavior in heavily doped Si single-crystals growing from a melt:

- The incorporated total V and I (sum of free V or I and V or I around the dopants) concentration at melting temperature depend on the type and concentration of dopant. This is due to the change in the formation energies of V and I around the dopant atoms, which is caused by the electrical state and magnitude of local strain depending on the types and sizes of the dopant.
- Most of the total V and I concentrations contribute to pair recombination at much higher temperatures than those at which voids are formed (1100 °C). This means that the values of $C_V^{eq,tot}(Tm)$ and $C_I^{eq,tot}(Tm)$ determine the impact of the dopant type and concentration on the dominant point defect (with v/G greater than the window of defect free Si) and also the critical $(v/G)_{crit}$.

The main strength of the proposed model is that it explains point defect behavior for all dopants and for all concentrations and is in excellent agreement with all experimental data known to the authors.

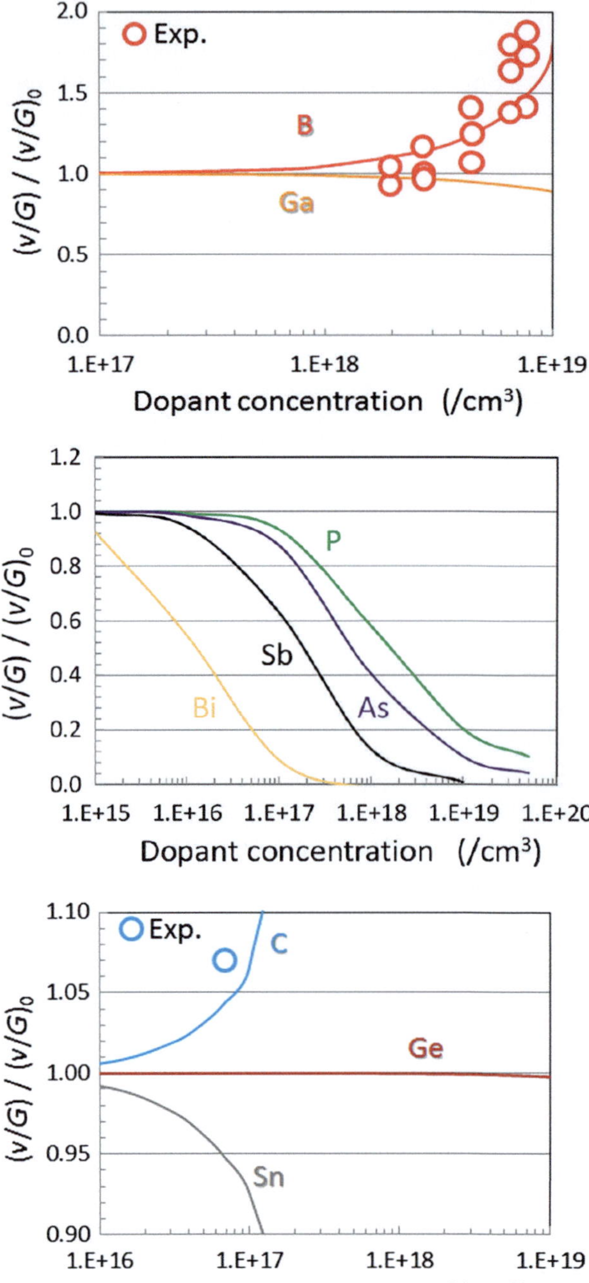

Fig. 4.44 Calculated dependence of $(v/G)_{crit}$ on dopant concentration, normalized with respect to the intrinsic value obtained with low doping. The *open symbols* are experimental data (Reprinted with permission from [55]. Copyright 2013, AIP Publishing LLC)

4.8 Open Questions: Impact of Fermi Level and Intrinsic Point Defect Formation Energy Near Crystal-Melt Interface

4.8.1 Impact of Fermi Level

Electrically active dopants influence the bandgap and the Fermi level and the bandgap and thus also the formation energy of charged intrinsic point defects as illustrated in Fig. 4.45 (based on DFT calculations at 0 K). In n^+ Si, the double negatively charged vacancy V^{2-} will have the lowest formation energy while in p^+ Si, it is the double positively charged self-interstitial I^{2+}. Close to melting temperature the situation changes as illustrated in Fig. 4.46 and only for very high n-type doping (well above a few times $10^{19}\,\mathrm{cm}^{-3}$), V^{2-} still has a slightly lower formation energy than the neutral vacancy V while For p-Si, I^{2+} has the lowest formation energy for all dopant concentrations. Figure 4.47 shows calculated $[v/G]_{crit}$ fitted to experimental data. Fermi level and bandgap effects were thereby taken into account. Figure 4.48 shows the data of Nakamura et al. and Dornberger et al. from the previous figure together with calculated curves showing that the data can be reproduced very well when assuming a different planar thermal stress (= hot zone) for both crystals.

4.8.2 Interstitial and Vacancy Formation Energy Near Crystal Surfaces

c(4 × 2) structure models of the Si (001) crystal surface were investigated to clarify the behavior of intrinsic point defects near crystal surfaces. Figure 4.49 shows the calculated dependence of the intrinsic point defect formation energy as function of the distance to the (001) surface. Regarding crystal growth from a melt, the most important result is the existence of the formation energy differences between the surface and the bulk for both types of intrinsic point defects. The presence of these energy differences supports the macroscopic model in which the generation and the recombination of Frenkel pairs is more important inside the bulk than at the surface. The obtained results also support that boundary conditions of the point defect concentrations at the surface in simulations can be set at fixed values. Namely, the existence of barriers makes it possible for the surface to act as a reservoir of intrinsic point defects. When simulating crystal growth from a melt, these fixed values for the boundary conditions should, however, be defined, taking into account the impact of the crystal crystal-melt interface.

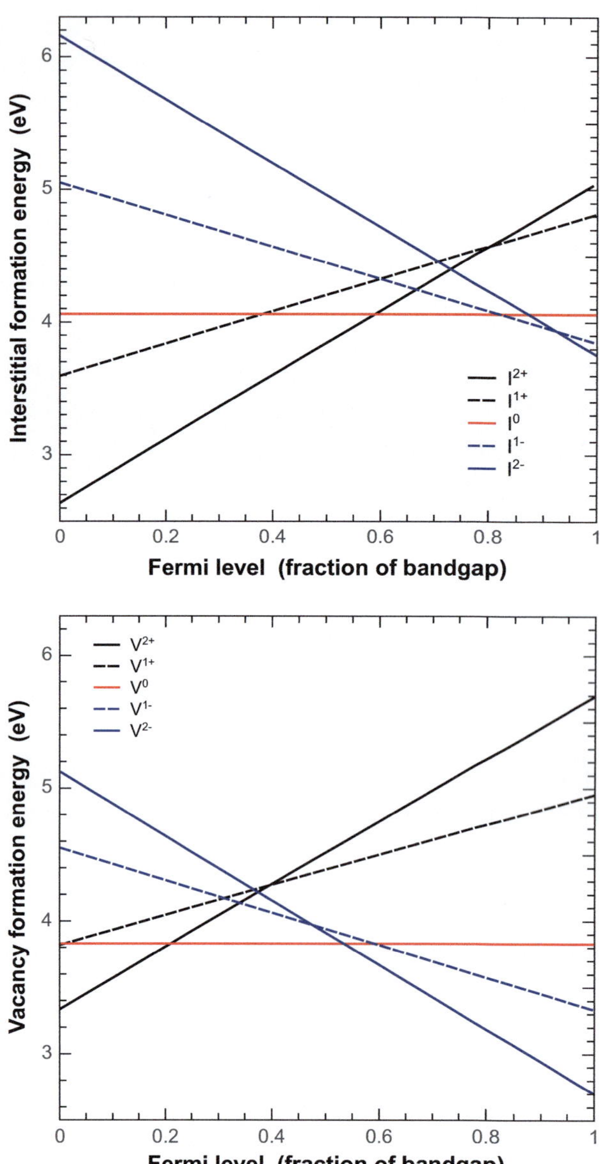

Fig. 4.45 *Top:* Calculated formation energies for the different charge states of the self-interstitial and *bottom:* the vacancy as a function of the Fermi level expressed as fraction of the bandgap (Reproduced with permission from [77]. Copyright 2013, The Electrochemical Society)

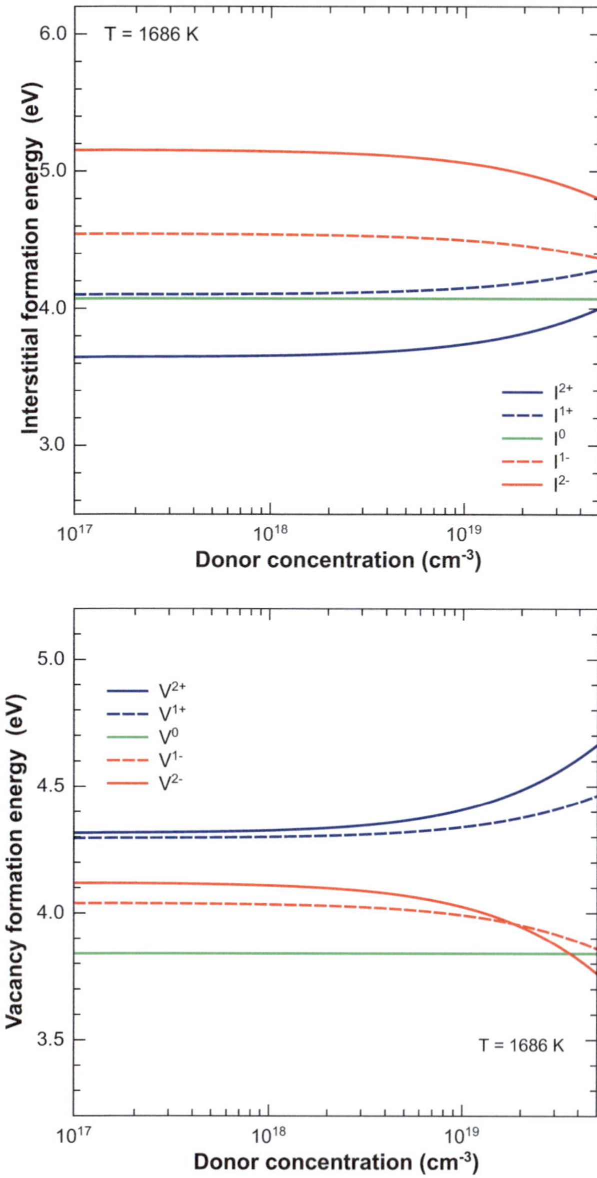

Fig. 4.46 *Top:* Calculated formation energies at 1413 °C of the different charge states of the self-interstitial and *bottom:* the vacancy as function of the donor concentration (Reproduced with permission from [77]. Copyright 2013, The Electrochemical Society)

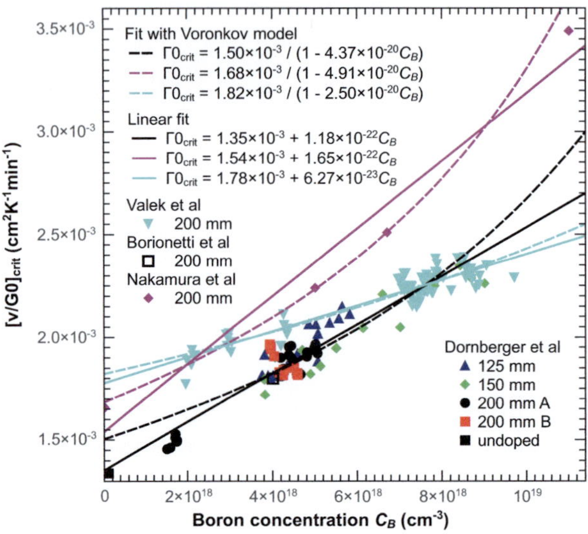

Fig. 4.47 Experimental and calculated curves obtained by taking into account dopant induced stress and Fermi level effects for a crystal grown with a thermal stress of 7.25 and 8 MPa (Reproduced with permission from [77]. Copyright 2013, The Electrochemical Society)

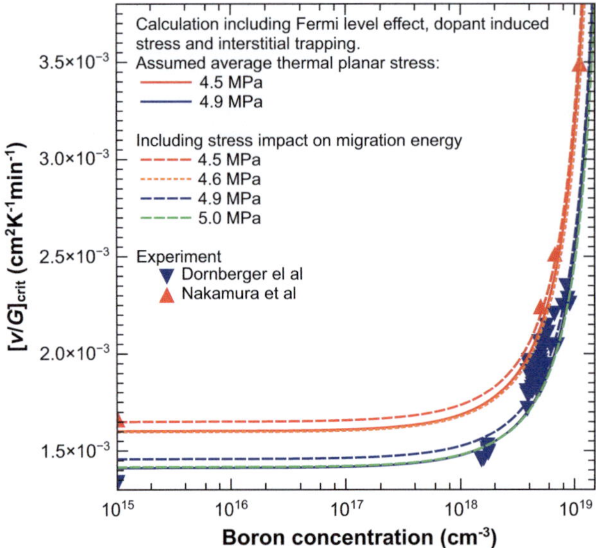

Fig. 4.48 Experimental data [15, 43] and calculated curves taking into account both dopant induced stress and Fermi level effect assuming double positively charged interstitials and neutral vacancies [77] and planar stress of 4.5 and 4.9 MPa [79]

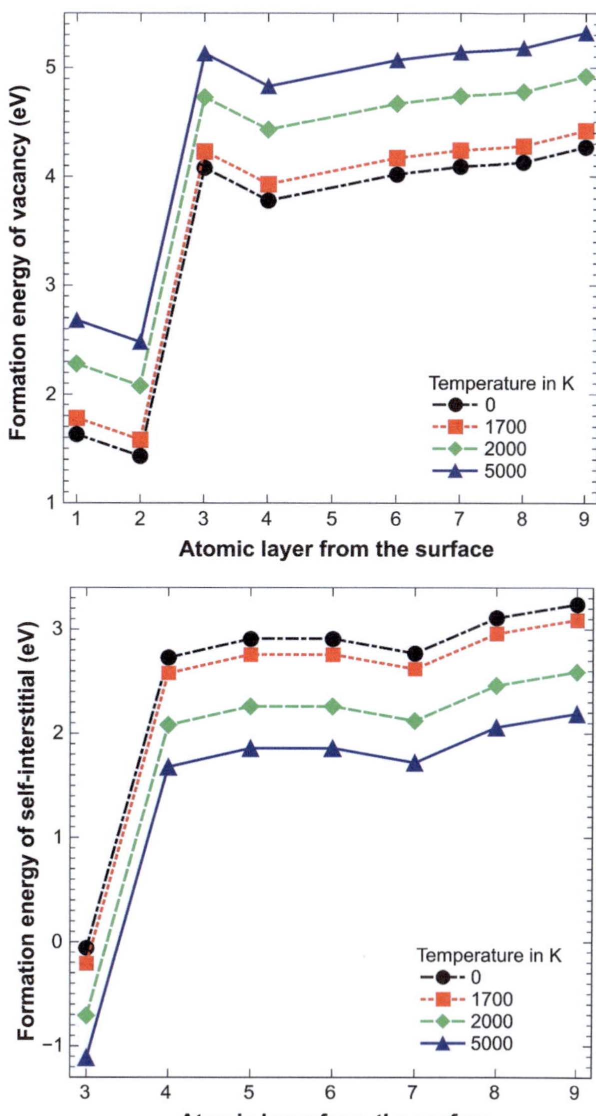

Fig. 4.49 DFT calculations reveal a decrease of the intrinsic point defect formation energy for the vacancy (*top*) and for the interstitial (*bottom*) in the first atomic layers near the Si surface (Reprinted with permission from [26]. Copyright 2012, AIP Publishing LLC)

4.9 Conclusions and Further Work

It was shown that the impact of thermal stress, substitutional dopants and Fermi level changes, when treated separately, are well understood and can be described quantitatively. The challenges for the near future are to:

- develop a unified model taking all effects into account simultaneously;

- clarify the thermal equilibrium intrinsic point defect concentrations at the crystal-melt interface;
- clarify the mechanisms behind the experimentally observed impact of interstitial oxygen, nitrogen and hydrogen doping using ab initio calculations.

These results will be very useful in the further development of economically viable pulling processes for 450 mm, defect-free Si single-crystals and even to improve those used at the moment for the production of 300 mm crystals.

Exploring the impact of doping on the intrinsic point defect properties in Ge will also allow to clarify if it will be possible to use a similar Voronkov criterion to pull defect-free Ge crystals.

References

1. Abe, T.: The formation mechanism of grown-in defects in CZ silicon crystals based on thermal gradients measured by thermocouplesnear growth interfaces. Mater. Sci. Eng. B **73**, 16 (2000)
2. Abe, T.: Generation and annihilation of point defects by doping impurities during FZ silicon crystal growth. J. Cryst. Growth **334**, 4 (2011)
3. Abe, T., Takahashi, T.: Point defects in silicon melt growth from the experimental results. ECS Trans. **335**(11), 75 (2010)
4. Abe, T., Takahashi, T.: Intrinsic point defect behavior in silicon crystals during growth from the melt: a model derived from experimental results. J. Cryst. Growth **334**, 16 (2011)
5. Abe, T., Harada, H., Chikawa, J.: Swirl defects in float-zoned silicon crystals. Physica B **116**, 139 (1983)
6. Abe, T., Harada, H., Chikawa, J.: Microdefects and impurities in dislocation-free silicon crystals. Mat. Res. Soc. Proc. **14**, 1 (1983)
7. Asayama, E., Ono, T., Takeshita, M., Hourai, M., Sano, M., Tsuya, H.: Radial distribution of thermally-induced defects in heavily boron-doped silicon wafers. Electrochem. Soc. Proc. **98–1**, 546 (1998)
8. Bender, H., Vanhellemont, J., Schmolke, R.: High resolution structure imaging of octahedral void defects in as-grown Czochralski silicon. Jpn. J. Appl. Phys. **36**, L1217 (1997)
9. Dash, W.C.: Growth of silicon crystals free from dislocations. J. Appl. Phys. **30**, 459 (1959)
10. De Gryse, O., Clauws, P., Van Landuyt, J., Lebedev, O., Claeys, C., Simoen, E., Vanhellemont, J.: Oxide phase determination in silicon using infrared spectroscopy and transmission electron microscopy techniques. J. Appl. Phys. **91**, 2493 (2002)
11. de Kock, A.J.R., Stacy, W.T., van de Wijgert, W.M.: The effect of doping on microdefect formation in as-grown dislocation-free Czochralski silicon crystals. Appl. Phys. Lett. **34**, 611 (1979)
12. de Kock, A.J.R., van de Wijgert, W.M.: The effect of doping on the formation of swirl defects in dislocation-free Czochralski-grown silicon crystals. J. Cryst. Growth **49**, 718 (1980)
13. Depuydt, B., De Jonghe, M., De Baets, W., Romandic, I., Theuwis, A., Quaeyhaegens, C., Deguet, C., Akatsu, T., Letertre, F.: Germanium materials. In: Claeys, C., Simoen, E. (eds.) Germanium-Based Technologies. From Materials to Devices. Elsevier (2007). ISBN-13:978-0-08-044953-1
14. Dornberger, E., von Ammon, W.: The dependence of ring-like distributed stacking faults on the axial temperature gradient of growing Czochralski silicon crystals. J. Electrochem. Soc. **143**, 1648 (1996)

15. Dornberger, E., Gräf, D., Suhren, M., Lambert, U., Wagner, P., Dupret, F., von Ammon, W.: Influence of boron concentration on the oxidation-induced stacking fault ring in Czochralski silicon crystals. J. Cryst. Growth **180**, 343 (1997)
16. Dornberger, E., von Ammon, W., Virbulis, J., Hanna, B., Sinno, T.: Modeling of transient point defect dynamics in Czochralski silicon crystals. J. Cryst. Growth **230**, 291 (2001)
17. Dupret, F., Van den Bogaert, N., Assaker, R., Regnier, V.: Mathematical modeling of the growth of large diameter Czochralski silicon crystals considering melt dynamics. Electrochem. Soc. Proc. **98–1**, 396 (1998)
18. Ebe, T.: Factors determining the saturation of point defects in growing silicon crystals. J. Crystal Growth **203**, 387 (1999)
19. Föll, H.: http://www.tf.uni-kiel.de/matwis/amat/semi_en/index.html
20. Gräf, D., Suhren, M., Lambert, U., Schmolke, R., Ehlert, A., Ammon, W.v., Wagner, P.: Characterization of crystal quality by delineation of COP and the impact on the silicon wafer surface. Electrochem. Soc. Proc. **96–13**, 117 (1996)
21. Harada, K., Tanaka, H., Watanabe, T., Furuya, H.: Defects in the oxidation-induced stacking fault ring region in Czochralski silicon crystal. Jpn. J. Appl. Phys. **37**, 3194 (1998)
22. Hens, S., Vanhellemont, J., Poelman, D., Clauws, P., Romandic, I., Theuwis, A., Holsteyns, F., Van Steenbergen, J.: Experimental and theoretical evidence for vacancy-clustering-induced large voids in Czochralski-grown germanium crystals. Appl. Phys. Lett. **87**, 061915 (2005)
23. Hourai, M., Kajita, E., Nagashima, T., Fujiwara, H., Ueno, S., Sadamitsu, S., Miki, S., Shigematsu, T.: Growth parameters determining the type of grown-in defects in Czockralski silicon crystals. Mater. Sci. Forum **196**, 1713 (1995)
24. Hourai, M., Kelly, G.P., Tanaka, T., Umeno, S., Ogushi, S.: Control of grown-in defects in Czochralski silicon crystals. Electrochem. Soc. Proc. **99–1**, 372 (1999)
25. Ikematsu, Y., Mizutani, T., Nakai, K., Fujinami, M., Hasebe, M., Ohashi, W.: Transmission electron microscope observation of grown-in defects detected by bright-field infrared-laser interferometer in Czochralski silicon crystals. Jpn. J. Appl. Phys. **37**, L196 (1998)
26. Kamiyama, E., Sueoka, K., Vanhellemont, J.: Ab initio study of vacancy and self-interstitial properties near single crystal silicon surfaces. J. Appl. Phys. **111**, 083507 (2012)
27. Kamiyama, E., Vanhellemont, J., Sueoka, K., Araki, K., Izunome, K.: Thermal stress induced void formation during 450mm defect free silicon crystal growth and implications for wafer inspection. Appl. Phys. Lett. **102**, 082108 (2013)
28. Kissinger, G., Vanhellemont, J., Gräf, D., Claeys, C., Richter, H.: IR-LST a powerful non-invasive tool to observe crystal defects in as-grown silicon, after device processing, and in heteroepitaxial layers. Inst. Phys. Conf. Ser. **149**, 19 (1996)
29. Kulkarni, M.S.: Lateral incorporation of vacancies in Czochralski silicon crystals. J. Cryst. Growth **310**, 3183 (2008)
30. Larsen, T.L., Jensen, L., Lüdge, A., Riemann, H., Lemke, H.: Numerical simulation of point defect transport in floating-zone silicon single crystal growth. J. Crystal Growth **230**, 300 (2001)
31. Lee, S.H., Song, D.W., Oh, H.J., Kim, D.H.: Modeling of defects generation in 300 mm silicon monocrystals during Czochralski growth. Jpn. J. Appl. Phys. **49**, 121303 (2010)
32. Lemke, H., Südkamp, W.: Analytical approximations for the distributions of intrinsic point defects in grown silicon crystals. Phys. Status Solidi A **176**, 843 (1999)
33. Lu, Z., Kimbel, S.: Growth of 450 mm diameter semiconductor grade silicon crystals. J. Cryst. Growth **318**, 193 (2011)
34. Menzel, R.: Growth Conditions for Large Diameter FZ Si Single Crystals. PhD Thesis, Technical University Berlin (2013)
35. Miyazaki, M., Miyazaki, S., Yanase, Y., Ochiai, T., Shigematsu, T.: Microstructure observation of "crystal-originated particles" on silicon wafers. Jpn. J. Appl. Phys. **34**, 6303 (1995)
36. Miyazaki, M., Miyazaki, S., Kitamura, T., Yanase, Y., Ochiai, T., Tsuya, H.: Influence of crystal-originated "particle" microstructure on silicon wafers on gate oxide integrity. Jpn. J. Appl. Phys. **36**, 6187 (1997)

37. Nakamura, K.: Control of point defects in growing single crystal Si. In: Proceedings of the 7th Forum on the Science and Technology of Silicon Materials, 19–22 Oct 2014, Hamamatsu, p. 237 (2014)
38. Nakamura, K., Saishoji, T., Tomioka, J.: Simulation of point defect distributions in silicon crystals during melt-growth. J. Cryst. Growth **210**, 49 (2000)
39. Nakamura, K., Saishoji, T., Tomioka, J.: Grown-in defects in silicon crystals. J. Cryst. Growth **237–239**, 1678 (2002)
40. Nakamura, K., Maeda, S., Togawa, S., Saishoji, T., Tomioka, J.: Effect of the shape of crystal-melt interface on point defect reaction in silicon crystals. Electrochem. Soc. Proc. **2000–17**, 31 (2000)
41. Nakamura, K., Saishoji, T., Togawa, S., Tomioka, J.: Influence of nitrogen on the point defect reaction in silicon crystal. In: Proceedings of Forum on the Science and Technology of Silicon Materials 2001, Japan, 26–28 Nov 2001, p. 109
42. Nakamura, K., Saishoji, T., Tomioka, J.: Simulation of the point defect diffusion and growth condition for defect free Cz silicon crystal. Electrochem. Soc. Proc. **2002–2**, 554 (2002)
43. Nakamura, K., Suewaka, R., Saishoji, T., Tomioka, J.: The effect of impurities on the grown-in defects in Cz-Si crystals, (B, C, N, O, Sb, As, P). In: Kaneta, H.Y., Sumino, K. (eds.) Proceedings of the Forum on the Science and Technology of Silicon Materials 2003, Shonan village center, Kanagawa, 25–27 Nov 2003, p. 161
44. Nakamura, K., Suewaka, R., Ko, B.-G.: Experimental study of the impact of stress on the point defect incorporation during silicon growth. ECS Solid State Lett. **3**, N5 (2014)
45. Nicolai, J., Burle, N., Pichaud, B.: Determination of silicon oxide precipitate stoichiometry using global and local techniques. J. Cryst. Growth **363**, 93 (2013)
46. Nishimoto, M., Nakamura, K., Hourai, M., Ono, T., Sugimura, W., Motooka, T.: Determination of physical properties for point defects during Cz silicon crystal growth by high-precision thermal simulations. J. Jpn. Inst. Met. Mater. **75**, 657 (2011, in Japanese)
47. Raskin, G., Romandic, I., Wouters, L., Hellin, D., Teerlinck, I., Meuris, M., Mijlemans, P.: Germanium Substrates for Micro-electronic Applications up to 300mm Diameter. Presented at Symposium B: High-Mobility Group-IV Materials and Devices, of the 2004 MRS Spring Meeting, San Francisco, 12–16 Apr 2004
48. Sadamitsu, S., Umeno, S., Koike, Y., Hourai, M., Sumita, S., Shigematsu, T.: Dependence of the grown-in defect distribution on growth rates in Czochralski silicon. Jpn. J. Appl. Phys. **32**, 3675 (1993)
49. Schmolke, R., Angelberger, W., von Ammon, W., Bender, H.: Characterization of interstitial-related bulk defects in p^+ silicon substrates by epitaxial deposition. Solid State Phenom. **82–84**, 231 (2002)
50. Shiraishi, Y., Maeda, S., Nakamura, K.: Prediction of solid-liquid interface shape during CZ Si crystal growth using experimental and global simulation. J. Cryst. Growth **266**, 28 (2004)
51. Sinno, T.: A bottom-up multiscale view of point-defect aggregation in silicon. J. Cryst. Growth **303**, 5 (2007)
52. Sinno, T., Brown, R.A.: Modeling microdefect formation in Czochralski silicon. J. Electrochem. Soc. **146**, 2300 (1999)
53. Sinno, T., Brown, R.A., Ammon, W.v., Dornberger, E.: Point defect dynamics and the oxidation-induced stacking-fault ring in Czochralskigrown silicon crystals. J. Electrochem. Soc. **145**, 302 (1998)
54. Sueoka, K., Akatsuka, M., Yonemura, M., Ono, T., Asayama, E., Katahama, H.: Effect of heavy boron doping on oxygen precipitation in Czochralski silicon substrates of epitaxial wafers. J. Electrochem. Soc. **147**, 756 (2000)
55. Sueoka, K., Kamiyama, E., Vanhellemont, J.: Density functional theory study on the impact of heavy doping on Si intrinsic point defect properties and implications for single crystal growth from a melt. J. Appl. Phys. **114**, 153510 (2013)
56. Sueoka, K., Kamiyama, E., Vanhellemont, J., Nakamura, K.: Impact of plane thermal stress near the melt/solid interface on the v/G criterion for defect-free large diameter single crystal Si growth. ECS Solid State Lett. **3**, P69 (2014)

57. Sugimura, W., Ono, T., Umeno, S., Hourai, M., Sueoka, K.: Defect formation behaviors in heavily doped Czochralski silicon. ECS Trans. **2**, 95 (2006)
58. Takada, K., Yamagishi, H., Imai, M.: Research and development of super silicon wafers. Electrochem. Soc. Proc. **98–1**, p. 376 (1998)
59. Tan, T.Y.: Mass transport equations unifying descriptions of isothermal diffusion, thermomigration, segregation, and position-dependent diffusivity. Appl. Phys. Lett. **73**, 2678 (1998)
60. Tanahashi, K., Kikuchi, M., Higashino, T., Inoue, N., Mizokawa, Y.: Concentration of point defects changed by thermal stress in growing CZ silicon crystal: effect of the growth rate. J. Cryst. Growth **210**, 45 (2000)
61. Tracy, C.J., Fejes, P., Theodore, N.D., Maniar, P., Johnson, E., Lamm, A.J., Paler, A.M., Malik, I.J., Ong, P.: Germanium-on-insulator substrates by wafer bonding. J. Electron. Mat. **33**, 886 (2004)
62. Trauwaert, M.-A., Vanhellemont, J., Lambert, U., Gräf, D., Kenis, K., Mertens, P.W., Heyns, M.: Differential interference contrast microscopy of defects in as-grown and annealed Si wafers. Solid State Phenom. **57–58**, 387 (1997)
63. Valek, L., Lysacek, D., Sik, J.: OISF pattern and grown-in precipitates in heavily boron doped silicon. J. Electrochem. Soc. **154**, H904 (2007)
64. Vanhellemont, J., Claeys, C.: A theoretical study on the critical radius of precipitates and its application to silicon oxide in silicon. J. Appl. Phys. **62**, 3960 (1987)
65. Vanhellemont, J.: Diffusion limited oxygen precipitation in silicon: Precipitate growth kinetics and phase formation. J. Appl. Phys. **78**, 4297 (1995)
66. Vanhellemont, J.: Intrinsic point defect incorporation in silicon single crystals grown from a melt, revisited. J. Appl. Phys. 110, 063519 (2011); Erratum: "Intrinsic point defect incorporation in silicon single crystals grown from a melt, revisited. J. Appl. Phys. 110, 063519 (2011)". J. Appl. Phys. 110, 129903 (2011); Response to "Comment on 'Intrinsic point defect incorporation in silicon single crystals grown from a melt, revisited'. J. Appl. Phys. 111, 116102 (2012)". J. Appl. Phys. 111, 116103 (2012)
67. Vanhellemont, J., Claeys, C.: A theoretical study of the critical radius of precipitates and its application to silicon oxide in silicon. J. Appl. Phys. **62**, 3960 (1987)
68. Vanhellemont, J., Dornberger, E., Graef, D., Esfandyari, J., Lambert, U., Schmolke, R., Ammon, W.v., Wagner, P.: Defects in as-grown Czochralski silicon: measurement, modeling and simulation. Proceedings of the Kazusa Akademia Park Forum on the Science and Technology of Silicon Materials, Chiba, pp. 173–196 (1997)
69. Vanhellemont, J., Senkader, S., Kissinger, G., Higgs, V., Trauwaert, M.-A., Gräf, D., Lambert, U., Wagner, P.: Measurement, modelling and simulation of defects in as-grown Czochralski silicon. J. Cryst. Growth **180**, 353 (1997)
70. Vanhellemont, J., De Gryse, O., Clauws, P.: Critical precipitate size revisited and implications for oxygen precipitation in silicon. Appl. Phys. Lett. **86**, 221903 (2005)
71. Vanhellemont, J., Simoen, E., Romandic, I., Theuwis, A.: Grown-in defects in germanium. In: Claeys, C., Simoen, E. (eds.) Germanium-Based Technologies. From Materials to Devices, pp. 41–66. Elsevier (2007). ISBN-13:978-0-08-044953-1
72. Vanhellemont, J., Śpiewak, P., Sueoka, K., Romandic, I., Simoen, E.: Intrinsic point defect properties and engineering in silicon and germanium Czochralski crystal growth. In: Proceedings of the 5th International Symposium on Advanced Science and Technology of Silicon Materials (JSPS Si Symposium), Kona, 10–14 Nov 2008 (2008). https://www.riam.kyushu-u.ac.jp/nano/hawaii2008/
73. Vanhellemont, J., Śpiewak, P., Sueoka, K., Simoen, E., Romandic, I.: A comparison of intrinsic point defect properties in Si and Ge. Mater. Res. Soc. Symp. Proc. **1070**, 1070-E06-05 (2008)
74. Vanhellemont, J., Van Steenbergen, J., Holsteyns, F., Roussel, P., Meuris, M., Młynarczyk, K., Śpiewak, P., Geens, W., Romandic, I.: On the characterisation of grown-in defects in Czochralski-grown Si and Ge. J. Mater. Sci.: Mater. Electron. **19**, S42 (2008)
75. Vanhellemont, J., Śpiewak, P., Sueoka, K., Romandic, I.: On intrinsic point defect cluster formation during Czochralski crystal growth. Phys. Status Solidi C **6**, 1906 (2009)

76. Vanhellemont, J., Zhang, X., Xu, W., Chen, J., Ma, X., Yang, D.: On the assumed impact of germanium doping on void formation in Czochralskigrown silicon. J. Appl. Phys. **108**, 123501 (2010)
77. Vanhellemont, J., Kamiyama, E., Sueoka, K.: Silicon single crystal growth from a melt: On the impact of dopants on the v/G criterion. ECS J. Solid State Sci. Technol. **2**, P166 (2013)
78. Vanhellemont, J., Kamiyama, E., Sueoka, K.: Experimental study of the impact of stress on the point defect incorporation during silicon growth. ECS Solid State Lett. **3**, N5 (2014). ECS Solid State Lett. **3**, X3 (2014)
79. Vanhellemont, J., Kamiyama, E., Nakamura, K., Sueoka, K.: Intrinsic point defect behavior close to silicon melt/solid interface. In: Proceedings of CSTIC 2015. doi:10.1109/CSTIC.2015.7153491
80. von Ammon, W.: FZ and CZ crystal growth: cost driving factors and new perspectives. Phys. Status Solidi A **211**, 2461 (2014)
81. von Ammon, W., Dornberger, E., Oelkrug, H., Weidner, H.: The dependence of bulk defects on the axial temperature gradient of silicon crystals during Czochralski growth. J. Cryst. Growth **151**, 273 (1995)
82. Voronkov, V.V.: The mechanism of swirl defects formation in silicon. J. Cryst. Growth **59**, 625 (1982)
83. Voronkov, V.V., Falster, R.: Vacancy-type microdefect formation in Czochralski silicon. J. Cryst. Growth **194**, 76 (1998)
84. Voronkov, V.V., Falster, R.: Vacancy and self-interstitial concentration incorporated into growing silicon crystals. J. Appl. Phys. **86**, 5975 (1999)
85. Voronkov, V.V., Falster, R.: Grown-in microdefects, residual vacancies and oxygen precipitation bands in Czochralski silicon. J. Cryst. Growth **204**, 462 (1999)
86. Watanabe, M., Kramer, S.: 450 mm silicon: an opportunity and wafer scaling. Electrochem. Soc. Interface **15**(4), 28 (2006)
87. Wünscher, M., Lüdge, A., Riemann, H.: Crucible-free pulling of germanium crystals. J. Cryst. Growth **318**, 1039 (2011)
88. Yamagishi, H., Fusegawa, I., Fujimaki, N., Katayama, M.: Recognition of d-defects in silicon single-crystals by preferential etching and effect on gate oxide integrity. Semicond. Sci. Technol. **7**, A135 (1992)
89. Zhang, X., Xu, W., Chen, J., Ma, X., Yang, D., Gong, L., Tian, D., Vanhellemont, J.: On the impact of heavy doping on grown-in defects in Czochralski-grown silicon. ECS Trans. **34**(1), 1151 (2011)
90. Zhang, X., Ma, X., Yang, D., Vanhellemont, J.: Scanning infrared microscopy study of thermal processing induced defects in low resistivity Si wafers. Semicond. Sci. Technol. **28**, 085013 (2013)
91. Zulehner, W.: Historical overview of silicon crystal pulling development. Mater. Sci. Eng. **B73**, 7 (2000)

Chapter 5
Numerical Analysis of Impurities and Dislocations During Silicon Crystal Growth for Solar Cells

Bing Gao and Koichi Kakimoto

Abstract Impurities and dislocations in silicon crystals can cause significant deterioration in the conversion efficiency of solar cells. For increasing solar cell efficiency, reduction of impurities and dislocations is necessary. Numerical simulation is a powerful tool for improving the quality of silicon crystal for solar cells. A set of numerical analysis system that includes all processes involved in crystal growth has been developed for studying the carbon and oxygen transport in global furnace, and a three-dimensional Alexander-Haasen model was developed for studying the dislocation multiplication. The simulation helped to reduce carbon and oxygen impurities by designing a simple crucible cover and to decrease the dislocation multiplication and residual stress by using a slow cooling process. Further quality improvements can be achieved using these solvers to optimize furnace structure and operating conditions at a low cost.

Keywords Reduction of carbon and oxygen impurities • Control of dislocation multiplication • Numerical simulations • Furnace structure and operating conditions optimization

5.1 Introduction

Impurities and dislocations in silicon crystals can cause significant deterioration in the conversion efficiency of solar cells. Light elements, such as carbon and oxygen, are two of the major impurities that occur in silicon materials. Carbon impurities can strongly affect the density and electrical activity of dislocations in crystalline silicon [1], and oxygen impurities can cause SiO_2 precipitation [2], dislocation [3], and stacking faults [4]. Dislocations have been identified as one of the most efficiency-relevant defect centers in crystalline silicon for photovoltaic applications [5]. For increasing solar cell efficiency, reduction of impurities and dislocations is necessary.

B. Gao (✉) • K. Kakimoto
Research Institute for Applied Mechanics, Kyushu University, Kasuga, Fukuoka, 816-8580 Japan
e-mail: gaobing@riam.kyushu-u.ac.jp; kakimoto@riam.kyushu-u.ac.jp

Y. Yoshida, G. Langouche (eds.), *Defects and Impurities in Silicon Materials*, Lecture Notes in Physics 916, DOI 10.1007/978-4-431-55800-2_5

Experimental exploration [6–8] toward reduction of oxygen and carbon impurities in a crystal has been carried out; however, the process is time consuming and incurs high cost. Analysis of the experimental data yielded is also complex. Developments in computer technology have made simulation of the global environments of crystal growth for improving the purity of crystals possible. Many simulations of impurity transport have been conducted [9–19]; however, most of them were local simulations [9–15] that neglect the gas transport of impurities. A few studies have used global simulations [16–19]. However, the oxygen and carbon impurities in the silicon melt were neglected in one of these studies [16], and the carbon impurities in both gas and silicon melt were neglected in the others [17–19]. There have been no simulations that took into account oxygen and carbon impurities in both cooling gas and silicon melt. Therefore, a set of analysis system has been developed that includes all processes involved in crystal growth. This set of analysis system incorporates the silicon melt flow into the global simulation done by Bornside and Brown [16]. The original boundary assumption of constant SiO concentration at the melt surface [16] is replaced by a dynamic update of SiO concentration. Therefore, this set of analysis system enables the prediction of oxygen impurity in a crystal. Another assumption of the equilibrium system in the melt [16], i.e., the carbon flux from the gas into the melt is equal to that from the melt into the crystal, is also replaced by a local nonequilibirum consideration. The carbon flux at the gas/melt interface is calculated locally, and thus, carbon accumulation in the melt is included. Therefore, the present simulation might be able to correct the difference between the simulation data and experimental data [16].

To reduce dislocations in a multicrystalline or seed-cast monocrystalline silicon ingot, optimization of the crystallization and cooling processes is required. It is known that the optimization of the crystallization and cooling processes depends on good control of the cooling flux [20–22]. To effectively control the cooling flux for the reduction of dislocations, the effect of cooling flux on the generation of dislocations inside the crystal must be understood. However, the relationship between the generation of dislocations and the cooling flux is very difficult to determine experimentally, since it is not possible to obtain the cooling flux inside the crystal in high-temperature opaque furnaces. Even if the temperature outside the crucible can be measured, it is impossible to measure the temperature inside the crystal. Therefore, numerical simulation provides a good alternative for obtaining the cooling flux and determining the relationship between the generation of dislocations and the cooling flux.

Numerical simulation is used to clarify the following problems: How to reduce carbon and oxygen impurities? How to reduce dislocations by controlling the cooling flux?

5.2 Simulation of Carbon and Oxygen Impurities

5.2.1 Mechanism of Carbon and Oxygen Incorporation

Before the numerical simulation, the mechanism of incorporation of carbon and oxygen impurities into the crystal has to be known. Based on the work of Bornside and Brown [16] in Cz furnace, a basic incorporation process of carbon and oxygen impurities into a crystal in a unidirectional solidification furnace is shown in Fig. 5.1. First, the quartz crucible (SiO_2) is dissolved, and the oxygen atoms enter into the Si melt. Second, the dissolved oxygen atoms get transported to the gas/melt interface and evaporate as SiO gas. Then, the SiO gas is carried away by the argon gas flow and reacts with all of the graphite components to produce gas-phase CO. After that, the resultant CO gets transported back to the melt surface by diffusion or convection and dissolves into the melt. Finally, the C and O atoms are segregated into the crystal [16].

5.2.2 Numerical Modeling for Global Heat Transfer, Gas Flow and Impurity Transport

To correctly simulate the impurity transport inside the furnace, the phenomena described in the above section should be included. For example, the incompressible

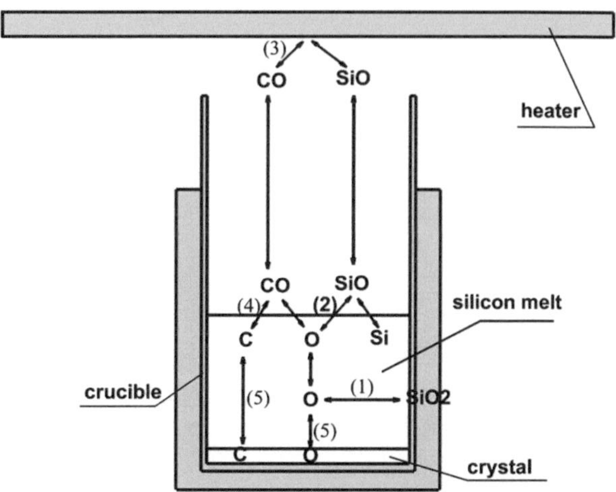

Fig. 5.1 Basic incorporation processes of carbon and oxygen impurities into a crystal (After Ref. [23] with permission from Elsevier)

flow of the silicon melt, the compressible flow of the argon gas, the heat transfer inside the solid components, the heat radiation inside all enclosures, the species transport inside the melt and gas flow, and all chemical reactions described in the above.

5.2.2.1 Numerical Modeling for Global Heat Transfer [24]

The flow of silicon melt inside crucible can be assumed to be incompressible and quasi-steady flow. The mass, momentum, and energy conservation equations for the melt can be written as follows:

$$\nabla \cdot \vec{V} = 0, \tag{5.1}$$

$$\rho \vec{V} \cdot \nabla \vec{V} = -\nabla p + \nabla \cdot \left[\mu (\nabla \vec{V} + \nabla \vec{V}^{T}) \right] - \rho \vec{g} \beta_T (T - T_m), \tag{5.2}$$

$$\rho c_p \vec{V} \cdot \nabla T = \nabla \cdot (k \nabla T). \tag{5.3}$$

where $\vec{V}$ is the flow velocity, ρ is the density of silicon melt, p is the pressure, μ is viscosity, $\vec{g}$ is the gravity acceleration, T_m is the temperature at the melting point of silicon crystal, c_p is the specific heat, and k is the thermal conductivity. The last term inside Eq. (5.2) is the thermal buoyancy force.

The calculations of melt flow require boundary conditions. For the calculation of melt flow, non-slip boundary conditions are applied on all of crucible walls and the crystal solidification front, and a free surface condition is enforced on the melt top surface. On the melt–crystal interface, the following relationship should be satisfied:

$$k_m (\nabla T)_m \cdot \vec{n} = k_s (\nabla T)_s \cdot \vec{n} + \rho_s \Delta H \vec{V}_g \cdot \vec{n}, \tag{5.4}$$

where k_m and k_s are the melt and solid thermal conductivities, respectively; $\vec{V}_g$ is the growth velocity of crystal; ΔH is the melting enthalpy of crystal silicon; and $\vec{n}$ is the normal direction of the melt-crystal interface pointing from melt to crystal.

The governing equation for heat transfer in the crystal and all of solid components inside furnace is

$$\rho_s c_s \vec{V}_s \cdot \nabla T = \nabla \cdot (k_s \nabla T) + Q_s, \tag{5.5}$$

where ρ_s, c_s, $\vec{V}_s$, and k_s are the density, specific heat, velocity and thermal conductivity of solid components, respectively; Q_s is the heat generate rate inside solid components.

The radiative heat transfer in radiative enclosures is modeled on the basis of an assumption of diffuse-gray surface radiation. Let ∂V stand for a radiative enclosure. The surface temperature $T\left(\vec{x}\right)$ and the net heat flux $q\left(\vec{x}\right)$ on ∂V are related by the following integral relationship:

$$\begin{aligned} &\frac{q(\vec{x})}{\varepsilon(\vec{x})} - \int_{\vec{x}^* \in \partial V} K(\vec{x}, \vec{x}^*) \frac{1-\varepsilon(\vec{x}^*)}{\varepsilon(\vec{x}^*)} q(\vec{x}^*) dS^*, \\ &= \sigma T^4(\vec{x}) - \int_{\vec{x}^* \in \partial V} K(\vec{x}, \vec{x}^*) \sigma T^4(\vec{x}^*) dS^* \end{aligned} \tag{5.6}$$

where $\vec{x}$ and $\vec{x}^*$ are infinitesimal radiative surface elements on ∂V, ε is radiative emissivity, dS^* is the area of the infinitesimal surface element $\vec{x}^*$, and $K(\vec{x}, \vec{x}^*)$ is the surface view factor between $\vec{x}$ and $\vec{x}^*$.

Whenever $\vec{x}$ and $\vec{x}^*$ face each other, it is given by the following formulation:

$$K(\vec{x}, \vec{x}^*) = -\frac{[(\vec{x} - \vec{x}^*) \cdot \vec{n}][(\vec{x} - \vec{x}^*) \cdot \vec{n}^*]}{\pi [(\vec{x} - \vec{x}^*) \cdot (\vec{x} - \vec{x}^*)]^2}, \tag{5.7}$$

where $\vec{n}$ and $\vec{n}^*$ are the unit normal to dS and dS^*.

Otherwise, when $\vec{x}$ and $\vec{x}^*$ do not face each other, it equals to zero.

5.2.2.2 Numerical Modeling for Argon Gas Flow

The flow of argon gas through the furnace is considered to be a compressible and axisymmetric flow. The compressible flow solver can accurately simulate the buoyancy-driven flow due to the large density variation in the furnace. Although the flow velocity in this furnace is low, yet the density variation is significant for this buoyancy-driven flow, which is similar as combustion problem. In present furnace, the temperature differences are rather large varying from 350 K at the gas inlet to 1722 K at the melt surface. Therefore, compressible flow solver is a good choice.

Full Navier-Stokes equations without any assumptions are used. The conservative compressible Navier-Stokes equations for axisymmetric geometry are given as follows:

$$\frac{\partial \rho}{\partial t} + \frac{1}{r}\frac{\partial (r\rho u)}{\partial x} + \frac{\partial (\rho v)}{\partial y} = 0, \tag{5.8}$$

$$\frac{\partial (\rho u)}{\partial t} + \frac{1}{r}\frac{\partial \left(r\rho u^2\right)}{\partial x} + \frac{\partial (\rho u v)}{\partial y} = -\frac{\partial p}{\partial x} + \frac{1}{r}\frac{\partial (r\tau_{xx})}{\partial x} + \frac{\partial \tau_{xy}}{\partial y} - \frac{\tau_{\theta\theta}}{r} + \rho g_x, \tag{5.9}$$

$$\frac{\partial (\rho v)}{\partial t} + \frac{1}{r}\frac{\partial (r\rho u v)}{\partial x} + \frac{\partial \left(\rho v^2\right)}{\partial y} = -\frac{\partial p}{\partial y} + \frac{1}{r}\frac{\partial \left(r\tau_{xy}\right)}{\partial x} + \frac{\partial \tau_{yy}}{\partial y} + \rho g_y, \tag{5.10}$$

$$\frac{\partial(\rho e)}{\partial t}+\frac{1}{r}\frac{\partial(r\rho eu)}{\partial x}+\frac{\partial(\rho ev)}{\partial y}$$
$$=-pD+\tau_{xx}\frac{\partial u}{\partial x}+\tau_{yy}\frac{\partial v}{\partial y}+\tau_{xy}(\frac{\partial u}{\partial y}+\frac{\partial v}{\partial x})+\frac{u}{r}\tau_{\theta\theta}, \tag{5.11}$$

where $D=\frac{1}{r}\frac{\partial(ru)}{\partial x}+\frac{\partial v}{\partial y}$, $\tau_{xx}=2\mu\frac{\partial u}{\partial x}+\lambda D$, $\tau_{yy}=2\mu\frac{\partial v}{\partial y}+\lambda D$, $\tau_{xy}=\mu(\frac{\partial u}{\partial y}+\frac{\partial v}{\partial x})$, $\tau_{\theta\theta}=\alpha\left(2\mu\frac{u}{r}+\lambda D\right)$, in which μ is the coefficient of viscosity, λ is the coefficient of dilatational viscosity. The radial position can be expressed as $r=\alpha x+1-\alpha$, where the value of α is zero for Cartesian coordinates and unity for Cylindrical coordinates.

The fluid pressure p is determined from an equation of state for an ideal gas as follows:

$$p=\rho RT, \tag{5.12}$$

where R is gas constant.

For numerical simulation of gas flow, the Navier-Stokes equation needs to be discretized. A second-order total variation diminishing (TVD) scheme is used for convection terms and Davis-Yee symmetric TVD is used for the flux limiter vectors. A fourth-order Runge-Kutta scheme is used for time marching and a central difference scheme is used for viscous terms.

The temperatures at all solid surfaces are set to the values computed from the global heat transfer. The velocities at all solid surfaces are set to zero. For the conditions at inlet and outlet, different methods can be used according to practical tests. For axisymmetric and subsonic inflow, three characteristic lines enter into the domain and, therefore, three analytical boundary conditions are specified. The remaining one is evaluated on the basis of information from the interior points near the inlet. For subsonic outflow, one characteristic line enters into the computation domain from the outside and, thus, one analytical boundary condition is given and others are evaluated from the interior points near the outlet. For the first simulation, inlet flow rate Q is set to 0.8 *liter*/min, inlet static temperature T is set to 350 K and inlet tangential velocity u is set to zero. The outlet static pressure p is set to 0.1 *atm*.

The flow field and temperature field inside the whole furnace is shown in Fig. 5.2. The top part of furnace is high temperature zone, where silicon raw material is first melted and then crystallized. The gas flow can bring some impurities from the top part to the bottom part of furnace or the outside of the furnace.

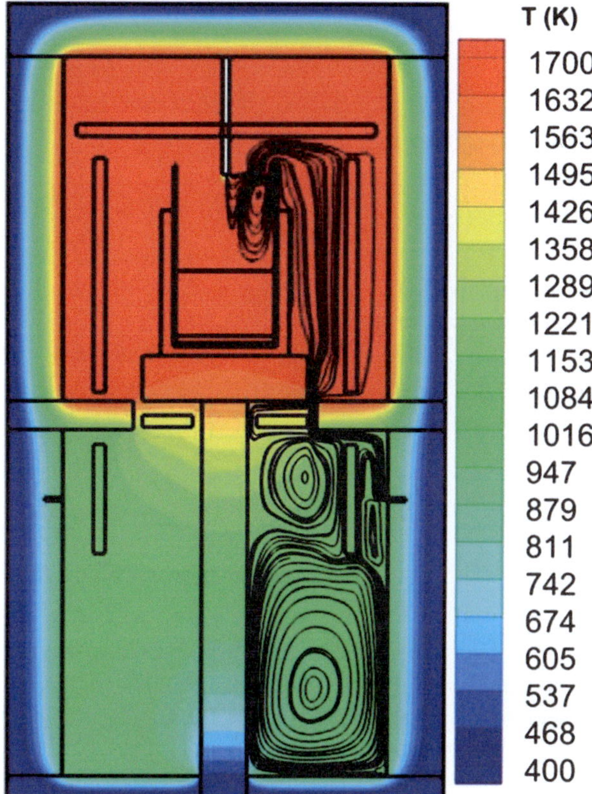

Fig. 5.2 Gas flow field and temperature field inside furnace

5.2.2.3 Numerical Modeling for Impurity Transport

In the argon gas, the concentration fields of SiO and CO gas are assumed to be dilute. The equations for the conservation of moles are written in dimensional form as

$$\frac{\partial c_{SiO}}{\partial t} + \nabla \cdot (c_{SiO}\vec{u}_{Ar}) = \nabla \cdot [c_{Ar}D_{SiO}\nabla(\frac{c_{SiO}}{c_{Ar}})], \tag{5.13}$$

$$\frac{\partial c_{CO}}{\partial t} + \nabla \cdot (c_{CO}\vec{u}_{Ar}) = \nabla \cdot [c_{Ar}D_{CO}\nabla(\frac{c_{CO}}{c_{Ar}})], \tag{5.14}$$

where c_{SiO} is the molar concentration of SiO gas, c_{CO} is the molar concentration of CO gas, and c_{Ar} is the molar concentration of argon gas. The vector, $\vec{u}_{Ar}$, is the flow vector of argon gas. The diffusivities D_{SiO} and D_{CO} in Eqs. (5.13) and (5.14) depend on the temperature and pressure [16, 25]:

$$D_{SiO} = a_{SiO}\frac{T^{1.75}}{p}, \tag{5.15}$$

$$D_{CO} = a_{CO}\frac{T^{1.75}}{p}, \tag{5.16}$$

where $a_{SiO} = 0.862611$ and $a_{CO} = 1.79548$ if D, T, and p are measured in cm^2/s, K, and dyn/cm^2, respectively.

Similarly, in the Si melt, the concentrations of oxygen and carbon atoms are modeled as

$$\frac{\partial c_C}{\partial t} + \nabla \cdot (c_C \vec{u}_{Si}) = \nabla \cdot [c_{Si} D_C \nabla (\frac{c_C}{c_{Si}})], \tag{5.17}$$

$$\frac{\partial c_O}{\partial t} + \nabla \cdot (c_O \vec{u}_{Si}) = \nabla \cdot [c_{Si} D_O \nabla (\frac{c_O}{c_{Si}})], \tag{5.18}$$

where c_C is the molar concentration of carbon atoms in the melt, c_O is the molar concentration of oxygen atoms in the melt, and c_{Si} is the molar concentration of silicon atoms. The vector, $\vec{u}_{Si}$, is the flow vector of the silicon melt. D_O and D_C are the diffusivities of oxygen and carbon in the melt, respectively. Their values are taken to be $5.0 \times 10^{-8}\ m^2/s$ [26].

For every chemical reaction, a detailed illumination is given. The first is the dissolution of the quartz crucible,

$$SiO_2(s) \leftrightarrow Si(m) + 2O(m). \tag{5.19}$$

Here the index symbols (s), (m), (g), (l), and (c) denote solid, melt, gas, liquid, and crystal, respectively. For a unidirectional solidification furnace, the equilibrium concentration of oxygen atoms, c_O, on the wall of the quartz crucible is expressed as [1]

$$c_O = \frac{b}{1-b} \times 0.5 \times 10^{23}\left[atom/cm^3\right],$$

$$b = 1.32 \times \exp\left(\frac{-7150}{T} - 6.99\right). \tag{5.20}$$

The oxygen atoms in the melt are then transported to the melt/gas interface by convection or diffusion. At the interface, they combine with silicon atoms to form

SiO gas according to the second reaction.

$$O(m) + Si(l) \leftrightarrow SiO(g). \tag{5.21}$$

The equilibrium relationship between the concentrations of SiO(g) and O(m) is given in [16]:

$$c_{SiO} = \frac{101325}{RT}\frac{c_O}{c_{Si}}e^{-21000/T+17.8}, \tag{5.22}$$

Where R is the universal gas constant, which is equal to $8.314\,J \cdot K^{-1} \cdot mol^{-1}$.

The third reaction is the transportation of SiO(g) in the argon flow. When SiO(g) comes into contact with hot carbon surfaces during the transportation, it reacts with them as

$$SiO(g) + 2C(s) \leftrightarrow CO(g) + SiC(s). \tag{5.23}$$

Temperature is an important parameter for chemical reactions, but the most important parameter for specifying the nature of a chemical reaction is the Gibbs free energy of formation ($\Delta G = \Delta H - T\Delta S$). If $\Delta G < 0$, the reaction proceeds spontaneously, and the net result is a large ratio of product to reactant; if $\Delta G > 0$, very little of the reactant may be converted to product without the input of additional energy into the system. All of reactions in the present model are assumed to be reversible. The reverse reaction rate constant, k_r, and the forward reaction rate constant, k_f, are functions of the Gibbs free energy and the equilibrium constant, k_{eq}. The temperature's effect on the reaction direction and degree is automatically considered if all of the Gibbs free energies are taken to be functions of temperature in the present calculations.

The free energy change, ΔG, of the reaction in Eq. (5.23) is expressed as given in [16]:

$$\Delta G = -81300 + 3.02T \ (J/mol) \ \text{for}\ T < 1640K,$$

$$\Delta G = -22100 - 33.1T \ (J/mol) \ \text{for}\ 1640K < T < 1687K,$$

$$\Delta G = -72100 - 3.44T \ (J/mol) \ \text{for}\ T > 1687K. \tag{5.24}$$

The equilibrium constant of the reaction in Eq. (5.23) can be obtained as $K = e^{-\Delta G/RT}$. The rate constant in the forward direction, $k^{forward}$, is assumed to be large ($k^{forward} = 10^6 cm/s$), and the rate constant for the reverse direction is computed from the relation $k^{reverse} = k^{forward}/K$ [16]. This extremely large forward rate constant corresponds to small carbon concentration in the crystal, as pointed out by Bornside and Brown [16]. The predicted carbon concentration will be greater if a low forward rate constant is considered. The reaction $SiO(g) + 2C(s) \leftrightarrow$

$CO(g) + SiC(s)$ is assumed to be a simple one-step, reversible reaction. The rate constant for the forward direction is set large enough such that SiO is consumed very fast once SiO is transported to the graphite wall. However, the concentration ratio of CO to SiO is nearly constant. Therefore, if the concentration of SiO is small, the concentration of CO is also small, indicating that the extremely large forward rate constant corresponds to the smallest carbon concentration in the crystal. Hence, the predicted concentration of carbon will be greater if a low forward rate constant is assumed.

After the rate constants are obtained, they can be used to obtain the flux relations at the hot carbon surfaces:

$$-\vec{n} \cdot \left[cD_{SiO} \nabla (\frac{c_{SiO}}{c}) \right] = k^{forward} c_{SiO} - k^{reverse} c_{CO}, \tag{5.25a}$$

$$-\vec{n} \cdot \left[cD_{CO} \nabla (\frac{c_{CO}}{c}) \right] = k^{reverse} c_{CO} - k^{forward} c_{SiO}, \tag{5.25b}$$

where the unit vector, $\vec{n}$, points outward from the gas field, and $k^{forward}$ and $k^{reverse}$ are the dimensional forward and reverse rate constants, respectively.

The fourth reaction is the transportation of the product CO back to the gas/melt interface in the argon flow. At the interface, the CO gas dissolves into the melt according to the reaction.

$$CO(g) \leftrightarrow C(m) + O(m). \tag{5.26}$$

The equilibrium relationship between the above reactants and resultants is given as [16].

$$c_{CO} = \frac{101325}{RT} \frac{c_O}{c_{Si}} \frac{c_C}{c_{Si}} e^{-5210/T + 14.5}. \tag{5.27}$$

The fifth reaction is the segregation of carbon and oxygen from the melt into the crystal. The equilibrium expression is

$$C(m) \leftrightarrow C(c), \tag{5.28}$$

$$O(m) \leftrightarrow O(c). \tag{5.29}$$

Here $C(c) = k_{Cseg} C(m)$ and $O(c) = k_{Oseg} O(m)$, with $k_{Cseg} = 0.07$ [16] and $k_{Oseg} = 0.85$ [27, 28].

Equations (5.22) and (5.27) can be regarded as boundary conditions for the gas/melt interface. Other than those, two extra boundary conditions are required for solving the four unknowns c_{SiO}, c_{CO}, c_O, and c_C at that interface. The conservation of moles for carbon and oxygen is introduced. This implies that the flux of oxygen and carbon at that interface should satisfy the following:

For oxygen

$$c_{Ar} D_{SiO} \nabla \left(\frac{c_{SiO}}{c_{Ar}} \right) + c_{Ar} D_{CO} \nabla \left(\frac{c_{CO}}{c_{Ar}} \right) = c_{Si} D_O \nabla \left(\frac{c_O}{c_{Si}} \right), \tag{5.30}$$

For carbon

$$c_{Ar} D_{CO} \nabla \left(\frac{c_{CO}}{c_{Ar}} \right) = c_{Si} D_C \nabla \left(\frac{c_C}{c_{Si}} \right). \tag{5.31}$$

Equations (5.22), (5.27), (5.30), and (5.31) are a set of nonlinear second-order equations. The concentrations c_{SiO}, c_{CO}, c_O, and c_C at the interface can be expressed by the values of the inner field points if Eqs. (5.30) and (5.31) are discretized and substituted into Eqs. (5.22) and (5.27). However, although those concentrations can be solved directly, extremely fine grids at the melt side of interface are still required for numerical stability. As has already been known, there is a dramatic difference between the diffusivity in the gas ($10^{-3} m^2/s$) and that in the melt ($10^{-8} m^2/s$). The flux of O(m) and C(m) is negligible without fine grids. In the present simulation, the minimum grid size in the axis direction is about 5 μm and that in the radius direction is about 0.7 mm. Fine grids require a small marching time step to maintain numerical stability. $\Delta t = 0.005 s$ was set for the impurity calculation in both the melt and gas field.

Beyond the above specifications, other boundary conditions in the gas and the melt are also given. For non-carbon walls in gas, zero fluxes of SiO(g) and CO(g) are applied; for the gas inlet, the concentrations of SiO(g) and CO(g) are set to zero; for the gas outlet, zero gradients of SiO(g) and CO(g) are used, and for the melt/crucible wall, zero flux of C(m) is used.

5.2.3 *Distribution of Impurities*

5.2.3.1 Distribution of SiO(*g*) in Gas and O(*m*) in Melt

The concentrations of SiO(g) and O(m) are shown in Figs. 5.3 and 5.4. The basic order of SiO(g) concentration in the gas is 10^{-9} *mol/cm*^{3} and the maximum concentration is about 10^{-8} *mol/cm*^{3}, which are almost the same as the simulation results of Smirnov and Kalaev [18] in a Czochralski furnace. The basic order of oxygen atom concentration in the melt is 10^{17} *atom/cm*^{3}, which is consistent with the experimental result found by Matsuo [30].

The concentration of SiO is large in the top half of the furnace and small in the bottom half. SiO(g) evaporates from the melt surface and is carried by the gas from the top of the furnace to the bottom of the furnace. During transportation, the SiO(g) reacts with the hot carbon walls and reduces gradually. Therefore, the distribution of the SiO(g) is consistent with the experiment if the chemical reaction is considered.

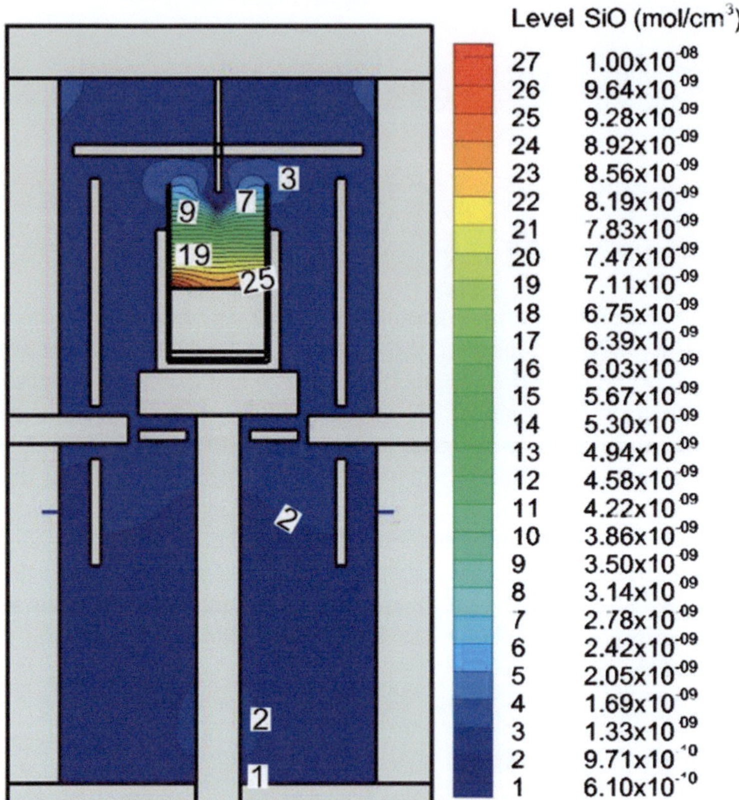

Fig. 5.3 SiO concentration in argon gas (After Ref. [29] reproduced with permission from The Electrochemical Society)

The concentration of O(*m*) is small at the center of the crucible and large at the wall. At the gas/melt interface, the concentration of O(*m*) is minimum. O(*m*) originates from the wall of the crucible and is transported to the center of the crucible by convection or diffusion. At the gas/melt interface, O(*m*) combines with Si(*m*) to form SiO gas. Thus, the distribution of the O(*m*) is also consistent with the experimental result.

5.2.3.2 Distributions of CO(*g*) in the Gas and C(*m*) in the Melt

The distributions of CO(*g*) in the gas and C(*m*) in the melt are shown in Figs. 5.5 and 5.6. The maximum CO concentration in the gas is about 10^{-9} mol/cm^3, which is one order of magnitude less than the maximum SiO concentration. The minimum CO concentration is about 10^{-10} mol/cm^3, which lies on the melt surface. Similar to SiO(g), the concentration of CO(g) is large in the top half of the furnace (except

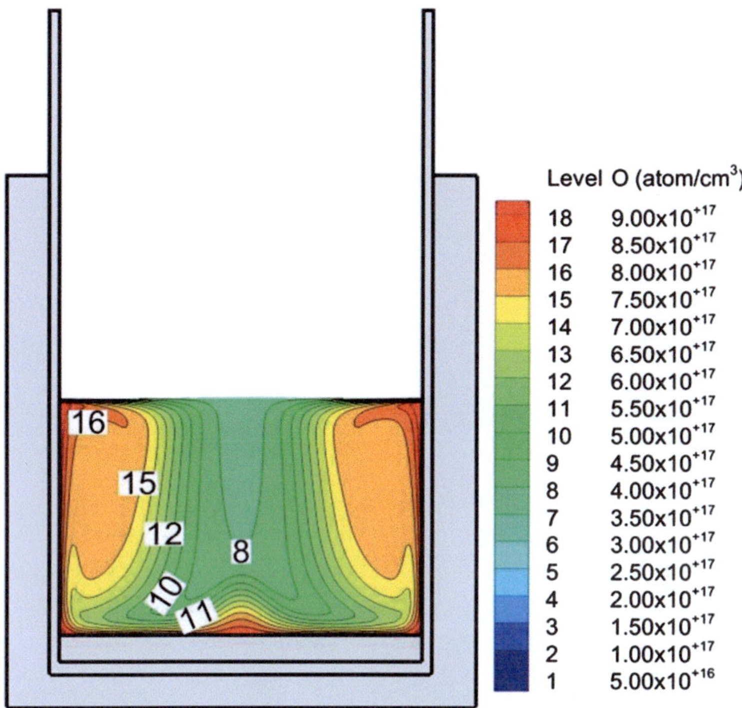

Fig. 5.4 Oxygen atom concentration in the melt (After Ref. [29] reproduced with permission from The Electrochemical Society)

inside the crucible) and small in the bottom half of the furnace. It is understood that CO(g) is a resultant of a reaction, and its concentration is proportional to the SiO(g) concentration.

The basic order of C(m) concentration is about 10^{18} *atom/cm*^{3}, which is the same as the experimental result reported by Ganesh et al. [31]. The concentration of C(m) in the melt is large at the top of the melt and small at the bottom. This distribution is consistent with the process by which CO(g) is first absorbed and dissolved in the melt and carbon is then transported to the bottom by convection or diffusion.

5.2.3.3 Comparison with Experiments

Results of the simulation were compared with experimental results. Figure 5.6 shows that the carbon concentration at the bottom of the crucible is 4.19×10^{18} *atom/cm*^{3}. After considering the segregation relation, the carbon concentration in the crystal can be expressed as

$$c_C(c) = 0.07 \times c_C(m) = 0.07 \times 4.19 \times 10^{18} = 3.0 \times 10^{17} atom/cm^3. \quad (5.32)$$

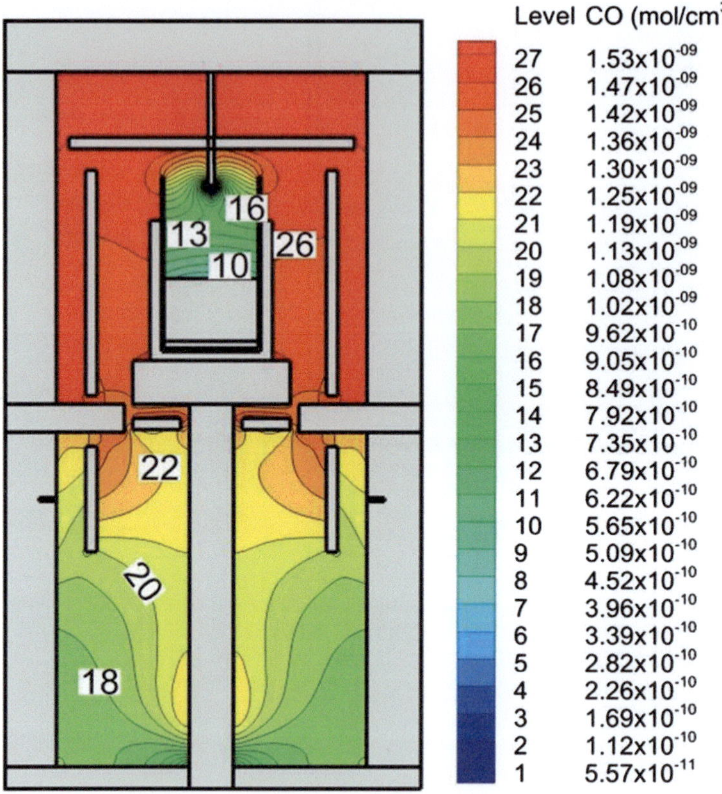

Fig. 5.5 CO concentration in gas (After Ref. [29] reproduced with permission from The Electrochemical Society)

Experimental results for carbon concentration [31] at normal freezing are in the range of 2.2×10^{17} *atom*/*cm*^{3} to 4.2×10^{17} *atom*/*cm*^{3}. The simulated carbon concentration falls into this range of experimental data.

Figure 5.4 shows that the oxygen concentration at the bottom of the crucible is 5.5×10^{17} *atom*/*cm*^{3}. The oxygen concentration in the crystal is given by

$$c_O(c) = 0.85 \times c_O(m) = 0.85 \times 5.5 \times 10^{17} = 4.6 \times 10^{17} atom/cm^3. \quad (5.33)$$

Experiment results for oxygen concentration [32] at normal freezing are in the range 4×10^{17} *atom*/*cm*^{3} to 5×10^{17} *atom*/*cm*^{3}. The simulated oxygen concentration also falls in the range of experimental data.

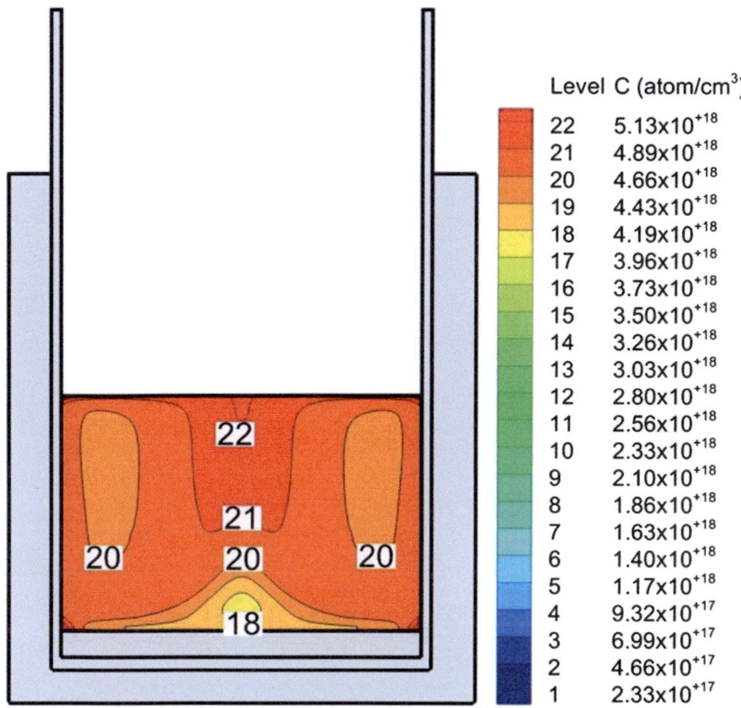

Fig. 5.6 Carbon atom concentration in the melt (After Ref. [29] reproduced with permission from The Electrochemical Society)

5.2.4 Reduction of Carbon and Oxygen Impurities

After the introduction of the mechanism of carbon and oxygen incorporation into a crystal and the global model for impurity transport, the reduction of carbon and oxygen impurities can be performed on the basis of previous simulations. Figure 5.7 shows the argon gas flow above the melt under the conditions of an inlet flow rate of 0.8 *liter*/min, inlet static temperature of 350 *K*, and outlet static pressure of 0.1 *atm*. There is a recirculation flow above the melt surface and a main convection flow passing through the top of the crucible. The back-diffusion of CO from the outside of crucible into the melt is hindered by the top convection flow. Therefore, if the top convection flow is strengthened, in addition to eliminating the recirculation flow, CO transportation into the melt might be dramatically reduced.

Figure 5.7 shows that CO is transported into the melt mainly by a diffusion flux that passes through the area AB. If the diffusion area is markedly reduced, the CO flux into the melt can be significantly reduced. An optimized design according to the above analysis is given in the following.

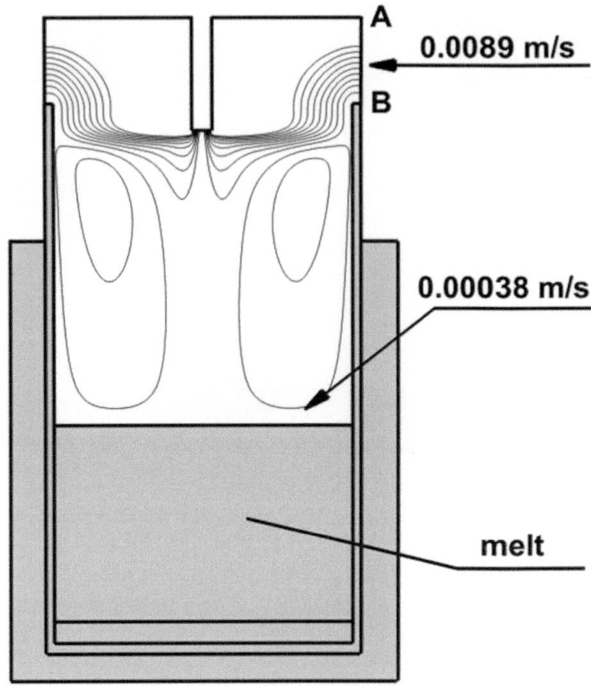

Fig. 5.7 Flow above the melt (After Ref. [29] reproduced with permission from The Electrochemical Society)

The gas tube is extended to form a cover as shown in Fig. 5.8. The cover material can be made of tungsten or molybdenum, both having high melting points. The distance, h, between the cover and the melt surface is variable. Global simulation of coupled oxygen and carbon transports was carried out for $h = 28$, 49, and 105 mm. The third case ($h = 105$ mm) corresponds to an old furnace that has no cover. All of the calculations have the same inlet and outlet conditions.

For the two new furnace cases with different h values, the C distributions are shown in Fig. 5.9a, b. The carbon concentrations in the melt show almost the same order as those in the gas. The cases with distances of 28 and 49 mm have basic orders of carbon concentration in the melt of 10^{-9} mol/cm^3; whereas the basic order of the carbon concentration in the old case is 10^{18} $atom/cm^3$, which is about 10^{-6} mol/cm^3. Thus, the two new furnaces provide smaller carbon concentrations than the old case.

If the segregation relation for carbon atoms is considered, the carbon concentration for a chosen position in the crystal is given in Fig. 5.10. The basic order of the carbon concentration in the crystal for the old furnace was 10^{17} $atom/cm^3$. Compared to that, the basic order for the improved furnace is $10^{13} \sim 10^{14}$ $atom/cm^3$, which is of a negligible quantity in the crystal growth.

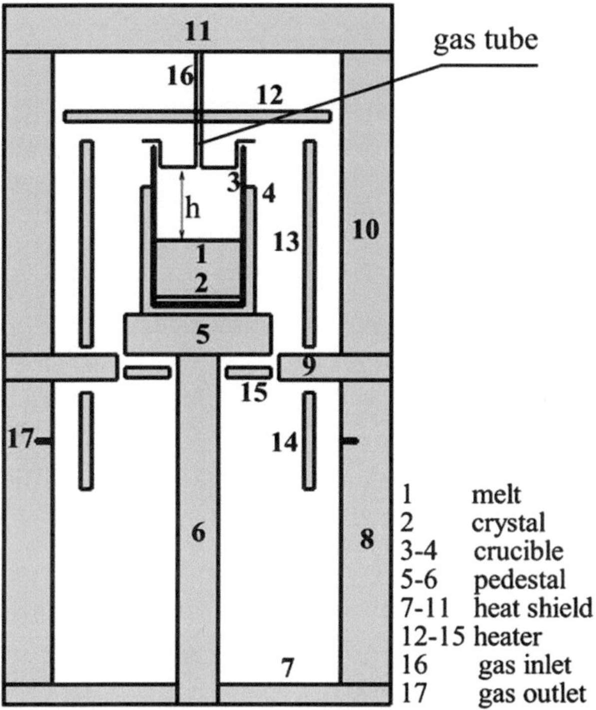

Fig. 5.8 Improved design of a unidirectional solidification furnace (After Ref. [23] with permission from Elsevier)

The oxygen distribution for the improved case with a distance of 49 *mm* is presented in Fig. 5.11. The oxygen concentration in most of the melt volume for the improved furnace is less than that for the old furnace (Fig. 5.4). In the old furnace, the oxygen first originates from the wall of the crucible, and then it is transported to the melt surface by convection flow. At the melt surface, the oxygen evaporates as SiO gas. In the improved furnace, however, more SiO is taken away because of stronger convection flow above the melt surface.

A comparison of the oxygen concentrations at a chosen position in the crystal is shown in Fig. 5.12. In the case of h=, the oxygen concentration is reduced by 30 % compared to that in the old furnace. A smaller distance corresponds to a stronger convection flow intensity, which brings away more oxygen from the melt surface. Therefore, strengthening the convection flow at the melt surface proves to also be favorable for transporting more oxygen away from the melt.

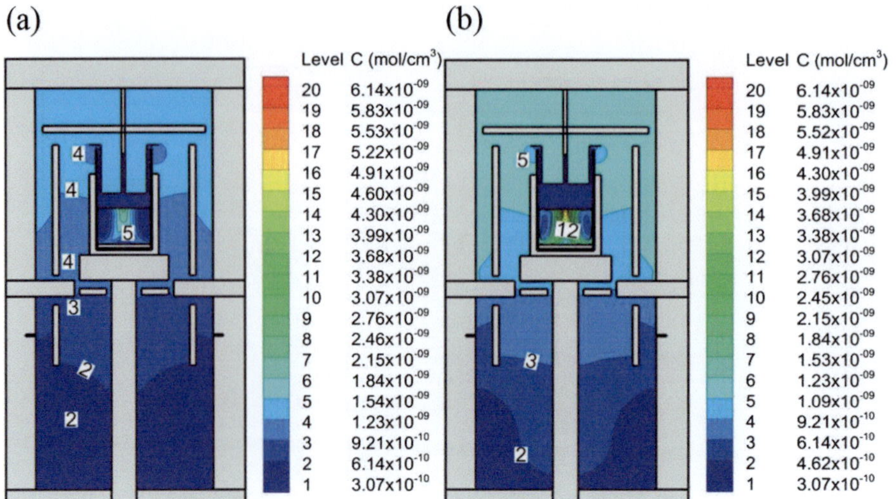

Fig. 5.9 (**a**) Carbon concentration distribution (CO in gas and C in melt) for the improved furnace at $h = 28\ mm$. (**b**) Carbon concentration distribution (CO in gas and C in melt) for the improved furnace at $h = 49\ mm$ (After Ref. [23] with permission from Elsevier)

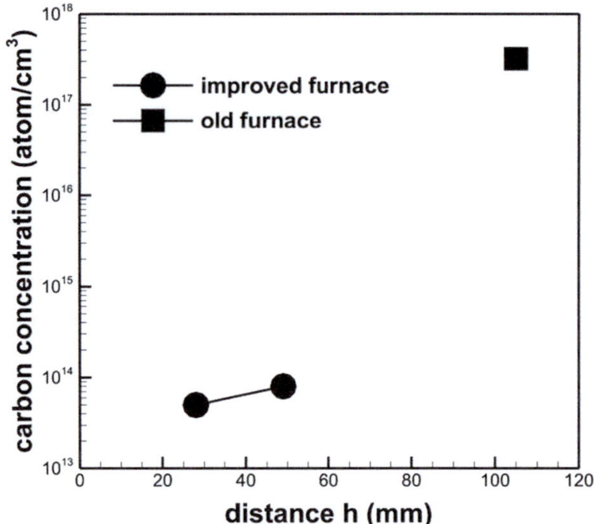

Fig. 5.10 Comparison of carbon concentrations in the crystal for the improved and old furnaces (After Ref. [23] with permission from Elsevier)

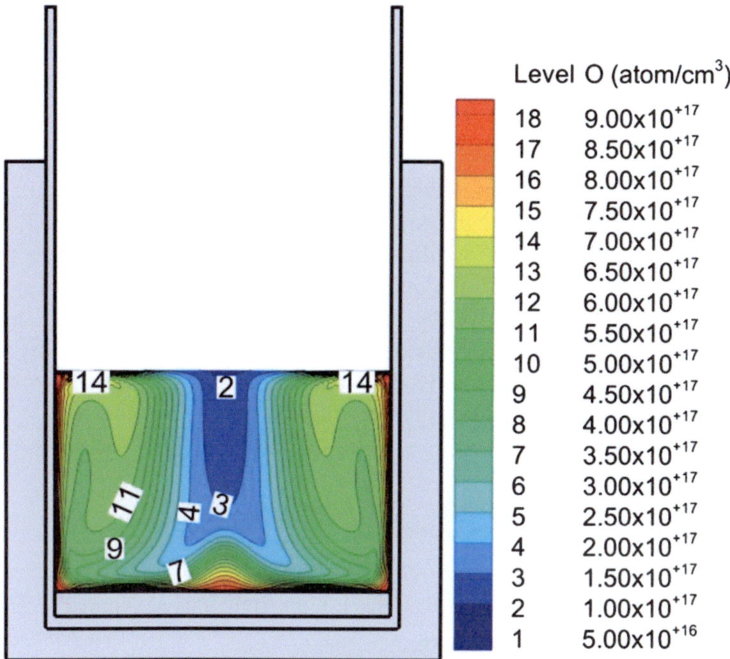

Fig. 5.11 Oxygen distribution inside melt for improved case (After Ref. [23] with permission from Elsevier)

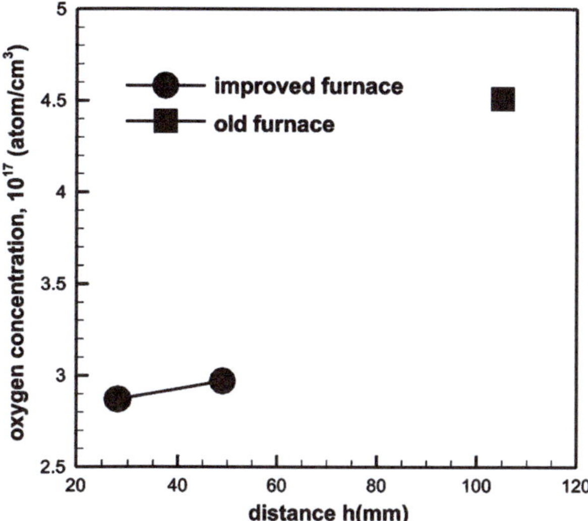

Fig. 5.12 Comparison of oxygen concentrations in the crystal for the improved and old furnaces (After Ref. [23] with permission from Elsevier)

5.2.5 *Summary*

Global simulation of oxygen and carbon impurities in a unidirectional solidification furnace for solar cells was implemented. Both gas transport and melt transport were included in the model. Five chemical reactions were considered. The impurity distributions in the gas and melt were given. The present simulation results showed good agreement with experimental data.

An improved unidirectional solidification furnace with a modified gas tube was designed. Global simulations of coupled oxygen and carbon transport showed that the carbon concentration in the crystal can be reduced from 10^{17} *atom*/*cm*^{3} to 10^{14} *atom*/*cm*^{3}. The oxygen concentration can also be reduced by at least 30 %. Therefore, the present improvement enables the production of a high-purity multicrystalline silicon crystal in a unidirectional solidification furnace.

5.3 Simulation of Dislocations

5.3.1 *Mechanism of Dislocation Generation*

In most works, thermal stress is regarded as the main reason for the generation of dislocations, because thermal stress drives the multiplication of existing dislocations. However, dislocation multiplication and dislocation nucleation must be distinguished from each other. Dislocation nucleation is not possible solely due to thermal stress, because the nucleation of dislocations requires a stress on the order of shear modulus of material, which is impossible to reach through thermal stress alone. The nucleation of dislocations can only be triggered by a large local stress, possibly caused by factors such as agglomeration of point defects, surface damage due to indentation, scratching, grinding, impingement of hard particles, the presence of a second phase in the form of foreign particles or precipitates, and chemical inhomogeneity due to doping or impurity atoms [33].

5.3.2 *Numerical Models for Dislocation Multiplication*

The dislocation generation or nucleation determines the first existing dislocations; however, the dislocation multiplication, driven by the thermal stress, determines the final dislocations in the as-grown crystal. Therefore, the thermal stress is the most important factor in determining the presence of final dislocations.

To model the dislocation multiplication due to thermal stress, one of two approaches is usually adopted based on whether the crystal is assumed to be elastic or viscoplastic. For the elastic crystal, an elastic problem is solved and the results yield information about the displacements, thermal and elastic strain, and elastic stress inside crystal. The obtained elastic stress is then used to estimate the plastic deformation, and whether a particular area of the crystal will have a high or low dislocation density [33]. Jordan [34] assumed that plastic deformation occurs if the resolved elastic shear stress on any slip system exceeds a critical value, called the critical resolved shear stress (CRSS). For the viscoplastic crystal, a direct simulation of the time evolution of the dislocation density during the crystal growth process, which considers stress relaxation and strain hardening by generated dislocation, is used. The advantage of this approach is that the final dislocation density and state of the as-grown crystal is a function of the initial state of crystal and the entire growth and cooling temperature histories [33]. This approach can also provide an estimate of the residual stress in the cooled crystal. Therefore, in the next section, a viscoplastic model, called the Alexander–Haasen (AH) model, is introduced and used. This model includes a multislip for multiplication of the dislocation, immobilization of mobile dislocations, jog formation between different slip systems and its influence on dislocation generation, and the internal stress due to short-range interactions from the total dislocation density. The formulations are given as follows:

First, the force equilibrium satisfies

$$\nabla \cdot \boldsymbol{\sigma} = 0, \tag{5.34}$$

where $\boldsymbol{\sigma}$ is the stress tensor. The stress tensor is determined by constitutive relation

$$\boldsymbol{\sigma} = \mathbf{C}\boldsymbol{\varepsilon}_{el}, \tag{5.35}$$

where $\mathbf{C}$ is the elastic moduli tensor and $\boldsymbol{\varepsilon}_{el}$ is the elastic strain tensor. The elastic strain tensor can be obtained from

$$\boldsymbol{\varepsilon}_{el} = \boldsymbol{\varepsilon} - \boldsymbol{\varepsilon}_{pl} - \boldsymbol{\varepsilon}_{th}, \tag{5.36}$$

where $\boldsymbol{\varepsilon}$, $\boldsymbol{\varepsilon}_{pl}$, and $\boldsymbol{\varepsilon}_{th}$ are the total, plastic, and thermal strain tensors, respectively. The total strain tensor is determined by displacement as follows:

$$\boldsymbol{\varepsilon} = \frac{1}{2}[\nabla\mathbf{u} + (\nabla\mathbf{u})]^T, \tag{5.37}$$

where $\mathbf{u}$ is the displacement vector. The thermal strain is given by

$$\boldsymbol{\varepsilon}_{th} = \beta^*\delta\,(T - T_0)\,, \tag{5.38}$$

where β^* is the thermal expansion coefficient.

The plastic strain is given by the Orowan relationship [35]:

$$\frac{d\boldsymbol{\varepsilon}_{pl}^{(\alpha)}}{dt} = N_m^{(\alpha)} v^{(\alpha)} b, \tag{5.39}$$

where the subscript m denotes the mobile dislocation, the superscript α denotes the slip direction, b is Burger's vector, N is the dislocation density, and v is the slip velocity of dislocation.

The rate of the mobile dislocation density, $dN_m^{(\alpha)}/dt$, in the slip direction, α, is given by

$$\frac{dN_m^{(\alpha)}}{dt} = KN_m^{(\alpha)} v^{(\alpha)} \tau_{eff}^{(\alpha)} + K^* N_m^{(\alpha)} v^{(\alpha)} \tau_{eff}^{(\alpha)} \sum_{\beta \neq \alpha} f_{\alpha\beta} N_m^{(\beta)} - 2r_c N_m^{(\alpha)} N_m^{(\alpha)} v^{(\alpha)}, \tag{5.40}$$

where τ_{eff} is the effective stress for dislocation multiplication, K and K^* are multiplication constants, and r_c is an effective dipole half-width. The values of $f_{\alpha\beta}$ are either 1 or 0, depending on whether a jog is formed on a screw dislocation of system α when cut by a forest dislocation of system β. On the right side of Eq. (5.40), the first term represents the increase in dislocation because of the glide on the slip plane, the second term represents the formation of jogs on screw dislocation and the consequent expansion through spiral formation, and the third term represents the sink for the mobile dislocation, i.e., immobilization. Thus, the immobilization rate, $dN_i^{(\alpha)}/dt$, can be expressed as [36].

$$dN_i^{(\alpha)}/dt = 2r_c N_m^{(\alpha)} N_m^{(\alpha)} v^{(\alpha)}. \tag{5.41}$$

The slip velocity of dislocation, v, is given by

$$v^{(\alpha)} = v_0 (\frac{\tau_{eff}^{(\alpha)}}{\tau_0})^m \exp(-\frac{U}{k_b T}), \tag{5.42}$$

where $v_0 = 5000 m/s$, $\tau_0 = 1 MPa$, $m = 1$, and $U = 2.2 eV$ are used here for crystalline silicon [36].

The values of K, K^*, and r_c are given by [36]

$$K(T) = 7.6 \times 10^{-1} e^{-0.72/k_b T} \, (m/N), \tag{5.43}$$

$$K^* = 8.0 \times 10^{-15} m^3/N, \tag{5.44}$$

$$r_c(T) = 7.2 \times 10^{-3} e^{-1.48/k_b T} (m). \tag{5.45}$$

Table 5.1 Long-range interaction coefficient $A_{\alpha\beta}$ and short-range interaction coefficients $\alpha_{\alpha\beta}$

Interaction	Self	Coplanar	Collinear	Hirth locks	Lomer lock	Glissile junction
$A_{\alpha\beta}$	0.21	1/16	1/20	1/12	1/12	1/12
$\alpha_{\alpha\beta}$	0.09	0.09	0.41	0.06	0.12	0.12

The effective stress, necessary for dislocation motion, is given by

$$\tau_{eff}^{(\alpha)} = \left\langle \tau^{(\alpha)} - \tau_i^{(\alpha)} - \tau_b^{(\alpha)} \right\rangle, \tag{5.46}$$

where $\tau^{(\alpha)}$ is the resolved shear stress, $\tau_i^{(\alpha)}$ is the stress required to overcome short-range obstacles, and $\tau_b^{(\alpha)}$ is the internal long-range elastic stress generated by mobile dislocations [36]. The value of $\langle x \rangle = x$ if $x > 0$, and 0 if $x \leq 0$.

The short- and long-range interactions are given by [36]

$$\tau_i^{(\alpha)} = \mu b \sqrt{\sum_\beta a_{\alpha\beta} (N_m^{(\beta)} + N_i^{(\beta)})}, \tag{5.47}$$

$$\tau_b^{(\alpha)} = \mu b \sum_\beta A_{\alpha\beta} \sqrt{N_m^{(\beta)}}. \tag{5.48}$$

where the coefficients $A_{\alpha\beta}$ and $a_{\alpha\beta}$ are shown in Table 5.1.

After the dislocation densities and creep strains are calculated for all of the slip directions, the total dislocation density and total creep strain can be expressed respectively as

$$N_m = \sum_{\alpha=1}^{12} N_m{}^{(\alpha)}, \tag{5.49}$$

$$\boldsymbol{\varepsilon}_{pl} = \sum_{\alpha=1}^{12} \boldsymbol{\varepsilon}_{pl}^{(\alpha)} \frac{1}{2} (\mathbf{n}^{(\alpha)} \otimes \mathbf{m}^{(\alpha)} + \mathbf{m}^{(\alpha)} \otimes \mathbf{n}^{(\alpha)}) sign(\tau^{(\alpha)}). \tag{5.50}$$

5.3.3 *Effect of Cooling Rate on the Generation of Dislocations*

After developing a model for dislocation multiplication, numerical simulations for obtaining the effect of the cooling rate on the generation of dislocations are performed.

The cubic single crystal has four slip planes, each of which has three slip directions. To simplify the description of the 12 slip systems, a number is assigned to every slip system, as shown in Table 5.2. Furthermore, to simplify the description

Table 5.2 Slip directions and slip planes for cubic single crystal

	1	2	3	4	5	6	7	8	9	10	11	12
$\sqrt{2}\mathbf{m}_i$	$\left[\bar{1}0\bar{1}\right]$	[011]	$\left[1\bar{1}0\right]$	$\left[\bar{1}\bar{1}0\right]$	[011]	$\left[10\bar{1}\right]$	$\left[\bar{1}0\bar{1}\right]$	[110]	$\left[0\bar{1}1\right]$	$\left[10\bar{1}\right]$	$\left[0\bar{1}1\right]$	$\left[\bar{1}10\right]$
$\sqrt{3}\mathbf{n}_i$	$\left(\bar{1}\bar{1}1\right)$	$\left(\bar{1}\bar{1}1\right)$	$\left(\bar{1}\bar{1}1\right)$	$\left(1\bar{1}1\right)$	$\left(1\bar{1}1\right)$	$\left(1\bar{1}1\right)$	$\left(\bar{1}11\right)$	$\left(\bar{1}11\right)$	$\left(\bar{1}11\right)$	(111)	(111)	(111)

of similar slip systems, the slip systems are classified into groups (A–E) with a subscript indicating the symmetry. In the study of [001] growth, the slip directions of 1, 2, 5, 6, 7, 9, 10, and 11 are classified as group A_{B2} (eightfold symmetry), and the slip directions of 3, 4, 8, 12 are classified as group B_{A1} (fourfold symmetry). The subscripts B2 and A1 indicate the symmetry in the global fourfold symmetry system (C_{4V}). In addition, for conciseness, the expression "the slip directions in groups A–E" is simplified to "groups A–E".

5.3.3.1 Furnace Structure and Cooling Settings

To check the effect of the cooling rate on the generation of dislocations, a basic configuration of a unidirectional solidification furnace [23, 29, 37] is used. The configuration near the crucible is shown in Fig. 5.13a. The crystal growth direction, i.e., the axial line direction, is set to be the [001] direction. The diameter of the cylindrical crystal is 10.6 cm. Both a top heater and a side heater are used. The power ratio of the side heater to the top heater is 3. For the numerical simulations, two cooling rates are set, as shown in Fig. 5.13b. The first is fast cooling with a decrease rate of 4.7 KW/h in total power, and the other is slow cooling with a decrease rate of 1.2 KW/h in total power. The temperature history at a monitoring point (P) is shown in Fig. 5.13c. From Fig. 5.13b and c, it can be seen that the history of the temperature is similar to that of the heater power. Therefore, the present furnace has small thermal inertia, which is helpful for controlling the growth and cooling processes.

5.3.3.2 Distribution of Dislocations and Residual Stress in Different Cooling Rates

The distributions of dislocation density inside the grown crystal at room temperature for fast and slow cooling are shown in Fig. 5.14, where Fig. 5.14a shows fast cooling and Fig. 5.14b shows slow cooling. Both cases show inhomogeneous distributions along the cylindrical surfaces. However, at the cylindrical surface under slow cooling, the dislocation density is less than that under fast cooling. Therefore, the cooling rate affects the generation of dislocations during the cooling process.

Since the distribution of the final dislocations determines the residual stress inside the grown crystal at room temperature, it is possible that the cooling rate can also affect the residual stress. Considering that residual stress inside the crystal is three-dimensional, an average along the peripheral direction is taken, and the average residual stresses for the fast and slow cooling rates are shown in Fig. 5.15, where Fig. 5.15a shows fast cooling and Fig. 5.15b shows slow cooling. It can be seen that the fast cooling rate causes much higher residual stress inside the crystal. Therefore, the slow cooling rate is beneficial for reducing residual stress.

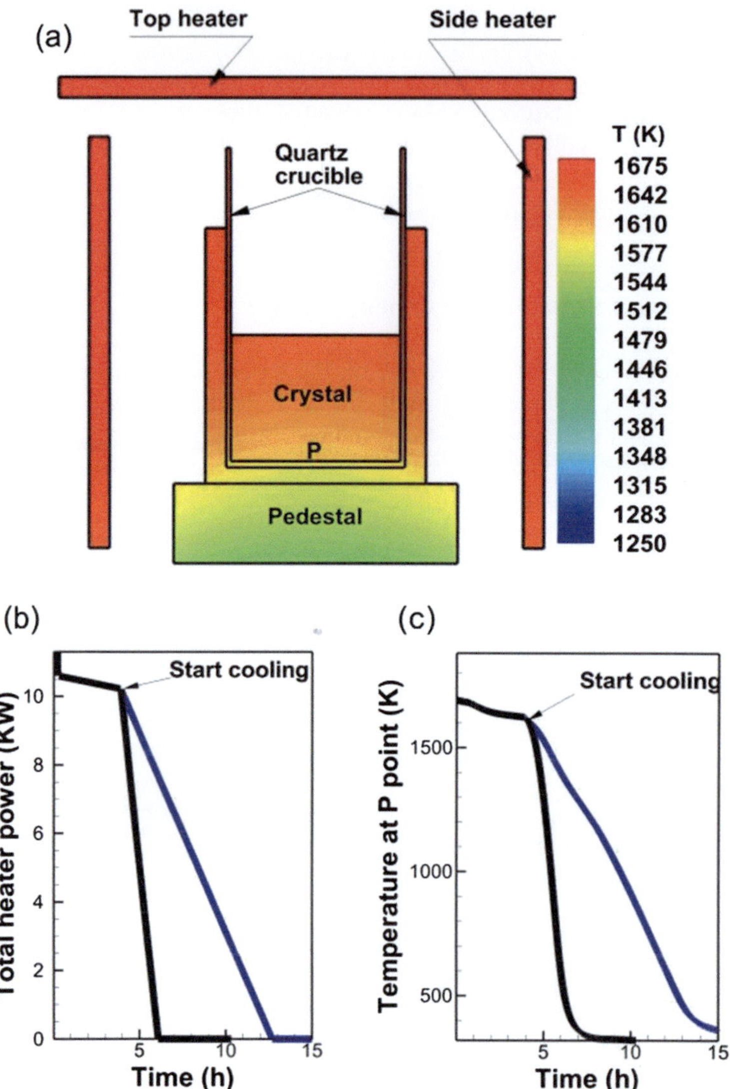

Fig. 5.13 Furnace structure and cooling settings after crystallization. (**a**) Configuration of furnace near the crucible, (**b**) the history of total heater power, and (**c**) the temperature history of a point (P point) at the *bottom* of the crystal (Reprinted with permission from Ref. [21], Copyright 2013, American Chemical Society)

5.3.3.3 Activation of Slip Systems in Different Cooling Rates

Since the cooling rate can affect the generation of dislocations, which mainly originate from the activation of slip systems, there must be a relationship between the cooling rate and the activation of slip systems. To determine the relationship, it is

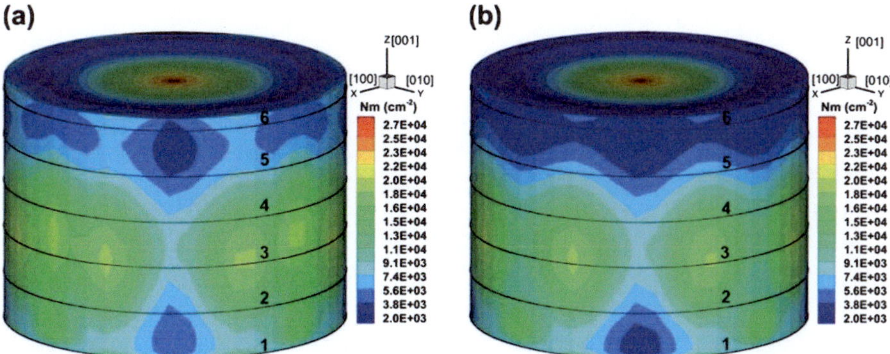

Fig. 5.14 The distribution of the dislocation density in a grown-crystal at room temperature for [001] growth, (**a**) fast cooling rate, (**b**) slow cooling rate

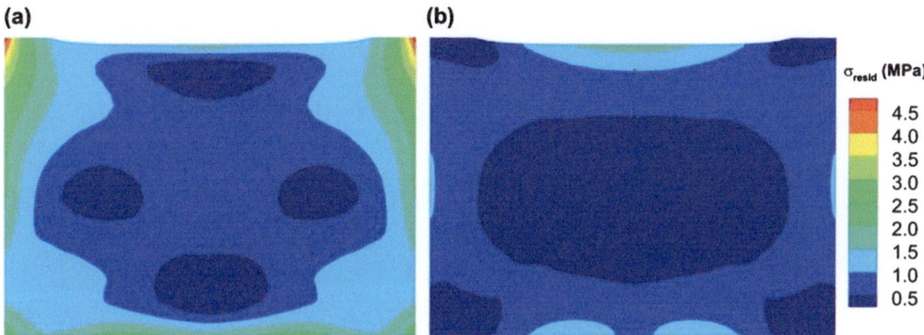

Fig. 5.15 Distribution of the average residual stress in a crystal at room temperature for [001] growth. (**a**) Fast cooling rate, and (**b**) slow cooling rate (Reprinted with permission from Ref. [21], Copyright 2013, American Chemical Society)

important to know what type of slip systems are activated at a specific cooling rate. This can be obtained by observing the dislocation density generated by every slip system. If the dislocation density is greater than a specific value, the corresponding slip system can be regarded as being activated. Figure 5.16 shows the dislocation density generated by every slip system for slice 1 in Fig. 5.14a. The slip systems of 1, 2, 5, 6, 7, 9, 10, and 11 (group A_{B2}) are mainly activated at points along the edge and in the interior, and the slip directions of 3, 4, 8, and 12 (group B_{A1}) are mainly activated at points along the edge. If all of the dislocations generated by the 12 slip systems are superposed, an eightfold symmetric distribution is obtained by group A_{B2} and a fourfold symmetric distribution is obtained by group B_{A1}, as shown in Fig. 5.17. The eightfold symmetric distribution exists at the edge and at the center, and the fourfold symmetric distribution exists only at the edge. Therefore, the activated slip systems in group A_{B2} mainly contribute an eightfold symmetric distribution at the center and the edge, whereas the activated slip systems in group B_{A1} mainly contribute a fourfold symmetric distribution only at the edge.

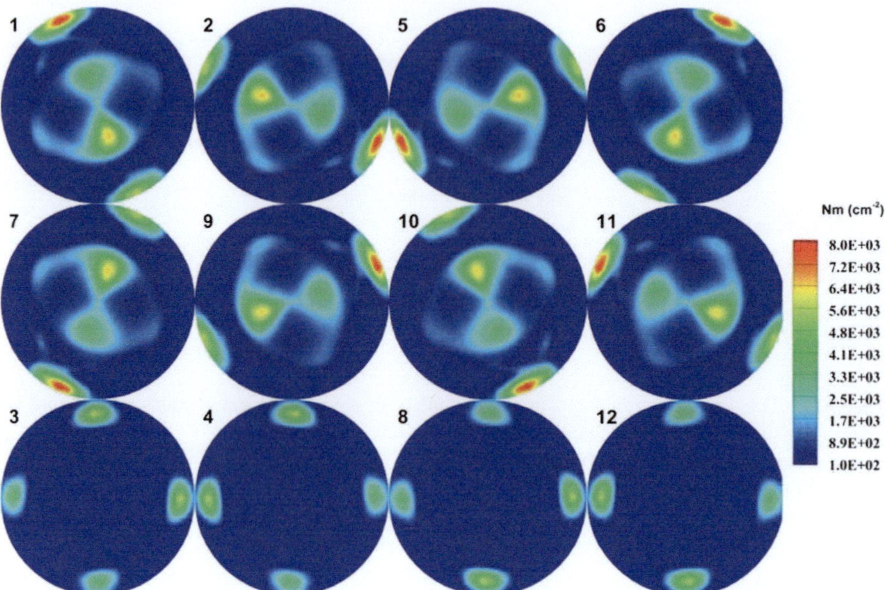

Fig. 5.16 Contribution of every slip direction on the increase of dislocation density in slice 1 in Fig. 5.14a. The number above the figure indicates the slip direction (Reprinted with permission from Ref. [21], Copyright 2013, American Chemical Society)

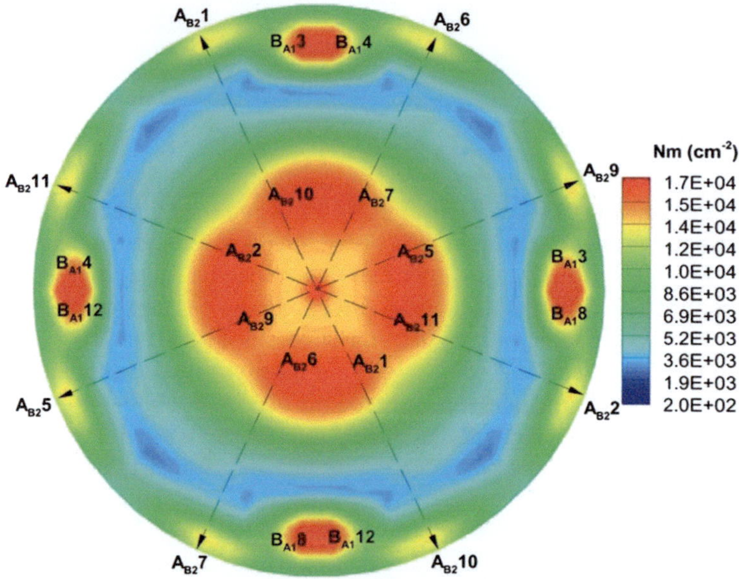

Fig. 5.17 Dislocation density and its corresponding activated slip system in slice 1 of Fig. 5.14a (Reprinted with permission from Ref. [21], Copyright 2013, American Chemical Society)

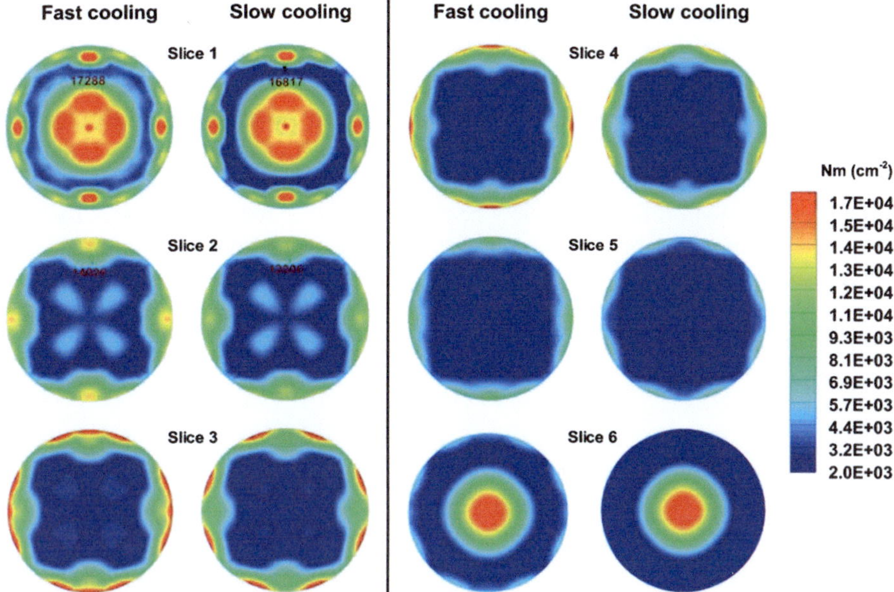

Fig. 5.18 Comparison of the dislocation distributions of (001) slices of a [001] growth crystal for fast cooling and slow cooling (Reprinted with permission from Ref. [21], Copyright 2013, American Chemical Society)

Since the corresponding relationship between the dislocation density and the activated slip system is known, the effect of the cooling rate on the activation of slip systems can be easily determined by calculating the final dislocation distributions under different cooling rates. Figure 5.18 shows comparisons of the dislocation densities obtained by fast cooling and slow cooling. The six slices are at the z positions of 0 % (0.0 m), 19 % (0.0148 m), 38 % (0.0296 m), 57 % (0.0443 m), 76 % (0.0591 m), and 95 % (0.0739 m), which are labeled in Fig. 5.14a. In slices 1 and 2, group B_{A1} is activated for both fast and slow cooling. However, the dislocation under the slow cooling rate is smaller than that under the fast cooling rate, as shown in Fig. 5.18. In slices 3–6, group B_{A1} is not activated under the slow cooling rate, but it is activated under the fast cooling rate. In summary, under slow cooling, group B_{A1} is weakly activated or inactive, whereas under fast cooling, group B_{A1} is strongly activated or activated. Therefore, in the [001] growth direction, the slip system in group B_{A1} under slow cooling is weakly activated or inactive, and that under fast cooling is strongly activated.

5.3.4 Summary

The cooling rate has a marked effect on the generation of dislocation and final residual stress inside the crystal. The slow cooling rate is beneficial for the reduction of dislocations and residual stress. The influence of cooling rate on the generation of dislocations originates from the activation of the slip systems under thermal stress. In the [001] growth direction, a slow cooling rate either weakly activates the slip systems in group B_{A1} or does not activate them at all, while a fast cooling rate strongly activates the slip systems in group B_{A1}. The differences between activated slip systems under slow or fast cooling rates are the main cause of the differences in dislocation and residual stress.

5.4 Conclusions

Numerical simulation is a powerful tool for improving the quality of silicon crystal for solar cells. The carbon and oxygen impurity transport and dislocation multiplication during crystal growth were studied by global simulation. The simulation helped to reduce carbon and oxygen impurities by designing a simple crucible cover and to decrease the dislocation multiplication and residual stress by using a slow cooling process. Further quality improvements can be achieved using these solvers to optimize furnace structure and operating conditions at a low cost.

Acknowledgments This work was partly supported by the Japan Society for the Promotion of Science (Grants-in-Aid for Scientific Research, Grant No. 24360012).

References

1. Pizzini, S., Sandrinelli, A., Beghi, M., Narducci, D., Allegretti, F., Torchio, S., Fabbri, G., Ottaviani, G.P., Demartin, F., Fusi, A.: Influence of extended defects and native impurities on the electrical properties of directionally solidified polycrystalline silicon. J. Electrochem. Soc. **135**, 155 (1988)
2. Moller, H.J., Long, L., Werner, M., Yang, D.: Oxygen and carbon precipitation in multicrystalline solar silicon. Phys. Status Solidi A **171**, 175 (1999)
3. Moller, H.J., Funke, C., Lawerenz, A., Riedel, S., Werner, M.: Oxygen and lattice distortions in multicrystalline silicon. Sol. Energy Mater. Sol. Cells **72**, 403 (2002)
4. Bolotov, V.V., Efremov, M.D., Babanskaya, I., Schmalz, K.: Raman study of mechanical stresses in processes of oxygen precipitation in silicon. Mater. Sci. Eng. B **21**, 49 (1993)
5. El Ghitani, H., Pasquinelli, M., Martinuzzi, S.: Influence of dislocations on photovoltaic properties of multicrystalline silicon solar cells. J. Phys. III **3**, 1941 (1993)
6. Machida, N., Suzuki, Y., Abe, K., Ono, N., Kida, M., Shimizu, Y.: The effects of argon gas flow rate and furnace pressure on oxygen concentration in Czochralski-grown silicon crystals. J. Cryst. Growth **186**, 362 (1998)

7. Machida, N., Hoshikawa, K., Shimizu, Y.: The effects of argon gas flow rate and furnace pressure on oxygen concentration in Czochralski silicon single crystals grown in a transverse magnetic field. J. Cryst. Growth **210**, 532 (2000)
8. Reimann, C., Friedrich, J., Müller, G., Wurzner, S., Möller, H.J.: Analysis of the formation of SiC and Si_3N_4 precipitates during directional solidification of multicrystalline silicon for solar cells. 22nd European Photovoltaic Solar Energy Conference, Milan (2007)
9. Kobayashi, N.: Oxygen transport under an axial magnetic field in Czochralski silicon growth. J. Cryst. Growth **108**, 240 (1991)
10. Hirata, H., Hoshikawa, K.: Three-dimensional numerical analyses of the effects of a cusp magnetic field on the flows, oxygen transport and heat transfer in a Czochralski silicon melt. J. Cryst. Growth **125**, 181 (1992)
11. Kinney, T.A., Brown, R.A.: Application of turbulence modeling to the integrated hydrodynamic thermal-capillary model of Czochralski crystal growth of silicon. J. Cryst. Growth **132**, 551 (1993)
12. Kakimoto, K., Yi, K.W., Eguchi, M.: Oxygen transfer during single silicon crystal growth in Czochralski system with vertical magnetic fields. J. Cryst. Growth **163**, 238 (1996)
13. Yi, K.W., Kakimoto, K., Eguchi, M., Noguchi, H.: Oxygen transport mechanism in Si melt during single crystal growth in the Czochralski system. J. Cryst. Growth **165**, 358 (1996)
14. Watanabe, M., Yi, K.W., Hibiya, T., Kakimoto, K.: Direct observation and numerical simulation of molten silicon flow during crystal growth under magnetic fields by x-ray radiography and large-scale computation. Prog. Cryst. Growth Charact. Mater. **38**, 215 (1999)
15. Reimann, C., Jung, T., Friedrich, J., Müller, G.: The importance of convective heat and mass transfer for controlling material properties in ingot casting of multi-crystalline-silicon for photovoltaic applications. Proceedings of the 33rd IEEE Photovoltaic Specialists Conference, ISBN:978-1-4244-1641-7 (2008)
16. Bornside, D.E., Brown, R.A.: The effects of gas-phase convection on carbon contamination of Czochralski-grown silicon. J. Electrochem. Soc. **142**(8), 2790 (1995)
17. Li, Y.R., Li, M.W., Imaishi, N., Akiyama, Y., Tsukada, T.: Oxygen-transport phenomena in a small silicon Czochralski furnace. J. Cryst. Growth **267**, 466 (2004)
18. Smirnov, A.D., Kalaev, V.V.: Development of oxygen transport model in Czochralski growth of silicon crystals. J. Cryst. Growth **310**, 2970 (2008)
19. Smirnov, A.D., Kalaev, V.V.: Analysis of impurity transport and deposition processes on the furnace elements during Cz silicon growth. J. Cryst. Growth **311**, 829 (2009)
20. Gao, B., Kakimoto, K.: Relationship between the locations of activated dislocations and the cooling flux direction in monocrystalline-like silicon grown in [001] and [111] directions. J. Appl. Cryst. **46**, 1771–1780 (2013)
21. Gao, B., Nakano, S., Harada, H., Miyamura, Y., Kakimoto, K.: Effect of cooling rate on the activation of slip systems in seed cast-grown monocrystalline silicon in the [001] and [111] directions. Cryst. Growth Des. **13**, 2661–2669 (2013)
22. Gao, B., Kakimoto, K.: Numerical investigation of the influence of cooling flux on the generation of dislocations in cylindrical mono-like silicon growth. J. Cryst. Growth **384**, 13–20 (2013)
23. Gao, B., Chen, X.J., Nakano, S., Kakimoto, K.: Crystal growth of high-purity multicrystalline silicon using a unidirectional solidification furnace for solar cells. J. Cryst. Growth **312**, 1572–1576 (2010)
24. Liu, L.J., Kakimoto, K.: Partly three-dimensional global modeling of a silicon Czochralski furnace. I. Principles, formulation and implementation of the model. Int. J. Heat Mass Transf. **48**, 4481 (2005)
25. Reid, R.C., Prausnitz, J.M., Sherwood, T.K.: The Properties of Gases and Liquids, 3rd edn. McGraw-Hill, Inc., New York (1987)
26. Liu, L.J., Nakano, S., Kakimoto, K.: Investigation of oxygen distribution in electromagnetic CZ-Si melts with a transverse magnetic field using 3D global modeling. J. Cryst. Growth **299**, 48 (2007)

27. Hoshikawa, K., Huang, X.: Oxygen transportation during Czochralski silicon crystal growth. Mater. Sci. Eng. **B 72**, 73 (2000)
28. Gao, B., Kakimoto, K.: Global simulation of coupled carbon and oxygen transport in a Czochralski furnace for silicon crystal growth. J. Cryst. Growth **312**, 2972 (2010)
29. Gao, B., Nakano, S., Kakimoto, K.: Global simulation of coupled carbon and oxygen transport in a unidirectional solidification furnace for solar cells. J. Electrochem. Soc. **157**(2), H153–H159 (2010)
30. Matsuo, H., Ganesh, R.B., Nakano, S., Liu, L.J., Kangawa, Y., Arafune, K., Ohshita, Y., Yamaguchi, M., Kakimoto, K.: Analysis of oxygen incorporation in unidirectionally solidified multicrystalline silicon for solar cells. J. Cryst. Growth **310**, 2204 (2008)
31. Ganesh, R.B., Matsuo, H., Kawamura, T., Kangawa, Y., Arafune, K., Ohshita, Y., Yamaguchi, M., Kakimoto, K.: Estimation of growth rate in unidirectionally solidified multicrystalline silicon by the growth induced striation method. J. Cryst. Growth **310**, 2697 (2008)
32. Matsuo, H., Bairava Ganesh, R., Nakano, S., Liu, L.J., Kangawa, Y., Arafune, K., Ohshita, Y., Yamaguchi, M., Kakimoto, K.: Thermodynamical analysis of oxygen incorporation from a quartz crucible during solidification of multicrystalline silicon for solar cell. J. Cryst. Growth **310**, 4666 (2008)
33. Prasad, V., Pendurti, S.: Models for stress and dislocation generation in melt based compound crystal growth. In: Dhanaraj, G., Byrappa, K., Prasad, V., Dudley, M. (eds.) Handbook of Crystal Growth. Springer, Berlin/Heidelberg (2010)
34. Jordan, A.S., Caruso, R., Von Neida, A.R.: A thermo-elastic analysis of dislocation generation in pulled GaAs crystals. Bell Syst. Tech. J. **59**, 593–637 (1980)
35. Orowan, E.: Problems of plastic gliding. Philos. Trans. R. Soc. London, Ser. A **52**, 8 (1940)
36. Cochard, J., Yonenaga, I., Gouttebroze, S., M'Hamdi, M., Zhang, Z.L.: Constitutive modeling of intrinsic silicon monocrystals in easy glide. J. Appl. Phys. **107**, 033512 (2010)
37. Gao, B., Nakano, S., Kakimoto, K.: Influence of reaction between silica crucible and graphite susceptor on impurities of multicrystalline silicon in a unidirectional solidification furnace. J. Cryst. Growth **314**, 239 (2011)

Chapter 6
Oxygen Precipitation in Silicon

Gudrun Kissinger

Abstract This chapter starts with the basic features of interstitial oxygen which are crucial for further considerations. Nucleation and growth of oxygen precipitates are described from the viewpoint of classical nucleation theory. The initial states of oxygen precipitation as suggested by ab initio calculation are also shown. Results about the impact of intrinsic point defects, doping, and co-doping on oxygen precipitation are presented. The most important methods for detection and characterization of oxygen precipitates with their possibilities and limitations are comprehensively described. A second focus of this chapter is directed towards the impact of grown-in oxygen precipitate nuclei in silicon wafers on creation of high quality defect denuded zones and oxygen precipitation during device processing. Conventional and modern methods of thermal processing and their impact on oxygen precipitation are discussed. Methods to determine the getter efficiency of oxygen precipitates and their results are described.

Keywords Oxygen precipitation • Gettering of metallic impurities • Nucleation of oxygen precipitates • Grown-in nuclei • Annealing • Denuded zone • Preferential etching • Getter test

6.1 Introduction

Oxygen is the most important impurity in Czochralski (CZ) silicon. It is incorporated in the interstitial form during crystal pulling as the result of the partial dissolution of the walls of the quartz crucible which contains the molten silicon. During cooling of the CZ silicon ingot, the interstitial oxygen supersaturates and starts precipitating in the form of SiO_x with x ranging from one to two whereby according to recently published results the exact stoichiometry is rather that of SiO_2. First oxygen precipitate nuclei can be found already after cooling of the ingot. Their size is below the detection limit of common tools used for studying oxygen precipitates but their characterization is the way to understand the impact of

G. Kissinger (✉)
IHP-Innovations for High Performance Microelectronics, Frankfurt (Oder), Germany
e-mail: gkissinger@ihp-microelectronics.com

Y. Yoshida, G. Langouche (eds.), *Defects and Impurities in Silicon Materials*, Lecture Notes in Physics 916, DOI 10.1007/978-4-431-55800-2_6

the thermal history of the silicon ingot on oxygen precipitation during subsequent thermal processing of the silicon wafers when the majority of the supersaturated interstitial oxygen precipitates.

Oxygen precipitates located in the device active region of microelectronic devices can degrade the integrity of thin gate oxides, cause leakage currents or even shortcuts if decorated with metals (see e.g. Refs. [1–3]). If located in the bulk of silicon wafers they can act as gettering sinks for metallic impurities and thus improve the device characteristics especially in case of undesired metal contamination.

In 1994, Shimura has edited a book on oxygen in silicon which contains the knowledge gained until this stage and which is very comprehensive [4]. This chapter describes the new knowledge generated during the following 20 years summarizing the role of oxygen precipitation from crystal growth to device processing with its many facets ranging from theory to applications.

6.2 Basic Features of Interstitial Oxygen in Silicon

In general, oxygen precipitation in silicon is a diffusion limited process. The most important features necessary to describe it are the temperature T dependent solubility $C_{Oi}^{eq}(T)$ and diffusivity $D_{Oi}(T)$ of the interstitial oxygen. The most widely accepted values were determined by Mikkelsen [5] as follows

$$C_{Oi}^{eq}(T) = 9 \times 10^{22} \exp\left(\frac{-1.52eV}{kT}\right) \tag{6.1}$$

where k is Boltzman's constant. Figure 6.1 shows the solubility as a function of the temperature together with concentrations typical for Czochralski silicon using new ASTM standard. The meaning of the standards is explained in Sect. 6.3. It can be seen that for common temperatures of device processing the interstitial oxygen is always supersaturated.

The interstitial oxygen is a slow diffusor. Its temperature dependent diffusivity can be calculated as follows [5]

$$D_{Oi}(T) = 0.13 \cdot \exp\left(\frac{-2.53eV}{kT}\right). \tag{6.2}$$

In the precipitation process of oxygen beneath the interstitial oxygen also the intrinsic point defects are involved. Figure 6.2 compares the diffusivities of all these species. The diffusivity of the intrinsic point defects is considerably higher than that of the interstitial oxygen. We use here the diffusivities according to Frewen et al. [6] as an example. Many other authors determined diffusivities of intrinsic point defects. Details can be found in Chap. 1.

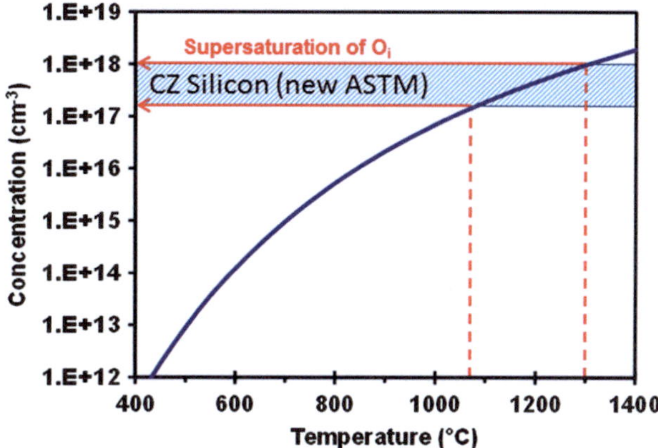

Fig. 6.1 Solubility of interstitial oxygen plotted as a function of temperature given in new ASTM standard [5]. The hatched area marks the concentration of interstitial oxygen typical for Czochralski silicon and the *dashed lines* indicate the onset of supersaturation for decreasing temperature

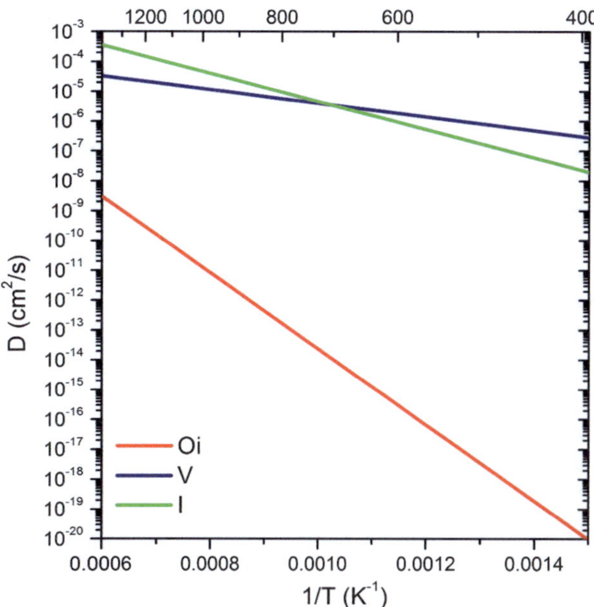

Fig. 6.2 Diffusivities of interstitial oxygen (O_i), silicon vacancies (V), and silicon interstitials (I) all plotted as a function of reciprocal temperature [5, 6]

A second oxygen species which is increasing in importance with decreasing temperature is the oxygen dimer O_{2i}. The oxygen dimer is known to diffuse much faster than interstitial oxygen [7]. Theoretical models of the diffusion path of the dimer were developed which are in good agreement with the experiments [8, 9]. The

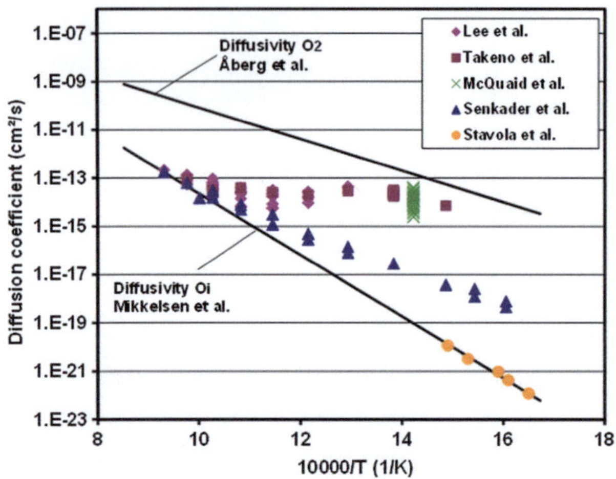

Fig. 6.3 Effective diffusivities from Refs. [10–14] plotted as a function of the reciprocal temperature. Additionally, the diffusivities of interstitial oxygen [5] and oxygen dimers [7] are shown as *straight lines* (Reproduced with permission from [15]. Copyright 2008, Trans Tech Publications Ltd)

higher diffusivity of O_{2i} results in deeper out-diffusion of oxygen in the temperature range below 700 °C than it would be expected from the interstitial oxygen diffusivity extrapolated from high temperature diffusivities [5, 10, 11]. It also results in a higher growth rate of oxide precipitates below 700 °C [12].

Figure 6.3 summarizes the effective diffusivities of oxygen from the most extensive studies in this temperature range [10–14] as a function of the reciprocal temperature. Obviously, all experiments which determine the diffusivity from oxygen out-diffusion [10, 11] and from growth of oxide precipitates [12] result in quite similar effective diffusivities while the diffusivities obtained from dislocation locking experiments [13] are clearly lower and the diffusivities from relaxation of stress induced dichroism [14] do not show any deviation from the extrapolation from high temperature diffusivities because by this method only O_i is measured. In the first two cases, oxygen depth profiles were established which result from diffusion of the oxygen to either the surface or the oxide precipitates. The contribution of the interstitial monomer to the out-diffusion or precipitate growth is very small because of its low transport capacity at temperatures below 700 °C. The creation of the oxygen out-diffusion profiles is first of all the result of the slow conversion of interstitial oxygen into dimers [11]. This means also that precipitate growth at temperatures below 700 °C would be strongly influenced by the kinetics of ongoing dimer formation in the oxygen depleted region around the growing oxide precipitate. The effective diffusivities obtained from dislocation locking experiments are lower but the kinetics of dimer formation in the oxygen depleted region near the dislocation core cannot be neglected either, although dislocations have a limited capacity for the capture of oxygen atoms and the capture process is reversible.

6.3 Measurement of Interstitial Oxygen in Silicon

Commonly, the concentration of interstitial oxygen is determined by Fourier transform infrared spectroscopy (FTIR). Figure 6.4 shows a typical absorption spectrum of interstitial oxygen in silicon recorded at room temperature in the wave number range from 400 to 1800 cm^{-1}. The absorbance A at the peak position of the main absorption band at 1106 cm^{-1} is determined from the incident radiant power I_0 and the transmitted radiant power I as follows:

$$A = \ln\left(\frac{I_0}{I}\right). \tag{6.3}$$

Knowing the thickness of the sample d, the absorption coefficient α with

$$\alpha = \frac{A}{d} = \frac{1}{d}\ln\left(\frac{I_0}{I}\right) \tag{6.4}$$

can be calculated. From the absorption coefficient of the main absorption band at 1106 cm^{-1}, the O_i concentration is obtained via multiplication with a conversion factor. The problem is that several conversion factors exist. They are listed in Table 6.1. Although the IOC-88 factor is regarded as the most accurate one nowadays [16], all conversion factors are still in use and it is very important to mention the conversion factor in connection with a concentration of interstitial oxygen measured by FTIR because otherwise the value is useless.

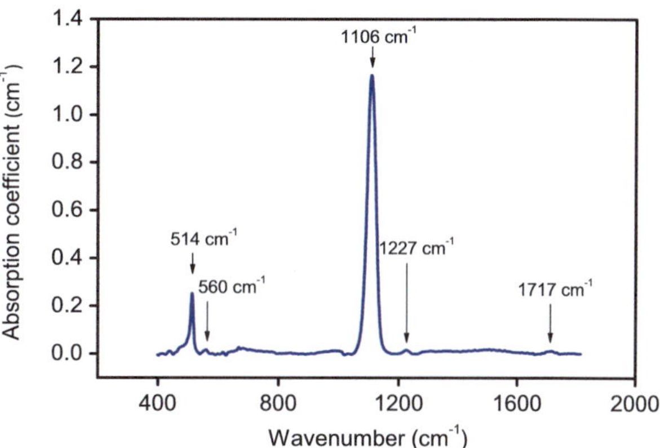

Fig. 6.4 Typical absorption spectrum of interstitial oxygen in silicon recorded at room temperature in the wave number range from 400 to 1800 cm^{-1}

Table 6.1 Conversion factors for measurement of the concentration of interstitial oxygen by FTIR at room temperature

Conversion factor (cm^{-2})	Source	Short name
2.45×10^{17}	ASTM F 121-83	New ASTM
2.45×10^{17}	DIN 50 438/1	DIN
3.03×10^{17}	JEIDA	JEIDA
4.81×10^{17}	ASTM F 121-79	Old ASTM
3.14×10^{17}	ASTM F 1188-88	IOC-88

6.4 Oxygen Precipitation Described by Classical Nucleation Theory

6.4.1 Volumetric Considerations

Classical nucleation theory is the easiest way to describe nucleation of oxygen precipitates and it helps to understand the important parameters and the basic physics involved. Most models for homogeneous nucleation of oxide precipitates are based on nucleation of incoherent amorphous SiO_x particles [see e.g. Refs. 17 and 18].

In the beginning, we will make some volumetric considerations about oxygen precipitation which are demonstrated by the three steps in Fig. 6.5. If we assume that a certain amount of silicon atoms is taken from the silicon matrix and transformed into SiO_x, a spherical cavity with the volume V_m and the radius r_m would be left by these silicon atoms. Because the density of SiO_x is lower than the density of silicon, the volume V_p and radius r_p of the unconstrained precipitate are larger than volume and radius of the cavity. In the third step, we merge the cavity and the precipitate. For the cavity's and unconstrained precipitate's volumes and radii and the resulting volume V_f and radius r_f of the final constrained precipitate, the following relationship holds: $V_m < V_f < V_p$ and $r_m < r_f < r_p$.

This means that the precipitates would be always compressively strained. However, there are mechanisms to relieve the stain. One possibility is the ejection of silicon self-interstitials I from the precipitate/matrix interface which diffuse away or aggregate by forming secondary defects like extrinsic stacking faults or punched out dislocation loops. Another possibility for strain relieve is the absorption of silicon lattice vacancies V by the precipitate. Finally, we can formulate the general equilibrium reaction of oxygen precipitation as follows:

$$P_n + \left(\frac{1}{x} + \gamma\right) Si + O + \beta V \leftrightarrow P_{n+1} + \gamma I + \text{strain energy} \tag{6.5}$$

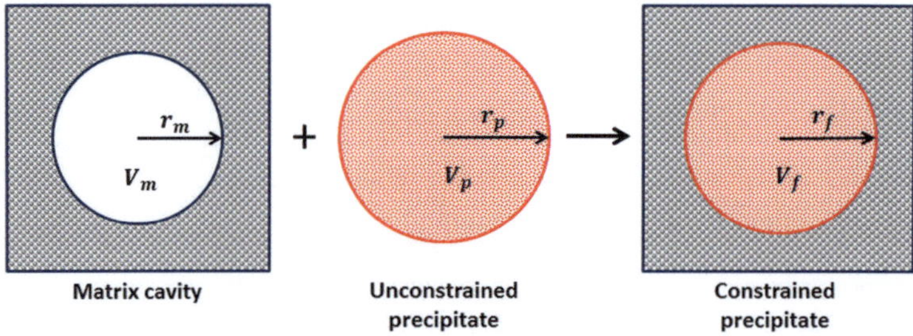

Fig. 6.5 Volumetric considerations of oxygen precipitate formation

where P_n and P_{n+1} are precipitates containing n and $n+1$ oxygen atoms, respectively , γ is the number of interstitials ejected, and β is the number of vacancies absorbed both per precipitated oxygen atom. The equilibrium reaction suggests that oxygen precipitation is enhanced in the environment of vacancy supersaturation and is retarded by interstitial supersaturation. Effective vacancy supersaturation can be created e.g. by rapid thermal annealing (RTA) at temperatures above 1150 °C or below silicon nitride layers. Whereas the oxidation of silicon leads to the supersaturation of self-interstitials below the oxide layer. However, the precipitation of interstitial oxygen itself can also cause a supersaturation of interstitials and thus retard itself.

The relationship between the volumes of the unconstrained precipitate and the matrix cavity is defined via the misfit δ and the relationship between the volumes of the constrained precipitate and the matrix cavity is defined via the strain ε as follows:

$$\frac{V_p}{V_m} = (1+\delta)^3 = \frac{\Omega_{SiOx}}{\Omega_{Si}\left[1+(\beta+\gamma)\right]} \quad \text{and} \tag{6.6}$$

$$\frac{V_f}{V_m} = (1+\varepsilon)^3 \text{ with } \varepsilon = \frac{\delta}{1+\left(\frac{4\mu}{3K}\right)} \tag{6.7}$$

whereby the misfit can be obtained from the molecular volumes of the SiO_x precipitate Ω_{SiOx} and of the Si matrix Ω_{Si}, which is 2×10^{-23} cm^3, taking into account also the number of vacancies absorbed and interstitials generated to release the strain. The concentration of oxygen in the precipitate C_p can be calculated from the molecular volumes Ω by $C_p = x/\Omega$. Typical values for the molecular volumes of oxygen precipitates applied are 3.448×10^{-23} cm^3 for SiO and 4.348×10^{-23} cm^3 for unstrained SiO_2 [19]. For the calculation of the strain in Eq. 6.5 the shear modulus μ of silicon (6.41×10^{10} Pa) and the bulk modulus K of the precipitate (3.47×10^{10} Pa) are needed.

6.4.2 Homogeneous Nucleation of Spherical Precipitates

Considering spherical oxygen precipitate nuclei is the easiest approach in classical nucleation theory. The total free energy of formation of a spherical precipitate nucleus ΔG_{hom}^{sphere} consisting of n oxygen atoms contains three major contributions. These are the free energy of formation of the precipitated oxide phase ΔG_v, the free energy necessary to create the new surface of the sphere ΔG_σ^{sphere}, and the elastic strain energy $\Delta G_\varepsilon^{sphere}$ caused by the volume deficiency between the spherical silicon cavity and the spherical volume of the precipitated oxide phase. We are neglecting here any influence of possible temperature dependence of the elastic moduli and of different thermal expansion. Most classical considerations of nucleation follow this approach. The total free energy can be obtained as follows:

$$\begin{aligned} \Delta G_{hom}^{sphere} &= -\Delta G_v + \Delta G_\sigma^{sphere} + \Delta G_\varepsilon^{sphere} \quad \text{with} \\ \Delta G_v &= n \cdot k \cdot T \cdot \ln\left[\frac{C_{Oi}}{C_{Oi}^{eq}} \cdot \left(\frac{C_V}{C_V^{eq}}\right)^\beta \cdot \left(\frac{C_I^{eq}}{C_I}\right)^\gamma\right], \\ \Delta G_\sigma^{sphere} &= (36\pi)^{\frac{1}{3}} \cdot \left(\frac{\Omega_p \cdot n}{x}\right)^{\frac{2}{3}} \cdot \sigma \cdot \left(\frac{1+\delta}{1+\varepsilon}\right)^{-2} \quad \text{and} \\ \Delta G_\varepsilon^{sphere} &= \left(\frac{\Omega_p \cdot n}{x}\right) \cdot 6 \cdot \mu \cdot \delta \cdot \varepsilon \cdot (1+\delta)^{-3}. \end{aligned} \tag{6.8}$$

Here C^{eq} denotes the equilibrium concentrations and C the concentrations of vacancies, interstitials, and interstitial oxygen as indicated by V, I, and O_i, respectively. The strain energy according to Ref. [20] was used here with a correction factor respecting the volume misfit because the strain energy originally refers to the volume of the cavity in the matrix. The surface energy σ cannot be measured. This parameter is often used as a fit parameter when a model is adjusted to experimental data. Typical values of the surface energy are in the range of 10^{-5} J/cm^2 [21].

Figure 6.6 shows a plot of the free energy of an oxygen precipitate as a function of the precipitated oxygen atoms. Because the positive and negative components of

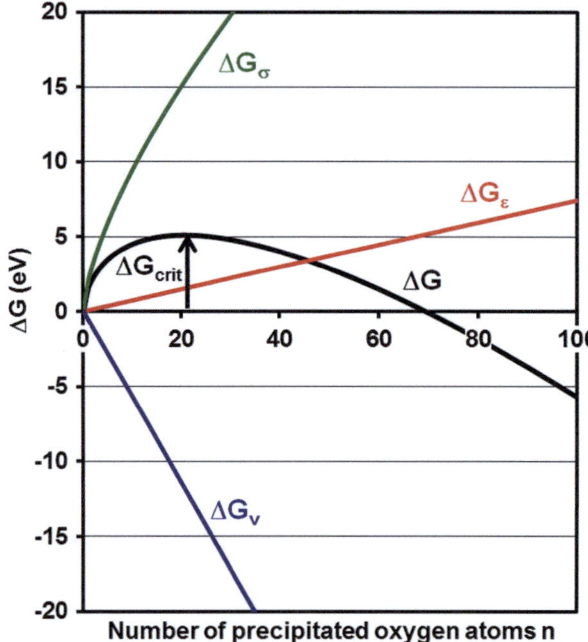

Fig. 6.6 Components of the free energy of formation of an oxygen precipitate plotted as a function of the number of oxygen atoms per precipitate

ΔG are involved which have different dependency on the number of oxygen atoms per precipitate, the free energy of formation of a precipitate nucleus has a maximum for $n = n_{crit}$. Precipitates with $n \geq n_{crit}$ are stable because the attachment of further oxygen atoms leads to a decrease of free energy whereas precipitates with $n < n_{crit}$ are unstable.

The number of oxygen atoms in a critical nucleus n_{crit} can be obtained by setting the first derivative of $d\Delta G/dn$ to zero. This leads to

$$n_{crit} = \left(\frac{32\pi \cdot x \cdot \sigma^3}{3 \cdot \Omega_p}\right) \cdot \left(\frac{1+\delta}{1+\varepsilon}\right)^{-6} \cdot \left\{\frac{k \cdot T \cdot x}{\Omega_p} \cdot ln\left[\frac{C_{Oi}}{C_{Oi}^{eq}} \cdot \left(\frac{C_V}{C_V^{eq}}\right)^{\beta} \cdot \left(\frac{C_I^{eq}}{C_I}\right)^{\gamma}\right] -6\mu \cdot \delta \cdot \varepsilon \cdot (1+\delta)^{-3}\right\}^{-3} \tag{6.9}$$

The thermodynamic equilibrium concentration of oxide precipitate nuclei of critical size N_{crit} can be determined from Boltzman's statistics as

$$N_{crit} = N_0 \cdot \exp\left(-\frac{\Delta G_{crit}}{k \cdot T}\right). \tag{6.10}$$

Here, ΔG_{crit} is the total free energy for formation of a critical nucleus and N_0 is the density of nucleation sites. Homogeneous nucleation models based on agglomeration of interstitial oxygen assume that N_0 is equal to the initial concentration of interstitial oxygen [22].

The nucleation rate J for spherical oxide precipitate nuclei of critical size can be expressed by

$$J = N_0 \cdot \exp\left(-\frac{\Delta G_{crit}}{k \cdot T}\right) \cdot C_{Oi} \cdot 4\pi \cdot \left(\frac{n_{crit} \cdot 3 \cdot \Omega_p}{4\pi \cdot x}\right)^{\frac{2}{3}} \cdot \left(\frac{1+\delta}{1+\varepsilon}\right)^{-2} \cdot D_{Oi} \cdot d^{-1} \cdot Z \tag{6.11}$$

where the first two terms represent the equilibrium concentration of critical nuclei as given in Eq. 6.10. The following terms except the last one represent the impingement frequency of an oxygen atom onto a critical nucleus and Z is the Zeldovich factor. The Zeldovich factor Z can be determined according to Ref. [23] as follows:

$$Z = \sqrt{\frac{\Delta G_{crit}}{3\pi \cdot k \cdot T \cdot n_{crit}^2}}. \tag{6.12}$$

The variable n_{crit} is the number of oxygen atoms in a nucleus of critical size and d is the atomic distance in silicon (0.235 nm). Because all calculations refer to the strained precipitate radius a correction factor for the volume of the precipitate has to be taken into account which contains the volume misfit δ and the elastic strain ε of the precipitated oxide phase.

6.4.3 Homogeneous Nucleation of Plate-Like Precipitates

We first consider again the free energy of formation of a circular plate-like precipitate ΔG^{plate} and its components. The free energy of formation of the precipitated oxide phase ΔG_v is the same as for spherical precipitate nuclei. The free energy necessary for the creation of the platelet surface ΔG_σ^{plate} is well approximated by taking twice the area of the circular platelet. In order to obtain the strain energy, the platelet is regarded as an intrinsic loop with the displacement vector b. For the calculation of the strain energy, we followed the approach adapted in Ref. [24] employing the energy of the dislocation loop surrounding the platelet $\Delta G_\varepsilon^{plate}$. This results in the following set of equations for the total free energy:

$$\begin{aligned} \Delta G_{hom}^{plate} &= -\Delta G_v + \Delta G_\sigma^{plate} + \Delta G_\varepsilon^{plate} \quad \text{with} \\ \Delta G_v &= n \cdot k \cdot T \cdot \ln\left[\tfrac{C_{Oi}}{C_{Oi}^{eq}} \cdot \left(\tfrac{C_V}{C_V^{eq}}\right)^\beta \cdot \left(\tfrac{C_I^{eq}}{C_I}\right)^\gamma\right], \\ \Delta G_\sigma^{plate} &= \left(\tfrac{\Omega_a \cdot 2n}{x}\right) \cdot \sigma = 2\pi \cdot r^2 \cdot \sigma, \quad \text{and} \\ \Delta G_\varepsilon^{plate} &= \tfrac{\mu \cdot b^2}{8\pi \cdot (1-\nu)} \cdot \ln\left(\tfrac{64 \cdot n \cdot \Omega_a}{x \cdot \pi \cdot b^2 \cdot e^2}\right) \cdot \sqrt{\tfrac{4\pi \cdot n \cdot \Omega_a}{x}} = \tfrac{\mu \cdot b^2 \cdot r}{2(1-\nu)} \ln\left(\tfrac{8r}{b \cdot e}\right) \end{aligned} \tag{6.13}$$

with $\nu = 0.28$ being Poisson's ratio of silicon [25], e being the base of the natural logarithm, and Ω_a being the molecular area. The molecular area is related to the radius r as follows:

$$r^2 = \frac{n \cdot \Omega_a}{x \cdot \pi}. \tag{6.14}$$

Again, setting the first derivative $d\Delta G/dn$ to zero delivers the critical number of oxygen atoms. However, for the plate-like precipitate the solution for n_{crit} is a Lambert function. Therefore, in practice an iterative solution is recommended.

The nucleation rate of the plate-like precipitates J_{plate} can be obtained according to the following equation:

$$J_{plate} = N_0 \cdot \exp\left(-\frac{\Delta G_{crit}}{k \cdot T}\right) \cdot C_{Oi} \cdot \left(\frac{2n_{crit} \cdot \Omega_a}{x}\right) \cdot D_{Oi} \cdot d^{-1} \cdot Z \tag{6.15}$$

If we compare the nucleation rate as a function of temperature that means the so-called nucleation curves for spherical and plate-like precipitates we will see that they are relatively broad and symmetric for the spherical precipitates whereas the nucleation curves for plate-like precipitates decrease abruptly at the side towards higher temperature [26].

6.4.4 Heterogeneous Nucleation

Heterogeneous nucleation of oxygen precipitates at secondary defects like dislocations or stacking faults can be often observed in silicon after thermal treatments. The energy barrier for heterogeneous nucleation at dislocations is lower than for homogeneous nucleation because of the energy gained by the annihilation of a part of the dislocation. This leads to an additional negative free energy contribution. As an example, the total free energy for heterogeneous nucleation at edge dislocation loops ΔG_{het}^{loop} is described here. It can be obtained if we assume that the nucleus is spherical and the part of the surrounding dislocation loop that is annihilated corresponds to the diameter of the sphere. The free energy gained by annihilation of part of an edge dislocation loop of radius R by a precipitate nucleus of radius r containing n oxygen atoms ΔG_{loop} can be calculated using the dislocation energy according to [27] assuming that the radius of the dislocation core corresponds to the Burgers vector b. Then, the free energy for heterogeneous nucleation at an edge dislocation loop can be obtained from

$$\begin{aligned} \Delta G_{het}^{loop} &= -\Delta G_v + \Delta G_\sigma^{sphere} + \Delta G_\varepsilon^{sphere} - \Delta G_{loop} \quad \text{with} \\ \Delta G_{loop} &= \frac{\mu \cdot b^2}{2\pi \cdot (1-\nu)} \cdot \ln\left(\frac{8 \cdot R}{b \cdot e}\right) \cdot \left(\frac{1+\delta}{1+\varepsilon}\right)^{-1} \cdot \sqrt[3]{\frac{3n \cdot \Omega_p}{4\pi \cdot x}} = \frac{\mu \cdot b^2}{2\pi \cdot (1-\nu)} \cdot \ln\left(\frac{8 \cdot R}{b \cdot e}\right) \cdot 2r. \end{aligned} \tag{6.16}$$

The other free energy contributions remain the same as in Eq. 6.8.

6.4.5 Early Stages of Nucleation

Vacancies enhance oxygen precipitation as it was demonstrated by classical nucleation theory in the previous sections. Now, the question is what is going on at the atomistic level in the initial stages of oxygen precipitation. A second issue regards the coherency of the interface between the oxygen precipitate and the silicon matrix. The interfacial energy of coherent precipitates is much lower because there is only the chemical contribution resulting from the bonds between the precipitate and matrix atoms [28]. Incoherent interfaces have additional energy contributions to the interface energy resulting from structural distortions. Therefore, the nucleation barrier ΔG_{crit} would be much lower for nucleation of coherent oxygen precipitates. From investigations of oxygen precipitates by transmission electron microscopy (TEM) it was found that the oxygen precipitates in silicon consist of amorphous silicon oxide [29, 30]. However, it cannot be excluded that in the initial stage fully coherent phases comparable to Guinier-Preston zones [28], well known from metal physics, are formed. Ab initio calculation based on density functional theory as described in Chap. 2 is a good way of getting insight into the initial stages of oxygen precipitation.

Table 6.2 Comparison of binding energies given in eV for O_n and VO_n complexes as obtained by different authors [31]

Reaction	Kissinger et al. [31]	Torres et al. [32]	Sueoka [33]	Pesola et al. [34]	Casali et al. [35]
$V+O \rightarrow VO$	1.48	1.50	1.29	0.80	1.70
$VO+O \rightarrow VO_2$	1.31	1.56	1.37	1.00	1.00
$VO_2+O \rightarrow VO_3$	1.28	1.41	1.44		
$VO_3+O \rightarrow VO_4$	1.23	1.37	1.43		
$VO_4+O \rightarrow VO_5$	1.20	1.14			
$VO_5+O \rightarrow VO_6$	1.10	1.08			
$VO_6+O \rightarrow VO_7$	0.57				
$VO_7+O \rightarrow VO_8$	0.62				
$O+O \rightarrow O_2$	0.30	0.52	0.23		
$O_2+O \rightarrow O_3$	0.43	0.71	0.31		
$O_3+O \rightarrow O_4$	0.54	0.80	0.28		
$O_4+O \rightarrow O_5$	0.61	0.88			
$O_5+O \rightarrow O_6$	0.65	0.95			
$O_6+O \rightarrow O_7$	0.66				

We first consider the binding energies E_b obtained from the formation energies E_f^n of O_n and VO_n complexes which are formed by the attachment of oxygen atoms to either a chain of oxygen atoms or a vacancy E_f^{n-1}, respectively, with $E_b = \left(E_f^{n-1} + E_f^{Oi}\right) - E_f^n$. Table 6.2 compares the values of different authors and most of them are in quite good agreement. For low n, the binding energies of VO_n complexes are clearly higher than the binding energies of the O_n chains. However, the more oxygen atoms are attached the lower the difference of the binding energy becomes and for $n > 6$ they are at the same level.

In Fig. 6.7, the binding energies and the formation energies of O_n and VO_n are plotted as a function of the oxygen atoms and Fig. 6.8 shows some examples of the VO_n complexes. In three steps, the VO_n complexes change their structure qualitatively. This is indicated also by dashed lines in Fig. 6.7 left. First the vacancy bonds are oxidised up to $n = 2$. Then short $\langle 110 \rangle$ chains of oxygen interstitials are built up to $n = 6$. After this, the second neighbours are oxidised up to $n = 10$ with one additional oxygen atom each. This would be possible up to $n = 14$ until the second neighbours are fully oxidized. Figure 6.7 left demonstrates that if the number of oxygen atoms in the final complex is less than 7, new oxygen interstitials are preferably attached to the oxygen atoms surrounding the vacancy at the center. The energy gain gradually decreases and for larger complexes there is no remarkable difference between VO_n and O_n defects. One might try to interpret this result as reflecting the fact that at a certain point the negative internal pressure exerted on

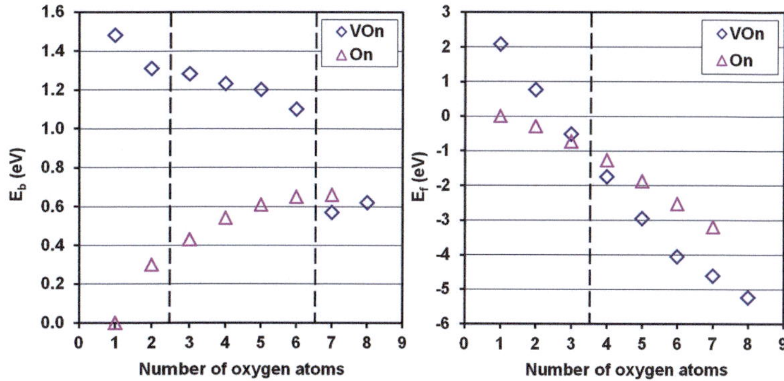

Fig. 6.7 Binding energies $E_b = \left(E_f^{n-1} + E_f^{Oi}\right) - E_f^n$ (*left graph*) and formation energies (*right graph*) of interstitial oxygen atoms attached to either O_{n-1} chains (*triangles*) or VO_{n-1} complexes (*diamonds*); E_f^n is the total energy of the complex containing *n* oxygen atoms, and E_f^{Oi} is the total energy of the oxygen interstitial. *Left graph* reprinted with permission from [31] (Copyright 2011, The Electrochemical Society)

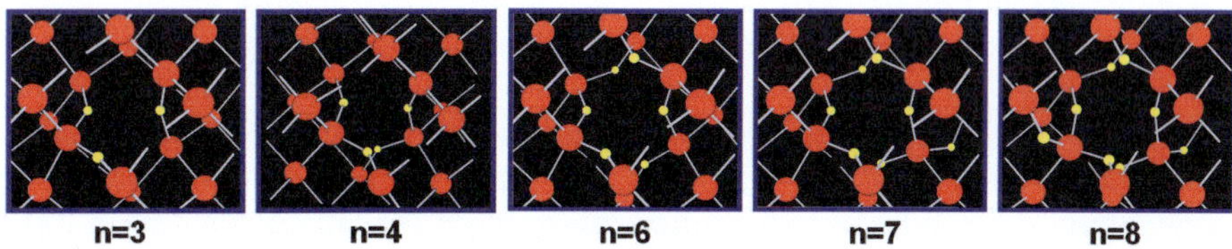

Fig. 6.8 Atomic structures of several VO_n complexes, for n = 3, 4, 6, 7, and 8. Si atoms are large *red spheres*, oxygen atoms are small *yellow spheres*; the vacancy is in the center (Reprinted with permission from [31]. Copyright 2011, The Electrochemical Society)

the Si matrix by the vacancy is compensated by the positive pressure exerted by the oxygen interstitials and the "attractive power" of the vacancy comes thus to an end. At $n > 6$ there is a remarkable jump in the binding energy to VO_n complexes. For up to 6 oxygen atoms, the oxidized vacancy consists of two planar, ⟨110⟩-like chains of oxygen interstitials, with the vacancy Si-Si bond being the center of each of them (see Fig. 6.8 for $n = 6$). The seventh atom is preferably attached to one of the chains out of the plane as seen in Fig. 6.8 for $n = 7$. The apparent reason for that is that in this way another component of the strain field is reduced with the maximum energy gain. This component is extending along another class of [110] chains, which does not contain the vacancy Si-Si bond. Now, the strain is comparable to that around an O_n chain. This remains true for the attachment of further oxygen atoms. If we regard the vacancy as the nucleus for oxygen precipitation we can interpret the result in the way that for $n = 7$ a second vacancy is needed to increase the binding energy again to a level as for lower *n*.

If we look at the formation energies of the O_n and VO_n complexes in Fig. 6.7 right we can see that for $n < 4$ the O_n complexes possess the lower formation energies and for $n > 3$ the VO_n complexes are the more stable ones with the lower formation energy. This change is indicated by a dashed line bin Fig. 6.7 right.

The binding energies can be used for modeling the kinetics of complex formation during thermal treatments. This was done for RTA pre-treated silicon wafers in Ref. [31]. Rate equation modeling for I, V, O_n with $n = (1–4)$ and VO_n with $n = (1–8)$ based on the ab initio calculated binding energies without fitting parameters yielded O_i, O_2, O_3 and VO_4 to appear with nearly the same concentration as it was determined by highly sensitive FTIR in RTA pre-treated silicon wafers [36].

In the following, we will describe silicon oxide phases which can serve as coherent oxygen precipitate nuclei the so-called seed-oxides. The features of these artificial phases were determined by ab initio calculation. Two dimensional and three dimensional seed-oxides were proposed [15, 26, 37, 38]. We describe here the seed-oxides which were suitable to model the experimental nucleation curves described in the next section. Most of them are planar configurations on (100) consisting of either single or double layers of oxygen dimers or of layers of VO_2 as can be seen in Fig. 6.9. They could be formed by attachment of O_i, O_2, V, VO, or VO_2. All these species are diffusive [39] and can thus contribute to formation of such platelets.

In Table 6.3, the parameters of these planar seed-oxides which are needed for simulating the nucleation rates by classical nucleation theory are summarized. The Burgers vector b results from the difference between the lattice constant in [001] direction with and without the defect. The planar precipitates consisting of oxygen atoms practically do not need vacancies for nucleation. While these nuclei continue growing, coherency is lost due to coherency strains resulting from lattice misfit between precipitate and matrix.

The first seed oxide which was ever proposed is a three dimensional $m(VO_2)$ cluster so-called seed-SiO_2 where m is the number of VO_2 complexes involved in its configuration [37]. Figure 6.10 shows a $4(VO_2)$ cluster which comprises the smallest SiO_2 precipitate because just one silicon atom in the center is fully oxidized. Table 6.4 summarizes the parameters of this artificial phase which would be necessary for calculation of nucleation rates by classical nucleation theory. It can be seen that the molecular volume of seed–SiO_2 is lower than the molecular volume of silicon. In contrast to any other real SiO_2 phase considered for oxygen precipitate nucleation, these coherent nuclei are tensile strained with a lattice mismatch of about 10 %. Unlimited coherent growth of such nuclei would not be possible because of the increasing strain energy due to the misfit and because the concentration of VO_2 complexes is orders of magnitude smaller than the concentration of interstitial oxygen O_i. However, interstitial oxygen can be attracted by the tensile strain field of the nucleus and oxidizes the Si–Si bonds in the immediate vicinity of the mVO_2

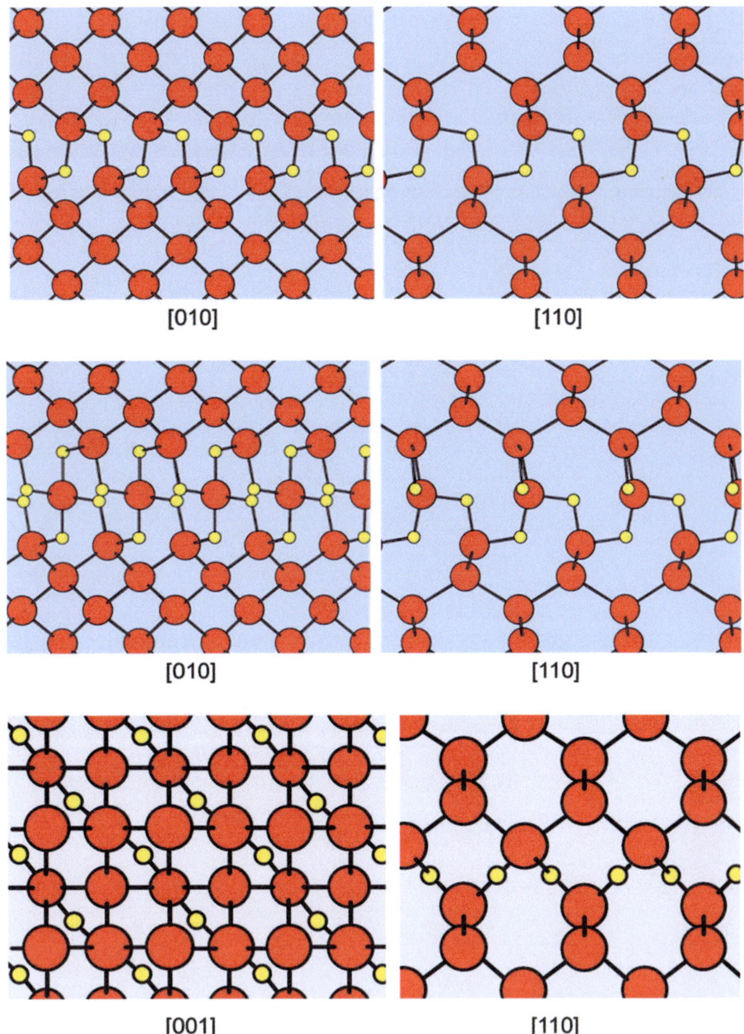

Fig. 6.9 Ball-and-stick models of a planar oxygen precipitate consisting of a single plane of oxygen dimers (*top*) and a planar oxygen precipitate consisting of a double plane of oxygen dimers, (Oi)2-p2. The silicon atoms are marked in *red* and the oxygen atoms are smaller *yellow* atoms (Reproduced with permission from [15]. Copyright 2008, Trans Tech Publications Ltd. Ball-and-stick model of a planar oxygen precipitate consisting of a single plane of VO_2 (*bottom*). Reprinted with permission from [26]. Copyright 2008, The Electrochemical Society)

cluster thus forming an amorphous SiO_2 mantle. In the initial phase, the compressive strain in the amorphous SiO_2 mantle could be annihilated by the tensile strain in the seed–SiO_2 core.

Table 6.3 Parameters of seed-oxide plates determined by ab initio calculation

Type of nucleus	Ω_p (cm^{-2})	b (cm)	x	β	σ (eV/cm^2)
Single layer $(O_i)_2$-p1 (001) platelet	7.23×10^{-16}	4.70×10^{-9}	1	0	4.30×10^{14}
Double layer $(O_i)_2$-p2 (001) platelet	7.23×10^{-16}	1.03×10^{-8}	2	0	4.20×10^{14}
Single layer (VO_2)(001) platelet	7.23×10^{-16}	7.05×10^{-9}	1	0.5	3.80×10^{14}

Reproduced with permission from [15]. Copyright 2008, Trans Tech Publications Ltd. Reprinted with permission from [26]. Copyright 2008, The Electrochemical Society

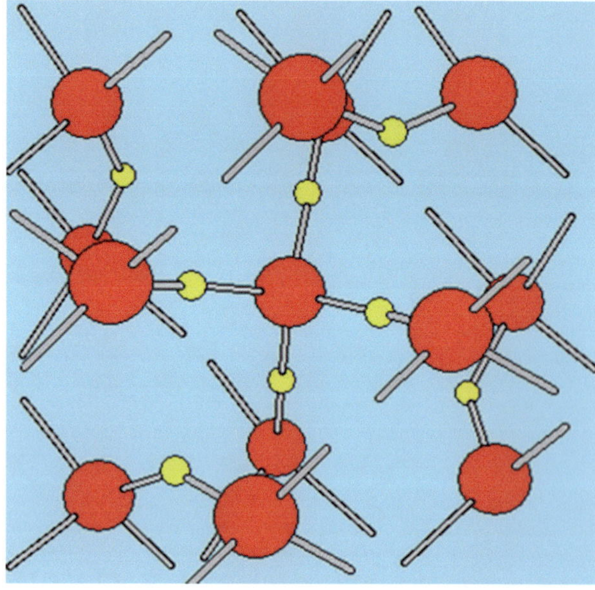

Fig. 6.10 Ball-and-stick model of a $4(VO_2)$ cluster with silicon and oxygen atoms marked in *red* and *yellow color*, respectively (Reprinted with permission from [37]. Copyright 2007, The Electrochemical Society)

Table 6.4 Parameters of a three-dimensional seed oxide determined by ab initio calculation

Type of nucleus	Ω_p (cm^{-3})	K (Pa)	δ	x	β	σ (eV/cm^2)
$m(VO_2)$	2.95×10^{-23}	1.47×10^{11}	−0.097	2	0.5	4.40×10^{14}

Reprinted with permission from [26]. Copyright 2008, The Electrochemical Society

6.4.6 Nucleation Curves

Experimentally, nucleation curves are generated by at least three step annealing. The first step is needed to reduce the influence of the thermal history on the results which can be tremendous [40] because different crystal growth conditions result in different size and density of grown-in oxygen precipitate nuclei. One possibility to reduce their impact is the so-called tabula rasa pre-treatment [41]. Such pretreatments require rapid heating of the samples to temperatures in the range from 1000 °C to 1100 °C and keeping them there for 15 min–1 h [41]. During such pre-treatments, the grown-in oxide precipitate nuclei shrink to nearly equal sizes [42]. Another possibility is “overwriting” the thermal history by RTA pre-treatments at temperatures in the range 1150–1250 °C in quasi inert atmosphere. By this pre-treatment, well defined vacancy-profiles are installed in the wafers [43]. These lead to an enhanced nucleation of oxygen precipitates which is superseding the impact of the grown-in nuclei. The second method is of the advantage that the impact of vacancies on the nucleation curves can be studied as well.

The second step in the generation of nucleation curves are the so called nucleation anneals themselves which are heat treatments for a certain time at different temperatures. After this step, the precipitate nuclei are too small to be detectable by common methods of investigation. Therefore, as the third step additional growth annealing is needed to increase their size above the detection limit of the method of investigation. Growth of oxygen precipitates is diffusion limited but as shown in the first section of this chapter the diffusivity is relatively low. For an efficient growth rate, temperatures of at least 900 °C are needed. A typical precipitate growth treatment used by many authors is annealing at 1000 °C for 16 h.

The temperature step between nucleation and growth anneals should not be too high because otherwise many nuclei generated during the nucleation anneal will not survive ramping towards the quite high growth temperature. To overcome this problem, so-called stabilization treatments are inserted after nucleation. Typical stabilization treatments known from the literature are anneals at 800 °C for 4 h or at 780 °C for 3 h (please see e.g. [44, 45]).

Now, let's have a look at nucleation curves for oxygen precipitates which were generated for different RTA pre-treatments and 8 h of nucleation time. The higher the RTA temperature the higher is the vacancy supersaturation prior to the nucleation anneal. This effect is described in detail later in Sect. 6.10.3. The nucleation curves in Fig. 6.11 clearly demonstrate that the nucleation rate increases with increasing vacancy supersaturation. This is well in agreement with classical nucleation theory described in the beginning of this chapter. Furthermore, there are three peaks in the nucleation curves, a peak at about 450 °C, 650 °C, and 800 °C. Nucleation curves generated after tabula rasa pre-treatment at 1000 °C for 15 min also exhibit the peaks at 450 °C and 650 °C but not the peak at 800 °C [44]. The latter

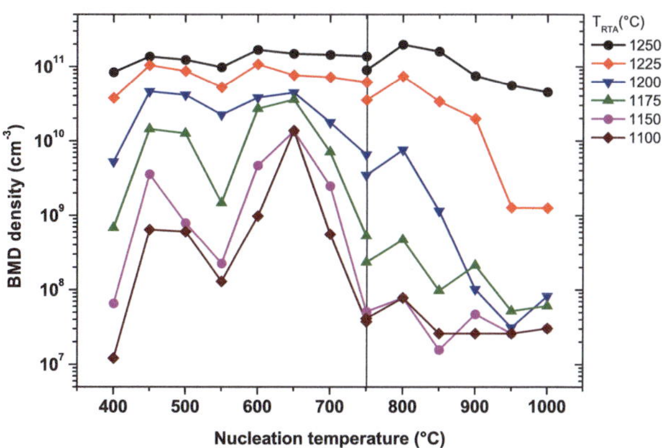

Fig. 6.11 Nucleation curves for B-doped (10 Ωcm) CZ silicon wafers ($C_{Oi} = 6.5 \times 10^{17}$ cm^{-3}) pre-treated by RTA for 30 s in Ar/1000 ppm O_2 atmosphere. Nucleation time was 8 h in N_2 followed by either 780 °C 3 h and 1000 °C 16 h in N_2 for nucleation at 400–750 °C (*left part*) or only 1000 °C 16 h in N_2 for nucleation at 750–1000 °C (*right part*)

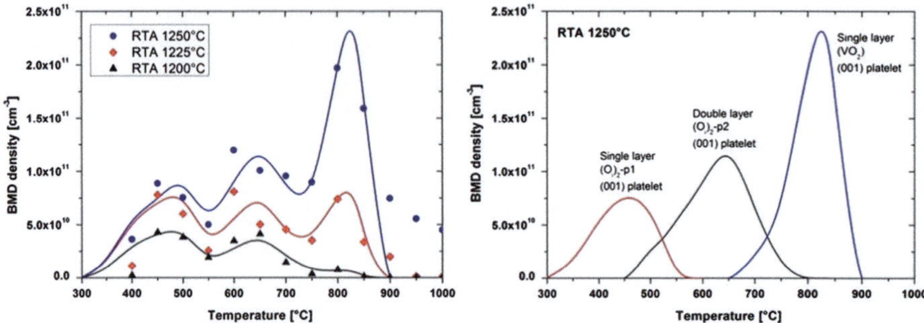

Fig. 6.12 *Left graph*: Comparison of experimentally obtained BMD densities (data points) and simulated BMD densities (*full lines*) for RTA pre-treatments at 1250 °C, 1225°°C, and 1200 °C. *Right graph*: Contributions of the different types of nuclei used for the simulation by classical nucleation theory (Reprinted with permission from [45]. Copyright 2008, The Electrochemical Society)

peak strongly depends on the vacancy supersaturation. Because there is no distinct vacancy supersaturation, the curves with pre-treatment at 1000 °C for 15 min are in good agreement with the curves after RTA pre-treatment at 1100 °C which actually is a tabula rasa pre-treatment as well.

It is still not known what causes the three maxima of the nucleation curves. All modeling attempts assuming one type of nuclei failed to explain this phenomenon [44, 46]. Therefore, it is very likely that different types of nuclei exist. With the help of the "seed oxides" described in Sect. 6.4.5 it is possible to model the nucleation maxima. The single layer $(O_i)_2$-p1 and double layer $(O_i)_2$-p2 platelets are suitable to model the peaks at 450 °C and 650 °C, respectively [15]. The single layer (VO_2) platelet is most suitable to model the 800 °C peak [45] but with the three-dimensional $m(VO_2)$ phase this was also possible [45]. Figure 6.12 shows an example.

6.5 Rate Equation Modeling of Oxygen Precipitation

Advanced models for precipitation of oxygen are usually based on rate equations describing the kinetics of growth or shrinkage of precipitates by attachment or detachment of single oxygen atoms, respectively (see e.g. Sch95, Sen96, [47–50]). In this section, we just shortly describe the basics of this approach because otherwise it would exceed the scope of this chapter.

Growth and shrinkage of a precipitate O_n containing n oxygen atoms can be described by the following quasi chemical reactions in a one dimensional model:

$$O_n + O_i \overset{g(n,x,t)}{\rightarrow} O_{n+1} \tag{6.17}$$

$$O_n \overset{d(n,x,t)}{\rightarrow} O_{n-1} + O_i \tag{6.18}$$

where $g(n,x,t)$ and $d(n,x,t)$ are the growth and dissolution rates, respectively, at time t and position x. Rate means the number of oxygen atoms attached or detached per precipitate in a certain period of time.

In order to describe the density of oxygen precipitates, we need to define a distribution function $f(n,x,t)$ which describes the concentration of precipitates with n atoms at time t in an infinitesimal volume of size $(\Delta x)^3$ which is large compared to the precipitate size. At least, the concentration $f(n,x,t)$ of O_n is changed by four processes and we can write

$$\frac{\partial f(n,x,t)}{\partial t} = g(n-1,x,t)\cdot f(n-1,x,t) - d(n,x,t)\cdot f(n,x,t) - g(n,x,t)\cdot f(n,x,t) + d(n+1,x,t)\cdot f(n+1,x,t)\ . \tag{6.19}$$

If we summarize the first two and the last two contributions we get the two precipitate formation fluxes I

$$I(n,x,t) = g(n-1,x,t)\cdot f(n-1,x,t) - d(n,x,t)\cdot f(n,x,t) \quad \text{and} \tag{6.20}$$

$$I(n+1,x,t) = g(n,x,t)\cdot f(n,x,t) - d(n+1,x,t)\cdot f(n+1,x,t)\ . \tag{6.21}$$

Now, we have the kinetic system of equations to be solved with

$$\frac{\partial f(n,x,t)}{\partial t} = I(n,x,t) - I(n+1,x,t) \text{ for } (n = 2,\ 3,\ldots) \tag{6.22}$$

and taking the balance of the oxygen atoms in the volume $(\Delta x)^3$ into account

$$\frac{\partial f(1,x,t)}{\partial t} + \frac{\partial J(x,t)}{\partial x} = -I(2,x,t) - \sum_{n=2}^{n_{max}} I(n,x,t) \text{ for } (n = 1)\,. \tag{6.23}$$

Now, the problem is that such models would require solving of thousands of coupled differential equations which can be very time-consuming. Therefore for larger precipitates starting in the range of about 20–50 atoms commonly Fokker-Planck equations are used [50] which are described as follows:

$$\frac{\partial f(n,x,t)}{\partial t} = -\frac{\partial I(n,x,t)}{\partial n} \text{ with} \tag{6.24}$$

$$I(n,x,t) = -B\frac{\partial f(n,x,t)}{\partial n} + A\cdot f(n,x,t)\,, \tag{6.25}$$

$$B(n,x,t) = \frac{g(n,x,t) + d(n,x,t)}{2}, \text{ and} \tag{6.26}$$

$$A(n,x,t) = g(n,x,t) - d(n,x,t) - \frac{\partial}{\partial n}B(n,x,t)\,. \tag{6.27}$$

Last but not least, we need the growth and dissolution rates g and d, respectively. These are similar to the nucleation rate in classical nucleation theory often neglecting Z and $(1+\delta)/(1+\varepsilon)$ which are both close to one. The following growth and dissolution rates hold for spherical clusters [48]:

$$g(n,x,t) = \exp\left(-\frac{\Delta G_{nn+1}}{k \cdot T}\right) \cdot C_{Oi}^{if} \cdot 4\pi \cdot \left(\frac{n \cdot 3 \cdot \Omega_p}{4\pi \cdot x}\right)^{\frac{2}{3}} \cdot D_{Oi} \cdot d^{-1} \tag{6.28}$$

$$d(n,x,t) = \exp\left(-\frac{\Delta G_{nn+1}}{k \cdot T}\right) \cdot C_{Oi}^{if,eq} \cdot 4\pi \cdot \left(\frac{n \cdot 3 \cdot \Omega_p}{4\pi \cdot x}\right)^{\frac{2}{3}} \cdot D_{Oi} \cdot d^{-1}. \tag{6.29}$$

Here C_{Oi}^{if} and $C_{Oi}^{if,eq}$ are the concentration and the equilibrium concentration, respectively, of interstitial oxygen at the precipitate/matrix interface and ΔG can be calculated as in classical nucleation theory.

In the literature, many more or less complicated rate equation models can be found which additionally involve secondary defect formation, vacancy and interstitial generation, or reaction limited processes (see e.g. [46–50]). At the end, the models have to be adjusted by experimental data and the model which can describe the oxygen precipitation under many different conditions accurately will turn out as the best one. At present, there is still work to be done to achieve this.

At the end of this section, let us have a look at an example of rate equation modeling. In Fig. 6.13, we can see the evolution of the size distribution during nucleation, stabilization and growth of oxygen precipitate nuclei. At the end of nucleation, we have a certain amount of stable nuclei ready to grow (blue curve). After stabilization, their size is already increased (green curve) and they grow further during the growth anneal (red curve). One of the advantages of rate equation modeling is that it shows the size distribution of the oxygen precipitates.

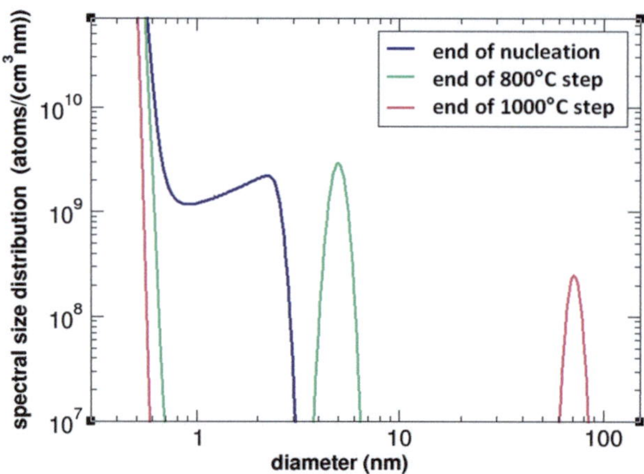

Fig. 6.13 Evolution of the size distribution of oxygen precipitates during a thermal treatment consisting of nucleation at 550 °C for 32 h, stabilization at 800 °C for 4 h, and growth at 1000 °C for 16 h as modeled by a rate equation model using $C_{Oi} = 6.1 \times 10^{17}$ cm^{-3}

6.6 Methods for Characterization of Oxygen Precipitates

6.6.1 Transmission Electron Microscopy for Characterization of Oxygen Precipitate Morphology

Because of its high resolution, transmission electron microscopy (TEM) is the best method for the investigation of the morphology of oxygen precipitates. Thin foils of <300 nm thickness prepared by ion milling or etching are suitable for common TEM investigation.

For a long time, only two basic morphologies of oxygen precipitates were reported, octahedral and plate-like [29, 30]. Whereby, the octahedral precipitates were formed at temperatures >900 °C and the platelets were mainly found after annealing at temperatures < 900 °C [29, 30]. The octahedral morphology usually consists of {111} planes. Sometimes, octahedrons were observed, which were truncated by {100} planes or even polyhedral precipitates [29, 30]. Commonly, platelets are located on {100} planes. They are bound by {110} planes. An important parameter of the platelets is their aspect ratio which is the ratio between height and side length. Figure 6.14 shows examples of plate-like and octahedral precipitates. It is important to hint here on the danger of erroneous identification of the morphology of the oxygen precipitates in the [110] view which becomes obvious in Fig. 6.14. Plate-like precipitates can be located perpendicular with one side plane in parallel (Fig. 6.14 middle) or tilted by 45° (Fig. 6.14 right) with respect to the (110) plane. In the latter case, the shape of a platelet is identical to the shape of an octahedron (Fig. 6.14 right and left). Energy dispersive x-ray spectroscopy (EDX) can be used to distinguish the morphologies because the oxygen signal of an octahedron would be much stronger than the oxygen signal of a thin platelet. A further hint is that

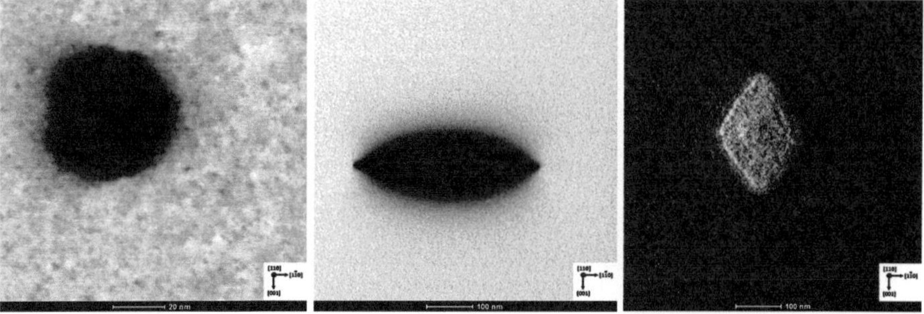

Fig. 6.14 Octahedral (*left*) and plate-like (*middle*) oxygen precipitates observed by bright field (BF) scanning transmission electron microscopy (STEM). Plate-like oxygen precipitate (*right*) observed by STEM in the High angle annual dark field (HAADF) mode. Observation direction is always [110]. *Left image* reprinted with permission from [51] (Copyright 2014, The Electrochemical Society)

octahedrons usually exhibit much sharper borders and characteristic strain fields in bright field observation.

Early nineties, a new type of oxygen precipitates was detected after annealing at 1000 °C [52]. In regions where the oxygen precipitated quickly, dendritic precipitates with main branches in ⟨110⟩ direction were found. In regions with lower precipitation rate, starfish-like dendritic precipitates were found extending in ⟨110⟩ direction with concentric circular growth stripes. Later, such dendritic oxygen precipitates were detected in RTA pre-treated silicon wafers already after annealing at 800 °C [51]. These dendritic precipitates were often three-dimensional consisting of two intersecting dendritic platelets. Figure 6.15 shows examples of dendritic oxygen precipitates. On the left side, the bright field image of a three-dimensional dendritic precipitate (top) and the corresponding EDX map of oxygen (bottom) are shown. The oxygen precipitate on the right side is of the type called starfish-like.

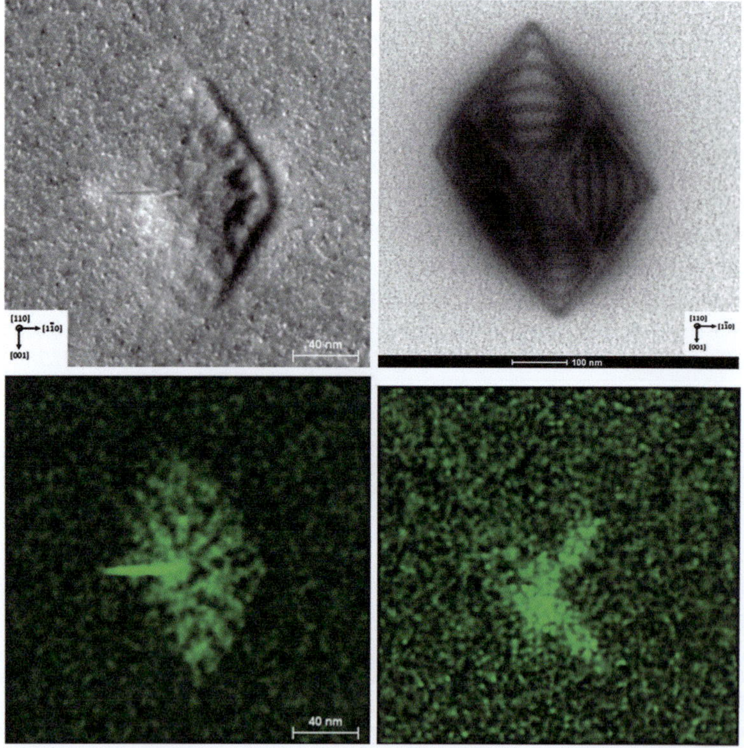

Fig. 6.15 HAADF and bright field STEM images of a three-dimensional dendritic oxygen precipitate (*top left*) and a starfish-like dendritic oxygen precipitate (*top right*), respectively. The EDX maps of oxygen in the lower row always belong to the images above. Observation direction is always [110]

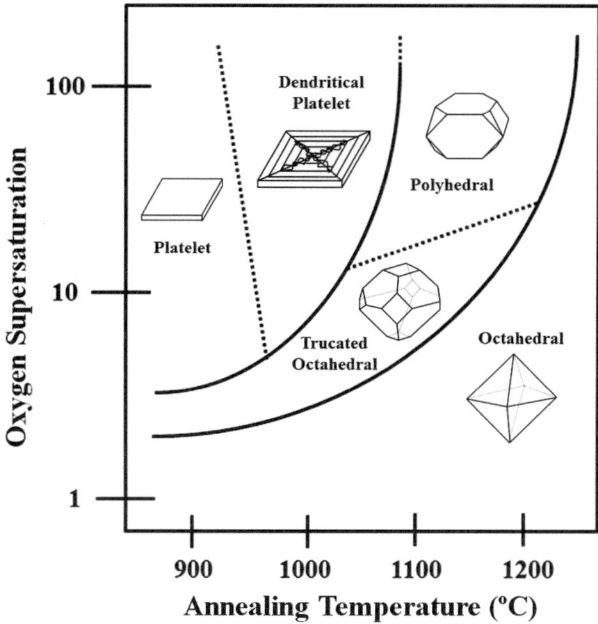

Fig. 6.16 Morphology of oxygen precipitates as function of annealing temperature and supersaturation of interstitial oxygen (Reprinted with permission from [53]. Copyright 1997, The Electrochemical Society)

The appearance of the different morphologies of oxygen precipitates depends on the supersaturation of interstitial oxygen and on the annealing temperature as demonstrated in Fig. 6.16 [53]. However, the graph in Fig. 6.16 neglects the impact of the supersaturation of intrinsic point defects on the morphology of the oxygen precipitates. It was shown theoretically that also the supersaturation of intrinsic point defects impacts the morphology of oxygen precipitates [54].

In the course of precipitate growth, the supersaturation of interstitial oxygen decreases. In this case, it can happen that the preferred morphology changes from plate-like to octahedral. In practice, this means that two-dimensional growth turns into three-dimensional growth. As the result, the platelets can start to develop so-called fins as one is shown in Fig. 6.17.

6.6.2 Preferential Etching of Oxygen Precipitates

The most common and easily to apply method for the determination of the density of oxygen precipitates is preferential etching of the surface, of cleavage planes, or of polished bevels of silicon wafers. A preferential etchant usually consists of three main ingredients, an oxidizer which oxidizes the silicon surface, a complexer which

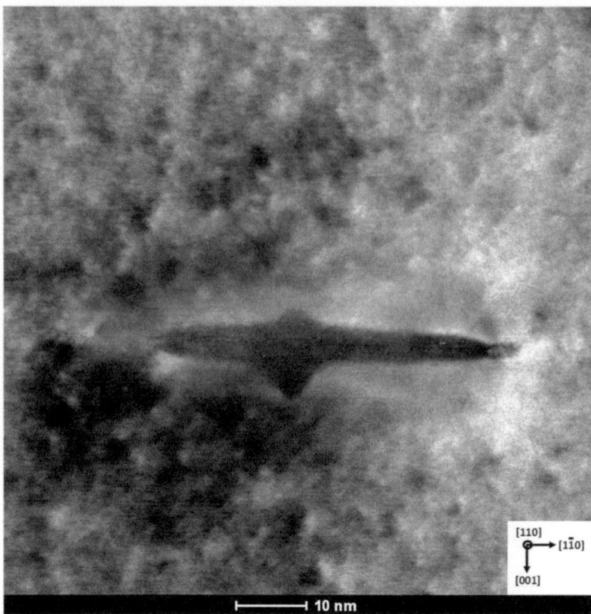

Fig. 6.17 HAADF STEM image of a plate-like oxygen precipitate with a fin demonstrating the change from two-dimensional to three-dimensional growth mode

Table 6.5 Examples for oxidizers, complexers, and diluents used in preferential etchants

Oxidizer	Complexer	Diluent
HNO_3	HF	H_2O
H_2O_2	HCl	CH_3COOH
CrO_3	H_2SO_4	CH_3O_4
$K_2Cr_2O_7$		
Br_2		

dissolves and complexes the oxidized silicon, and a diluent which influences the etch rate. Table 6.5 summarizes some examples of these ingredients. The etching process in a preferential etch is reaction controlled in contrast to a polishing etch which is diffusion controlled [55, 56]. However, it is also possible that an etching solution turns from polishing to preferential by change of temperature or composition during etching [56].

An important parameter of an etchant is its selectivity which is the ratio between the removal at the position of the defect and the removal at the perfect crystal [55]. Etchants with higher selectivity can reveal smaller oxygen precipitates. In other words, their detection limit is lower. However, the disadvantage of the highly selective etchants is their very low etch rate of just a few nm per minute which results in etching times of several hours.

Table 6.6 Recipes of classical preferential etchants for delineation of defects in silicon [57]

Etchant	Composition	Etch rate (23 °C) (μm/min)	Reference
Secco	(a) 44 g $K_2Cr_2O_7$ in 1000 ml H_2O (b) HF (48 %) (a):(b) = 1:2	1.5	[58]
Wright	(a) 45 g CrO_3 in 90 ml H_2O (b) 6 g $Cu(NO_3)_2$°$3H_2O$ in 180 ml H_2O (c) 90 ml HNO_3 (69 %) 180 ml CH_3COOH (98 %) 180 ml HF (49 %) (a) and (b) then add (c)	1	[59]
Yang	(a) 200 g CrO_3 in 1000 ml H_2O (b) HF (49 %) (a):(b) = 1:1	1.5	[60]
Sirtl	(a) 50 g CrO_3 in 100 ml H_2O (b) HF (48 %) (a):(b) = 1:1	1.3	[61]
Schimmel	(a) 75 g CrO_3 in 1000 ml H_2O (b) HF (48 %) (a):(b) = 1:2	1	[62]

According to Ref. [55], two main factors are contributing to dislocation etch pit formation:

1. the strain field produced by a dislocation or an extended lattice defect and
2. impurities, in particular metals, segregated to the defect.

In most cases, the etch rate at the defect is higher than the etch rate at the perfect silicon lattice and more or less characteristic etch pits are formed. This is due to the reduced activation energy at the defect because of the increased surface potential of the distorted lattice.

In Table 6.6, the recipes of some of the most widely used preferential etchants can be found. Etchants are often named according to person or the company which published the recipe. The Sirtl etchant developed first is suitable especially for (111) surfaces. Later modified recipes were developed to make CrO_3 containing etchants applicable to (100) surfaces.

All of the classical etchants contain Cr(VI) compounds which are the best oxidizers for silicon but Cr(VI) is carcinogenic. Because the use of Cr(VI) compounds is restricted in more and more countries Cr-free etchants were developed. In Table 6.7, some recipes of Cr-free etchants are given. The variety of published recipes containing HF, HNO_3, and CH_3COOH is huge but the quality of the defect delineation is not always sufficient.

Now, let's have a look at the typical results of preferential etching. The best tool for observing etched defects is a Nomarski Microscope. It is not possible to deduce information about type and size of the defect directly from the etch pit except stacking faults and systems of punched out dislocations. Usually, further information or measurements are necessary to identify the corresponding defect. One of the common aims of preferential etching is to determine the BMD density

Table 6.7 Chromium free preferential etchants for delineation of defects in silicon

Etchant	Composition	Etch rate (23 °C)	Reference
MEMC	36 Vol% HF (49 %) 25 Vol% HNO_3 (70 %) 18 Vol% CH_3COOH (98 %) 21 Vol% H_2O Add 1 g of $Cu(NO_3)_2 \cdot 3H_2O$ per 100 ml of mixed acid	1.5 μm/min	[63]
JEITA	3.59 Vol% HF (49 %) 53.57 Vol% HNO_3 (69 %) 21.42 Vol% CH_3COOH (100 %) 21.42 Vol% H_2O 0.0083 g potassium iodide (KI) added to 100 ml H_2O	1 μm/min	[64]
Toshiba	25 ml HF (49 %) 100 ml HNO_3 (70 %) 300 ml CH_3COOH (99 %) 0.1 g $CuNO_3$ 5 ml $AgNO_3$ (1 % aqueous solution)	1–2 μm/min	[65]
OPE:C	50 ml H_2O_2 50 ml HF 100 ml CH_3COOH	1.34 nm/min	[66]

which can be easily obtained by counting the etch pits per area and dividing it by the etched depth which can be measured after masking part of the sample. BMD densities are usually given in cm^{-3}.

In Fig. 6.18, an example of a Secco etched cleavage plane is shown. Among the typical round etch pits of BMDs which are caused by oxygen precipitates with and without secondary defects also the line shaped etch pits of bulk staking faults can be observed. These lines appear only in well-defined directions because stacking faults only exist on $\{111\}$ planes. The etch pit delineates a cut through such a circular stacking fault plane.

Figure 6.19 was derived from an oxidized silicon wafer which was preferentially etched with the Yang etchant after removal of the oxide layer with HF. Here, we can observe very large oxidation induced stacking faults and oxygen precipitates.

Systems of punched out dislocations appear as lined-up pits in $\langle 110 \rangle$ directions as can be seen in Fig. 6.20. Usually, not all of the possible $\langle 110 \rangle$ directions exhibit pits. In the center of these punching systems, the etch pit of the originating oxygen precipitate can be found. The lined-up pits are originated by the etched punched out dislocation loops.

In this section, we focused the attention to etchants most suitable for etching of oxygen precipitates and their secondary defects. More comprehensive overviews about the high variety of preferential etchants for various defects can be found in Refs. [55, 57, 67–69].

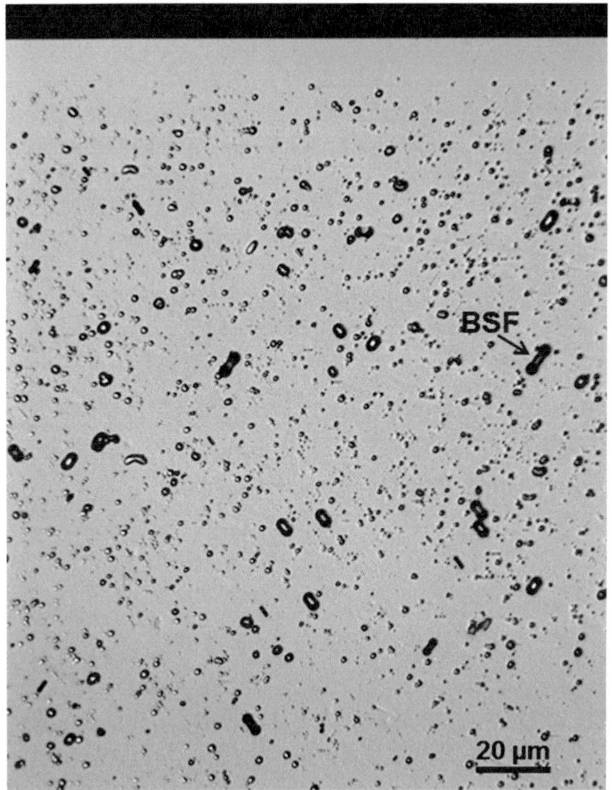

Fig. 6.18 Secco etched (110) cleavage plane showing a defect denuded zone below the surface (*top*) and a bulk defect zone with *black etch* pits generated by bulk microdefects. One of the characteristic *line-shaped etch* pits caused by bulk stacking faults (BSF) is marked by an *arrow*

6.6.3 Infrared Laser Scattering Tomography

The infrared laser scattering tomography (IR-LST) is the second most widely applied technique for quick and easy investigation of the density of oxygen precipitates. Except the generation of a cleavage plane, it is non-destructive. So, it would be possible to investigate a cleaved sample before and after a thermal treatment at the same position [70].

IR-LST is based on the principle of ultramicroscopy [71–74]. Particles which are much smaller than the resolution of an optical microscope can be observed due to light scattering. Figure 6.21 demonstrates the two basic optical arrangements for investigation of bulk defects in silicon wafers by IR-LST. An incident infrared laser beam with the unit vector s_0 penetrates the silicon wafer perpendicular to the observation direction of the scattered light with the unit vector s. The incident beam can either penetrate a (110) cleavage plane or the (100) surface of the silicon wafer. In the first case, the plan view, the defects are observed on a plane in a certain

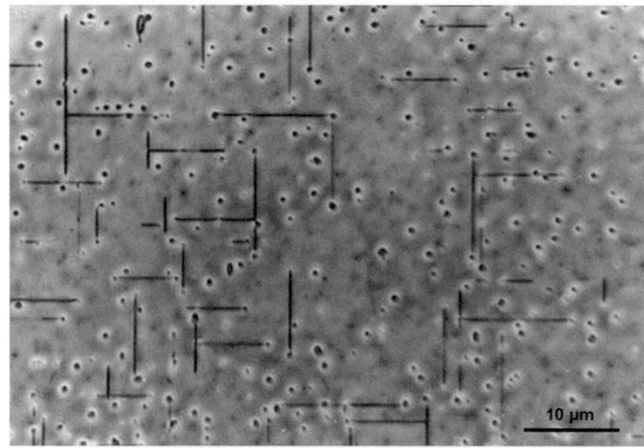

Fig. 6.19 Yang etched surface of a (100) silicon wafer showing etch pits of oxygen precipitates (*black dots*) and oxidation induced stacking faults (*black lines*)

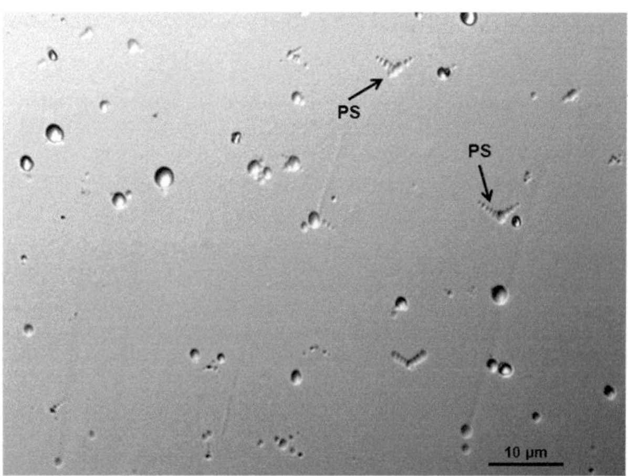

Fig. 6.20 BMDs (pits) and systems of punched out dislocations (PS) appearing as lined-up pits observed after Secco etching of a (100) silicon surface. Two of the systems of punched out dislocations are marked by an *arrow*

depth perpendicular to the surface. In the second case, the cross-sectional view, the defects on the cross-section perpendicular to the wafer surface can be investigated. The incident laser beam has to be scanned in the direction perpendicular to s and s_0 for a complete image. The scanned volume corresponds to the diameter of the laser beam multiplied with the area of the scanned plane.

Depending on the tool, oxygen precipitates of radii larger than 10–20 nm are detectable by IR-LST. The oxygen precipitates appear as light points in the LST image as can be seen in the examples in Fig. 6.22. Sometimes it is also possible to identify secondary defects like stacking faults or punched out dislocation loops by IR-LST [75]. This is only possible if the defect structures are larger than the

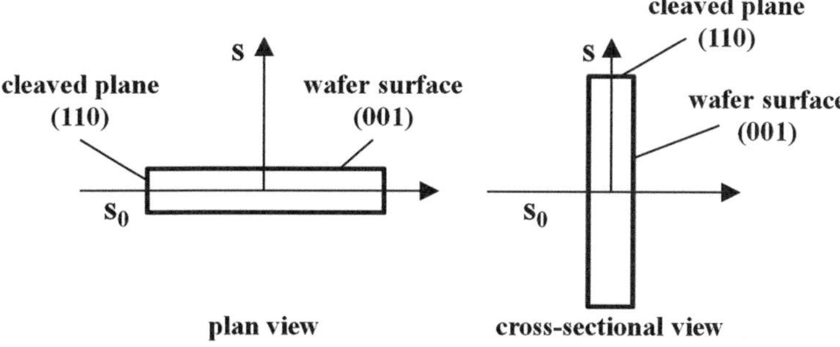

Fig. 6.21 Basic optical arrangements for investigation of bulk defects in silicon wafers by infrared laser scattering tomography

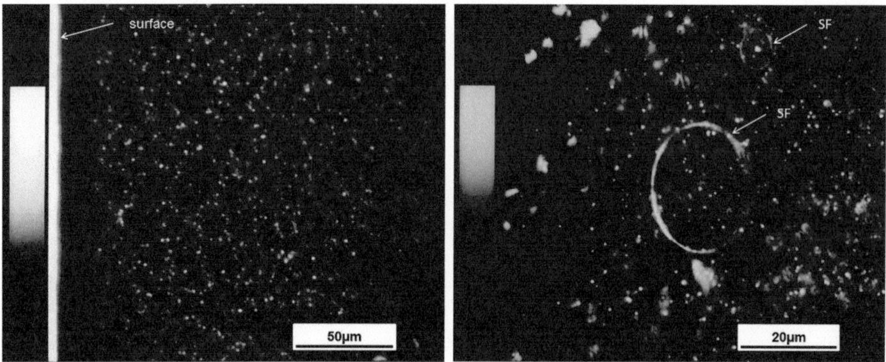

Fig. 6.22 *Left*: IR-LST image of oxygen precipitates (*white dots*) in a silicon wafer with a defect denuded zone below the surface. *Right*: IR-LST image of bulk microdefects in a silicon wafer showing oxygen precipitates as *white dots* and two stacking faults (SF)

resolution of the optical microscope and if dislocations are decorated by impurities. Examples of stacking faults observed by IR-LST can be found in Fig. 6.22 right.

After calibration, it is also possible to determine the size of the oxygen precipitates via the intensity of the scattered light. The intensity I of the light scattered perpendicular to the scattering plane of a particle is proportional to the square of the volume V that means to the sixth power of its radius r [72]. It can be approximated by the following formula

$$I \cong \left(\frac{\nu^2}{4\pi} \cdot \Delta\varepsilon \cdot V \right)^2 I_0 \tag{6.30}$$

for $r \leq 0.01\lambda$ that means if the radius of the particle is sufficiently small compared to the wavelength of the laser beam λ which corresponds to Rayleigh scattering. Here ν and $\Delta\varepsilon$ are the wave number of the laser beam and the difference of the dielectric constant between particle and matrix, respectively.

6.6.4 Infrared Absorption Spectra of Oxygen Precipitates

The infrared absorption bands of oxygen precipitates in silicon can be found in the spectral range from 1000 to 1300 cm^{-1}. Their exact location depends on their geometry. In most cases, two characteristic broad bands at 1095 and at 1225 cm^{-1} can be observed which can be attributed to spherical and plate-like precipitates, respectively [76–78]. Recently, it was found that dendritic oxygen precipitates give rise to a broad absorption band at 1040 cm^{-1} [51, 79].

The first model which theoretically describes the absorption bands of oxygen precipitates in silicon was published by Hu [80]. The model is based on a spheroidal precipitate geometry with the main axes a_1, a_2, *and* a_3 whereby $a_2 = a_3$. Spheroids are a common approximation for the shape of oxygen precipitates in modeling. By continuously increasing the ratio a_1/a_2, the shape can be varied from a disc (oblate spheroid with $a_1 < a_2$) via a sphere ($a_1 = a_2$) to a needle (prolate spheroid with $a_1 > a_2$). For precipitates with axes of less than 360 nm lengths, the electric field E_p inside the precipitate due to an incident electromagnetic wave can be regarded as being uniform. It is related to the electric field of the surrounding matrix E_m as follows:

$$E_p = g \cdot E_m. \tag{6.31}$$

The tensor g becomes diagonal if the axes of the spheroid are chosen to coincide with the axes of the coordinate system. If the complex dielectric functions of the precipitate ε_p and the matrix ε_m are isotropic, the elements of g can be obtained via

$$g_i = \frac{\varepsilon_m}{\left[L_i \varepsilon_p + (1 - L_i)\,\varepsilon_m\right]} \tag{6.32}$$

with $i = 1$–3. The depolarization factors L_i along the axis a_i of the spheroid are related to each other as follows:

$$L_1 + L_2 + L_3 = 1 \tag{6.33}$$

and it can be approximated with good accuracy that

$$L_1 : L_2 : L_3 = \frac{1}{a_1} : \frac{1}{a_2} : \frac{1}{a_3} \tag{6.34}$$

It is useful to assume that the principal axes of the spheroids coincide with the $\langle 100 \rangle$ axes of the silicon axes. For oxygen precipitates oriented randomly in the three possible orientations, g can be averaged and becomes the scalar

$$g = \frac{1}{3}\left(g_1 + g_2 + g_3\right) \tag{6.35}$$

When effective medium theory [81, 82] is applied, the average electric field of the composite medium E_{avg} and the average polarization P_{avg} are given by

$$E_{avg} = (1-f)\,E_m + fE_p \text{ and} \tag{6.36}$$

$$P_{avg} = (1-f)\,P_m + fP_p, \tag{6.37}$$

respectively, with f being the volume fraction occupied by the oxygen precipitates. Combining Eqs. 6.31, 6.36, and 6.37 using the relationship $P = E\,(\varepsilon - 1)/4\pi$ for the composite, the matrix, and the precipitate the following relationship is obtained which allows calculating the average dielectric constant of the composite:

$$\varepsilon_{avg} = 1 + \frac{(1-f)\,(\varepsilon_m - 1) + f\,(\varepsilon_p - 1)\,g}{(1-f) + fg}. \tag{6.38}$$

Now, the absorption coefficient of the composite α as a function of the wavenumber ν can be calculated as follows:

$$\alpha\,(\nu) = \frac{\mathrm{Im}\varepsilon_{avg}\,(\nu)}{\mathrm{Re}n_{avg}\,(\nu)} 2\pi\nu. \tag{6.39}$$

Two input functions are needed for the calculation of the infrared absorption spectra of the oxygen precipitates, the complex dielectric function of the silicon matrix ε_m and the complex dielectric function of the precipitates ε_p. The former can be well approximated using the index of refraction $n = 3.41626 + 1.443 \times 10^{-9}\,\nu^2$ from Ref. [83] and the extinction coefficient $k = 2.5 \times 10^5\,\nu^{-3.13}$ according to Ref. [80]. The complex dielectric function of the precipitated phase is not known but several approximations exist like the dielectric functions of SiO_2 glass [84] or amorphous SiO_2 [85, 86].

In Fig. 6.23, infrared absorption spectra of SiO_2 precipitates calculated for different aspect ratios ranging from needle to disc shapes are shown [80]. Actually, the different location of the absorption bands would make it possible to identify the geometry of the oxygen precipitates by FTIR. However, it is not yet clarified what the exact stoichiometry of the oxygen precipitates is. The stoichiometry influences the dielectric function [87] and thus the calculated spectra [76]. In Fig. 6.24, the impact of the stoichiometry on the absorption bands of spheres and ellipsoids with different aspect ratios is demonstrated. It becomes unambiguously clear that without the knowledge of x, the morphology, or the aspect ratio β of the precipitates it will be very difficult to interpret the absorption bands observed.

Several authors determined the stoichiometry of the oxygen precipitates by fitting their infrared spectra. The majority of them found that x is close or equal 2 [51, 79, 88–91] only in one case substoichiometric SiO_x with x close to 1 was found [92]. However, an x close to 1 was determined also by analyzing TEM data of many authors assuming diffusion limited growth [19] and by TEM analysis combined

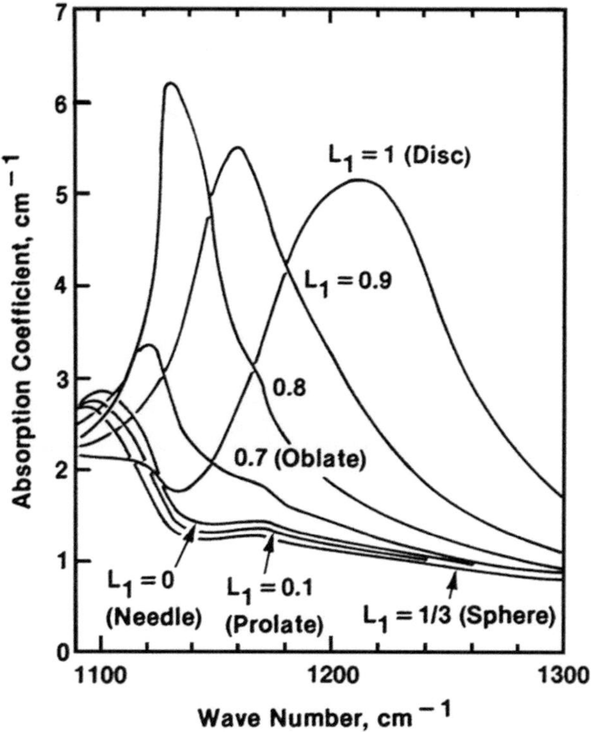

Fig. 6.23 Calculated infrared absorption spectra of silicon substrates imbedded with $f = 5 \times 10^{-5}$ of small SiO_2 particles of various shapes. L_l is the depolarization factor along the axis of revolution of spheroid. $L_l < 1/3$ for prolate spheroids; $L_l > 1/3$ for oblate spheroids (Reproduced with permission from [80]. Copyright 1980, AIP Publishing LLC)

with EDX [93]. The latest work combined EDX, electron energy loss spectroscopy (EELS), and FTIR analysis and confirmed $x = 2$ [94].

6.7 Characterization of Oxygen Precipitate Nuclei

Precipitation of oxygen happens already during cooling of the ingot in the crystal pulling process because the interstitial oxygen built in at melting temperature supersaturates. Tiny grown-in oxide precipitate nuclei are assumed to exist in as-grown CZ silicon wafers. They consist of just a few atoms and cannot be detected by commonly available techniques. A dedicated annealing technique combined with IR-LST was developed in order to gain information about the grown-in oxide precipitate nuclei and their evolution during processing steps of device production [42, 95, 96]. The method is based on the effect that during slowly ramped anneals with 1 K/min, only the grown-in nuclei which are larger than the critical radius at

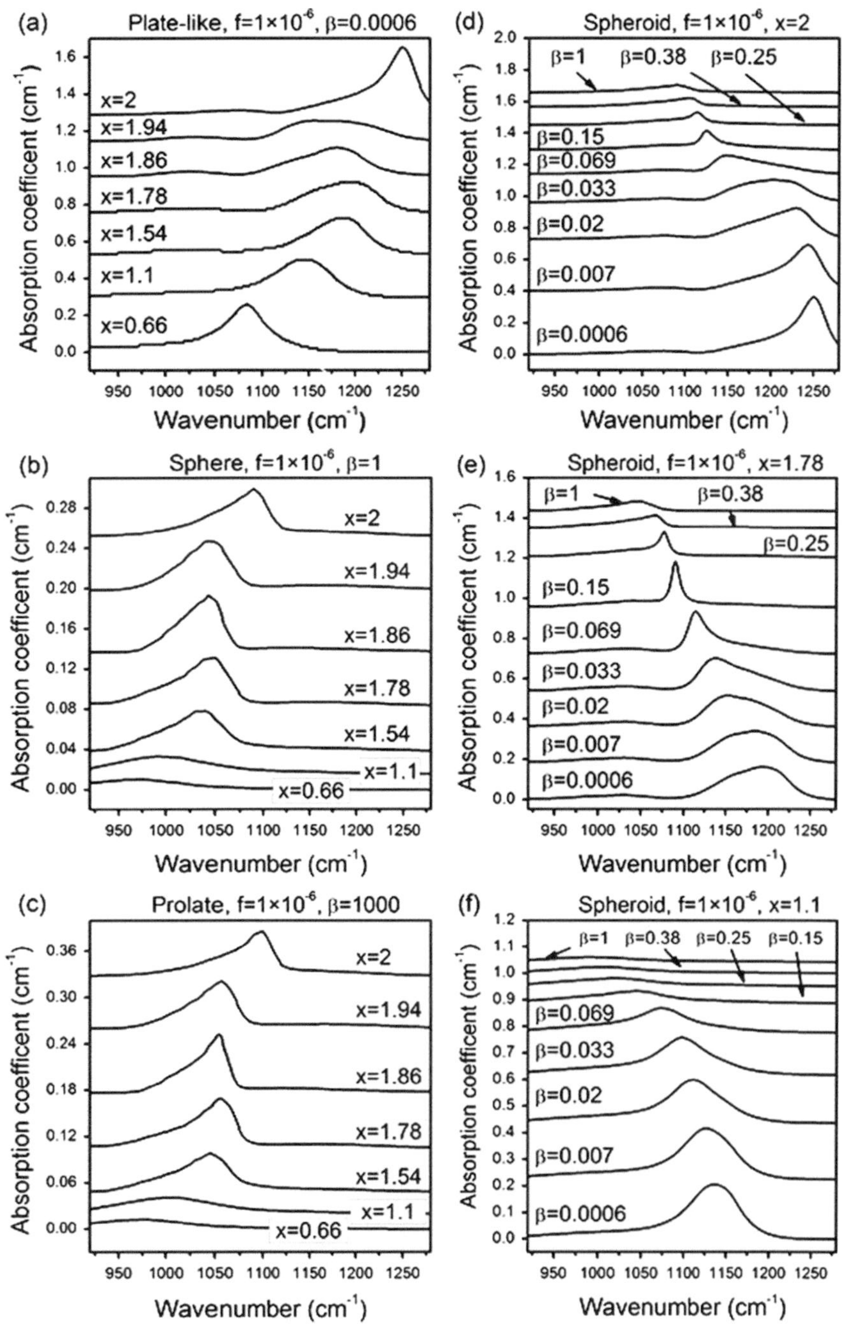

Fig. 6.24 Simulated absorption spectra of oxygen precipitates: for different x thin oblate spheroids (**a**), spheres (**b**), long prolate spheroids (**c**) and for oblate spheroids with different aspect ratios β and $x = 2$ (**d**), $x = 1.78$ (**e**), and $x = 1.1$ (**f**) (Reprinted with permission from [51]. Copyright 2014, The Electrochemical Society)

the corresponding starting temperature can grow to a detectable size. Nucleation of additional nuclei is strongly suppressed because these nuclei would become immediately unstable because of the steadily increasing critical radius during ramping.

In detail the method is carried out as described in the following. BMD densities are measured after ramped anneals with 1 K/min starting from 500 °C, 600 °C, 700 °C, 800 °C, or 900 °C combined with a final growth anneal at 1000 °C for 2–16 h. One sample receives only the growth anneal. The duration of annealing at 1000 °C depends on the detection limit of the method which is applied for the measurement of the BMD density. Originally, IR-LST was used to measure the BMD densities but preferential etching is suitable as well. The ramp start temperature has to be reached with the highest ramp rate possible. All anneals are carried out in nitrogen atmosphere. The method is suitable for moderately boron doped silicon. It does not work for highly boron doped silicon because the very high nucleation rate prevents the growth of oxygen precipitates for ramp starts at 500–700 °C [97].

Before we look at experimental results, we will use modeling to understand what is going on during the slowly ramped anneals. For this purpose, the BMD densities after the ramped anneals were simulated via rate equations and Fokker-Planck equations describing the clustering process of oxygen atoms during the thermal treatments as shown in Sect. 6.5. The parameters for the simulations were adjusted to the nucleation experiments published in Ref. [44] in a similar way as described in Ref. [21]. Modeling was carried out for different sizes of grown-in oxygen precipitate nuclei and without pre-existing nuclei. The values for the simulated BMD density for each temperature of ramp start were obtained from the accumulated density of oxygen precipitates being larger than the size detection limit of 20 nm. The results in Fig. 6.25 left demonstrate that without pre-existing oxygen

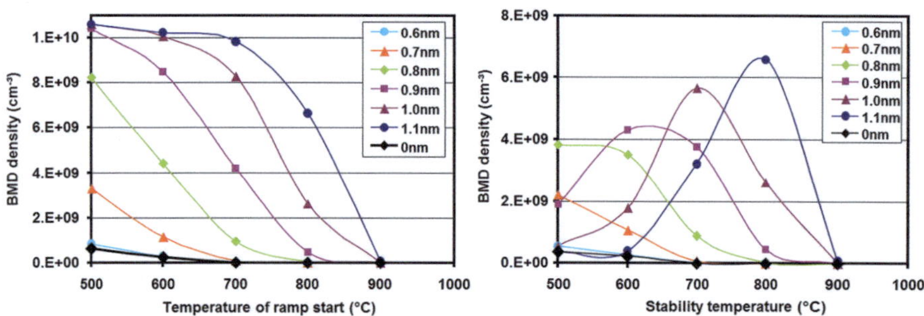

Fig. 6.25 *Left*: Simulated BMD density as a function of the temperature for different sizes of grown-in nuclei and a nuclei density of 1×10^{10} cm^{-3}. A value of 0 nm means that there are no grown-in nuclei. The initial concentration of interstitial oxygen is 7×10^{17} cm^{-3}. *Right*: Defect density spectrum obtained from the BMD values in the *left graph*. The data points in both graphs are connected by *splines* (Reprinted with permission from [38]. Copyright 2007, The Electrochemical Society)

precipitate nuclei the BMD density is negligibly small. This means the nucleation of fresh nuclei can be neglected and the resulting BMDs are mainly due to grown-in oxygen precipitate nuclei. It can be also seen in Fig. 6.25 left that the larger the grown-in nuclei the higher the temperature of ramp start which they survive for growing further above the detection limit.

Based on the BMD density values so-called grown-in defect density spectra can be obtained as one is shown in Fig. 6.25 right. For this purpose, the difference between the BMD density at the stability temperature T_{ST} and the BMD density at $T_{ST} + 100$ K is plotted against T_{ST}. So for each stability temperature the density of nuclei is shown which is stable at this temperature but becomes unstable at a temperature being 100 K higher. The resulting graph shows a size distribution of grown-in nuclei. The stability temperature is more useful here than a calculated radius because it allows directly estimating how many nuclei are able to grow during subsequent thermal processing. The resulting graph in Fig. 6.25 right demonstrates that larger grown-in nuclei possess higher stability temperatures.

Now, let's have a look at experimental results. Figure 6.26 shows the results of the test with 1 K/min ramping for wafers obtained from vacancy-rich ingots grown with different cooling rates. They look very similar to the simulated graphs in Fig. 6.25. It can be seen that with increasing cooling rate of the ingot the size distribution of the grown-in oxygen precipitate nuclei shifts to higher stability temperature that means to larger sizes [42]. This has consequences for defect generation during thermal processing as it will be described in detail in Sect. 6.10.1.

It is also possible to fit modeled data to the experimental data in order to extract the size of the grown-in nuclei. This was done for the data of wafer material A150. These wafers were also subjected to a tabula rasa treatment, already described in Sect. 6.4.6. In this case, the test with 1 K/min ramping was applied after the tabula rasa pre-treatment at 1100 °C for 15 min. It can be seen in Fig. 6.27 that the tabula rasa pre-treatment has caused shrinkage of the grown in nuclei. Best fit was obtained

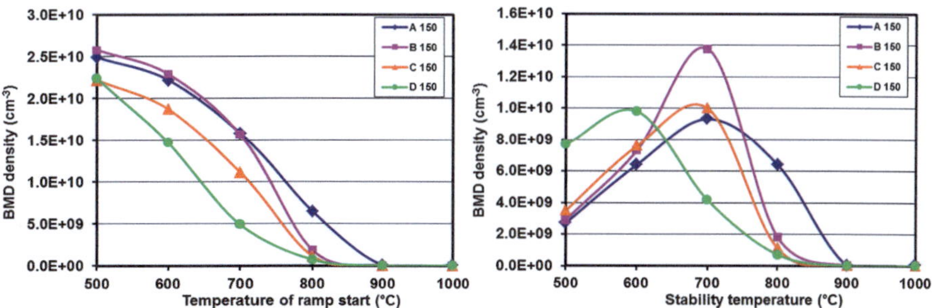

Fig. 6.26 *Left*: BMD density of four different 150 mm silicon wafer materials measured after ramped anneals with 1 K/min. The cooling rate of the ingots was decreasing from A to D. *Right*: Defect density spectra obtained from the BMD values in the *left graph*. The data points in both graphs are connected by *splines*

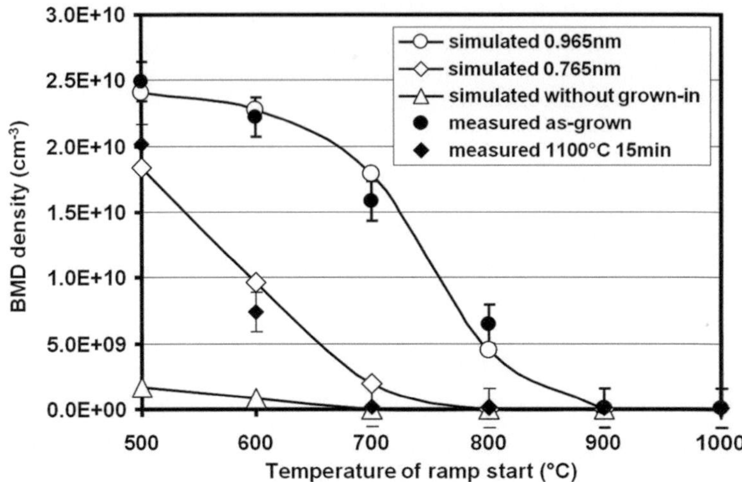

Fig. 6.27 BMD density of 150 mm silicon wafer materials with and without tabula rasa pre-treatment at 1100 °C for 15 min obtained from fast cooled vacancy-rich ingots measured after ramped anneals with 1 K/min. The simulated data points assuming oxygen precipitate nuclei of 0.965 and 0.765 nm size are connected by *splines* (Reprinted with permission from [38]. Copyright 2007, The Electrochemical Society)

for grown-in nuclei of 0.965 nm in the as-grown wafer and for nuclei of 0.765 nm after the tabula rasa pre-treatment.

In summary, it can be said that the ramped anneals with 1 K/min provide a powerful method for understanding what is going on below the detection limit of common detection methods.

6.8 Thermal Donors

The impact of oxygen on the resistivity of silicon has been intensively studied since the end of the fifties [98]. It is now well established that heat treatment in the 350–500 °C temperature range generates microdefects which act as shallow double donor centers, so-called thermal donors (TDs). TDs are double donors with ionization energies of about 70 and 150 meV [99, 100]. These defects, undesirable in commercial silicon wafers, are usually eliminated by thermal treatments above 500 °C [98, 101, 102]. In this way, the resistivity can be stabilized.

Upon annealing at high temperatures, a different type of donor states is observed. These so-called new donors (NDs) occur in the temperature range 600–900 °C with a maximum generation rate at 800 °C. Their ionization energy varies between 30 and 120 meV [103], their annihilation requires temperatures above 1000 °C and they are more difficult to cure than TDs.

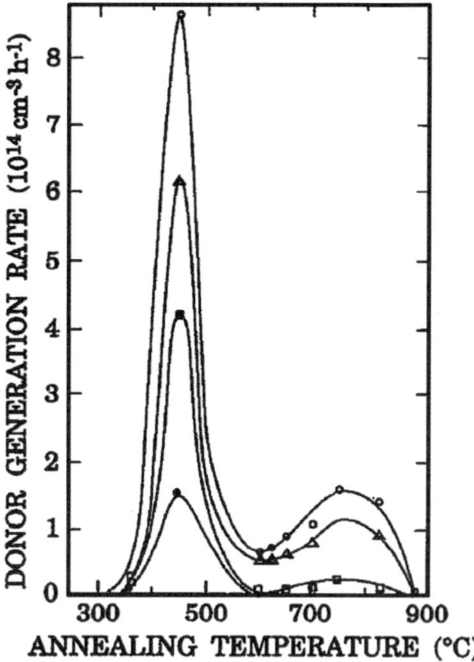

Fig. 6.28 Donor generation rate as a function of annealing temperature for different initial concentrations of interstitial oxygen: 1×10^{18} cm^{-3} (*empty circles*), 9.25×10^{17} cm^{-3} (*triangles*), 8×10^{17} cm^{-3} (*squares*), and 6×10^{17} cm^{-3} (*full circles*) according to [104] (Reproduced with permission from [104]. Copyright 1980, AIP Publishing LLC)

Figure 6.28 shows the donor generation rate for different concentrations of interstitial oxygen [104]. The two peaks related to TDs and NDs are clearly seen. The moderately B-doped material had an initial resistivity of 12–19 Ωcm. The donor concentration was measured with a collinear four-probe array according to ASTM procedure F-84-73 at room temperature. The number of carriers before and after thermal treatment in nitrogen was derived from the Irving curves. The difference corresponds to the number of donors generated or eliminated during the heat treatments.

The nature of TDs was discussed for a long time but it is well accepted now that they consist of chains of oxygen atoms only. The double donor character of the chains is due to two trifold bonded oxygen atoms involved in the chains [32, 105, 106]. Figure 6.29 compares the structure of an O_6 chain with and without a thermal double donor. In the first structure, the oxygen atoms only bridge first neighbor silicon atoms but in the second structure, two oxygen atoms also bridge second neighbor silicon atoms.

Pensl et al. have developed the SiO_x interface model which explains the origin of NDs [107]. They found a correlation of the formation and annihilation kinetics between NDs and oxygen related precipitates. The model is illustrated in Fig. 6.30. Two types of states are assumed to contribute to the continuous distribution of energy states. One type is interface states at the surface of oxygen precipitates in the entire range of measured trap states. The other type is bound states in the Coulombic wells of fixed positive charges which reside in the oxygen precipitate. These are

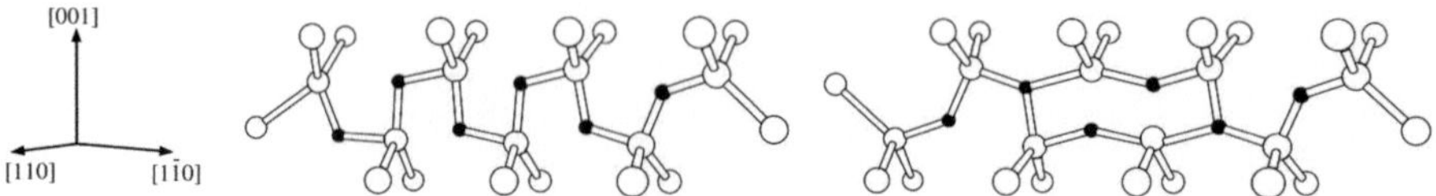

Fig. 6.29 Atomistic picture for the ground state structure of O_6^{1NN} (*left*) and O_6^{2NN} (*right*) [32]. Atomistic picture for the ground state structure of an O_6 chain (*left*) and a thermal double donor consisting of six oxygen atoms (*right*) (Reprinted from [32]. Copyright 2006, with permission from Elsevier)

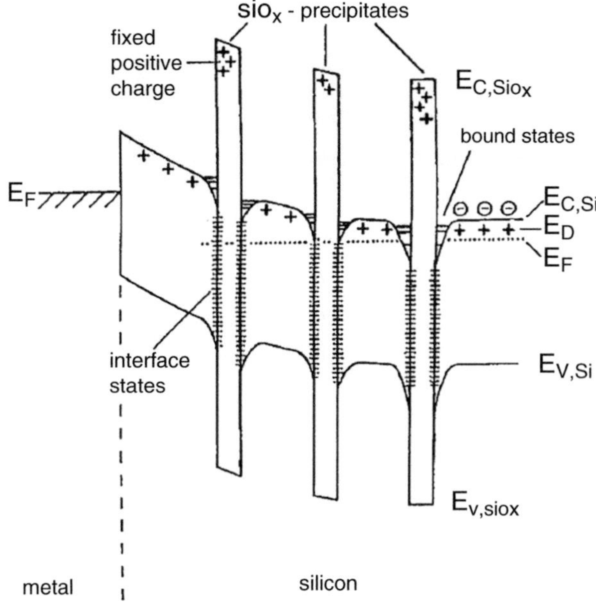

Fig. 6.30 Schematic of the Si-SiO_x (precipitate) band structure illustrating the "SiO_x Interface Model" for oxygen-related ND defects in silicon according to [107] (Reproduced with permission from [107]. Copyright 1989, Springer Science + Business Media)

limited to energetically shallow states (e.g. 250 meV below conduction band edge). Hydrogen passivation of ND traps provides experimental evidence for both types of energy states. The interface states at the surface of oxide precipitates are completely passivated whereas the fixed positive charge in the oxide and, consequently, also the bound states in the Coulombic wells remain unaffected or only slightly affected by the hydrogenation process [107].

All TD species possess a temperature and oxygen dependent equilibrium concentration which they reach after a certain time of annealing [108]. High interstitial concentrations increase the formation of TDs [109, 110] and high vacancy concentrations delay formation of TDs [111]. It was also found that rod-like defects and NDs are anti-correlated [112]. Thus, the precipitation of silicon self-interstitials in rod-like defects reduces the ND concentration. Generation of TDs by thermal pre-treatments was found to increases the generation of NDs [113, 114].

TDs consist of a family of TDs named TD1-TD16 which contain a different number of oxygen atoms increasing with the numbering [108, 115]. They evolve sequentially by attachment of oxygen atoms. Therefore, rate equation models are very suitable to describe their kinetics. A very detailed rate equation model to describe TD kinetics was developed by McQuaid et al. [116]. They showed that oxygen dimer formation is the rate limiting step in the loss of interstitial oxygen from solution during anneals at temperatures $\leq 500\,°C$. They also found that oxygen diffusion was enhanced in the early stages of TD formation but only by a factor of about 3 and they assumed that this was due to catalysis by atomic hydrogen. This catalysis is much larger when the material was pre-treated in hydrogen gas [116].

However, this was not the first rate equation model to describe TD formation. Already at a time when the donor structure was much more unclear, Kaiser et al. and Oehrlein used rate equation models [117, 118]. This model was the so-called KFR model.

Wada et al. have extended the KFR model by considering electrons emitted from TDs in mass action law [22, 119]. This was the first model which succeeded in expressing both the temperature dependence and carrier concentration dependence in addition the oxygen concentration dependence. The formation rate and concentration of TDs C_{TD} in this model are given by

$$\frac{dC_{TD}}{dt} = a \cdot D_{Oi} \cdot {C_{Oi}}^4 \cdot n^{-2} \cdot \exp\left(b \cdot D_{Oi} \cdot C_{Oi} \cdot t\right) \tag{6.40}$$

$$C_{TD}(t) = \frac{a}{b} \cdot {C_{Oi}}^3 \cdot n^{-2} \cdot \left[1 - \exp\left(-b \cdot D_{Oi} \cdot C_{Oi} \cdot t\right)\right]. \tag{6.41}$$

Here n is the electron concentration, $a = 2.75 \times 10^{-10}$ cm^4 and $b = 5.0 \times 10^{-5}$ cm^4 are constants. The electron concentration can be obtained from the electron-hole equilibrium. The maximum concentration of TDs just before a decrease in annealing time dependence can be obtained by

$$C_{TD}^{eq} = \frac{a}{b} \cdot {C_{Oi}}^3 \cdot n^{-2}. \tag{6.42}$$

Wijaranakula also developed a kinetic model which describes the reduction of interstitial oxygen during a TD anneal by the following solution of the differential equation in which neither enhanced diffusivity of interstitial oxygen nor best fitting parameters are required [120]

$$C_{Oi}(t) = C_{Oi}^0 \cdot \left[1 + \frac{2}{3} \cdot D_{Oi} \cdot t \cdot {C_{Oi}}^{2/3}\right]^{-3/2}. \tag{6.43}$$

The formation kinetics of thermal donors according to Wijaranakula is as follows

$$C_{TD}(t) = k_t \cdot \left\{ C_{Oi}^0 \cdot \left[1 + \frac{2}{3} \cdot D_{Oi} \cdot t \cdot {C_{Oi}}^{2/3} \right]^{-3/2} \right\}^c \cdot t^{1.02} \tag{6.44}$$

with $c = 3.44$ and $k_t = 4.61 \times 10^{-52}$ [120].

Ourmazd et al. developed a rate equation model for TD formation which assumes that oxygen clusters above a certain size (number of oxygen atoms) can transfer from electrically active state to electrically inactive state by the emission of a silicon self-interstitial [121]. They also assume an effective diffusivity for oxygen atoms which includes self-interstitial oxygen (IO) complexes as a fast diffusing species. This is similar to the model of Voronkov et al. [110] who also include IO as fast diffusing species in TD formation.

Lee et al. developed a quite different model to describe TD formation kinetics [122, 123]. Their model is based on the assumption that two migrating oxygen chains O_j (containing j O atoms) and O_k (containing k O atoms) associate into an O_{j+k} complex with a diffusion controlled reaction and dissolution rate. They found that TDs grow via consecutive reactions where fast diffusing oxygen dimers and all TDs capture interstitial oxygen atoms.

Mao et al. found that a pre-treatment at 1100 °C for 30 min reduces the number of TDs generated during an anneal at 450 °C [124]. The kinetics of both processes was well described by the Johnson-Mehl equation.

Most detailed investigations of TDs can be obtained by using FTIR spectrometry. Here, the different TD species from TD1 to TD16 can be distinguished and studied separately. Wagner et al. and Götz et al. have determined the infrared absorption bands and experimental binding energies for TD1-TD11 (neutral and singly ionized states) and TD12-TD16 (neutral state), respectively [108, 115]. However, with FTIR only absorption coefficients can be measured and it is necessary to find calibration coefficients in order to determine the concentration of the different species of TDs.

Hallberg et al. proposed a calibration coefficient K_m for the TD related vibrational bands $m = 975$–1015 cm^{-1} [125]. They fitted the vibrational bands to the total TD concentration determined from resistivity measurements. Using the integrated absorption coefficient of the bands, the best fit was obtained by using the same calibration coefficient for all the different bands. This calibration coefficient is $K_{975-1015} = 6(\pm 1) \times 10^{15}$ cm^{-1}. It is valid for the entire temperature range from 350 °C to 470 °C. This method assumes that the calibration coefficient of interstitial oxygen is applicable to the oxygen atoms of the TDx centers.

Wagner determined optical cross sections to calculate the concentrations of TD1-TD6 from the absorption coefficients [108]. They were evaluated by a combination of mathematical fitting and intuition to optical cross sections for the $2p_0$-transitions of neutral TDs from the single and total TD curves of samples with $O_i = 10^{18}$ cm^{-3}.

The NDs do not seem to have special FTIR lines because they are assumed to be related to oxygen precipitates. Murin et al. found in local vibration mode (LVM) studies that the annihilation at 650 °C of TDs formed at 450 °C for up to 240 h was found to result not only in a partial recovery of the interstitial oxygen, but also in the appearance of a number of new O-related LVM bands in the range 990–1110 cm^{-1} [126]. The positions of the lines and their shapes are identical to those observed for CZ-Si irradiated with electrons or neutrons and annealed at 600–700 °C. For irradiated samples these new lines appear upon annealing out of VO_3 and VO_4 defects and they were suggested to arise from VO_m ($m > 4$) defects. In both kinds of samples, pre-annealed and pre-irradiated, the new LVM bands disappear upon prolonged annealing at 650 °C while highly enhanced oxygen precipitation occurs.

The total concentration of TDs can be measured by four-point-probe measurements of the resistivity at room temperature. The measurement procedure is standardized (DIN 50431, ASTM F-84-73, ASTM F723-82). The conversion of the resistivity to the total TD concentration is based on the assumption that the TDs are the dominant donors in the samples and that all the TDs are fully ionized at room temperature. For the determination of the TD concentration from the resistivity measurements, it has to be taken into account that TDs are double donor centers [127]. Therefore, the electron concentration equals twice the TD concentration.

TD depth profiles in [128] were determined after forming Au Schottky barrier diodes by capacitance-voltage measurements at room temperature.

TDs can be also detected by DLTS in n-type silicon. A peak can be found around 55 K. This peak shifts with increasing annealing time at 450 °C and increases in height [102]. This means that the energy levels become shallower during the formation of larger clusters.

NDs exhibit a continuous DLTS spectrum [129]. The density of states determined increases towards the conduction band edge. Enhanced values of the density of states are observed for increased oxygen content, for a prolonged duration of the annealing at 650 °C, and for pre-annealing at 500 °C prior to annealing at 650 °C. The density of states ranges from 10^{14} to 10^{16} cm^{-3} eV^{-1}. Kamiura et al. have also identified a NTD in the temperature range where only TDs are usually generated [130].

In this section just some very basic features of TDs and NDs are described because there is a strong relationship to oxygen precipitation. The number of publications in this field is huge and we leave it to the reader to explore these into detail.

6.9 Impact of Dopants and Impurities on Oxygen Precipitation

Oxygen precipitation is strongly influenced by most dopants and impurities. Because the majority of all silicon wafers produced worldwide is boron doped we focus our attention on the impact of boron on oxygen precipitation. Among the impurities, nitrogen is also an important co-dopant. Hydrogen is also used as a co-dopant but can be also incorporated via the wafer surface during certain steps of electronic device fabrication. Co-dopants are introduced intentionally in silicon ingots in order to influence defect generation. The impact of these three species on oxygen precipitation is reported in detail here.

6.9.1 Boron

High boron concentrations were found to affect the oxygen transport in silicon. It was found that high boron concentration enhances the oxygen diffusivity [131]. This results in a higher growth rate of oxygen precipitates.

Murphy et al. also found an enhanced diffusivity of oxygen in highly B-doped silicon in dislocation locking experiments at 350–550 °C [132]. In wafers containing 5.4×10^{18} cm^{-3} boron, they determined an effective diffusivity of

$$D_{Oi(eff)} = 2.7 \times 10^{-6} \cdot \exp\left(-\frac{1.4\text{eV}}{kT}\right) \text{cm}^{-2}\text{s}^{-1}. \tag{6.45}$$

The activation energy is nearly the same as for the effective oxygen diffusivity found in dislocation locking experiments for moderately B-doped wafers. Only the pre-factor increased.

Sugimura et al. studied the temperature dependence of enhanced diffusivity in highly B-doped wafers by using SIMS profiles fitted to error functions [133]. Accelerated diffusion was observed for boron concentrations higher than 3.8×10^{18} cm^{-3} and the acceleration increased with increasing boron concentration.

Sueoka et al. have extensively studied oxygen precipitation in epi-wafers with different B-doping by isothermal anneals in the temperature range 700–1000 °C [131, 134]. With increasing B-concentration, the precipitate density increases and the precipitates nucleate at higher temperature. A fourth power dependence of the nucleation rate on the boron concentration was determined for annealing at 700 °C and 800 °C. It was also found that the precipitate growth rate in highly boron doped wafers is increased. Sueoka et al. found that the precipitate growth process in highly B-doped wafers is reaction limited which is different from the diffusion limited growth in moderately B-doped silicon.

Sugimura et al. [133] studied the BMD generation in p/p$^+$ epi wafers with different B-concentration. The epi-process has shrunken all BMD nuclei. At highest B-concentration nucleation of oxide precipitates is very fast at 650 °C and 700 °C. Sugimura et al. have fitted their data to nucleation rates of classical nucleation theory applying enhanced diffusivity. However, the enhanced diffusivity alone could not explain the strongly increased nucleation rate. Therefore, they believe that the dependence of the formation energy of silicon self-interstitials from the doping conditions could change the equilibrium constant of the precipitation reaction. Sueoka has discussed the impact of the dependence of self-interstitial formation energy on dopant concentration and of the binding energy of self-interstitials with dopants on oxygen precipitation kinetics [33]. The formation energy of self-interstitials decreases in p^+ silicon, while increases in n^+ silicon. The reaction $2Si + 2O_i \leftrightarrow SiO_2 + I_{Si}$ is driven to the right side for p^+ and to the left side for n^+ silicon.

Takeno developed a nucleation model based on the reaction $(1 + \gamma)Si + 2O_i + \beta V + \alpha B \leftrightarrow SiO_2B_\alpha + \gamma I +$ strain [135]. It is assumed that B is incorporated into the precipitate thus reducing the strain due to the different volume between precipitate and silicon matrix. α is the fraction of B atoms absorbed to the precipitate.

Wijaranakula has carried out isochronal anneals at different nucleation temperatures for very long periods (100 h) followed by growth anneals for very long periods (50 h) [136]. In Fig. 6.31, the nucleation curves are shown for different B-concentrations and medium oxygen concentration. In heavily B-doped silicon, the two peaks shift to higher temperatures, 700 °C and 800 °C, whereas at the highest B-concentration, the precipitation is strongly retarded. The latter result is well in agreement with Ref. [137].

For the aggregation of oxygen in heavily doped silicon, the following reactions are proposed by Wijaranakula as being important:

$$Si_s + 2\,(VO_i) \leftrightarrow SiO_2 + 2V$$
$$(VO_i) + O_i \leftrightarrow (VO_2) + Si_s \leftrightarrow SiO_2 + V$$

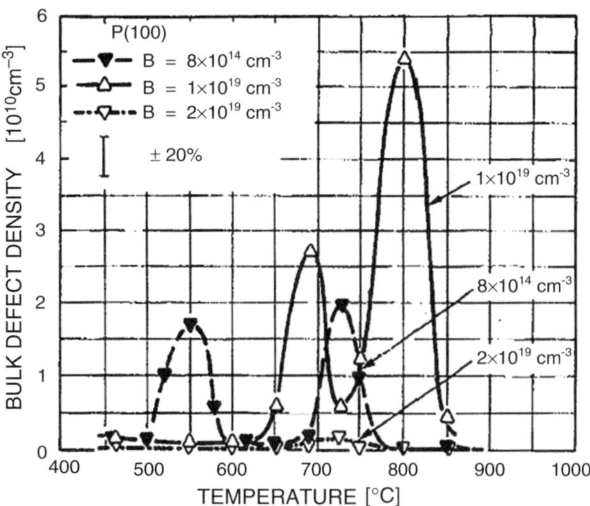

Fig. 6.31 The effect of interstitial boron doping concentration on the isochronal transformation of the bulk defects after a 100 h isochronal anneal and a 50 h anneal at 1050 °C [136]. The oxygen concentrations were 7.7×10^{17}, 6.8×10^{17}, and 6.17×10^{17}, for 8×10^{14}, 1×10^{19}, and 2×10^{19} cm^{-3}, respectively (Reproduced with permission from [136]. Copyright 1992, AIP Publishing LLC)

where SiO_2 is the oxide nucleus. Additional reactions are proposed for highly B-doped silicon:

$$Si_s + 2\,(B_iO_i) \leftrightarrow SiO_2 + 2B_s$$
$$(B_iO_i) + O_i \leftrightarrow (BO_2) + Si_s \leftrightarrow SiO_2 + B_s$$

where the impurity oxygen pairs are regarded as mobile species. The presence of small amounts of boron dissociated from boron-oxygen pairs in the vicinity of the precipitate can relax the lattice strain and reduce surface energy of the precipitate. This can explain the shift of the nucleation peaks to higher temperature for high B-doping. The reaction at the precipitate/matrix interface is described by:

$$\gamma Si_s + 2O_i \leftrightarrow SiO_2 + (\gamma - 1)\,I$$

with γ was suggested to be 2.25 for strain-free precipitation. The reduced precipitation at highest B-concentration, which is much higher that the oxygen concentration, is explained by a significant decrease of interstitial oxygen due to binding of oxygen to boron forming BO_2 and more complex defects like B_2O_3.

Yonemura measured the lattice strain around plate-like precipitates in highly B-doped wafers by the convergent beam electron diffraction (CBED) technique in TEM [138]. The comparison with moderately B doped wafers showed that the lattice strain around the platelets in highly B-doped wafers is far relaxed compared to moderately B-doped wafers.

6.9.2 Hydrogen

It is known that hydrogen reduces the migration barrier for oxygen and thus enhances the diffusion of oxygen in silicon [139, 140]. A hydrogen concentration as low as 10^8 cm^{-3} can significantly enhance oxygen diffusivity. A diffusion coefficient for the hydrogen enhanced diffusion of interstitial oxygen of

$$D_{Oi(enhanced)} = 7.1 \times 10^{-4} \cdot \exp\left(-\frac{2.0\text{eV}}{kT}\right) \text{cm}^{-2}\text{s}^{-1} \tag{6.46}$$

was determined in silicon given a heat treatment at 900 °C for 2 h in hydrogen [139]:

The hydrogen enhanced diffusivity of interstitial oxygen leads to an enhanced growth rate of oxygen precipitates and to an enhanced rate of thermal donor formation. During thermal donor formation, hydrogen acts as a catalyzer for forming electrically active ring structures with three fold coordinated oxygen from oxygen chains [141]. The experimentally determined activation energy for hydrogen enhanced thermal donor formation amounts to 1.7±0.3 eV [140]. Simoen et al. concluded from combined DLTS and FTIR measurements that hydrogen not only plays a catalytic role but actively takes part in donor formation [142]. According to their results, the H-related shallow thermal donors are a different species than the usual oxygen thermal donors. Hara et al. confirmed that hydrogen is incorporated in crystals during pulling and they estimated the hydrogen concentration in silicon ingots using hydrogen enhanced thermal donor formation [143].

The hydrogen, which can be incorporated in silicon ingots from polysilicon, quartz crucibles, and Ar gas during pulling, was shown to affect oxygen precipitation also [143, 144]. Annealing of thick bulk crystal generates a much higher BMD (bulk microdefect) density than annealing of wafers. This behavior can be explained by out-diffusion of hydrogen from the wafers during annealing. In the bulk crystals, only a hydrogen denuded zone is formed below the surfaces.

It was also found that hydrogen can be introduced during high temperature annealing from hydrogen contaminated gases [144]. There was found a correlation between the hydrogen content measured after annealing at 1270 °C in nitrogen or oxygen and the amount of precipitated oxygen.

6.9.3 Nitrogen

It is well investigated that nitrogen doping enhances oxygen precipitation. The reason for this is first of all a larger size of the grown-in oxide precipitate nuclei which are supposed to be stable up to 1100 °C, while most of the nuclei in crystals which are not nitrogen doped shrink at this temperature [145].

Several authors investigated the grown-in size distribution of oxide precipitate nuclei in NCZ (nitrogen-doped CZ) silicon using the slow ramping procedure with 1 or 1.5 K/min from different start temperatures described in Sect. 6.7 [146, 147]. There is very good agreement of the results. In all cases, the size of grown-in oxide precipitate nuclei increases with increasing nitrogen concentration of the material. Figure 6.32 provides an example of such investigations comparing the defect density spectra of NCZ and CZ silicon.

An interesting phenomenon being unique in NCZ wafers is the fact that the BMD density generated during ramped anneals with 1 K/min from 500 °C to 1000 °C

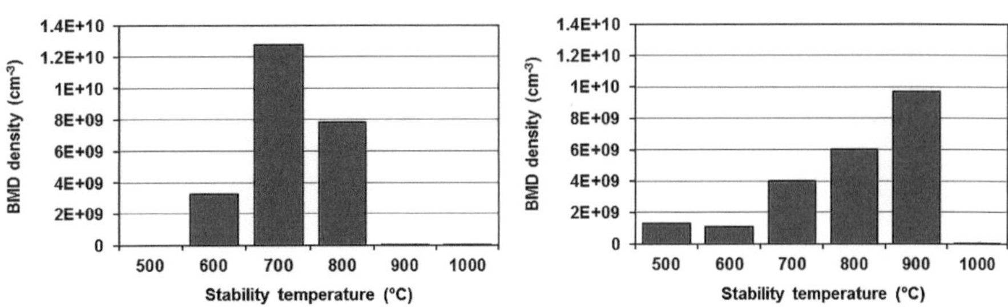

Fig. 6.32 Grown-in defect density N_{LST} as a function of stability temperature for an as-grown CZ silicon wafer with an initial concentration of interstitial oxygen in the range of 8.4–9.1 × 10^{17} cm^{-3} (*left*) and for an as-grown CZ silicon wafer doped with nitrogen (2 × 10^{14} cm^{-3}) and an initial concentration of interstitial oxygen of 7.4 × 10^{17} cm^{-3} (*right*). *Right figure* reproduced with permission from [147] (Copyright 2005, Trans Tech Publications Ltd)

followed by 2 h at 1000 °C correlates to the radial *v/G* (pulling speed/thermal gradient) profile of the wafer [147].

Nakai et al. [145] found plate-like precipitates with a strong strain field in NCZ crystals with a nitrogen concentration of 3×10^{15} cm^{-3}. They detected oxygen and nitrogen in these precipitates and found that the precipitation process is diffusion limited similar to CZ silicon. Further authors have also analyzed precipitates in N-doped silicon and confirmed that precipitates in NCZ silicon contain oxygen and nitrogen [148–151].

Yonemura [138] measured the lattice stress around plate-like precipitates in N-doped wafers by the convergent beam electron diffraction (CBED) technique in TEM and found that the strain is nearly the same as in wafer without N-doping.

Yu et al. [152] studied the oxygen precipitation on NCZ wafers with an oxidation induced stacking fault (OSF) ring region. The precipitation in the OSF ring is different from the precipitation in the central V-region. For two step anneals with a nucleation step, it is higher than in the V-region and for a one-step treatment it is lower. Similar results were obtained already by von Ammon et al. for two-step anneals [153].

The density of grown-in oxide precipitate nuclei in N-doped wafers cut from ingots optimized with respect to voids by advanced pulling methods is distributed radial very inhomogeneous according the variation of the vacancy supersaturation. Wafers cut from these ingots contain a central OSF region and an outer perfect vacancy region. Two step thermal treatments with a nucleation at 780 °C 3 h or an oxygen out-diffusion anneal at 1100 °C 2 h both followed by 1000 °C 16 h have shown that nitrogen doping of ingots can result in a homogeneous BMD density on the whole wafer [154]. A threshold nitrogen concentration of at least 3×10^{13} cm^{-3} is required for the nitrogen enhanced oxygen precipitation indicating that diffusive nitrogen in the form of N_2 complexes is prerequisite for the increase of the nucleation rate [154]. This is demonstrated in Fig. 6.33.

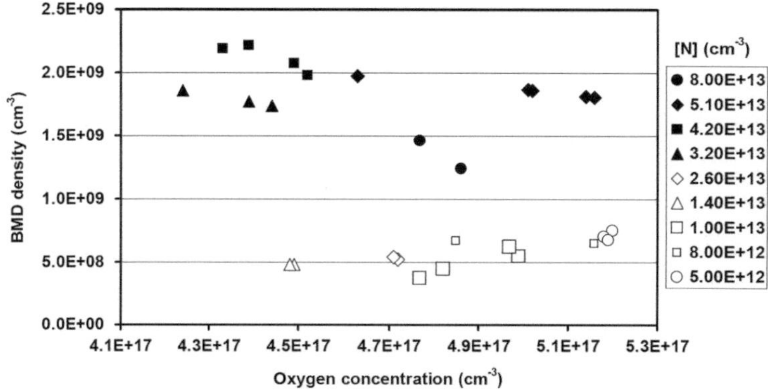

Fig. 6.33 Average BMD density after annealing for 2 h at 1100 °C followed by 1000 °C 16 h shown as a function of the initial concentration of interstitial oxygen in wafers optimized with respect to voids containing different initial concentrations of nitrogen [154] (Reprinted from [154]. Copyright 2012, with permission from Elsevier)

6.10 Oxygen Precipitation in Device Processing

Control of oxygen precipitation in device processing becomes increasingly important for increasing wafer diameters and decreasing feature sizes of devices. Warpage of silicon wafers due to plastic deformation caused by high densities of oxygen precipitates with secondary defects is an important issue which can result in overlay problems in photolithography steps of device processing [155]. Furthermore, oxygen precipitates located in the device active zone are known to degrade the device function (see e.g. [3]). However, bulk defect zones consisting of carefully controlled densities of oxygen precipitates can efficiently getter metallic contamination accidentally introduced during device processing and thus increase the device yield (see e.g. [4]). Therefore in this section, the device processing relevant issues of oxygen precipitation are considered into detail.

6.10.1 Impact of Grown-In Oxygen Precipitate Nuclei on Defect Generation in Device Processing

As shown in Sect. 6.7, there is a characteristic size distribution of grown-in oxygen precipitate nuclei which depends among others on the crystal pulling process conditions. So it was found that high cooling rates lead to larger grown-in oxygen precipitate nuclei than low cooling rates. Co-dopants are important as well as we have seen already in Sect. 6.9.3 where it was shown that nitrogen doping strongly increases the size of grown-in nuclei.

In order to find out what happens to the grown-in oxygen precipitate nuclei during device processing, the method described in Sect. 6.7 can be also used after the respective processing steps [42, 95–97, 156, 157]. This was done for typical processing steps at the beginning of a CMOS process and the results can be found in Fig. 6.34. It is clearly seen that already during the first thermal step, the pad oxidation, the grown-in size distribution splits in two parts, a growing part and a shrinking part. This split is maintained in the second thermal step. The BMD density in the growing part depends on the grown-in size distribution. In the faster cooled materials with the larger grown-in nuclei the density is the highest and it is decreasing with decreasing cooling rate of the ingot. This means that the size distribution of the grown-in oxide precipitate nuclei is important for the BMD density which is found after device processing.

A well-known technique for reduction of the BMD density in the device active region is out-diffusion of interstitial oxygen at high temperatures prior to nucleation and growth of oxygen precipitates [158, 159]. Such an oxygen out-diffusion step can also be carried out at the beginning of CMOS processing. Now it is interesting, what happens to the depth distribution of grown-in nuclei if the oxygen was out-diffused in the surface near region? The method described in Sect. 6.7 is of the further advantage that depth distributions of defects with different stability temperature can

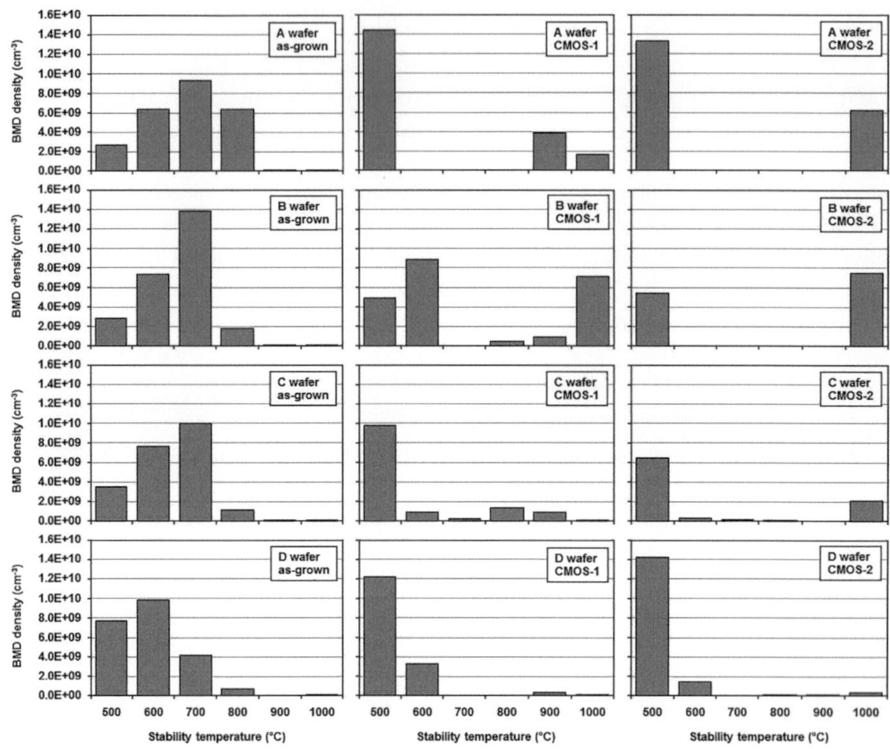

Fig. 6.34 Defect density spectra of wafers obtained from ingots with different cooling rate decreasing from A to D measured in the as-grown state, after a pad oxidation at 900 °C for 40 min (CMOS-1), and after a well drive anneal at 1150 °C for 2 h in nitrogen (CMOS-2) (Reproduced with permission from [95]. Copyright 1997, IOP Publishing, all rights reserved)

be generated with the 1 K/min ramps after different processing steps. This more advanced method is applied now to wafers which received an out-diffusion anneal at 1150 °C for 1 h prior to pad oxidation at 900 °C for 40 min, well drive anneal at 1150 °C for 2 h, and a summary of steps at 900 °C for 8 h. Figure 6.35 shows the profiles for different stability temperatures after the last step for wafers obtained from slowly and fast cooled ingots. We find again the shrunken and grown fractions. The shrunken fractions stable up to 600 °C possess a defect denuded zone below the wafer surface. However, the grown fraction stable at $\geq$1000 °C can be found up to the surface in a concentration which again depends on the cooling rate of the ingot. These defects are harmful because they also exist in the defect denuded zone and thus in the device active region. It becomes clear that it is possible to influence the density of denuded zone defects that means the quality of the defect denuded zone via the pulling conditions of the silicon crystal.

We have seen in the beginning of this chapter that the first thermal step of device processing is decisive for growth and shrinkage of grown-in oxygen precipitate nuclei. Therefore, there should be the opportunity to modify denuded zone defects

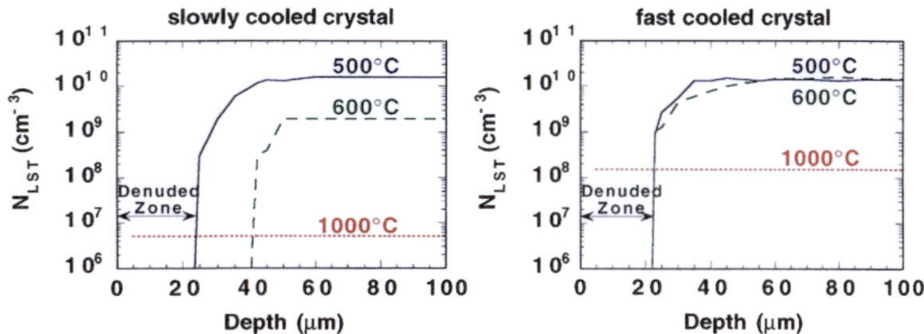

Fig. 6.35 Depth distribution of defects measured by IR-LST and classified according to their stability temperature for wafers obtained from vacancy-rich slowly and fast cooled ingots after the following processing steps: oxygen out-diffusion at 1150 °C for 1 h, pad oxidation at 900 °C for 40 min, well drive anneal at 1150 °C for 2 h and annealing at 900 °C for 8 h (Reprinted with permission from [157]. Copyright 1999, The Electrochemical Society)

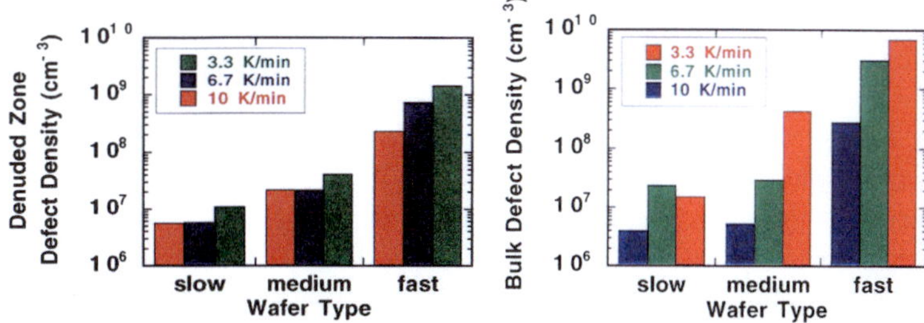

Fig. 6.36 Denuded zone defect density (*left*) and bulk defect density (*right*) measured on wafers obtained from vacancy-rich ingots with slow, medium and high cooling rate after a thermally simulated CMOS process with the following steps: oxygen out-diffusion at 1150 °C for 1 h, pad oxidation at 900 °C for 40 min, well drive anneal at 1150 °C for 2 h and annealing at 900 °C for 8 h. *Left graph* reprinted with permission from [156] (Copyright 1998, The Electrochemical Society)

and bulk defect density by changing the ramp rates of the first thermal step. This was done for the oxygen out-diffusion step and the results at the end of the following steps of a thermally simulated CMOS process described in detail above (pad oxidation, well drive anneal, and summary of steps) can be found in Fig. 6.36. Typical processing ramp rates of 3.3, 6.7, and 10 K/min were applied. It can be clearly recognized that both denuded zone defect density and BMD density decrease with increasing ramp rate for a given type of material.

In summary, it can be said that oxygen precipitation during device processing and quality of the defect denuded zone can be influenced by the pulling conditions of the crystal and by the ramp rate of the first thermal step.

For high-end low thermal budget device processing however, it has to be taken into account that prolonged thermal treatment at nucleation temperatures (see Sect. 6.4.6) can result in regrowth of previously shrunken oxide precipitate nuclei.

6.10.2 Internal Gettering

Copper, nickel, and iron are the most important impurities in device processing. All these metals are very fast diffusors which possess a marked diffusivity even at room temperature. The supersaturated metals easily precipitate forming metal silicides. The preferred places of metal silicide precipitation are the surfaces of the wafer. After preferential etching, the metal silicide precipitates can be observed as surface haze. Moreover, metal silicide precipitates can be also formed at crystal defects by heterogeneous nucleation e. g. dislocations or stacking faults are decorated by metal silicide precipitates.

Metal silicides precipitated in the surface-near device active region are detrimental for electronic devices [3]. Therefore in the eighties, the multi-zone concept was proposed [2, 160, 161]. It consists of a defect denuded zone below the front surface including the device active region, an internal gettering zone, and an external gettering zone at the back surface. In this way, the metallic impurities are forced to precipitate at the internal bulk defects or at the defects at the back surface because precipitation at crystal defects is preferred compared to precipitation at the surface. In other words, the metallic impurities are gettered to a place where they are not harmful to the devices anymore.

Gettering is the attraction of metallic impurities by so-called gettering sinks. The physics of gettering can be various depending on the technique applied. A detailed description can be found in Ref. [162]. Myers et al. proposed different types of gettering according to their physics which are shortly mentioned here. Precipitation gettering is based on metal silicide precipitation at heterogeneous nucleation centers deliberately introduced. The minimum impurity concentration theoretically achievable corresponds to the solid solubility at temperature of last processing step. Segregation gettering is due to segregation into second phases with the impurity incorporated as dilute constituent. Gettering of single metal atoms at defects which provide open space in the lattice and unsaturated bonds is possible as well. This trapping is a saturable process that remains active at all concentrations but it is reversible. Interaction with dopants is also used for gettering where charged metals and dopant atoms with opposite charge form pairs as e. g. gettering of Fe in highly B doped silicon by forming B-Fe pairs. The last type, phosphorus gettering is based on the enrichment of metallic impurities in phosphorus diffused regions.

Recently, it was discovered by electron energy loss spectrometry (EELS) that the 2–3 nm thick interface layer between a SiO_2 precipitate and the surrounding silicon matrix consists of SiO [163]. It was shown that it is similar to the interface between a thermal oxide layer and silicon. Copper was found to segregate to this SiO interface of the thermal oxide layer and it was concluded that copper should be gettered by segregation to the SiO interface layer of the SiO_2 precipitates [163].

The most important and most widely used gettering technique in microelectronic manufacturing is internal gettering by oxygen precipitates [158, 159, 164]. This technique is easily to apply in Czochralski silicon by exploiting the inherent interstitial oxygen. In the previous section, we described already that out-diffusion of interstitial oxygen is the way of creating a defect denuded zone in the surface-near region. This is also the first step of the traditional method of internal gettering (see e.g. [4]). The second step is the formation of the bulk defect zone by oxygen precipitation. Various processes consisting of one or more steps often including a nucleation step were proposed for bulk defect generation (see e.g. [165]). However, in modern device technologies for economic reasons the device process itself should generate the bulk defects for internal gettering. The defect denuded zone and the microdefect density in the bulk defect zone should be tailored with respect to the device process. High densities of oxygen precipitates can cause plastic deformation of silicon wafers which results in overlay problems in photolithography steps [155, 166]. Therefore, density and size of oxygen precipitates need to be carefully tuned to efficiently getter the metals without causing geometry problems.

The International Technology Roadmap for Semiconductors (ITRS) requests less metal contamination than the defined critical surface concentration of 10^{10} cm^{-2}. Modern high end processing lines can meet the high standards and yield losses due to metal contamination are restricted to occasional events. In order to avoid the costs of a lot lost in such cases, insurance in the form of internal gettering the unintentional metal contamination is still needed.

6.10.3 Generation of Defect Denuded Zones

The timescale for creation of defect denuded zones has changed from hours over seconds to milliseconds over the decades. In the 1970s and 1980s, furnace annealing for oxygen out-diffusion was applied. These denuded zones are based on interstitial oxygen profiles. Because of the low diffusivity of interstitial oxygen this took one or more hours at temperatures above 1000 °C. In the 1990s, the so-called magic denuded zone (MDZ) effect was discovered [41, 43]. Now, it was possible to generate defect denuded zones via vacancy profiles installed by rapid thermal annealing (RTA). Such processes take less than a minute. In the new century, even more rapid processes like flash lamp annealing were developed [167, 168] being able to generate denuded zones in milliseconds of thermal treatment [169]. In the following, these possibilities to generate defect denuded zones will be described into detail.

Oxygen out-diffusion profiles can be calculated by solving the one dimensional diffusion equation

$$\frac{\partial C_{Oi}}{\partial t} = \frac{\partial}{\partial x} D_{Oi} \frac{\partial C_{Oi}}{\partial x} \tag{6.47}$$

numerically whereby the surface concentration C_{Oi}^{surf} can be assumed to be equal to the solubility which is valid for oxidized surfaces or set to zero in any other case. However, there exists also the following analytical solution for calculation of oxygen profiles.

$$C_{Oi}(x,t) = C_{Oi}^{surf} + \left(C_{Oi}^{0} - C_{Oi}^{surf}\right) \mathrm{erf}\left(\frac{x}{2\sqrt{D_{Oi}t}}\right). \tag{6.48}$$

Figure 6.37 compares a calculated oxygen out-diffusion profile with the depth profiles of the BMD density measured by IR-LST N_{LST}. The depth profiles were obtained after ramped anneals with 1 K/min as described in detail in Sect. 6.7. It can be seen that in the region where the oxygen is diffused out, the oxygen denuded zone, the grown-in oxide precipitate nuclei are dissolved except a few large grown-in nuclei which survived the ramping. This problem was discussed already in Sect. 6.10.1. The defect denuded zone (DZ) is in good agreement with the oxygen denuded zone but there is a range where the grown-in oxygen precipitate nuclei were just shrunken. Therefore, the exact depth of the defect denuded zone measured depends on subsequent processing. This becomes obvious already in the right graph

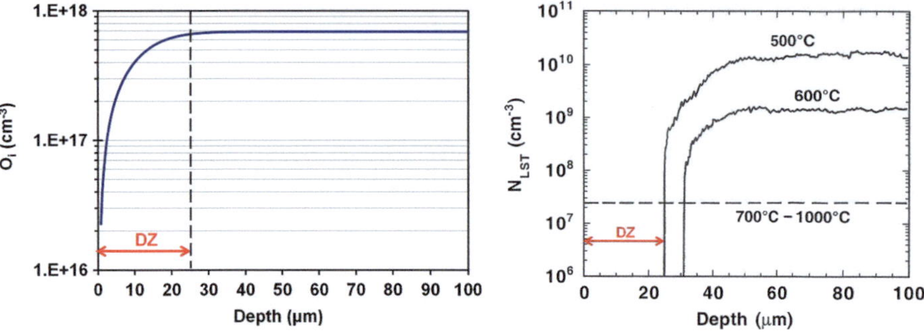

Fig. 6.37 Oxygen out-diffusion profile calculated for annealing of a silicon wafer with an initial concentration of interstitial oxygen of 6.9×10^{17} cm^{-3} at 1150 °C for 1 h in N_2 (*left graph*) compared to the BMD density depth profiles of a wafer of the same type subjected to the same out-diffusion anneal followed by ramped anneals with 1 K/min from different starting temperatures (as indicated in the *right graph*) to 1000 °C with soak for 2 h. *Right figure* reprinted from [170] (Copyright 2000, with permission from Elsevier)

in Fig. 6.37 where two different temperatures of ramp start lead to different depths of the denuded zone. So for any application in device processing, the conditions of oxygen out-diffusion must be adjusted to the subsequent thermal process in order to obtain the required depth of denuded zone and bulk microdefect density.

A much faster way for creation of denuded zones is based on the so-called magic denuded zone (MDZ) effect discovered by Falster et al. [43]. Rapid thermal annealing (RTA) is used to create a well-defined internally gettering bulk defect zone and a defect denuded zone below the surface with negligible out-diffusion of interstitial oxygen. This is possible because profiles of vacancy supersaturation are installed in the wafer and as it was shown in Sect. 6.4 the precipitation of interstitial oxygen strongly depends on the supersaturation of intrinsic point defects. Hereby, vacancy supersaturation enhances oxygen precipitation while interstitial supersaturation suppresses oxygen precipitation. It was also demonstrated that during RTA the supersaturated vacancies facilitate the formation of V_mO_n complexes which are the initial stages of oxygen precipitation [31]. However, the strength of the effect depends on the temperature and cooling rate of the RTA pre-treatment [171].

The MDZ effect can be well explained considering both diffusivity and solubility of the intrinsic point defects as a function of temperature. Figure 6.38 shows both plotted as a function of temperature. It can be seen that in the region of temperature being important for thermal treatments of silicon wafers, the diffusivity of interstitials is higher than the diffusivity of vacancies but the solubility of vacancies is higher than the solubility of interstitials. Two important physical processes have to be taken into account, diffusion and the Frenkel pair equilibrium [172]. Then, the basics of the MDZ effect can be modeled using the following

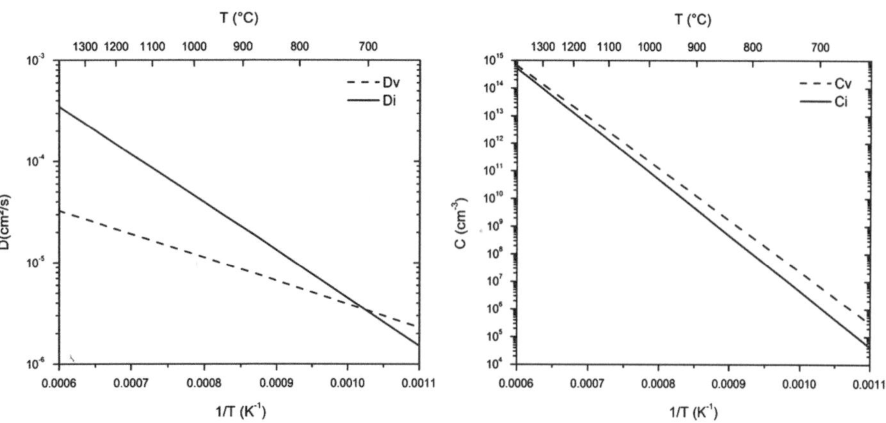

Fig. 6.38 Diffusivity (*left graph*) and solubility (*right graph*) of neutral intrinsic point defects in silicon both plotted as a function of temperature according to Ref. [6]

equations which describe the change of the concentrations of the intrinsic point defects as a function of time t and depth below the surface of a silicon wafer x:

$$\frac{\partial}{\partial t}C_V(x,t)=\frac{\partial}{\partial x}\left[D_V\frac{\partial}{\partial x}C_V(x,t)\right]-k_{IV}(x,t)\left[C_I(x,t)\,C_V(x,t)-C_I^{eq}(x,t)\,C_V^{eq}(x,t)\right] \tag{6.49}$$

$$\frac{\partial}{\partial t}C_I(x,t)=\frac{\partial}{\partial x}\left[D_I\frac{\partial}{\partial x}C_I(x,t)\right]-k_{IV}(x,t)\left[C_I(x,t)\,C_V(x,t)-C_I^{eq}(x,t)\,C_V^{eq}(x,t)\right]. \tag{6.50}$$

Here, D_I and D_V are the diffusivities of interstitials and vacancies, respectively, C_I and C_V are their concentrations, and C_I^{eq} and C_V^{eq} are their equilibrium concentrations. The Frenkel pair reaction constant k_{IV} can be obtained by

$$k_{IV}(x,t)=\frac{4\pi r_C}{\Omega C_{Si}}(D_I+D_V) \tag{6.51}$$

with the atomic density of silicon $C_{Si} = 5\times10^{22}\text{cm}^{-3}$, the capture radius $r_C = 10^{-7}\text{cm}$, and the volume of the silicon unit cell $\Omega = 2.002\times10^{-23}\text{cm}^3$. It was further assumed that the energy barrier of the Frenkel pair reaction is zero.

The plots in Fig. 6.39 demonstrate what happens to interstitials and vacancies during RTA. During heating Frenkel pairs are formed maintaining the Frenkel pair equilibrium condition $C_I\cdot C_V = C_I^{eq}\cdot C_V^{eq}$. Because after Frenkel pair formation, the concentration of vacancies is lower than the equilibrium concentration of vacancies and the concentration of interstitials is higher than their equilibrium concentration the former diffuse in from the surface and the latter diffuse out towards the surface. The time of the soak at 1250 °C is long enough to nearly reach equilibrium concentrations in the bulk. So at the end of RTA soak there is a difference between vacancy concentration, which is higher, and interstitial concentration. Then during cooling, Frenkel pairs try to recombine, however, the concentration of interstitials is not sufficient leading to a supersaturation of the residual vacancies. Close to the surfaces supersaturated vacancies can diffuse out. At the end, a profile of supersaturated vacancies is found as can be seen in Fig. 6.39.

A defect denuded zone generated in this way, is based on the installed vacancy profile. This needs even more adjustment with subsequent processing than oxygen out-diffusion because intrinsic point defects are very fast diffusors and the profiles could be destroyed again and because the grown-in oxygen precipitate nuclei are just shrunken in most RTA pre-treatments [170, 173].

The MDZ effect is of the advantage that it allows the installation of a well-defined vacancy concentration in the wafer via the choice of the RTA temperature which allows control of the bulk microdefect density as it is demonstrated in Fig. 6.40. It can be seen that the RTA temperature should amount at least 1200 °C to generate enough vacancies in order to enhance oxygen precipitation. In these cases, the so-called thermal history is practically overwritten and it is possible to

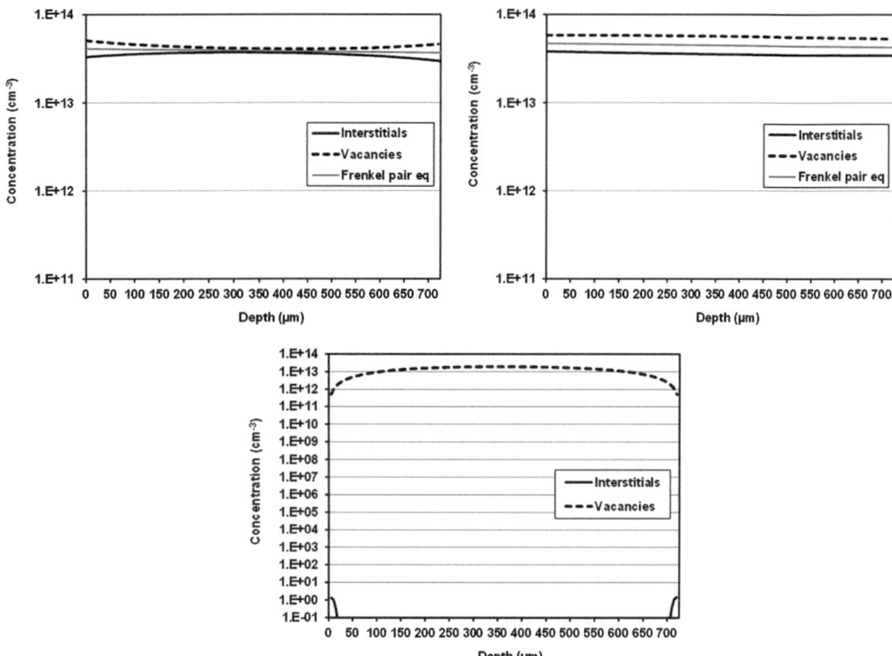

Fig. 6.39 Modeling of interstitial and vacancy depth profiles during RTA at 1250 °C with 30 s soak and cooling rates of about 75 K/s. The *top left* plot shows the profiles after RTA heating, the *top right* after RTA soak, and the *bottom* after RTA cooling. *Bottom graph* reprinted with permission from [169]. (Copyright 2012, The Electrochemical Society)

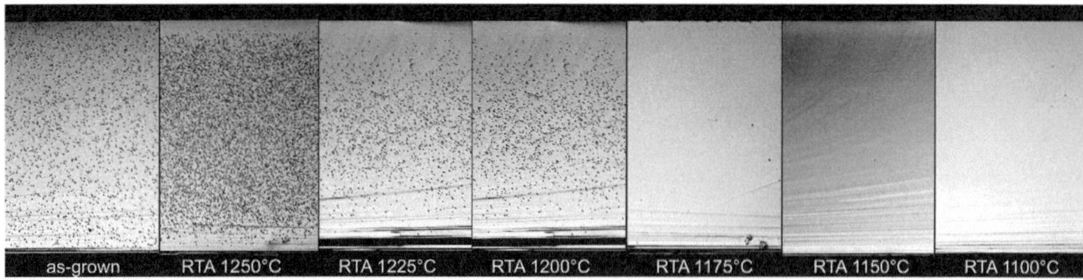

Fig. 6.40 Micrographs of the preferentially etched cleavage planes after different RTA pre-treatments for 30 s in Ar/1000 ppm oxygen atmosphere followed by 780 °C 3 h + 1000 °C 16 h anneals. BMDs appear as *black dots* and each micrograph shows the whole wafer thickness (725 µm) from top to bottom. As-grown means no RTA pre-treatment before annealing (Reprinted with permission from [169]. Copyright 2012, The Electrochemical Society)

generate a homogeneous distribution of BMDs in a wafer independent of the local concentration of intrinsic point defects generated during crystal growth [174].

Further, the defect denuded zone can be controlled via the RTA cooling rate [171]. Higher cooling rates lead to smaller defect denuded zones. Further effects can be caused by the ambient during RTA treatment which either can introduce more interstitials like oxygen or introduce even more vacancies like NH_3 or remain neutral like argon.

Especially, RTA in an ambient containing Ar and NH_3 has gained a lot of interest for proximity gettering of metal contamination in device manufacturing [174]. Due to the introduction of additional vacancies, bands of high concentration of oxygen precipitates are formed very close to the surface. The defect denuded zone is very narrow in this case and gettering of impurities in low thermal budget device processing is possible.

During the last years, thermal processing on the millisecond scale was developed first of all for the creation of shallow dopant profiles in high end electronic device technologies and for the thermal treatment of layers [167, 168]. Flash lamp anneals or laser anneals are able to heat the silicon wafer within a few milliseconds to temperatures up to the melting point. During such treatments temperature profiles are generated within the silicon wafer [168]. These temperature profiles lead to profiles of intrinsic point defects which are different from the point defect profiles installed by RTA [169]. Now, the different diffusivity of vacancies and interstitials in the temperature gradients is the most important physical effect leading to the interstitial becoming the dominating intrinsic point defect in most parts of the wafer. Thus, enhancement of oxygen precipitation does not take place anymore for anneals in the millisecond range. If the BMD formation is completely suppressed because oxygen precipitate nuclei were shrunken in the whole wafer depth or a defect denuded zone is formed below the front surface depends on the flash duration and irradiance. Figure 6.41 presents the results of 20 and 3 ms flash lamp annealing

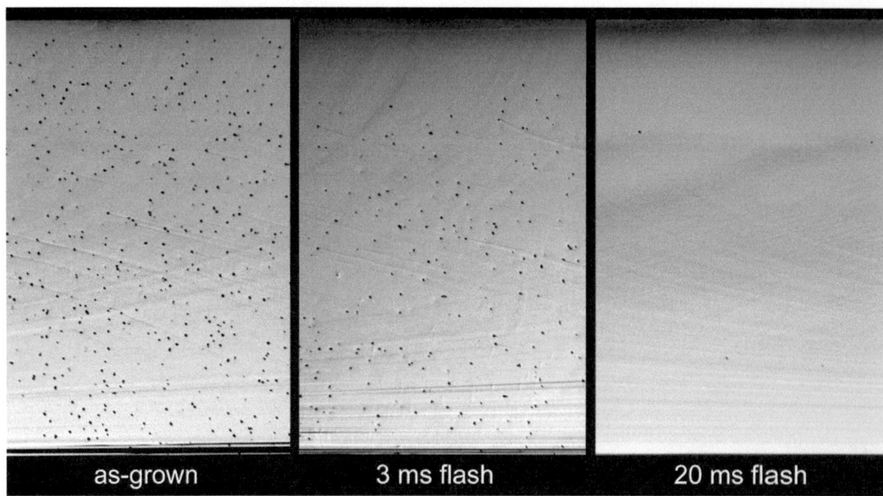

Fig. 6.41 Micrographs of the etched cleavage planes after flash lamp anneal for 3 ms and 20 ms both followed by 780 °C 3 h + 1000 °C 16 h anneals. BMDs appear as *black dots* and each micrograph shows the whole wafer thickness (725 μm) from top to bottom. As-grown means no flash pre-treatment before annealing. The irradiance normalized with respect to the irradiance required for melting of the surface of the 3 and 20 ms flash anneals were 0.95 and 0.98, respectively (Reprinted with permission from [169]. Copyright 2012, The Electrochemical Society)

where in the former case BMD formation was completely suppressed and in the second case a defect denuded zone was created below the front surface.

6.10.4 Getter Tests

In order to create a suitable internal gettering bulk microdefect zone, getter tests are needed which can proof that it can getter sufficient metal impurities or better measure the getter efficiency of the bulk microdefects. Such getter tests usually contain a sequence of steps starting with intentional contamination of a full wafer or with pieces of a silicon wafer. Among others, this can be done by spinning on of a contaminated solution, placing the sample in a contaminated solution, or scratching a metal piece on the back surface. The latter method leads to high metal contamination which cannot be controlled but the other two methods allow contamination with well-defined levels.

Next step is measurement of the initial surface contamination. Many getter tests work with inductively coupled plasma mass spectrometry (ICP-MS), total reflection x-ray fluorescence (TXRF), or graphite furnace atomic absorption spectroscopy (GF-AAS) measurement of the contaminant concentration. These methods require the collection of the surface concentration in a droplet. Several methods were developed which are compared by Shabani et al. [175]. Special methods are drop sandwich etching (DSE) and room temperature acid vapor phase decomposition (RT-AVPD) because they remove layers of the silicon wafers and can thus be used for layer-by-layer measurements of impurity profiles below the surface of silicon wafers or measurement of bulk concentrations [176]. Another wet chemical layer by layer etching method called ultra-trace profiling (UTP) was developed by Hölzl et al. [177]. Shabani et al. also compared the collection efficiency of Cu by vapor phase decomposition (VPD) and drop etching (DE) methods using different solutions for recovery [175]. It seems that it is useful to add H_2O_2 to the HF solution for a marked increase of the collection efficiency.

A further measurement method which allows localized measurements of surface concentrations as well as depth profiling is time of flight secondary ion mass spectrometry (ToF-SIMS). For this method, the surface pre-treatment prior to the intentional contamination is important. It was found for copper contamination that the surface must be hydrophobic because at hydrophilic surfaces the metal contamination values measured are too low because the conversion factors for silicon are not valid in the environment of water and hydroxyl groups [89].

The third step is a procedure to drive-in the metal contamination. Two different drive-in methods where applied for intentional contamination. These are drive-in anneals [176–180] or polishing with spiked slurries [178, 181]. Drive-in anneals should be done at temperatures relevant for the process where the wafers with the internal gettering zone are used for. Most getter tests are applied to front end device processing and drive-in temperatures are usually in the range 800–900 °C. Hölzl et al. mainly used anneals at 800 °C for 30 min in Ar with a cooling rate of 50 K/min.

Fabry et al. reported an anneal at 900 °C for 30 min. Shabani et al. used drive-in anneals at 900 °C for 2 h cooled with 5 K/min to 800 °C and pulled the wafers out of the furnace.

During polishing with spiked slurries a reversed out-diffusion takes place [182]. During out-diffusion near-surface band bending repels Cu_i^+ from the surface to the bulk and slows down out-diffusion. However, during chemo-mechanical polishing near-surface band bending drives Cu into the wafer. This makes the kinetics of in-diffusion much faster than out-diffusion. When polishing is finished and the flux into the wafer stops, out-diffusion becomes dominant again [182].

The getter efficiency η is calculated from the initial metal concentration C_{met}^0 and the final metal contamination measured after drive-in C_{met}^{final} as follows:

$$\eta = \left(1 - \frac{C_{met}^{final}}{C_{met}^{0}}\right) \cdot 100\%. \tag{6.52}$$

The initial contamination can be measured either at the contaminated sample or at a reference wafer if the detection method is destructive. It also has to be taken into account whether the contamination was done on both the front and back surface or just on one of the surfaces. Same consideration has to be done for the final contamination value.

Copper has a special behavior in silicon which requires special getter tests. After drive-in annealing, it slowly returns to the surface even at room temperature. This effect was demonstrated by Shabani after removal of a surface oxide layer from a p-type silicon wafer [176]. This special behavior lead to the development of storage getter tests [89]. In such tests, the final measurement of the Cu surface concentration is carried out 7 days after the drive-in anneal. During this storage time, the Cu which was not gettered returns to the surface. Other tests which take this special behavior of Cu into account use up to 400 °C hot plates for 2 h [183, 184] or infrared irradiation heating at 300 °C for 3 h [185] to accelerate low temperature out-diffusion of Cu.

A very special getter test which is rather a qualitative assessment of the getter efficiency is the so-called haze test developed by Graff [186]. It is the oldest among the getter tests but it is very quick and inexpensive. A haze forming metal like Cu, Ni, Fe, or Pd is scratched on the back surface of the wafer. Then the wafer is annealed at about 1050 °C for several minutes followed by quick cooling and preferential etching of the front surface. During cooling the metals supersaturate and precipitate either at BMDs or if these are not available at the surface forming the so-called surface haze which becomes visible after preferential etching. No haze means high getter efficiency and haze formation indicates that the getter efficiency is too low. The haze test can be also applied semi-quantitatively. In this case, the haze forming etch pits are counted and a getter efficiency is calculated from the pit density at the sample and a reference sample without gettering defects [90, 187].

Now, the question is which contamination level is necessary in order to measure a getter efficiency being relevant for device processing. Nearly all the advanced getter test applications described in the beginning work with low contamination levels in the range 10^{11}–10^{13} cm^{-3} which are device relevant levels. The contamination level of the haze test is very high because the metal contamination is in the range of the solubility at very high temperature. For copper contamination, the results of the haze test were compared with the results of a ToF-SIMS based getter test and it turned out that the results are different [90]. The reason is that high concentrations of metals are much easier to getter because the supersaturation is higher and metal silicide precipitation occurs earlier that means already at higher temperature during cooling.

6.10.5 Criteria for Efficient Gettering of Metal Impurities

Based on the results of getter tests, criteria for efficient gettering of metal impurities by oxygen precipitates were proposed by Sueoka et al. and Hölzl et al. [134, 188]. The concept of Sueoka's criterion for Ni gettering is based on size and density of oxygen precipitates. The Ni contamination level was 5×10^{11} cm^{-2} and the size of the plate-like oxygen precipitates was calculated assuming diffusion limited growth and a ratio between diagonal length and thickness of 0.01. In Fig. 6.42, Sueokas's criterion is demonstrated. The critical size L_{cri} as a function of the precipitate density N_p was determined as follows:

$$L_{crit}(nm) = 2.0 \times 10^6 \cdot N_p^{-\frac{1}{2}} \left(cm^{-3}\right). \tag{6.53}$$

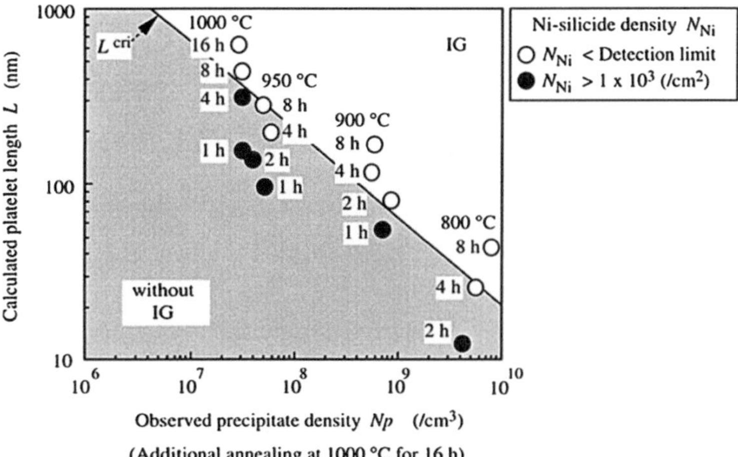

Fig. 6.42 Ni-silicide density as a function of precipitate density and size. The oxide precipitates have an internal gettering (IG) effect when their size is larger than the critical size L_{cri} (Reprinted with permission from [134]. Copyright 2000, The Electrochemical Society)

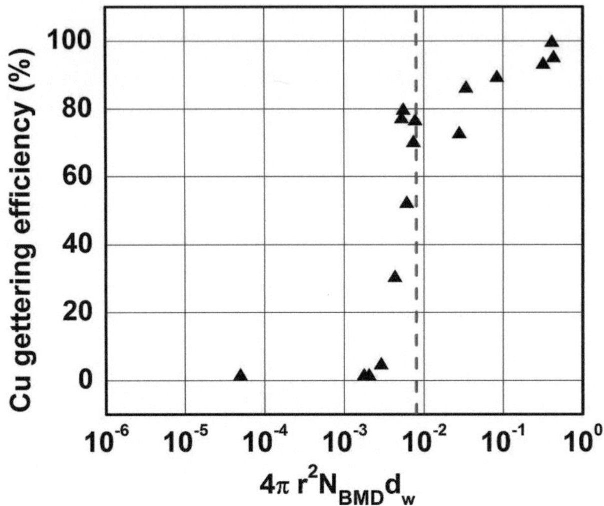

Fig. 6.43 Copper gettering efficiency as a function of the normalized inner surface of the oxygen precipitates. A threshold level of 8×10^{-3} must be obtained for an effective Cu gettering (Reprinted with permission from [188]. Copyright 2002, The Electrochemical Society)

Hölzl's criterion is based on the assumption that the total surface of the oxygen precipitates is proportional to the getter efficiency and he defined the normalized inner surface S as

$$S = 4\pi \cdot r^2 \cdot N_{BMD} \cdot d_w \text{ with } r = \left(2\frac{C_{Oi}-C_{Oi}^{eq}}{C_p}\right) \cdot (D_{Oi} \cdot t)^{0.5} \quad (6.54)$$

which can be calculated from the BMD density N_{BMD} and the thickness of the wafer d_w [188]. The model assumes spherical precipitates with the radius r to be calculated via diffusion limited growth. For both Ni and Cu, critical normalized inner surfaces for efficient gettering were found. These are 8×10^{-3} for Cu and 2×10^{-3} for Ni [188]. In Fig. 6.43, Hölzl's criterion is demonstrated.

6.11 Summary

With the knowledge provided to the reader of this chapter about oxygen precipitation she or he will be able to understand what is going on during thermal processing of silicon wafers with the interstitial oxygen and how it influences defect generation and device performance. This basic understanding is crucial for defect engineering in semiconductor manufacturing and crystal growth. The measurement and characterization techniques of interstitial oxygen and oxygen precipitates described in this chapter can be applied for defect control.

Acknowledgments The author would like to thank Dr. Dawid Kot very much for his engaged work in the investigation of oxygen precipitation during his work as a PhD student and as a post-doc at IHP. She also thanks her colleague Dr. Jaroslaw Dabrowski very much for this collaboration in the field of ab initio calculation. Many of the results presented here were obtained in common research projects between IHP and Siltronic AG and the author wants to thank all the colleagues from Siltronic AG for the fruitful collaboration. Especially acknowledged are Dr. Wilfried von Ammon, Dr. Andreas Sattler, Dr. Timo Müller, Dr. Ulrich Lambert, and Dr. Dieter Gräf. Many thanks also to Prof. Jan Vanhellemont for the inspiring discussions during his time at IMEC and Wacker Siltronic.

References

1. Hölzl, R., Huber, A., Fabry, L., Range, K.-J., Blietz, M.: Integrity of ultrathin gate oxides with different oxide thickness, substrate wafers and metallic contaminations. Appl. Phys. A: Mater. Sci. Process. **72**, 351 (2001)
2. Lawrence, J.E., Huff, H.R.: Silicon material properties for VLSI circuitry. In: Einspruch, N.G. (ed.) VLSI Electronics: Microstructure Science, vol. 5, pp. 51–102. Academic, New York (1982)
3. Tsuya, T.: Oxygen effect on electronic device performance. In: Shimura, F. (ed.) Semiconductors and Semimetals, vol. 42, pp. 619–667. Academic, New York (1994)
4. Shimura, F.: Oxygen in silicon. In: Shimura, F. (ed.) Semiconductors and Semimetals, vol. 42. Academic, San Diego (1994)
5. Mikkelsen Jr., J.C.: The diffusivity and solubility of oxygen in silicon. Mater. Res. Soc. Symp. Proc. **59**, 19 (1986)
6. Frewen, T.A., Kapur, S.S., Häckl, W., von Ammon, W., Sinno, T.: A microscopically accurate continuum model for void formation during semiconductor silicon processing. J. Cryst. Growth **279**, 258 (2005)
7. Åberg, D., Svensson, B.G., Hallberg, T., Lindström, J.L.: Kinetic study of oxygen dimer and thermal donor formation in silicon. Phys. Rev. B **58**, 12944 (1998)
8. Ramamoorthy, M., Pantelides, S.T.: Enhanced modes of oxygen diffusion in silicon. Solid State Commun. **106**, 243 (1998)
9. Snyder, L.C., Corbett, J.W., Deak, P., Wu, R.: On the diffusion of oxygen dimer in a silicon crystal. Mat. Res. Soc. Proc. **104**, 179 (1987)
10. Lee, S.-T., Fellinger, P., Chen, S.: Enhanced and wafer-dependent oxygen diffusion in CZ-Si at 500–700 °C. J. Appl. Phys. **63**, 1924 (1988)
11. McQuaid, S.A., Johnson, B.K., Gambaro, D., Falster, R., Ashwin, M.J., Tucker, J.H.: The conversion of isolated oxygen atoms to a fast diffusing species in Czochralski silicon at low temperatures. J. Appl. Phys. **86**, 1878 (1999)
12. Takeno, H., Hayamizu, Y., Miki, K.: Diffusivity of oxygen in Czochralski silicon at 400–750 °C. J. Appl. Phys. **84**, 3113 (1998)
13. Senkader, S., Wilshaw, P.R., Falster, R.J.: Oxygen-dislocation interactions in silicon at temperatures below 700 °C: dislocation locking and oxygen diffusion. J. Appl. Phys. **89**, 4803 (2001)
14. Stavola, M., Patel, J.R., Kimerling, L.C., Freeland, P.E.: Diffusivity of oxygen in silicon at the donor formation temperature. Appl. Phys. Lett. **42**, 73 (1983)
15. Kissinger, G., Dabrowski, J., Sattler, A., Müller, T., von Ammon, W.: Two paths of oxide precipitate nucleation in silicon. Solid State Phenom. **131–133**, 293 (2008)
16. Baghdadi, A., Bullis, W.M., Croarkin, M.C., Yue-zhen Li, Scace, R.I., Series, R.W., Stallhofer, P., Watanabe, M.: Interlaboratory determination of the calibration factor for the measurement of interstitial oxygen content if silicon by infrared absorption. J. Electrochem. Soc. **136**, 2015 (1989)

17. Vanhellemont, J., Claeys, C.: A theoretical study of the critical radius of precipitates and its application to silicon oxide in silicon. J. Appl. Phys. **62**, 3960 (1987), Erratum J. Appl. Phys. **71**, 1073 (1992)
18. Voronkov, V.V., Falster, R.: Nucleation of oxide precipitates in vacancy-containing silicon. J. Appl. Phys. **91**, 5802 (2002)
19. Vanhellemont, J.: Diffusion limited oxygen precipitation in silicon: precipitate growth kinetics and phase formation. J. Appl. Phys. **78**, 4297 (1995)
20. Nabarro, F.R.N.: The strain produced by precipitation in alloys. Proc. Roy. Soc. A **175**, 519 (1940)
21. Zschorsch, M., Hölzl, R., Rüfer, H., Möller, H.J., von Ammon, W.: Optimized parameters for modeling oxygen nucleation in silicon. Solid State Phenom. **95–96**, 71 (2004)
22. Wada, K., Inoue, N.: Thermal double donors in silicon. Electrochem. Soc. Proc. **86–4**, 778 (1986)
23. Becker, R., Döring, W.: Kinetische Behandlung der Keimbildung in übersättigten Dämpfen (German). Ann. Phys. **416**, 719 (1935)
24. Hu, S.M.: Growth law for disk precipitates, and oxygen precipitation in silicon. Appl. Phys. Lett. **48**, 115 (1986)
25. Wortman, J.J., Evans, R.A.: Young's modulus, shear modulus, and Poisson's ratio in silicon and germanium. J. Appl. Phys. **36**, 153 (1965)
26. Kissinger, G., Dabrowski, J.: Oxide precipitation via coherent seed-oxide phases. J. Electrochem. Soc. **155**, H448 (2008)
27. Kroupa, F.: Circular edge dislocation loop. Czechoslovak J. Phys. B **10**, 284 (1960)
28. Porter, D.A., Easterling, K.E.: Phase Transformations in Metals, 2nd edn, pp. 143–155. Nelson Thornes Ltd, Cheltenham (2001)
29. Bergholz, W.: Grown-in and process-induced defects. Semiconductors and semimetals. Oxygen in Silicon. **42**, 513 (1994)
30. Bergholz, W.: Grown-in and process-induced defects. In: Shimura, F. (ed.) Semiconductors and Semimetals, vol. 42, pp. 513–575. Academic, New York (1994)
31. Kissinger, G., Dabrowski, J., Kot, D., Akhmetov, V., Sattler, A., Von Ammon, W.: Modeling the early stages of oxygen agglomeration. J. Electrochem. Soc. **158**, H343 (2011)
32. Torres, V.J.B., Coutinho, J., Jones, R., Barroso, M., Öberg, S., Briddon, P.R.: Early SiO_2 precipitates in Si: vacancy-oxygen versus interstitial-oxygen clusters. Phys. B **376–377**, 109 (2006)
33. Sueoka, K.: Oxygen precipitation in lightly and heavily doped Czochralski silicon oxygen, nitrogen and hydrogen in silicon. ECS Trans. **3**(4), 71 (2006)
34. Pesola, M., Von Boehm, J., Mattila, T., Nieminen, R.M.: Computational study of interstitial oxygen and vacancy-oxygen complexes in silicon. Phys. Rev. B **60**, 11449 (1999)
35. Casali, R.A., Rücker, H., Methfessel, M.: Interaction of vacancies with interstitial oxygen in silicon. Appl. Phys. Lett. **78**, 913 (2001)
36. Akhmetov, V., Kissinger, G., von Ammon, W.: Interaction of oxygen with thermally induced vacancies in Czochralski silicon. Appl. Phys. Lett. **94**, 092105 (2009)
37. Kissinger, G., Dabrowski, J., Sattler, A., Seuring, C., Müller, T., Richter, H., von Ammon, W.: Analytical modeling of the interaction of vacancies and oxygen for oxide precipitation in RTA treated silicon wafers. J. Electrochem. Soc. **154**, H454 (2007)
38. Kissinger, G., Sattler, A., Dabrowski, J., von Ammon, W.: Verification of a method to detect grown-in oxide precipitate nuclei in Czochralski silicon. ECS Trans. **11**(3), 161 (2007)
39. Quemener, V., Raeissi, B., Herklotz, F., Murin, L.I., Monakhov, E.V., Svensson, B.G.: Kinetics study of vacancy-oxygen-related defects in monocrystalline solar silicon. Phys. Stat. Sol. B **251**, 2197 (2014)
40. Swaroop, R., Kim, N., Lin, W., Bullis, M., Shive, L., Rice, A., Castel, E., Christ, M.: Testing for oxygen precipitation in silicon wafers. Solid State Technol. **3**, 85–89 (1987)
41. Falster, R., Cornara, M., Gambaro, D., Olmo, M., Pagani, M.: Effect of high temperature pre-anneal on oxygen precipitates nucleation kinetics in Si. Solid State Phenom. **57–58**, 123 (1997)

42. Kissinger, G., Gräf, D., Lambert, U., Richter, H.: A method for studying the grown-in defect density spectra in Czochralski silicon wafers. J. Electrochem. Soc. **144**, 1447 (1997)
43. Falster, R., Pagani, M., Gambaro, D., Cornara, M., Olmo, M., Ferrero, G., Pichler, P., Jacob, M.: Vacancy-assisted oxygen precipitation phenomena in Si. Solid State Phenom. **57–58**, 129 (1997)
44. Kelton, K.F., Falster, R., Gambaro, D., Olmo, M., Cornara, M., Wei, P.F.: Oxygen precipitation in silicon: experimental studies and theoretical investigations within the classical theory of nucleation. J. Appl. Phys. **85**, 8097 (1999)
45. Kissinger, G., Kot, D., Dabrowski, J., Akhmetov, V., Sattler, A., von Ammon, W.: Analysis of the nucleation kinetics of oxide precipitates in Czochralski silicon. ECS Trans. **16**(6), 97 (2008)
46. Wei, P.F., Kelton, K.F., Falster, R.: Coupled-flux nucleation modeling of oxygen precipitation in silicon. J. Appl. Phys. **88**, 5062 (2000)
47. Esfandyari, J., Vanhellemont, J., Obermeier, G.: Computer simulation of oxygen precipitation by considering a temperature dependent interfacial energy. Electrochem. Soc. Proc. **99–1**, 437 (1999)
48. Schrems, M.: Simulation of oxygen precipitation. In: Shimura, F. (ed.) Semiconductors and Semimetals, vol. 42, pp. 391–447. Academic, New York (1994)
49. Trzynadlowski, B.C., Dunham, S.T.: A reduced moment-based model for oxygen precipitation in silicon. J. Appl. Phys. **114**, 243508 (2013)
50. Senkader, S., Hobler, G., Schmeiser, C.: Determination of the oxide-precipitate-silicon-matrix interface energy by considering the change of precipitate morphology. Appl. Phys. Lett. **69**, 2202 (1996)
51. Kot, D., Kissinger, G., Schubert, M.A., Sattler, A.: Morphology of oxygen precipitates in RTA pre-treated Czochralski silicon wafers investigated by FTIR spectroscopy and STEM. ECS J. Solid State Sci. Technol. **3**(11), P370 (2014)
52. Aoki, S.: Morphology of oxide precipitates in silicon crystals. Mater. Trans. JIM **34**, 746 (1993)
53. Fujimori, H.: Dependence on morphology of oxygen precipitates upon oxygen supersaturation in Czochralski silicon crystals. J. Electrochem. Soc. **144**, 3180 (1997)
54. Wang, Z., Brown, R.A.: Simulation of oxide formation and point defect dynamics in silicon: the role of oxide morphology. In: Claeys, C.L., Watanabe, M., Rai-Choudhury, P., Stallhofer, P. (eds.) Proceedings semiconductor silicon 2002, vol. 2002-20, p. 49. The Electrochemical Society, Pennington (2002)
55. Kolbesen, B.O.: Defect delineation in silicon materials by chemical etching techniques. In: Kissinger, G., Pizzini, S. (eds.) Silicon, Germanium and Their Alloys, Growth, Defects, Impurities, and Nanocrystals, pp. 289–322. CRC Press, Boca Raton/London/New York (2015)
56. Kulkarni, M.S., Libbert, J., Keltner, S., Mulestagno, L.: A theoretical and experimental analysis of macrodecoration of defects in monocrystalline silicon. J. Electrochem. Soc. **149**, G153 (2002)
57. Kolbesen, B.O., Possner, D., Mähliß, J.: Delineation of crystalline defects in semiconductor substrates and thin films by chemical etching techniques. ECS Trans. **11**, 195–206 (2007)
58. Secco d'Aragona, F.: Dislocation etch for (100) planes in silicon. J. Electrochem. Soc. **119**, 948 (1972)
59. Wright Jenkins, M.: A new preferential etch for defects in silicon crystals. J. Electrochem. Soc. **124**, 757 (1977)
60. Yang, K.H.: An etch for delineation of defects in silicon. J. Electrochem. Soc. **131**, 1140 (1984)
61. Sirtl, E., Adler, A.: Z. f. Metallkunde **52**, 529 (1961)
62. Schimmel, D.: Defect etch for <100> silicon evaluation. J. Electrochem. Soc. **126**, 479 (1979)
63. Chandler, T.C.: MEMC etch—A chromium trioxide-free etchant for delineating dislocations and slip in silicon. J. Electrochem. Soc. **137**, 944 (1990)

64. JEITA EM-3603 E 3.0
65. Saito, Y., Matsushita, Y.: European Patent EP 0281115B1, 20 Sept 1994
66. Possner, D., Kolbesen, B.O., Cerva, H., Klüppel, V.: Organic Peracid Etches: a new class of chromium free etch solutions for the delineation of defects in different semiconducting materials. ECS Trans. **10**(1), 21 (2007)
67. Abbadie, A., Bedell, S.W., Hartmann, J.M., Sadana, D.K., Brunier, F., Figuet, C., Cayrefourcq, I.: Study of HCl and secco defect etching for characterization of thick sSOI. J. Electrochem. Soc. **154**(8), H713 (2007)
68. Bogenschütz, A.F.: Ätzpraxis für Halbleiter. Hanser, Munich (1967)
69. Rozgonyi, G.A.: In: Mahajan, S. (ed.) Encyclopedia of Materials: Science and Technology, pp. 8524–8533. Elsevier Science Ltd, Amsterdam (2001)
70. Furukawa, J., Furuya, H.: Annealing behavior of a light scattering tomography detecting defect near the surface of Si wafers. Jpn. J. Appl. Phys. **34**, L156 (1995)
71. Moriya, K., Ogawa, T.: Observation of lattice defects in GaAs and heat-treated Si crystals by infrared light scattering tomography. Jpn. J. Appl. Phys. **22**, L207 (1983)
72. Moriya, K., Kashima, K., Takasu, S.: Development of a bulk microdefect analyzer for Si wafers. J. Appl. Phys. **66**, 5267 (1989)
73. Ogawa, T.: Dislocation lines in indium-doped GaAs crystals observed by infrared light scattering tomography of about 1 μm wavelength radiation. J. Cryst. Growth **88**, 332 (1988)
74. Taijing, L., Toyoda, K., Nango, N., Ogawa, T.: Observation of microdefects and microprecipitates in Si crystals by IR scattering tomography. J. Cryst. Growth **108**, 482 (1991)
75. Kissinger, G., Vanhellemont, J., Claeys, C., Richter, H.: Observation of stacking faults and prismatic punching systems in silicon by light scattering tomography. J. Cryst. Growth **158**, 191 (1996)
76. Borghesi, A., Sassela, A., Geranzani, P., Porrini, M., Pivac, B.: Infrared characterization of oxygen precipitates in silicon wafers with different concentrations of interstitial oxygen. Mater. Sci. Eng. B **73**, 145 (2000)
77. Hallberg, T., Lindström, J.L.: Enhanced oxygen precipitation in electron irradiated silicon. J. Appl. Phys. **72**, 5130 (1992)
78. Sassella, A., Borghesi, A., Garanzani, P., Borionetti, G.: Infrared response of oxygen precipitates in silicon: experimental and simulated spectra. Appl. Phys. Lett. **75**, 1131 (1999)
79. Kot, D., Kissinger, G., Schubert, M.A., Sattler, A.: Influence of RTA pre-treatment on the morphology of oxygen precipitates in Czochralski silicon wafers observed by FTIR spectroscopy and STEM. Proceedings Forum of the Science and Technology of Silicon Materials, The 145th Committee of the JSPS (Japanese Society for the Promotion of Science), Hamamatsu, 19–22 Oct 2014, p. 81 (2014)
80. Hu, S.M.: Infrared absorption spectra of SiO_2 precipitates of various shapes in silicon: calculated and experimental. J. Appl. Phys. **51**, 5945 (1980)
81. Genzel, L., Martin, T.P.: Infrared absorption in small ionic crystals. Phys. Stat. Sol. B **51**, 91 (1972)
82. Genzel, L., Martin, T.P.: Infrared absorption by surface phonons and surface plasmons in small crystals. Surf. Sci. **34**, 33 (1973)
83. Štoudek, R., Humlíček, J.: Infrared spectroscopy of oxygen interstitials and precipitates in nitrogen-doped silicon. Phys. B: Condens. Matter. **376–377**, 150 (2006)
84. Philip, H.R.: Part II Critiques Subpart 3 Insulators. In: Palik, E.D. (ed.) Handbook of Optical Constants of Solids, p. 749. Academic, San Diego (1985)
85. Henning, T., Mutchke, H.: Low-temperature infrared properties of cosmic dust analogues. Astron. Astrophys. **327**, 743–754 (1997)
86. Zolotarev, V.M.: Optical constants of amorphous SiO_2 and GeO_2 in the region of valence band. Optika Spectrosc. **29**, 66 (1970)
87. Tsu, D.V., Lucovsky, G., Davidson, B.N.: Effects of the nearest neighbors and the alloy matrix on SiH stretching vibrations in the amorphous SiOr:H (0<r<2) alloy system. Phys. Rev. B **40**, 1795 (1989)

88. Borghesi, A., Piaggi, A., Sassella, A., Stella, A., Pivac, B.: Infrared study of oxygen precipitate composition in silicon. Phys. Rev. B **46**, 4123 (1992)
89. Kot, D., Kissinger, G., Sattler, A., Müller, T.: Development of a storage getter test for Cu contaminations in silicon wafers based on ToF-SIMS measurements. Acta Phys. Pol. A **125**, 965 (2014)
90. Kot, D., Kissinger, G., Schubert, M.A., Sattler, A., Müller, T.: Influence of Cu concentration on the getter efficiency of dislocations and oxygen precipitates in silicon wafers. Solid State Phenom. **205–206**, 278 (2014)
91. Meduňa, M., Caha, O., Buršík, J.: Studies of influence of high temperature preannealing on oxygen precipitation in CZ Si wafers. J. Cryst. Growth **348**, 53 (2012)
92. De Gryse, O., Clauws, P., Van Landuyt, J., Lebedev, O., Claeys, C., Simoen, E., Vanhellemont, J.: Oxide phase determination in silicon using infrared spectroscopy and transmission electron microscopy techniques. J. Appl. Phys. **91**, 2493 (2002)
93. Nicolai, J., Burle, N., Pichaud, B.: Determination of silicon oxide precipitate stoichiometry using global and local techniques. J. Cryst. Growth **363**, 93 (2013)
94. Kot, D., Kissinger, G., Schubert, M.A., Klingsporn, M., Huber, A., Sattler, A.: Composition of oxygen precipitates in Czochralski silicon wafers investigated by STEM with EDX/EELS and FTIR spectroscopy. Phys. Status Solid RRL **9**, 405 (2015)
95. Kissinger, G., Gräf, D., Lambert, U., Grabolla, T., Richter, H.: Key influence of the thermal history on process-induced defects in Czochralski silicon wafers. Semicond. Sci. Technol. **12**, 933 (1997)
96. Kissinger, G., Gräf, D., Vanhellemont, J., Lambert, U., Richter, H.: The role of grown-in defects in advanced silicon technology. Solid State Phenom. **57–58**, 337 (1997)
97. Kissinger, G., Vanhellemont, J., Morgenstern, G., Blietz, M., Tittelbach-Helmrich, K., Obermeier, G., Wahlich, R.: Influence of boron doping on oxide precipitate nucleation and gettering of iron impurities in low thermal budget processing of Czochralski silicon. Electrochem. Soc. Proc. **99–1**, 268 (1999)
98. Fuller, C.S., Logan, R.A.: Effect of heat treatment upon the electrical properties of silicon crystals. J. Appl. Phys. **28**, 1427 (1957)
99. Benton, J.L., Kimerling, L.C., Stavola, M.: The oxygen related donor effect in silicon. Physica B **116**, 271 (1983)
100. Wruck, D., Gaworzewski, P.: Electrical and infrared spectroscopic investigations of oxygen-related donors in silicon. Phys. Stat. Sol. (a) **56**, 557 (1979)
101. Stein, H.J., Hahn, S.K., Shatas, S.C.: Rapid thermal annealing and regrowth of thermal donors in silicon. J. Appl. Phys. **59**, 3495 (1986)
102. Tokuda, Y., Kobayashi, N., Usami, A., Inoue, Y., Imura, M.: Thermal donor annihilation and defect production in n-type silicon by rapid thermal annealing. J. Appl. Phys. **66**, 3651 (1989)
103. Schmalz, K., Gaworzewski, P.: On the donor activity of oxygen in silicon at temperatures from 500 to 800 °C. Phys. Stat. Sol. (a) **64**, 151 (1981)
104. Cazcarra, V., Zunino, P.: Influence of oxygen on silicon resistivity. J. Appl. Phys. **51**, 4206 (1980)
105. Coutinho, J., Jones, R., Murin, L.I., Markevich, V.P., Lindström, J.L., Öberg, S., Briddon, P.R.: Thermal double donors and quantum dots. Phys. Rev. Lett. **87**, 235501 (2001)
106. Pesola, M., Lee, Y.J., von Boehm, J., Kaukonen, M., Nieminen, R.M.: Structures of thermal double donors in silicon. Phys. Rev. Lett. **84**, 5343 (2000)
107. Pensl, G., Schulz, M., Hölzlein, K., Bergholz, W., Hutchison, J.L.: New oxygen donors in silicon. Appl. Phys. A **48**, 49 (1989)
108. Wagner, P., Hage, J.: Thermal double donors in silicon. Appl. Phys. A **49**, 123 (1989)
109. Voronkov, V.V., Voronkova, G.I., Batunina, A.V., Falster, R., Golovina, V.N., Guliaeva, A.S., Tiurina, N.B., Milvidski, M.G.: The sensitivity of thermal donor generation in silicon to self-interstitial sinks. J. Electrochem. Soc. **147**, 3899 (2000)
110. Voronkov, V.V., Voronkova, G.I., Batunina, A.V., Falster, R., Golovina, V.N., Guliaeva, A.S., Tiurina, N.B., Milvidski, M.G.: Evolution of thermal donors in silicon enhanced by self-interstitials. Solid State Phenom. **131–133**, 387 (2008)

111. Tajima, M., Warashina, M., Takeno, H., Abe, T.: Effect of point defects on oxygen aggregation in Si at 450 °C. Appl. Phys. Lett. **65**, 222 (1994)
112. Kot, D., Mchedlidze, T., Kissinger, G., von Ammon, W.: Characterization of deep levels introduced by RTA and by subsequent anneals in n-type silicon. ECS J. Solid State Sci. Technol. **2**, P9 (2013)
113. Cadeo, S., Pizzini, S., Acciarri, M., Cavallini, A.: Oxygen precipitate precursors and low temperature gettering processes. I. Segregation of oxygen and thermal donor generation in the 600–850 °C range. Mat. Sci. Semicond. Proc. **2**, 57 (1999)
114. Tokuda, Y., Shimokata, T., Inoue, Y., Usami, A., Imura, M.: Donor formation at 650 degrees C in oxygen-rich silicon after rapid thermal annealing of thermal donors. Semicond. Sci. Technol. **6**, 66 (1991)
115. Götz, W., Pensl, G., Zulehner, W.: Observation of five additional thermal donor species TD12 to TD16 and of regrowth of thermal donors at initial stages of the new oxygen donor formation in Czochralski-grown silicon. Phys. Rev. B **46**, 4312 (1992)
116. McQuaid, S.A., Binns, M.J., Londos, C.A., Tucker, J.H., Brown, A.R., Newman, R.C.: Oxygen loss during thermal donor formation in Czochralski silicon: new insights into oxygen diffusion mechanisms. J. Appl. Phys. **77**, 1427 (1995)
117. Kaiser, W., Frisch, H.L., Reiss, H.: Mechanism of the formation of donor states in heat-treated silicon. Phys. Rev. **112**, 1546 (1958)
118. Oehrlein, G.S.: Silicon–oxygen complexes containing three oxygen atoms as the dominant thermal donor species in heat-treated oxygen-containing silicon. J. Appl. Phys. **54**, 5453 (1983)
119. Wada, K.: United model for formation kinetics of oxygen thermal donors in silicon. Phys. Rev. B **30**, 5884 (1984)
120. Wijaranakula, W.: Formation kinetics of oxygen thermal donors in silicon. Appl. Phys. Lett. **59**, 1608 (1991)
121. Ourmazd, A., Schröter, W., Bourret, A.: Oxygen-related thermal donors in silicon: a new structural and kinetic model. J. Appl. Phys. **56**, 1670 (1984)
122. Lee, Y.J., von Boehm, J., Nieminen, R.M.: Interstitial oxygen loss and the formation of thermal double donors in Si. Appl. Phys. Lett. **79**, 1453 (2001)
123. Lee, Y.J., Von Boehm, J., Pesola, M., Nieminen, R.M.: Aggregation kinetics of thermal double donors in Silicon. Phys. Rev. Lett. **86**, 3060 (2001)
124. Mao, B.-Y., Lagowski, J., Gatos, H.C.: Kinetics of thermal donor generation in silicon. J. Appl. Phys. **56**, 2729 (1984)
125. Hallberg, T., Lindström, J.L.: Infrared vibrational bands related to the thermal donors in silicon. J. Appl. Phys. **79**, 7570 (1996)
126. Murin, L.I., Lindström, J.L., Markevich, V.P., Misiuk, A., Londos, C.A.: Thermal double donor annihilation and oxygen precipitation at around 650 °C in Czochralski-grown Si: local vibrational mode studies. J. Phys. Condens. Matter **17**, S2237 (2005)
127. Claybourn, M., Newman, R.C.: Thermal donor formation and the loss of oxygen from solution in silicon heated at 450 °C. Appl. Phys. Lett. **52**, 2139 (1988)
128. Tokuda, Y., Katayama, M., Hattori, T.: Depth profiles of thermal donors formed at 450 degrees C in oxygen-rich n-type silicon. Semicond. Sci. Technol. **8**, 163 (1993)
129. Hölzlein, K., Pensl, G., Schulz, M.: Trap spectrum of the "new oxygen donor" in silicon. Appl. Phys. A **34**, 155 (1984)
130. Kamiura, Y., Hashimoto, F., Yoneta, M.: A new family of thermal donors generated around 450 °C in phosphorus-doped Czochralski silicon. J. Appl. Phys. **65**, 600 (1989)
131. Sueoka, K., Akatsuka, M., Yonemura, M., Ono, T., Asayama, E., Katahama, H.: Effect of heavy boron doping on oxygen precipitation in Czochralski silicon substrates of epitaxial wafers. J. Electrochem. Soc. **147**, 756 (2000)
132. Murphy, J.D., Wilshaw, P.R., Pygall, B.C., Senkader, S.: Enhanced oxygen diffusion in highly doped p-type Czochralski silicon. J. Appl. Phys. **100**, 103531 (2006)
133. Sugimura, W., Ono, T., Umeno, S., Hourai, M., Sueoka, K.: Defect formation behaviors in heavily doped Czochralski silicon 300 mm. ECS Trans. **2**(2), 95 (2006)

134. Sueoka, K., Sadamitsu, S., Koike, Y., Kihara, T., Katahama, H.: Internal gettering for Ni contamination in Czochralski silicon wafers. J. Electrochem. Soc. **147**, 3074 (2000)
135. Takeno, H., Aihara, K., Hayamizu, Y., Kitagawara, Y.: Influence of heavy boron doping on oxygen precipitation characteristics of Czochralski silicon crystals and its computer simulation. Electrochem. Soc. Proc. **98–1**, 1012 (1998)
136. Wijaranakula, W.: Oxygen precipitation and defects in heavily doped Czochralski silicon. J. Appl. Phys. **72**, 2713 (1992)
137. Ono, T., Asayama, E., Horie, H., Hourai, M., Sano, M., Tsuya, H., Nakai, K.: Behavior of defects in heavily boron doped Czochralski silicon. Jpn. J. Appl. Phys. **36**, L249 (1997)
138. Yonemura, M., Sueoka, K., Kamei, K.: Lattice strain around platelet oxide precipitates in C- and N-doped silicon epitaxial wafers. J. Electrochem. Soc. **148**, G630 (2001)
139. Newman, R.C., Tucker, J.H., Brown, A.R., McQuaid, S.A.: Hydrogen diffusion and the catalysis of enhanced oxygen diffusion in silicon at temperatures below 500 °C. J. Appl. Phys. **70**, 3061 (1991)
140. Stein, H.J., Hahn, S.: Hydrogen introduction and hydrogen-enhanced thermal donor formation in silicon. J. Appl. Phys. **75**, 3477 (1994)
141. Tsetseries, L., Wang, S., Pantelides, S.T.: Thermal donor formation processes in silicon and the catalytic role of hydrogen. Appl. Phys. Lett. **88**, 051916 (2006)
142. Simoen, E., Huang, Y.L., Ma, Y., Lauwaert, J., Clauws, P., Rafi, J.M., Ulyashin, A., Claeys, C.: J. Electrochem. Soc. **156**, H434 (2009)
143. Hara, A., Koizuka, M., Aoki, M., Fukuda, T., Yamada-Kaneta, H., Mori, H.: Influence of grown-in hydrogen on thermal donor formation and oxygen precipitation in Czochralski silicon crystals. Jpn. J. Appl. Phys. **33**, 5577 (1994)
144. Hara, A., Aoki, M., Fukuda, T., Ohsawa, A.: Hydrogen effects on oxygen precipitation in Czochralski silicon crystals. J. Appl. Phys. **74**, 913 (1993)
145. Nakai, K., Inoue, Y., Yokota, H., Ikari, A., Takahashi, J., Tachikawa, A., Kitahara, K., Ohta, Y., Ohashi, W.: Oxygen precipitation in nitrogen-doped Czochralski-grown silicon crystals. J. Appl. Phys. **89**, 4301 (2001)
146. Aihara, K., Takeno, H., Hayamizu, Y., Tamatsuka, M., Masui, T.: Enhanced nucleation of oxide precipitates during Czochralski silicon crystal growth with nitrogen doping. J. Appl. Phys. **88**, 3705 (2000)
147. Kissinger, G., Müller, T., Sattler, A., Häckl, W., Weber, M., Lambert, U., Huber, A., Krottenthaler, P., Richter, H., von Ammon, W.: Oxygen precipitation in nitrogen doped CZ silicon. Solid State Phenom. **108–109**, 17 (2005)
148. Karoui, A., Sahtout Karoui, F., Kvit, A., Rozgonyi, G.A.: Role of nitrogen related complexes in the formation of defects in silicon. Appl. Phys. Lett. **80**, 2114 (2002)
149. Karoui, A., Rozgonyi, G.A.: Oxygen precipitation in nitrogen doped Czochralski silicon wafers. II. Effects of nitrogen and oxygen coupling. J. Appl. Phys. **96**, 3264 (2004)
150. Rozgonyi, G.A., Karoui, A., Kvit, A., Duscher, G.: Nano-scale analysis of precipitates in nitrogen-doped Czochralski silicon. Microel. Eng. **66**, 305 (2003)
151. Yang, D., Yu, X.: Nitrogen in silicon. Defect Diffus. Forum **230–232**, 199 (2004)
152. Yu, X., Yang, D., Ma, X., Yang, J., Li, L., Que, D.: Grown-in defects in nitrogen-doped Czochralski silicon. J. Appl. Phys. **92**, 188 (2002)
153. von Ammon, W., Dreier, P., Hensel, W., Lambert, U., Köster, L.: Influence of oxygen and nitrogen on point defect aggregation in silicon single crystals. Mat. Sci. Eng. B **36**, 33 (1996)
154. Kissinger, G., Raming, G., Wahlich, R., Müller, T.: 300 mm Czochralski silicon wafers optimized with respect to voids with laterally homogeneous oxygen precipitation. Phys. B **407**, 2993 (2012)
155. Izunome, K.: In: Proceedings of the 6th International Symposium on Advanced Science and Technology of Silicon Materials (JSPS Symposium), Kona, 19–23 Nov 2012, pp. 9–13 (2012)
156. Kissinger, G., Vanhellemont, J., Lambert, U., Dornberger, E., Sorge, R., Morgenstern, G., Grabolla, T., Gräf, D., von Ammon, W., Wagner, P., Richter, H.: Curriculum vitae of oxide precipitates: from nucleation during crystal growth to their final destination in processed wafers. Electrochem. Soc. Proc.. **98–1**, 1095 (1998)

157. Kissinger, G., Grabolla, T., Morgenstern, G., Richter, H., Gräf, D., Vanhellemont, J., Lambert, U., von Ammon, W.: Grown-in oxide precipitate nuclei in Czochralski silicon substrates and their role in device processing. J. Electrochem. Soc. **146**, 1971 (1999)
158. Rozgonyi, G.A., Deysher, R.P., Pearce, C.W.: The identification, annihilation, and suppression of nucleation sites responsible for silicon epitaxial stacking faults. J. Electrochem. Soc. **123**, 1910 (1976)
159. Tan, T.Y., Gardner, E.E., Tice, W.K.: Intrinsic gettering by oxide precipitate induced dislocations in Czochralski Si. Appl. Phys. Lett. **30**, 175 (1977)
160. Richter, H.: Gettering in the silicon device technology - an overview. In: Proceedings 1st International Autumn School Gettering and Defect Engineering in the Semiconductor Technology (GADEST), 8–18 Oct 1985, GDR, Garzau, p. 1 (1985)
161. Takasu, S.: VLSI science and technology/1984. In: Bean, K.E., Rozgonyi, G.A. (eds.) The Electrochemical Society Proceedings, vol. 84–7, pp. 490 The Electrochemical Society, Pennington (1984)
162. Myers, S.M., Seibt, M., Schröter, W.: Mechanisms of transition-metal gettering in silicon. J. Appl. Phys. **88**, 3795 (2000)
163. Kissinger, G., Kot, D., Klingsporn, M., Schubert, M.A., Sattler, A., Müller, T.: Investigation of the copper gettering mechanism of oxide precipitates in silicon. ECS J. Solid State Sci. Technol. **4**(9), N124–N129 (2015)
164. Rozgonyi, G.A., Pearce, C.W.: Gettering of surface and bulk impurities in Czochralski silicon wafers. Appl. Phys. Lett. **32**, 747 (1978)
165. Borghesi, A., Pivac, B., Sassella, A., Stella, A.: Oxygen precipitation in silicon. J. Appl. Phys. **77**, 4169 (1995)
166. Hirano, Y., Yamazaki, K., Inoue, F., Imaoka, K., Tanahashi, K., Yamada-Kaneta, H.: Impact of defects in silicon substrate on flash memory characteristics. J. Electrochem. Soc. **154**, H1027 (2007)
167. Lanzerath, F., Buca, D., Trinkaus, H., Goryll, M., Mantl, S., Knoch, J., Breuer, U., Skorupa, W., Ghyselen, B.: Boron activation and diffusion in silicon and strained silicon-on-insulator by rapid thermal and flash lamp annealings. J. Appl. Phys. **104**, 044908 (2008)
168. Smith, M., McMahon, R.A., Voelskow, M., Skorupa, W.: Modeling and regrowth mechanisms of flash lamp processing of SiC-on-silicon heterostructures. J. Appl. Phys. **96**, 4843 (2004)
169. Kissinger, G., Kot, D., von Ammon, W.: Comparison of the impact of thermal treatments on the second and on the millisecond scales on the precipitation of interstitial oxygen. ECS J. Solid State Sci. Technol. **1**, P269 (2012)
170. Kissinger, G., Vanhellemont, J., Obermeier, G., Esfandyari, J.: Denuded zone formation by conventional and rapid thermal anneals. Mat. Sci. Eng. B **73**, 106 (2000)
171. Akatsuka, M., Okui, M., Morimoto, N., Sueoka, K.: Effect of rapid thermal annealing on oxygen precipitation behavior in silicon wafers. Jpn. J. Appl. Phys. **40**, 3055 (2001)
172. Frenkel, J.: Kinetic Theory of Liquids. Oxford University Press, Oxford (1946)
173. Müller, T., Kissinger, G., Krottenthaler, P., Seuring, C., Wahlich, R., von Ammon, W.: Precipitation enhancement of "so-called" defect-free Czochralski silicon material. Solid State Phenom. **108–109**, 11 (2005)
174. Park, J.-G., Lee, G.-S., Lee, J.-S., Kurita, K., Furuya, H.: Extremely proximity gettering for semiconductor devices. Mat. Sci. Eng. B **134**, 249 (2006)
175. Shabani, M.B., Shiina, Y., Kirscht, F.G., Shimanuki, Y.: Recent advanced applications of AAS and ICP-MS in the semiconductor industry. Mat. Sci. Eng. B **102**, 238 (2003)
176. Shabani, M.B., Yoshimi, T., Abe, H.: Low-temperature out-diffusion of Cu from silicon wafers. J. Electrochem. Soc. **143**, 2025 (1996)
177. Hölzl, R., Fabry, L., Range, K.-J., Pech, R.: MeV-boron implanted layer, oxygen precipitates and poly-silicon back side combined in one silicon wafer: at what defect will Cu and Ni be gettered? Appl. Phys. A **74**, 545 (2002)
178. Fabry, L., Hölzl, R., Andrukhiv, A., Matsumoto, K., Qiu, J., Koveshnikov, S., Goldstein, M., Grabau, A., Horie, H., Takeda, R.: Test methods for measuring bulk copper and nickel in heavily doped p-type silicon wafers. J. Electrochem. Soc. **153**, G566 (2006)

179. Hölzl, R., Range, K.J., Fabry, L.: Comparison of different gettering techniques for Cu-p+ versus polysilicon and oxygen precipitates. Appl. Phys. A **75**, 591 (2002)
180. Shabani, M.B., Okuuchi, S., Yoshimi, T., Shingyoji, T., Kirscht, F.G.: Effect of dopants and oxygen precipitation on low-temperature out-diffusion and gettering of Cu in silicon wafer. In: Claeys, C.L., Rai-Choudhury, P., Watanabe, M., Stallhofer, P., Dawson, H.J. (eds.) Proceedings high purity silicon V, vol. 98–13, p. 313. The Electrochemical Society, Pennington (1998)
181. Kim, K.-S., Lee, S.-W., Kang, H.-B., Lee, B.-Y., Park, S.-M.: Quantitative evaluation of gettering efficiencies below 1 × 1012 atoms/cm^3 in p-type silicon using a #2#1 tracer. J. Electrochem. Soc. **155**, H912 (2008)
182. Istratov, A.A., Weber, E.R.: Physics of copper in silicon. J. Electrochem. Soc. **149**, G21 (2002)
183. Hozawa, K., Isomae, S., Yugami, J.: Copper distribution near a SiO_2/Si interface under low-temperature annealing. Jpn. J. Appl. Phys. **41**, 5887 (2002)
184. Hozawa, K., Yugami, J.: Copper diffusion behavior in SiO_2/Si structure during 400 °C annealing. Jpn. J. Appl. Phys. **43**, 1 (2004)
185. Lee, S.-W., Kim, Y.-H., Kim, K.-S., Hong, B.-S., Lee, B.-Y.: Understanding the behaviors of Cu during a post-gate-oxidation device process by using an isotope tracking analysis. J. Korean Phys. Soc. **48**(6), 1548 (2006)
186. Graff, K.: Metal Impurities in Silicon-Device Fabrication. Springer Series Material Science, vol. 24. Springer-Verlag Berlin Heidelberg (1995)
187. Seacrist, M., Stinson, M., Libbert, J., Standley, R., Bins, J.: Determination of minimum oxygen precipitate growth conditions for gettering of copper and nickel. In: Huff, H.R., Fabry, L., Kishino, S, (eds.) Semiconductor silicon 2002, vol. 2002-2, p. 638. The Electrochemical Society, Pennington (2002)
188. Hölzl, R., Blietz, M., Fabry, L., Schmolke, R.: Gettering efficiencies and their dependence on material parameters and thermal processes: how can this be modeled. In: Huff, H.R., Fabry, L., Kishino, S. (eds.) Proceedings semiconductor silicon 2002, vol. 2002-2, p. 608. The Electrochemical Society, Pennington (2002)

Chapter 7
Defect Characterization in Silicon by Electron-Beam-Induced Current and Cathodoluminescence Techniques

Takashi Sekiguchi and Jun Chen

Abstract This chapter describes electron-beam-induced current (EBIC) and cathodoluminescence (CL) techniques, which have been used for the electrical/optical characterization of extended defects in Si. For these purposes, we use a scanning electron microscope (SEM) for electron beam irradiation. The electric current induced at the internal circuit and light emission from the specimen are used for the imaging of EBIC and CL, respectively. Using these techniques, we have succeeded in classifying the dislocations and grain boundaries (GBs). It is found that the clean dislocations are not so electrically active, but become active after metallic decoration. Large-angle (LA) GBs behave like dislocations. The coherency of GBs and the degree of contamination are the major factors determining the electrical activity of LA-GBs. Small-angle (SA) GBs are different from the former because they have certain carrier recombination activities at room temperature. We may attribute these activities to the bundle of dislocations at the GB plane. Due to such dislocation bundles, SA-GBs emit D-lines and are distinguished in the D-line imaging in CL mode. The SA-GBs are classified by D-lines according to the character and misorientation angle. Now, EBIC/CLs have been extensively used for multicrystalline Si for photovoltaic applications.

Keywords EBIC • Cathodoluminescence • Dislocation • Grain boundary

7.1 Introduction

In Si technology, extended defects, such as dislocations and grain boundaries, have been regarded as the classical defects because single crystalline Si wafers are now free from such extended defects. In most cases, we do not have to worry about

T. Sekiguchi (✉) • J. Chen
MANA Nanoelectromaterials Unit, National Institute for Materials Science, 1-1 Namiki, Tsukuba 305-0044, Japan
e-mail: SEKIGUCHI.Takashi@nims.go.jp

Y. Yoshida, G. Langouche (eds.), *Defects and Impurities in Silicon Materials*, Lecture Notes in Physics 916, DOI 10.1007/978-4-431-55800-2_7

these extended defects. In the early 2000s, however, Si solar cell technology has become important to solve the environmental problem. Since the majority of solar cells are made of multicrystalline (mc-) Si, the control of extended defects which suppress the conversion efficiency has come in the highlight again. It is known that such extended defects behave differently according to their morphology and metallic impurity decoration. Thus, special characterization techniques are indispensable for this study. Electron-beam-induced current (EBIC) and cathodoluminescence are suitable techniques because they can reveal the electrical and optical properties of semiconductor specimens on the scale of microns. This chapter describes the principles of these techniques and gives results about dislocations and grain boundaries in Si.

7.2 Principles

7.2.1 Electron Beam Excitation

Let us imagine the observation of a Si specimen using scanning electron microscopy (SEM). The Si specimen is irradiated with an electron beam of several kV to several tens of kV focused in a diameter of several tens of nm. As a result, various excitation signals appear from the specimen [1–3]. Figure 7.1 shows this situation. Reflected electrons (REs) and secondary electrons (SEs) are the electrons emitted from the specimen. X-rays or sometimes light (cathodoluminescence; CL) are also emitted

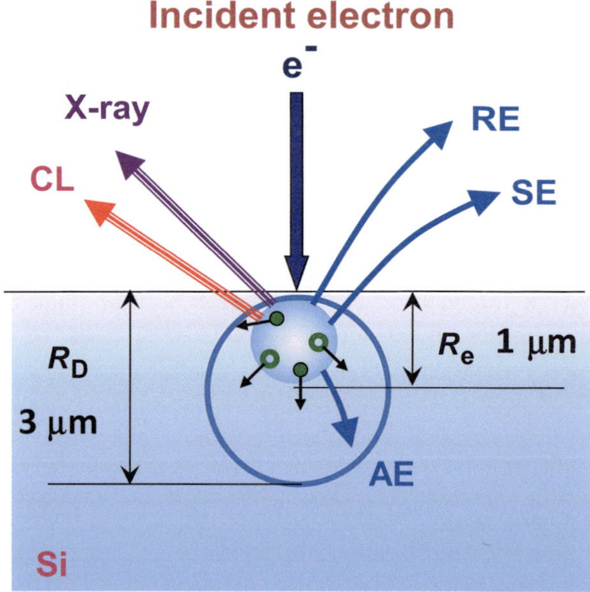

Fig. 7.1 Various excitations induced by an electron beam injection

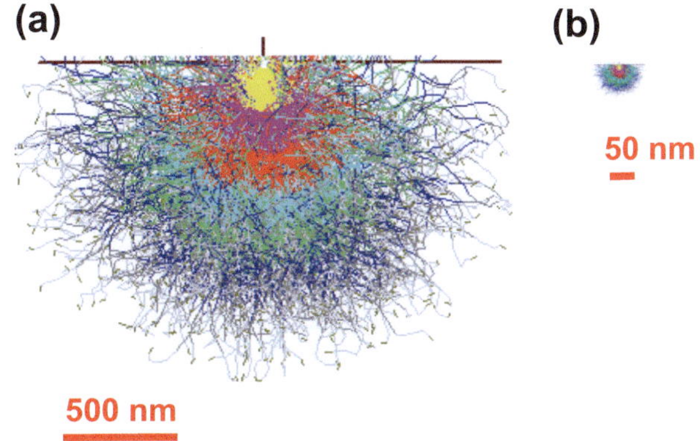

Fig. 7.2 Electron trajectories of (**a**) 10 kV and (**b**) 2 kV electrons in Si simulated by Monte Carlo method

from the specimen. Some fraction of the incident electrons are detected as absorbed electrons (AEs).

The incident electrons are elastically or inelastically scattered by the specimen and spread in a certain volume. The elastic scattering only deflects an electron in a certain direction, but does not modify its energy. Inelastic scattering, on the other hand, does not alter the direction but transfers energy to the specimen and eventually reduces its energy to zero. Thus, the electrons may spread in a certain volume, which is illustrated by Monte Carlo simulations [4]. Figure 7.2 shows the trajectories of 10 and 2 kV electrons in Si. The volume that incident electrons spread is defined as electron range. The electron range, R_e is empirically defined as [5]

$$R_e = \left(2.76 \times 10^{-2} A/\rho Z^{0.889}\right) E^{1.67}\ [\mu m]$$

where A is the mass number, ρ is the density [g/cm^3], Z is the atomic weight, E is the electron beam energy [kV]. This equation shows that R_e is roughly proportional to $E^{5/3}$. Giving the proper values in our case, the R_e in Si varies as shown in Fig. 7.3. The lower the electron beam energy is, the smaller the R_e becomes. Typical R_e s are 4.7 μm for 20 kV, 1.5 μm for 10 kV, 0.47 μm for 5 kV.

The transferred energy is mainly consumed to excite electrons from valence or core states to the conduction or higher excited levels. The highly excited electrons are then relaxed by Auger process or phonon excitation into the conduction band. Simultaneously, holes in the core state are also relaxed into the valence band. As the consequence, electrons are exited in the conduction band while holes are created in the valence band. The work to generate one electron hole pair in semiconductors is empirically known as about three times the bandgap energy [6]. Thus, an electron of 20 kV may yield c.a. 6000 electron–hole pairs in Si. Such electron–hole pairs are generated along the electron trajectory, so that the electron range also indicates the diameter of electron hole pair generation [7].

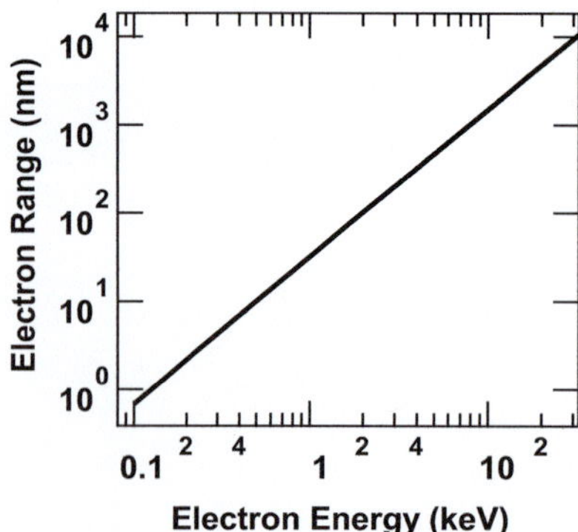

Fig. 7.3 Variation of electron range with electron beam energy in Si

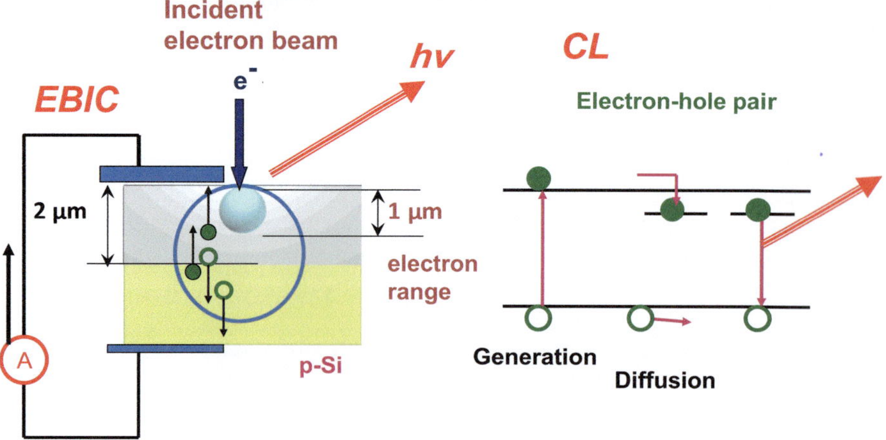

Fig. 7.4 Behavior of the generated carriers in Si. (**a**) Real space and (**b**) Reciprocal space

7.2.2 EBIC

The generated electrons and holes may diffuse in the specimen, and then be trapped in certain sites in the specimen and recombine. If an internal electric field exists, like in a Schottky or p-n junction, the electrons and holes drift in opposite directions, yielding an induced current. This current is the so-called EBIC (Electron-beam-induced current). Figure 7.4 represents these situations. If there exist electrically active defects, the EBIC current decreases due to recombination processes. Such defects are imaged as black contrast in EBIC image. Thus, we can study the distribution of electrically active defects in the specimen using EBIC. The strength

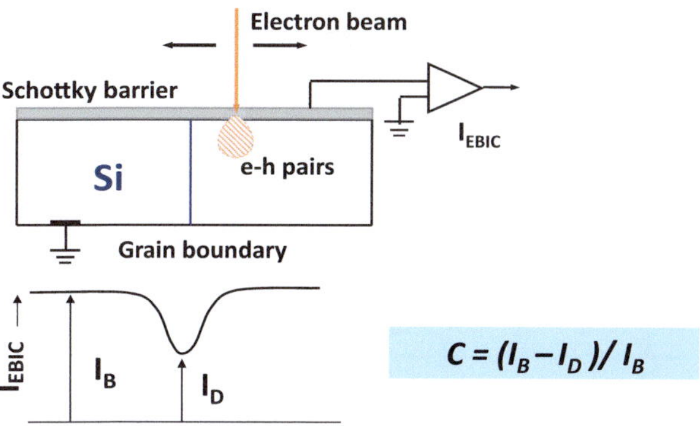

Fig. 7.5 EBIC contrast of electrical active defect

of carrier recombination at the defect is defined as the EBIC contrast C, which is determined as

$$C = \frac{I_B - I_D}{I_B} \times 100\ (\%)$$

where C (%) is the EBIC contrast, I_B and I_D are the EBIC currents of the background and defect, respectively, as shown in Fig. 7.5.

The EBIC image often changes with temperature. Since the carrier recombination takes place via an energy level of the defect, the carrier recombination strength may be changing with Fermi level position, thus with temperature. Such behavior will be discussed in Sect. 7.4.1.1 in connection with Shockley-Read-Hall (SRH) statistics.

7.2.3 CL

On the other hand, if the recombination is a direct transition, we may generate photons by electron beam irradiation. This light emission is so-called cathodo-luminescence (CL). The possible mechanism of carrier recombination in Si is schematically shown in Fig. 7.6. Since the band gap of Si is 1.1 eV, the light is infrared, even as band edge emission. Since Si is an indirect gap semiconductor, the possibility of light emission is small. Most of luminescence is observed at low temperature below 30 K. Among such luminescence, the dislocation related luminescence is known as D-lines. These lines were first found by Drozdov et al. [8, 9] from photoluminescence of plastically deformed Si. Many studies have been done afterwards. CL observation has significant advantage for this study due to its spatial resolution. The details will be discussed in Sect. 7.4.1.2.

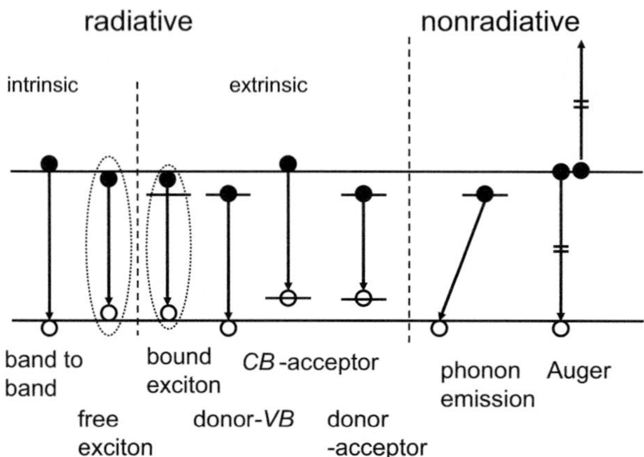

Fig. 7.6 Various carrier recombination processes

7.3 Instrumentation

Generally an EBIC/CL system is installed in a scanning electron microscope (SEM). The block diagram of a fully equipped system is shown in Fig. 7.7. The instrument is mainly composed of four components, namely the electron beam source, the electrical circuit for EBIC, the light detection unit for CL, and the specimen cooling system. Photographs of such an instrument are shown in Figs. 7.8 and 7.9. The details are described in Ref. [10].

7.3.1 Electron Beam Source

The electron beam energy and current should be selected according to the specimen. Generally we use an electron beam energy of 3–20 kV and a current of several tens of pA to several tens of nA for EBIC/CL observation. For this purpose, the thermal field emission SEM with a clean vacuum system is suitable as basic instrument. The cold field emission source sometimes limits the maximum electron beam current. On the other hand, a conventional thermionic filament cannot supply an electron beam of suitable size for high spatial resolution. A clean vacuum system avoiding oil contamination from the vacuum pump is necessary for long time observation at low temperatures. A modern in-lens secondary electron detector may allow to obtain useful information on the specimen. In such a case, we should avoid interference between the light detection system of CL and the secondary electron trajectory for in-lens SE.

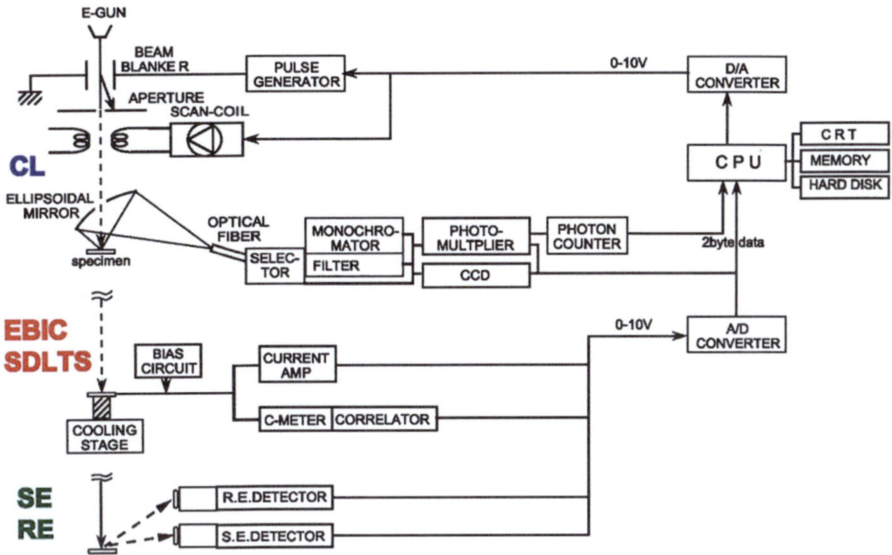

Fig. 7.7 Block diagram of fully equipped EBIC/CL system

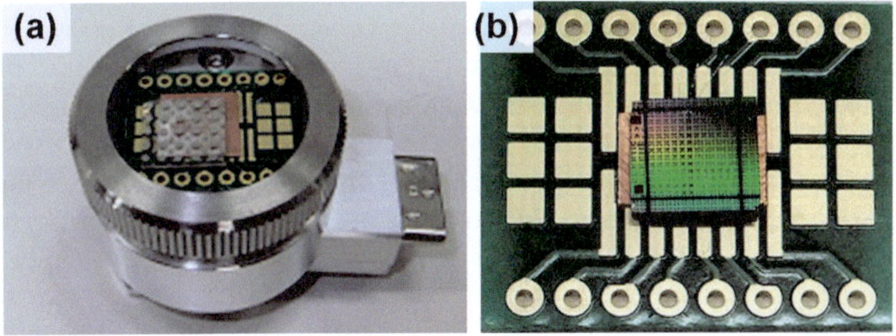

Fig. 7.8 Typical EBIC setting

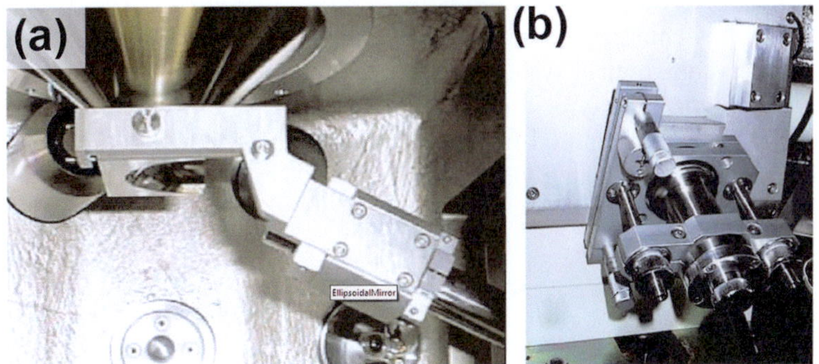

Fig. 7.9 Typical CL setting

7.3.2 Electric Circuit for EBIC

As shown in the previous section, the EBIC current is generally several thousand times larger than the current of the primary electron beam current. Thus, the current amplifier for the EBIC detection system is not necessary to be operating at high gain condition but the response should be fast enough to follow the electron beam scanning. The current amplifier used in this study is Model 427 (Keithley Inc.).

Since EBIC is the current going through the Schottky or p-n junction, it is necessary to prepare a circuit including this test element. Typically, the test element (Schottky or p-n junction) is fabricated at the sample surface and the ohmic contact is placed at the back surface of the specimen. This setting is shown in Fig. 7.8.

EBIC is often operated in different specimen geometriesdepending on the purposes. A variety of EBIC settings are described in Ref. [11].

7.3.3 Light Detection Unit for CL

Since the cathodoluminescence (CL) signal is often weak due to the small excitation volume, the light detection system for CL measurement should be designed with high efficiency. Several optical settings have been demonstrated so far Ref. [2]. Ellipsoidal mirrors and parabolic mirrors are recommended to collect the light emitted from the specimen. An ellipsoidal mirror can converge the light emitted from one focal point to another focal point. Thus, the light can be focused on the detector or on the slits of a monochromator. This system is suitable for the direct coupling or fiber transfer system. A parabolic mirror can convert the light from the focal point to parallel light travelling along an axis. This setting is typical for an optical light detection system.

When operated in low magnification, the electron beam may be deviated from the focal point. The light from the off focal point may be focused at the corresponding point in the vicinity of another focal point in ellipsoidal mirror. The distance from the focal point is enlarged according to the magnification of the mirror. Thus, this magnification of the optical system should be small.

Figure 7.9 shows the typical optical system using an ellipsoidal mirror and an optical fiber. The light emitted from the specimen is collected with an ellipsoidal mirror and focused on the inlet of an optical fiber.

7.3.4 Low Temperature System

A low temperature measurement is necessary for EBIC/CL observation. As for EBIC, the temperature dependence of EBIC contrast gives information on the energy level of the concerned defect. Since typical Si specimen becomes insulating below 50 K, the EBIC observation should be done from 50 K to room temperature.

A low temperature measurement is also essential in the CL study. The CL spectrum becomes sharp due to the suppression of thermal broadening, so that the identification of emission peaks become easy and more information becomes available at low temperatures. Moreover the suppression of nonradiative recombination may result in a significant increase of the CL signal.

The specimen cooling is usually realized a by liquid He flow unit. A simple He flow system has several advantages: it can cool the specimen down to 10 K and suppress the extra vibration of an artificial cooling unit. In any case, the cooling system should be designed to avoid the specimen drift due to thermal shrinkage of the stage.

7.4 Defects in Si

This section describes the EBIC/CL characterization of extended defects in Si. Although some information may be available for point defects like vacancies, interstitials and impurities, they are rather difficult to characterize by these techniques.

7.4.1 Dislocations in Single Crystalline Si

7.4.1.1 Electrical Activity

Dislocations are one of the most interesting subjects for EBIC/CL characterization. Since the image resolution of EBIC/CL is in the order of μm, these techniques are suitable to characterize the dislocation distribution. Moreover, we can study the electrical activity of dislocations by EBIC/CL.

In 1980s the electrical properties of dislocations in Si are one of the hottest topics for semiconductor technology. Although dislocation free Si wafers are available, some treatments of device processing may easily introduce dislocations. Some of them have acted as detrimental defects for device operation, but not always. A big dispute arose about the electrical activity of dislocations namely whether dislocations are accompanied by electrical active defects or not [12–16]. Until the early 1990s, a large part of R & D laboratories and Si device processes were not clean enough so that wafers could not be kept free from metallic impurities. Such environments have prevented the researchers to obtain reliable data. The introduction of clean processes has significantly reduced the metallic contamination.

Figure 7.10 shows the EBIC images of hexagonal dislocations introduced into Si by compression [17]. The FZ-Si specimen was compressively deformed at 800 °C in Ar atmosphere and was cut along a (111) primary slip plane. As shown in Fig. 7.10a, glide dislocations of hexagonal shape are visible in the EBIC image at 70 K. According to the specimen geometry, the Burgers vector was identified as shown in (c). Although the 60° dislocation component at the top region is darker

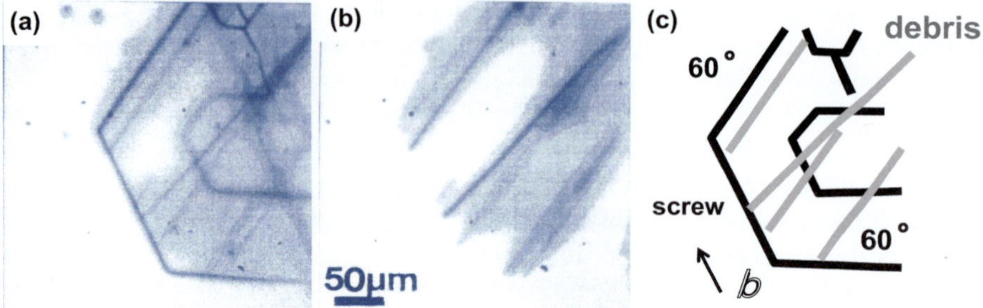

Fig. 7.10 EBIC images of hexagonal dislocations introduced into Si by compression taken at (**a**) 70 K and (**b**) 300 K, and (**c**) schematic of defect configuration

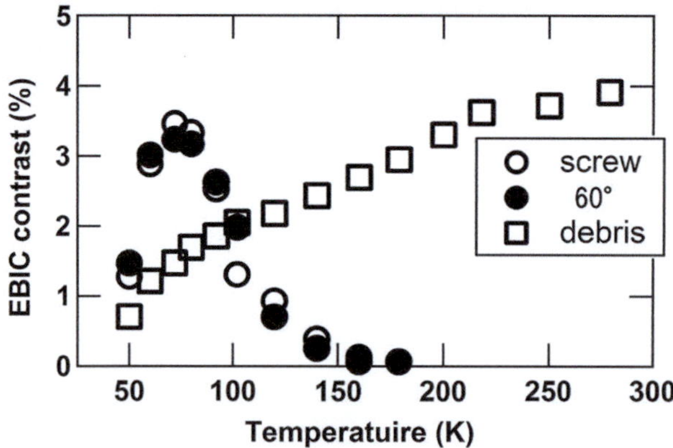

Fig. 7.11 Temperature dependence of EBIC contrasts of dislocations and debris in Si

than the others in this image, it is difficult to distinguish the difference between 60° and screw. This image has completely changed at 300 K (Fig. 7.10c). The hexagonal dislocations are no more visible but straight lines have significantly enhanced. These lines are so called debris, which are left over after dislocation glide. The main origin of these debris is the dislocation pinning. When a dislocation may be pinned by a certain obstacle, the dislocation may bow out and a faulted dipole may be left over. Such secondary defects are so called debris. The temperature dependence of EBIC contrasts of these defects are shown in Fig. 7.11. EBIC contrasts of 60° and screw dislocations are weak as 1 % at 50 K, increase until maximum of a few % around 70 K, and gradually decrease to zero around 150 K. It is found that the absolute value and peak temperature of the EBIC contrast of dislocations varies with the depth of dislocations, single or bundle, and also with the carrier concentration of the Si specimen. However, the EBIC contrasts of clean dislocations behave similarly and no significant difference has been found between 60° and screw dislocations. The debris, on the other hand, have faint contrast at 50 K and increase monotonically with temperature until 300 K.

These temperature dependences have been analyzed by Shockley-Read-Hall (SRH) statistics [18]. A defect acting as the carrier recombination center may be represented as an energy level E_d in the bandgap. Since the carrier recombination take place with this energy level, the recombination strength may be related to the occupation fraction f_T of this level. f_T is represented as

$$f_T = \frac{1}{1 + \exp\{(E_d - E_F)/kT\}}$$

where E_F is the Fermi level position. For the EBIC simulation, we only have to focus on the minority carrier lifetime under the electron beam excitation. For the simplicity, we will ignore the carrier diffusion and calculate the detailed balance of the carriers with this defect level. The SRH theory gives the minority carrier lifetime as

$$\tau = \frac{\tau_{p0}(n_0 + n_1 + \Delta n) + \tau_{n0}(p_0 + p_1 + \Delta n)}{n_0 + p_0 + \Delta n}$$

where τ_{n0}, τ_{p0} are the carrier lifetime of electrons and holes, respectively; n_0, p_0 are the electron and hole density under thermal equilibrium and n_1, p_1 are the electron and hole density when the Fermi level is pinned at the defect level. Δn is the excess carriers generated by the electron beam.

If we agree that the EBIC contrast is inversely proportional to this carrier lifetime, we can estimate the EBIC contrast. Putting typical values of the Si specimens, such as defect density $N_d = 1 \times 10^{14}$ cm^{-3}, $\sigma_n = \sigma_p = 1 \times 10^{-18}$ cm^{-2}, $\Delta n = 1 \times 10^{15}$ cm^{-3}, the temperature dependence of EBIC contrast ($1/\tau$) is calculated as Fig. 7.12. It is clear that, depending on the energy level E_d, the EBIC contrast varies differently with temperature.

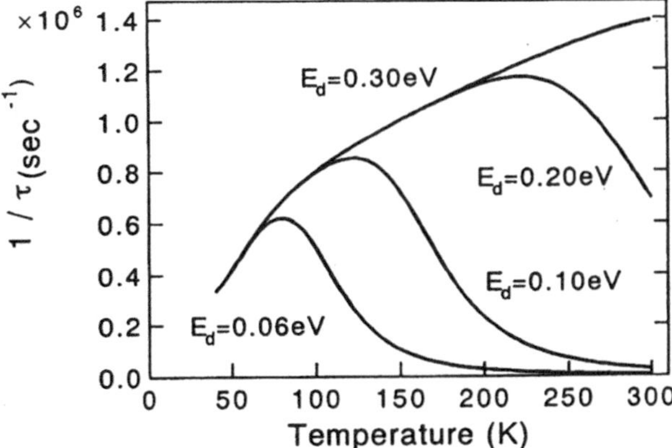

Fig. 7.12 Temperature dependence of recombination activity ($1/\tau$) of defects based on Shockley-Read-Hall statistics

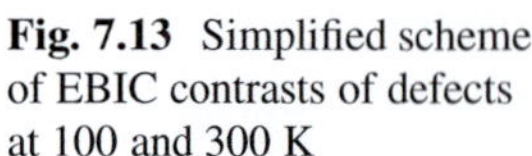

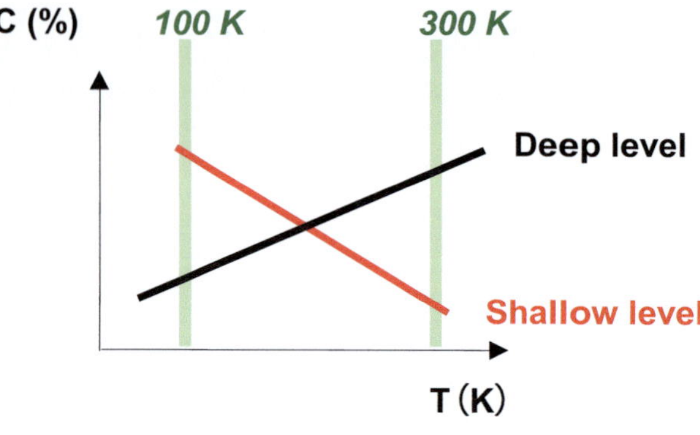

Fig. 7.13 Simplified scheme of EBIC contrasts of defects at 100 and 300 K

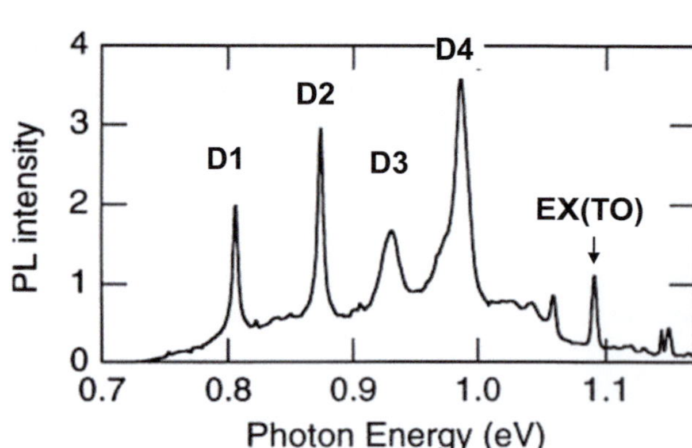

Fig. 7.14 Typical PL spectrum of plastically deformed Si

In early 2000s, Chen et al. [19] simplified these behaviors and proposed the EBIC imaging at 100 and 300 K. Figure 7.13 shows this idea. If the defect appears at 100 K and not at 300 K, it may have a shallow level. Alternately, if it appears at 300 K but not 100 K, it may have a deep level. This simple criterion works well with the later study.

7.4.1.2 Optical Activity

As stated in Sect. 7.2.3, dislocation related luminescence has been reported by Drozdov et al. in 1970s using photoluminescence (PL) [8, 9]. They have found four characteristic luminescence lines in deformed Si crystals and named them as D1–D4. Figure 7.14 shows the typical PL spectrum from plastically deformed Si. They also found that D3 and D4 are dominant in lightly deformed specimens, while D1 and D2 in heavily deformed ones. Various studies have been done to identify these emissions [20–24]. Since PL is rather difficult to clearly correlate D-line emissions to a certain defect, CL is expected to give decisive results.

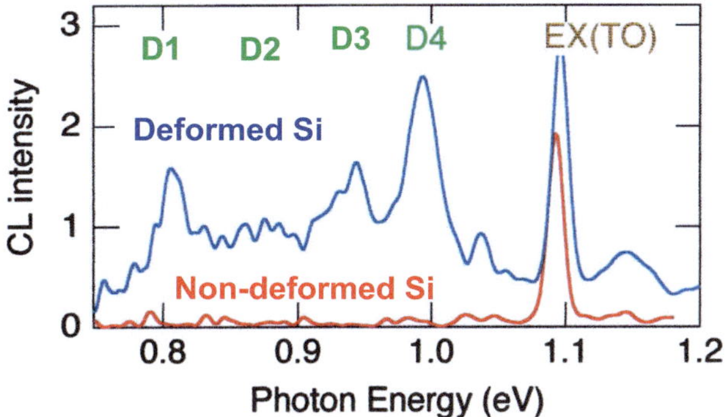

Fig. 7.15 CL spectra of deformed (*blue*) and non-deformed (*red*) Si crystal

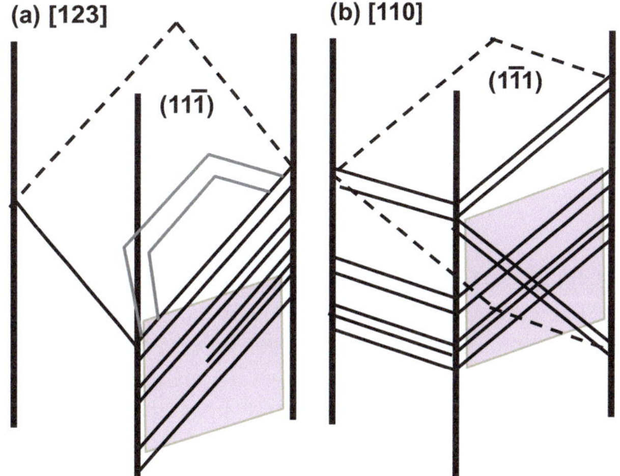

Fig. 7.16 Two geometries of plastic deformation

Figure 7.15 shows the CL spectra taken from deformed (blue) and non-deformed (red) Si crystals. The CL spectrum of deformed Si is identical with that of PL although the peaks are not so sharp. The peaks at 0.81, 0.88, 0.93, and 1.00 eV correspond to D1, D2, D3 and D4 emission, respectively. The peak at 1.09 eV is attributed to the transverse optical phonon replica of excitonic luminescence (EX(TO)). The non-deformed Si only gives the EX(TO). Since the CL intensity of Si is rather weak, it is difficult to detect D-line emission from individual dislocations. The observation of slip lines has been performed, instead [25]. Figure 7.16 illustrates the two geometries of plastic deformation. The Si rod (a) is compressed along the [123] direction, named "simple geometry". In this configuration, only one slip

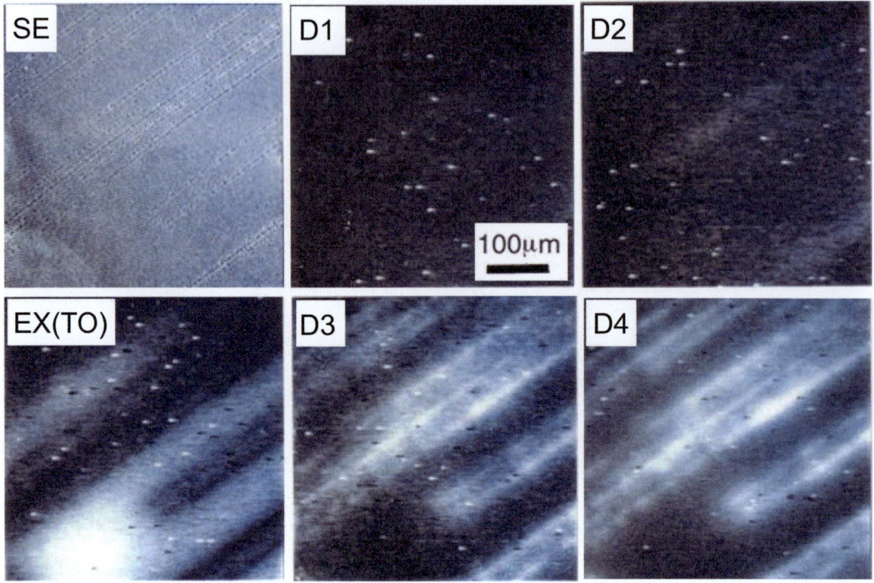

Fig. 7.17 Monochromatic CL images of deformed Si (simple geometry)

system of [1–10]/(11–1) becomes dominant and single slip lines, which are the arrays of glide dislocations, appear. On the other hand, the rod (b) is compressed along the [110] direction, named "complex geometry". Four equivalent slip systems are activated in this condition and the side (−110) plane has two sets of slip lines. Figure 7.17 shows the set of monochromatic CL images of the former case (simple geometry). Prior to the observation, the specimen surface was etched to reveal the dislocation arrays. Thus, a SE image can show us the distribution of slip lines. In an EX(TO) image, these slip lines become dark. D1 and D2 images do not give any significant emission related to slip lines. D3 and D4 images, however, give the bright lines corresponding to slip lines. It is clear that, in this simple deformation case, dislocations are generated and move along the slip planes and do not interact with each other. It is confirmed by TEM that most of the dislocations are straight and exist with a certain distance interval. Contrary to this case, a complex geometry specimen has all the D-lines as shown in Fig. 7.18. The configuration of slip lines corresponds to the schematic of Fig. 7.16. D3 and D4 emissions are observed along the two sets of slip lines. Contrarily, D1 and D2 are observed at the intersections of these slip lines. The CL spectra were taken at the three points indicated in D3 image and shown in Fig. 7.19. Point 1, namely the cross point, has D1 and D2 peaks. On the other hand, point 2 and 3, on the slip line, have D3 and D4. These observations confirm the idea that D3 and D4 are originating in glide dislocations, while D1 and D2 are related to the product of dislocation interactions. Thereafter, many studies have been devoted to this dislocation emission and some attempts have been performed to utilize this D-line emission for infrared LEDs [26, 27].

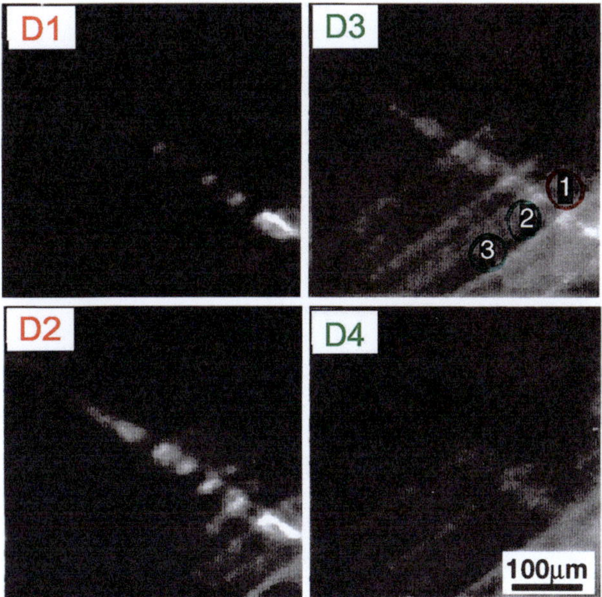

Fig. 7.18 Monochromatic CL images of deformed Si (complex geometry)

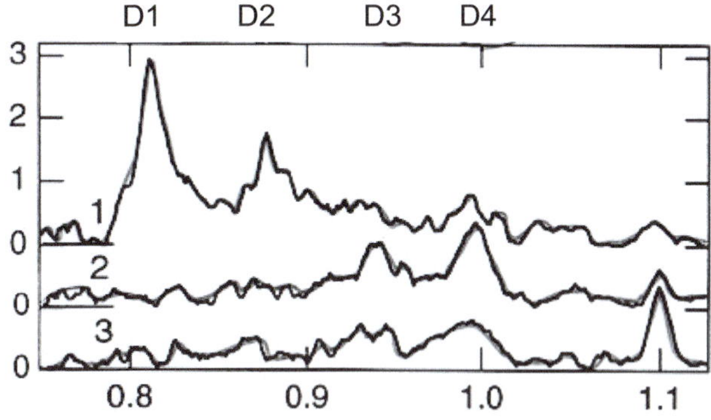

Fig. 7.19 CL spectra taken at the three points indicated in D3 image as shown in Fig. 7.19

7.4.2 *Misfit Dislocations in SiGe/Si Heterostructure*

Misfit dislocations (MDs) are generated at the hetero interface. Since they are confined at the interface, it is easy to analyze the carrier recombination mechanism. In early 2000s, SiGe virtual substrates have been extensively studied to supply strained Si wafer for advanced MOSFET [28]. The strained Si/SiGe layers were fabricated as shown in Fig. 7.20. First, a relaxed SiGe layer was grown on a Si substrate, which was composed of a graded SiGe layer (a few μm thick) and a

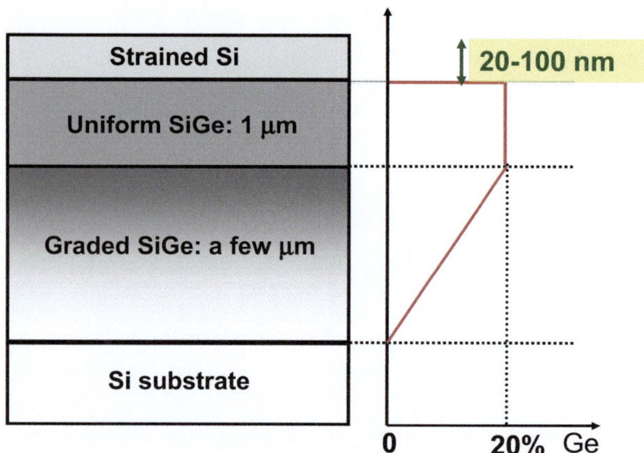

Fig. 7.20 Schematic of strained Si/SiGe layer structure

uniform SiGe layer (about 1 μm). The first graded SiGe layer relaxes the misfit dislocations between the Si substrate and SiGe. The next uniform SiGe layer separates the defect layer from the strained Si channel layer although a certain number of threading dislocations (TDs) remains. Then a thin Si layer was grown to form a strained Si layer. If this Si layer is below a critical thickness, no MDs should be created. However, there may exist MDs between the uniform SiGe and strained Si layers. Since such MDs may limit the carrier mobility enhancement, the suppression of MD generation is of great importance. The problem is how to observe MDs in this system by nondestructive methods. EBIC/CL is applicable for this purpose [29, 30].

Figure 7.21 illustrates (a) the dislocation distribution in a strained Si/SiGe system as well as (b) the cross sectional transmission electron micrograph inclined along the horizontal axis. A high density of tangled dislocations exists at the graded SiGe layer, while a small number of independent dislocations at the top interface.

EBIC observation under different accelerating voltage can distinguish these dislocation groups. Prior to the observation, an Al Schottky contact of 30 nm was deposited at the surface. Figure 7.22 shows the EBIC observation condition under 4 and 20 kV. The electron ranges of these voltages are 320 nm and 3.5 μm, respectively. Thus, the carriers are generated in the volume of the top interface at 4 kV, while in the volume of the graded SiGe layer at 20 kV.

Figure 7.23 shows the EBIC images taken at these voltages under the specimen temperature of 300 and 65 K. At 300 K, it is difficult to find any contrast in the 4 kV image. Only a faint cross hatch pattern is visible in the 20 kV image. At 65 K, however, dislocations are clearly distinguished. The EBIC image at 4 kV (Fig. 7.23b) has dark lines and dots. The dark lines run parallel to the <110> directions. They correspond to MDs and TDs, respectively. Their EBIC contrast

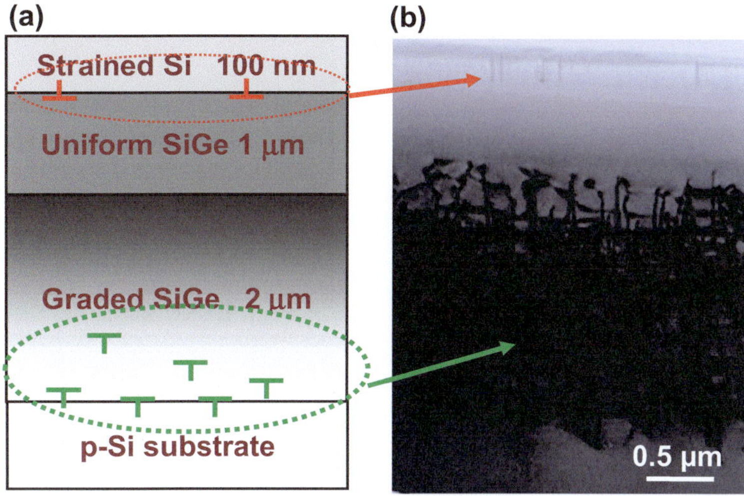

Fig. 7.21 (**a**) Structure and (**b**) bright field TEM image of strained Si/SiGe (cross section)

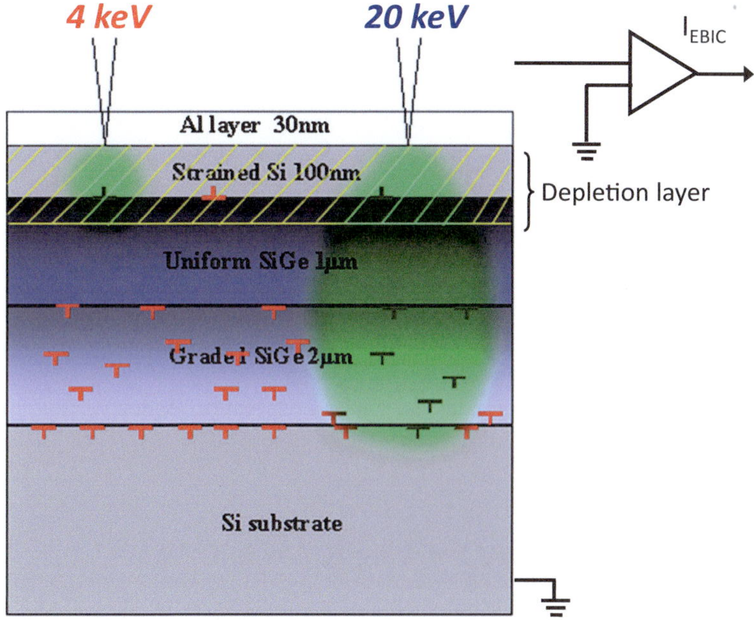

Fig. 7.22 EBIC configuration of strained Si/SiGe (cross section)

is rather weak showing small electrical activity. On the other hand, the EBIC image at 20 kV shows clear cross hatches (Fig. 7.23d). The contrast is about 10 %. The line directions are also along <110> s. Comparing Figs. 7.21 and 7.23, it is found that

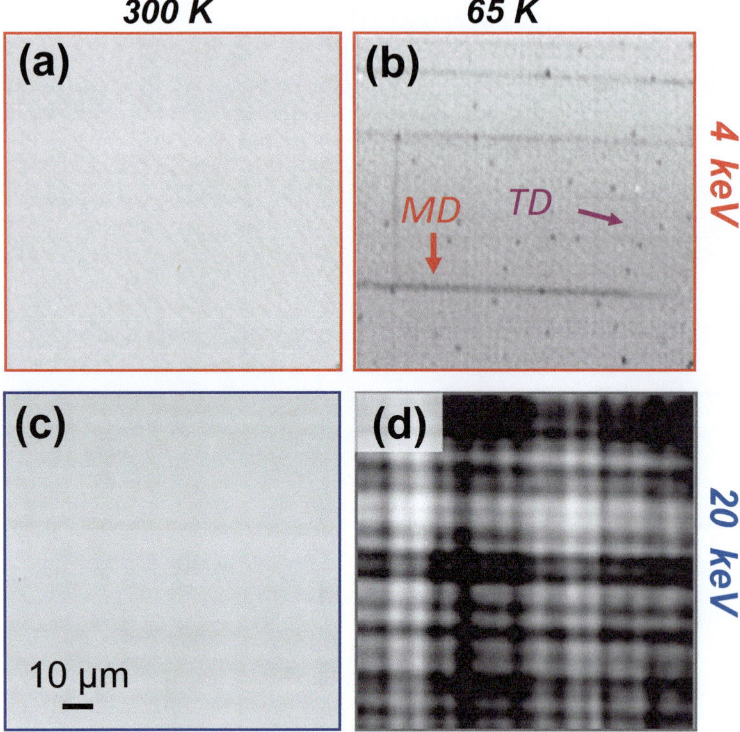

Fig. 7.23 Planar EBIC images of strained Si/SiGe taken at 4 kV (**a**, **b**) and 20 kV (**c**, **d**). Specimen temperature: 300 K (**a**, **c**) and 65 K (**b**, **d**)

individual dislocations are no more distinguished and only the density fluctuation of dislocations are imaged as a cross hatch pattern. Such a fluctuation at the graded layer was found to be the origin of TDs and MDs.

These observations demonstrate that EBIC observation under various voltages can show us the defect distribution in the depth region.

7.4.3 Oxidation-Induced Stacking Faults in CZ Si

Stacking faults, especially oxidation-induced stacking faults (OISF), are somewhat more ideal defects for characterization. OISFs are generated on the {111} plane by segregation of interstitial Si atoms which have been emitted from oxygen precipitates. Since OISFs are sessile and have their own geometry, it is easy to analyze their electrical activity [31]. The OISF are easily generated in CZ-Si by two-step annealing, namely at 750 °C for 15 h and 1100 °C for 20 h in Ar atmosphere. OISFs of 20–40 μm diameter exist in the ingot with a density of c.a. 10^5 cm^{-2}.

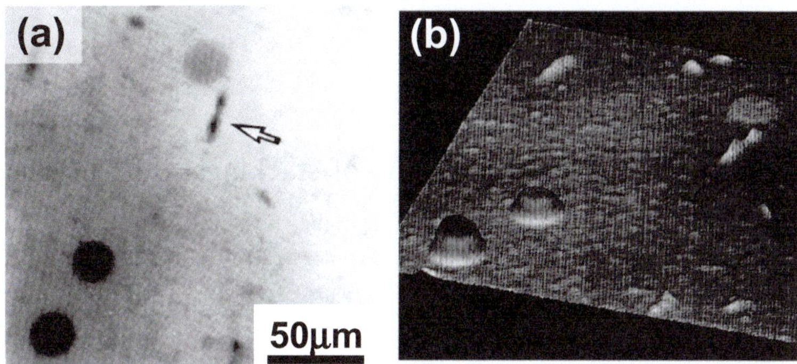

Fig. 7.24 (**a**) EBIC image and (**b**) 3-d plot of oxygen-induced stacking faults in Si

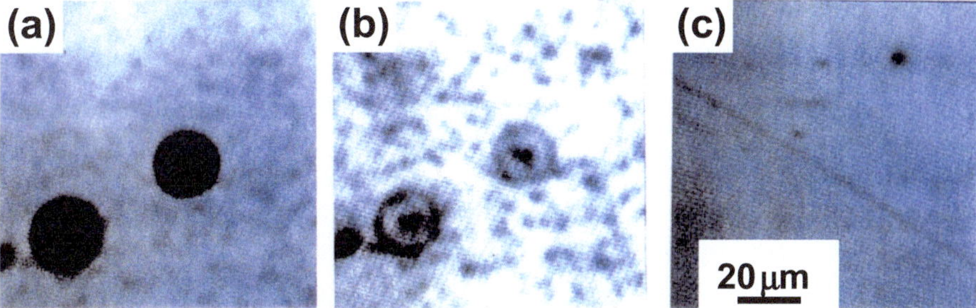

Fig. 7.25 EBIC images of oxygen-induced stacking faults in Si (**a**) 500 K, (**b**) 110 K, and (**c**) 290 K

After the OISF generation, specimens were sliced parallel to the (111) plane. Schottky contacts of Au or Al were fabricated at the surface for n- or p-type specimens. EBIC observations were conducted with a 20 kV electron beam at temperatures from 40 to 300 K [32]. Figure 7.24 shows an EBIC image of the (111) surface taken at 50 K as well as a bird's-eye view of an adjacent area. The black circle is an OISF lying parallel to the (111) surface, indicating that the stacking fault is electrically active in this condition. The EBIC contrast varies with the OISFs, suggesting that the depth from the surface is different. The OISFs lying other {111} planes are imaged as lines. Figure 7.25 shows the temperature dependence of OISF images. At 50 K, OISFs are seen as black circles. These EBIC contrasts decrease with temperature increase. Around 110 K, the periphery of OISF, namely Frank partial dislocations, become stronger than OISF itself. Thus, OISF changes its image from a solid circle at 50 K to an open circle at 110 K, then completely disappears at 300 K. Figure 7.26 shows the temperature dependence of EBIC contrasts of OISFs as well as the bounding Frank partial dislocations. The EBIC contrast of OISF significantly decreases with increase in temperature and becomes zero at 150 K. Frank partial dislocations first increase their contrast, become maximum at 70–80 K and decrease until 200 K. The Arrhenius plot of EBIC contrast of OISF

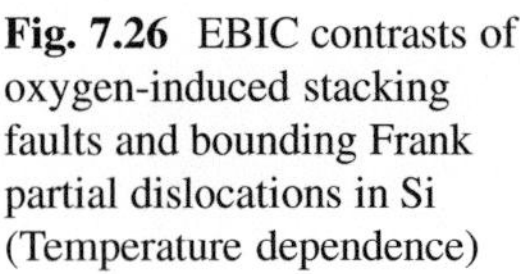

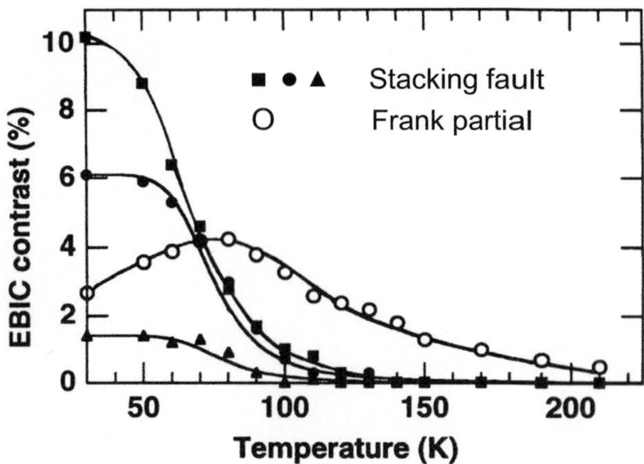

Fig. 7.26 EBIC contrasts of oxygen-induced stacking faults and bounding Frank partial dislocations in Si (Temperature dependence)

gives an energy value of 55 meV. It has been found that the EBIC contrast of OISF significantly decreases when the OISFs are inclined, namely when the specimen was slantingly cut from the (111) plane.

According to the above results, Sekiguchi et al. [32] have concluded that the stacking faults in Si are accompanied with an energy barrier of 55 meV and that the minority carrier diffusion through this barrier plane at lower temperature is suppressed. When the specimen temperature becomes high, the carriers no more feel this barrier and EBIC contrast disappears.

Shen et al. have studied the mechanism of Cu gettering on the OISF [33, 34]. It was found that the EBIC contrast of Frank partials significantly increases at room temperature but not OISF. The data suggest that Cu only decorates the Frank partials but not OISF.

7.4.4 Grain Boundaries in Multicrystalline Si

Photovoltaics are expected to solve the problem of energy consumption and sustainable environment. In 2013, more than 90 % of solar cells are made from Si: 60 % are multicrystalline (mc-) Si and 40 % monocrystalline (mono-) Si. Since their average conversion efficiency is 17 % and 19 %, at present, it is important to improve the efficiency of mc-Si solar cells. Generally, the minority carrier lifetime is small where many grain boundaries (GBs) exist. Thus, it has been believed that the GBs are detrimental for solar cells and should be avoided from the wafer as much as possible. Moreover, the purity of solar grade Si is much lower than semiconductor grade Si, which also reduces the solar cell efficiency. To clarify the killer defects of solar grade Si, we have applied EBIC technique for the characterization of mc-Si

Fig. 7.27 (**a**) Photograph of cast grown ingot of 50 cm square, and (**b**, **c**) photographs of horizontal cut wafers

grown by cast method. Figure 7.27a shows a photo of a cast grown ingot of 50 cm square. (b) and (c) show photographs of horizontally cut wafers. The (b) wafer has smaller grains than the (c) wafer. Such a difference appears depending on the growth conditions [35].

For the characterization of GBs, we have grown mc-Si ingots of rather high purity. Namely, it is important to avoid the effect of metallic impurities. For this purpose we have chosen semiconductor grade feed stock and cleaner furnace, and then performed cast Si growth.

7.4.4.1 Large Angle Grain Boundaries (LA-GBs)

First, we start the characterization of Large Angle Grain Boundaries (LA-GBs).

The EBIC images of this high-purity cast mc-Si are shown in Fig. 7.28. The top column shows the SE images. The grains are easily distinguished from the etch steps due to the different etching speed depending on the facet. The EBSD patterns in the second column clarify the GB characters. All GBs are LA-GBs, and dominant GBs in (a) are Σ 3, Σ 9, Σ 27 and random (denoted by R), while those in (b) are R, Σ 27 and Σ 3. The EBIC images at 300 K (third column) have no clear contrast related to LA-GBs. Only the R in (b) gives a weak contrast of less than 2 %. The 100 K images (bottom column) show faint contrast of LA-GBs and some grains become a little cloudy. Figure 7.29 shows the variation of EBIC contrast according to the GB character. These data indicate that the sigma (Σ 3, Σ 9, and Σ 27) and random GBs are intrinsically not so electrically active. This infers that the actual GBs in the conventional mc-Si are somewhat contaminated with metallic impurities.

To confirm the effect of metallic decoration on the electrical activity of LA-GBs, the test specimens were contaminated with Fe. The specimens were vacuum sealed in quarts tubes with Fe wires and annealed at 800, 900 and 1100 °C. These temperatures correspond the Fe concentrations of 3×10^{12}, 4×10^{13}, 3×10^{15}, respectively. They are named light, moderate and heavily Fe contaminated specimens. Figure 7.30 shows the set of EBIC images at 300 and 100 K, as well as SE

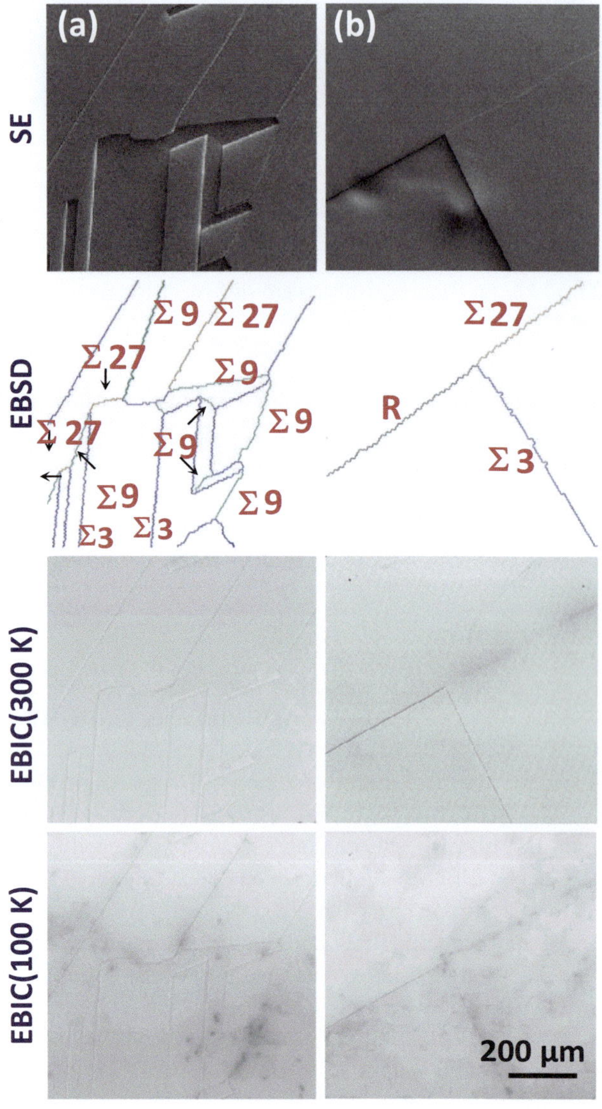

Fig. 7.28 SE, EBSD and EBIC images of high-purity cast mc-Si wafers

images and EBSD patterns. The left column corresponds to lightly contaminated one, the center to moderate, the right to heavily. The symbol Σ 3* in EBSD indicates Σ 3 with the habit plane other than {111}. As for the lightly contaminated specimen, R has the dark EBIC contrast, Σ 9 and Σ 3* have weaker contrast but not for Σ 3. In the EBIC image at 100 K, the EBIC contrasts of LA-GBs increase, and now, some of Σ 3 s become visible. As for the moderately contaminated specimen, R's and Σ 3*'s are clearly seen in EBIC image at 300 K, but not Σ 3. These contrasts become stronger at 100 K. In the heavily contaminated specimen, all the LA-GBs including

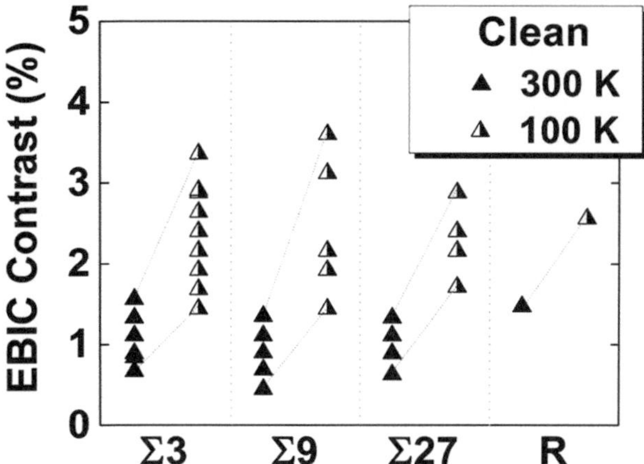

Fig. 7.29 Variation of EBIC contrast of large-angle grain boundaries in high-purity cast mc-Si wafers

Σ 3's are visible in EBIC image at 300 K. Not only LA-GBs but also dislocations in grains have given dark contrast in the image. It is noted that the vicinity of the GBs is brighter than the matrix. Such bright and dark contrasts are known as the dot (dark line) and halo pattern, indicating the impurity gettering of LA-GBs. Namely, interstitial Fe around LA-GBs have been precipitated at the some parts of GBs, which increases minority carrier lifetime at the vicinity of GBs. Interestingly, the EBIC contrasts of LA-GBs at 100 K has decreased with respect to those at 300 K. These changes are shown in Fig. 7.31. The discussion in Sect. 7.4.1.2 indicates that the LA-GBs under light or moderate contamination have shallow energy levels for carrier recombination, but heavily contaminated LA-GBs endow deep energy levels. These deep levels may be attributed to the Fe precipitations at certain positions at GBs. It is however difficult to identify the Fe precipitates at certain LA-GB by TEM. There exists a report of B. Chen et al. [36], who have found Fe atoms at LA-GB using high angle annular dark field (HAADF) technique. The EBIC activity of GBs according to GB character and Fe contamination levels are schematically shown in Fig. 7.32. The GBs are aligned by their coherency from left to right. This figure indicates that the GBs are intrinsically not electrically active. But, once decorated with Fe, they become electrically active. Such decoration starts with incoherent LA-GBs such as R under low contamination level. Then, Fe decoration proceeds to coherent LA-GBs. At last, under heavily contamination level, all the LA-GBs including Σ 3 become electrically active.

This figure is useful because if one knows the EBIC contrast of a certain GB, one can estimate the contamination level of this mc-Si wafer.

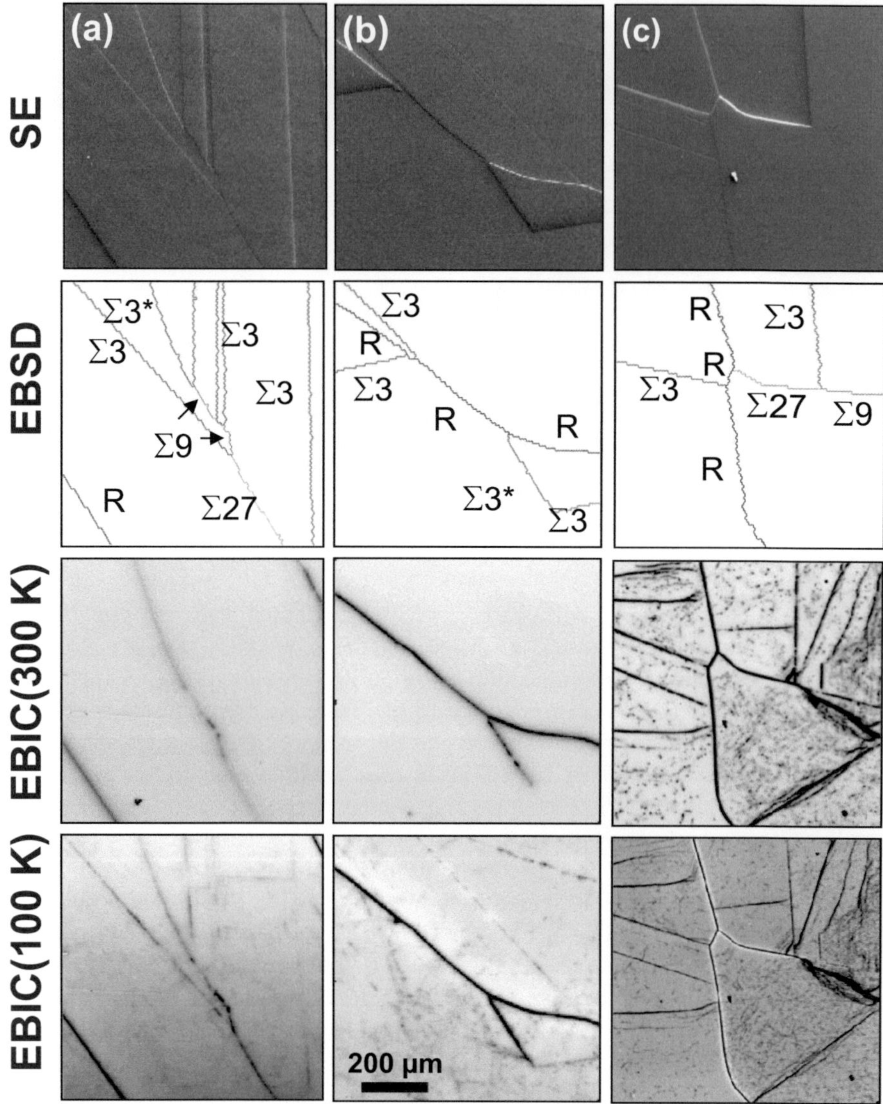

Fig. 7.30 SE, EBSD and EBIC images of Fe contaminated mc-Si wafers. (**a**) Light; [Fe] = 4×10^{13} cm^{-3}, (**b**) Moderate; [Fe] = 4×10^{14} cm^{-3}, (**c**) Heavy; [Fe] = 3×10^{15} cm^{-3}

Some attempts have been made to suppress electrical activity of Fe contaminated GBs. J. Chen et al. have made hydrogen passivation of these contaminated mc-Si specimens and measured EBIC contrast [37]. The effect of hydrogen passivation is not enough to passivate all the Fe at LA-GBs. The proper gettering procedure may be very promising to suppress the carrier recombination at contaminated GBs.

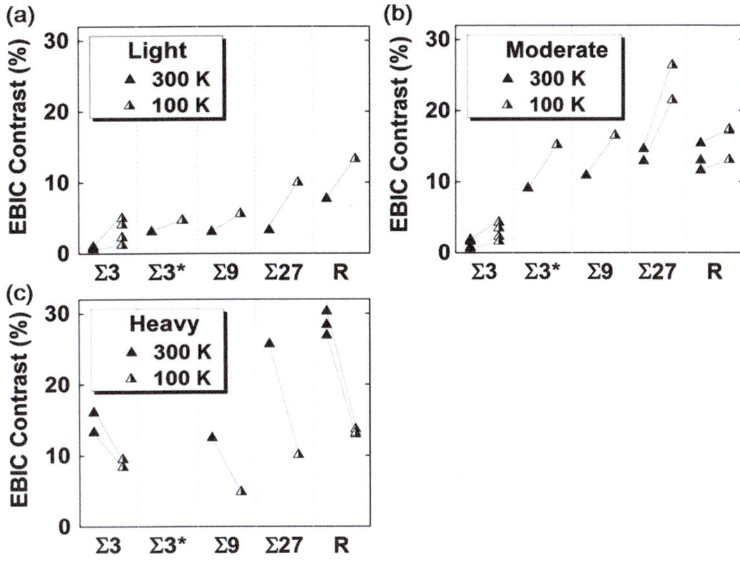

Fig. 7.31 Variation of EBIC contrast of LA-GBs in Fe contaminated mc-Si wafers

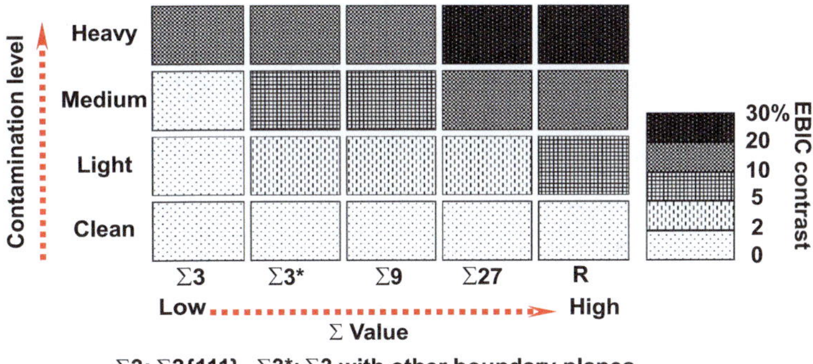

Fig. 7.32 Variation of EBIC activity of GBs according to GB character and Fe contamination level

7.4.4.2 Small Angle Grain Boundaries (SA-GBs)

The small angle grain boundaries (SA-GBs) are rather different from LA-GBs. Figure 7.33 shows the SE, EBSD, and EBIC images which contain SA-GBs. The image area (a) contains R, Σ 3, and SA1 and SA2 and (b) are SA1 to SA5 as well as Σ 3, Σ 9, Σ 11 and R. LA-GBs are not electrically active at both 300 and 100 K. The SA-GBs, on the other hand, have weak contrast in 300 K EBIC images. Their EBIC contrasts significantly increase at 100 K, ranging to 18–45 %. These values of SA-GBs are plotted in Fig. 7.34, which is very much different from those of LA-GBs in Fig. 7.29.

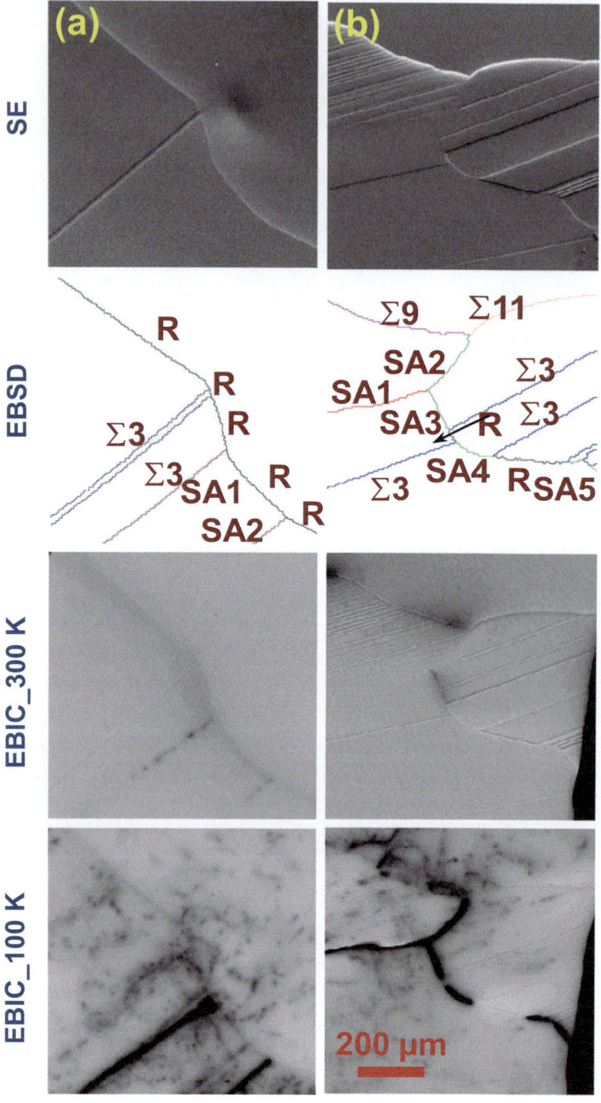

Fig. 7.33 SE, EBSD, and EBIC images of pure mc-Si wafer which contains SA-GBs

Then, TEM observation was performed to clarify such a significant difference. Figure 7.35 shows the bright field images of LA- and SA- GBs. The GBs diagonally intersect these thin specimens. The stripes along Σ 3 and R are the diffraction contours which appear when the fault plane diagonallys cut the specimens. Only a few dislocations are visible in Σ 3 image. They show that the LA-GBs have a rather smooth interface. On the other hand, the SA-GBs have lines perpendicular to GB. These lines are dislocations. When the misorientation angle is smaller than 1°, each dislocation is clearly distinguished. When it becomes large, dislocations are no more

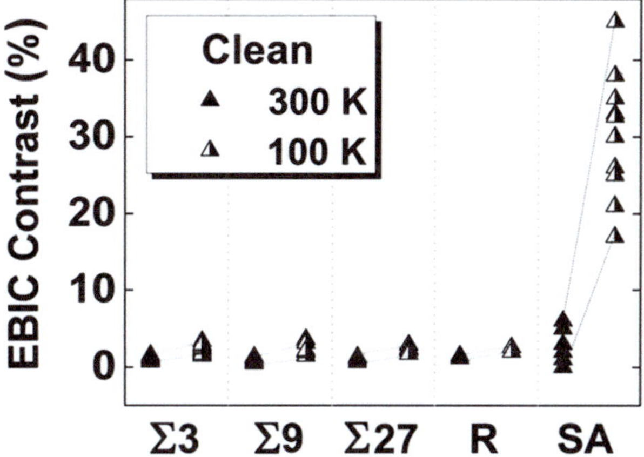

Fig. 7.34 Variation of EBIC contrast of SA-GBs in pure mc-Si wafers

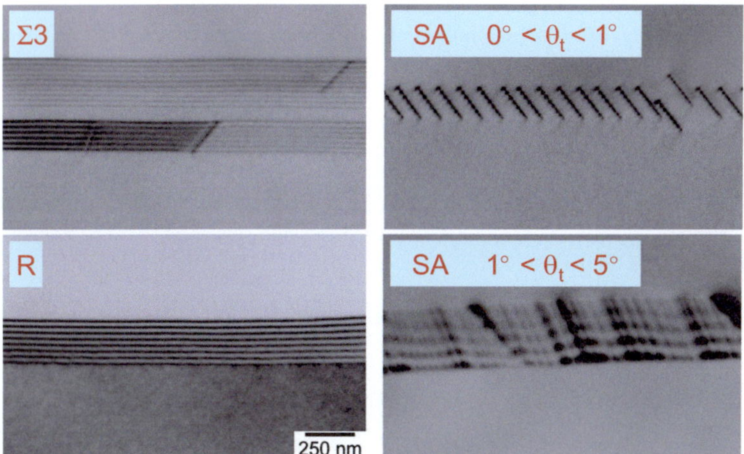

Fig. 7.35 TEM images of various GBs in mc-Si

resolved and a mixed pattern of diffraction contour and dislocations appears. These observations indicate that the SA-GBs are composed of dislocations. It is reasonable to attribute the electrical activity of SA-GBs to these dislocations. Because of the absence of dislocations, LA-GBs are not originally electrically active.

The fact that SA-GBs are composed of dislocations infers that D-line luminescence may be observed from SA-GBs [38]. Figure 7.36 shows the EBIC and CL images at the intersection of Σ 3 and SA-GB. The notation SA1° means the SA-GB with misorientation angle of 1°. EBIC images are the same as the above results.

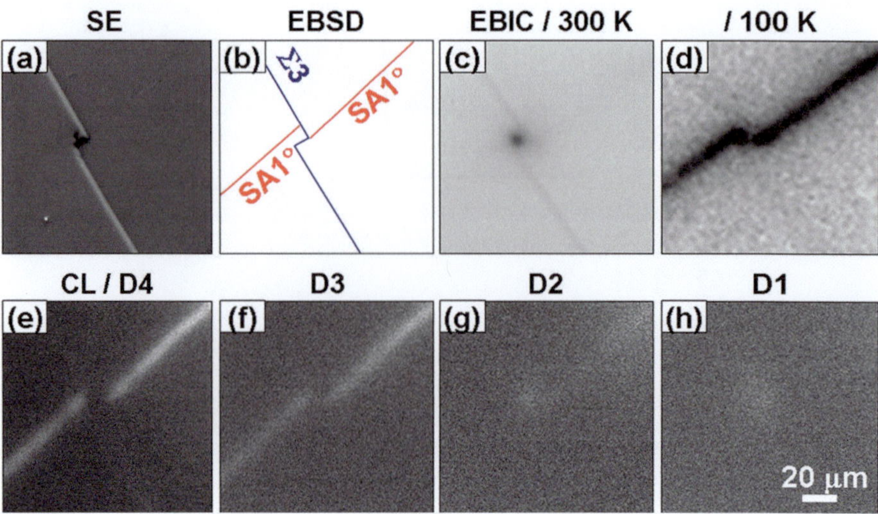

Fig. 7.36 SE, EBSD, EBIC and CL images of mc-Si wafer which contains Σ3 and general SA-GBs

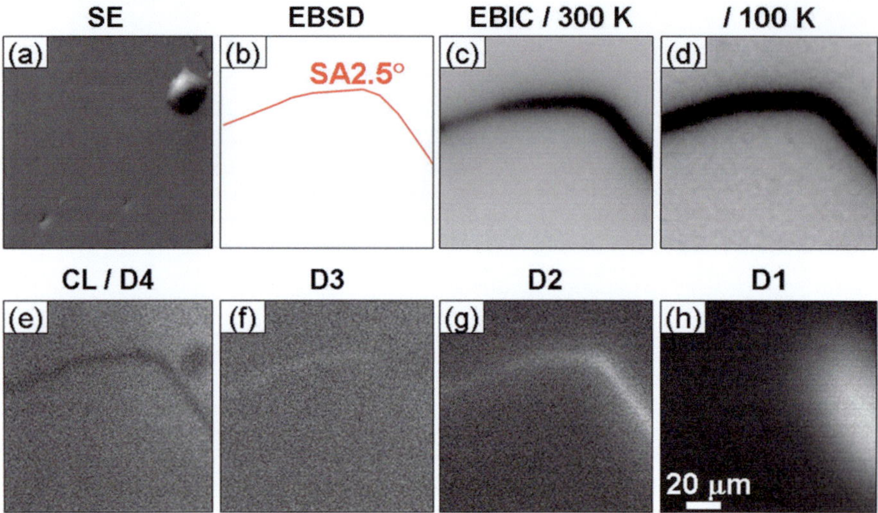

Fig. 7.37 SE, EBSD, EBIC and CL images of mc-Si wafer which contains special SA-GBs

It is noted that the intersection becomes EBIC active at 300 K, suggesting that the GB interaction has introduced some deep levels. The D-line CL images taken at 20 K show that SA1° gives D3 and D4 emissions. The upper-right part of SA1° is faintly bright in D2 image. The intersection of SA1° and Σ 3 is bright in D1 and D2 images. No D-line emission was observed from Σ 3. Figure 7.37 shows another set of EBIC/CL images of SA2.5°. This curved SA is EBIC active even at

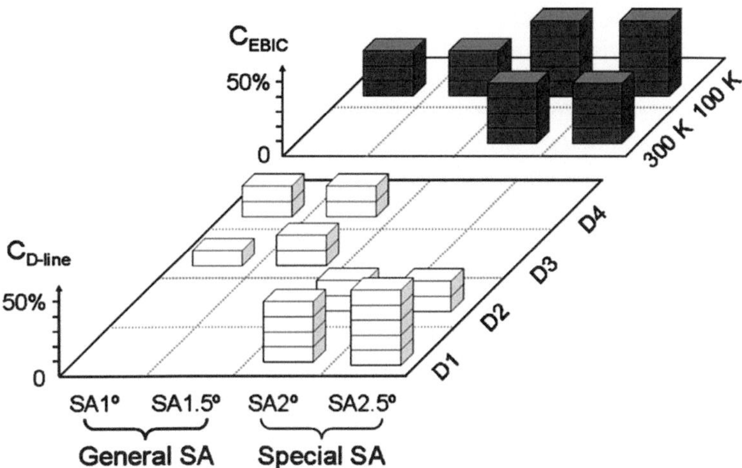

Fig. 7.38 Variation of EBIC contrasts and CL intensities of general and special SA-GBs

300 K, indicating the existence of deep levels. This SA2.5° is bright in D1, D2, and D3 images. The whole part of SA2.5° is bright in D2 image, while D1 is bright in right part and D3 in left. Moreover, the D1 emission of SA2.5° becomes wider than the others. These observations confirm the discussion of Sect. 7.4.1.2 (D-line emission from dislocations). The variation of D-lines suggests that the dislocation configuration of SA-GBs varies significantly not only with the misorientation angle but also with the shape of SA-GB. For the classification, we have distinguished these SA-GBs into two categories, namely "general" and "special". The former SA1° is classified in general SA, which has D3 and D4 emissions and EBIC contrast only at 100 K. The latter SA2.5° is classified in special SA, which has D1 and D2 emissions and EBIC contrast both at 100 and 300 K. This correspondence between EBIC contrasts and D-lines is schematically shown in Fig. 7.38.

7.5 Summary

We have reviewed the electrical and optical characterization of extended defects in Si by using EBIC and CL. The dislocations free from metallic contamination are not so electrical active and possess shallow levels, while the debris generated by dislocation interaction are active and have deep levels. The straight glide dislocations show D3 and D4 emissions in CL, the products of dislocation interaction show D1 and D2 emissions. Stacking faults have shallower levels than dislocations. Misfit dislocations at Si/SiGe interface and Frank partial dislocations bounding oxide induced stacking faults behave similar to glide dislocations. The large angle grain boundaries (LAGBs) are intrinsically not electrically active at room temperature.

They become electrically active after Fe contamination. Incoherent GBs are more easily decorated with Fe than coherent GBs. The small angle grain boundaries (SAGBs) are composed of dislocations and categorized into two groups, namely general and special. The general ones are somewhat electrical active at room temperature and show D3 and D4 emissions. The special ones are more electrical active and show D1 and D2 emissions.

Thus, the electrical and optical activity of dislocations and GBs are similar and well understood in terms of their structure and degree of metallic decoration.

In conclusion, EBIC and CL are very useful to characterize extended defects in semiconductors.

References

1. Goldstein, J., Newbury, D.E., Joy, D.C., Lyman, C.E., Echlin, P., Lifshin, E.: Scanning Electron Microscopy and X-ray Microanalysis, 3rd edn. Springer, New York (2007). ISBN 13: 978–0306472923
2. Joy, D.C.: Ultramicroscopy, **37**, 216 (1991). doi:10.1016/0304-3991(91)90020-7; 10.1016/0304-3991(91)90020-7#doilink
3. Yacobi, B.G., Holt, D.B.: Cathodoluminescnece Microscopy of Inorganic Solids. Springer, New York (2007). ISBN 13: 978–0306433146
4. Casino, http://www.gel.usherbrooke.ca/casino/What.html
5. Kanaya, K., Okayama, S.: J. Phys. D **5**, 43 (1972)
6. Clein, C.A.: J. Appl. Phys. **26**, 380 (1955)
7. Holt, D.B., Joy, D.C. (eds.): SEM Microcharacterization of Semiconductors. Academic, New York (1989). ISBN 0-12-353855-6
8. Drozdov, N.A., Patlin, A.A., Tkachev, V.D.: Pis'ma Zh. Eksp. Teor. Fiz **23**, 651 (1976) [JETP Lett. 23, 597 (1976)]
9. Drozdov, N.A., Patlin, A.A., Tkachev, V.D.: Phys. Status Solidi B **83**, K137 (1977)
10. Sekiguchi, T., Sumino, K.: Rev. Sci. Instrum. **66**, 4277 (1995). doi:10.1063/1.1145382
11. Leamy, H.J.: J. Appl. Phys. **53**, R51 (1982). doi:10.1063/1.331667
12. Wilshaw, P.R., Fell, T.S., Coteau, M.D.: J. Phys. **IV1**, C6-3 (1991)
13. Kittler, M., Seifert, W.: Phys. Stat. Sol. A **138**, 687 (1993)
14. Kusanagi, S., Sekiguchi, T., Sumino, K.: Appl. Phys. Lett. **61**, 792 (1992)
15. Radzimski, Z.J., Zhou, T.Q., Buczkovski, A., Rozgonyi, G.A., Flinnm, D., Hellwig, L.G., Ross, J.A.: Appl. Phys. Lett. **60**, 1096 (1992)
16. Kittler, M., Seifert, W.: Mater. Sci. Eng. B. **42**, 260 (1996)
17. Kusanagi, S., Sekiguchi, T., Sumino, K.: Mater. Sci. Technol. **11**, 685 (1995)
18. Blakemore, J.S.: Semiconductor Statistics. Dover, New York (2007). ISBN 0-486-65362-5
19. Chen, J., Sekiguchi, T., Yang, D., Yin, F., Kido, K., Tsurekawa, S.: J. Appl. Phys. **96**, 5490 (2004)
20. Suezawa, M., Sasaki, Y., Sumino, K.: Phys. Status Solidi A **79**, 173 (1983)
21. Sauer, R., Weber, J., Stolz, J., Weber, E.R., Kusters, K.-H., Alexander, H.: Appl. Phys. A **36**, 1 (1985)
22. Lerikov, Y., Rebane, Y., Ruvimov, S., Tarhin, D., Sitnikova, A., Shreter, Y.: Mater. Sci. Forum **83–87**, 1321 (1992)
23. Higgs, V., Lightowlers, E.C., Norman, C.E., Kightley, P.C.: Mater. Sci. Forum **83–87**, 1309 (1992)
24. Sekiguchi, T., Kveder, V., Sumino, K.: J. Appl. Phys. **79**, 3253 (1996)
25. Sekiguchi, T., Sumino, K.: J. Appl. Phys. **79**, 3253 (1996)

26. Kittler, M., Aguirov, T., Seifert, W., Yu, X., Gia, G., Vivenko, O.F., Mchedlidze, T., Reiche, M., Sha, J., Yang, D.: Mater. Sci. Eng. C **27**, 1252 (2007). doi:10.1016/j.msec.2006.09.034
27. Yu, X., Aguirov, T., Kittler, M., Seifert, W., Ratzke, M., Reiche, M.: Mater. Sci. Semicond. Process. **9**, 96 (2006). doi:10.1016/j.mssp.2006.01.070
28. Takagi, S., Hoyt, J.L., Welser, J.J., Gibbons, J.F.: J. Appl. Phys. **80**, 1567 (1996). doi:10.1063/1.362953
29. Yuan, X.L., Sekiguchi, T., Li, S.G., Ito, S.: Appl. Phys. Lett. **84**, 3316 (2004). doi:10.1063/1.1734688
30. Yuan, X.L., Sekiguchi, T., Niitsuma, J., Sakuma, Y., Ito, S., Li, S.G.: Appl. Phys. Lett. **86**, 162102 (2005). doi:10.1063/1.1905802
31. Kawado, S.: Jpn. J. Appl. Phys. **19**, 1591 (1980)
32. Sekiguchi, T., Shen, B., Watanabe, T., Sumino, K.: Mater. Sci. Eng. B. **42**, 235 (1996). doi:10.1016/S0921-5107(96)01713-8
33. Shen, B., Sekiguchi, T., Jabronski, J., Sumino, K.: J. Appl. Phys. **76**, 4540 (1996). doi:10.1063/1.357285
34. Shen, B., Zhang, R., Shi, Y., Sheng, Y.D., Sekiguchi, T., Sumino, K.: Appl. Phys. Lett. **68**, 214 (1996). doi:10.1063/1.116464
35. Nara, S., Sekiguchi, T., Chen, J.: Eur. Phys. J. Appl. Phys **27**, 389 (2004)
36. Chen, B., Chen, J., Sekiguchi, T., Saito, M., Kimoto, K.: J. Appl. Phys. **105**, 113502 (2009)
37. Chen, J., Yang, D.R., Xi, Z.Q., Sekiguchi, T.: Physica B **364**, 162 (2005). doi:10.1016/j.physb.2005.04.008
38. Lee, W., Chen, J., Chen, B., Kang, J., Sekiguchi, T.: Appl. Phys. Lett. **94**, 112103 (2009). doi:10.1063/1.3099001

Chapter 8
Nuclear Methods to Study Defects and Impurities in Si Materials

Guido Langouche and Yutaka Yoshida

Abstract The hyperfine interaction between the atomic nucleus and its surrounding charge and electromagnetic field distribution is extremely sensitive to the atomic and electronic configuration of this atom. In the field of defects and impurities in semiconductors, the study of their hyperfine interaction can therefore contribute substantially to their identification and characterization. The introduction of radioactive isotopes as impurity atoms allows to probe the hyperfine interaction of extremely low quantities of such impurities. Several dedicated nuclear methods such as Mössbauer Spectroscopy, Perturbed Angular Correlations and Low Temperature Nuclear Orientation allow to measure the hyperfine interaction at the nuclear site of the impurity atom by analysing the radiation emitted by these probe nuclei. The Emission Channelling technique, on the other hand, allows studying the precise lattice site location of the probe atoms from the channelling behaviour of the particles emitted by these probe nuclei.

In the first part of this chapter we will describe the basic principles of these nuclear methods and illustrate them with a few relevant examples.

In the second part of this chapter we will focus on a series of recent Mössbauer spectroscopy studies which have been set up to study ^{57}Fe solute atoms in Si wafers and solar cells under special conditions such as under applied voltage and external illumination. Finally, we will describe the basic principles of a Mössbauer microscope, allowing to map the presence of Fe across a Si wafer.

Keywords Hyperfine interactions • Mössbauer Spectroscopy • Perturbed Angular Correlations • Emission channelling technique • Fe in Si materials • Substitutional and interstitial Fe impurities

G. Langouche
University of Leuven (KU Leuven), Leuven, Belgium
e-mail: guido.langouche@kuleuven.be

Y. Yoshida (✉)
Shizuoka Institute of Science and Technology, Fukuroi, Japan
e-mail: yoshida@ms.sist.ac.jp

Y. Yoshida, G. Langouche (eds.), *Defects and Impurities in Silicon Materials*, Lecture Notes in Physics 916, DOI 10.1007/978-4-431-55800-2_8

8.1 Hyperfine Interactions

8.1.1 Introduction to Hyperfine Interactions

The term "hyperfine interactions" originated in the nineteenth century when superimposed on the fine structure of atomic spectra, attributed to the splitting of electronic levels due to spin-orbit coupling, a much finer substructure was experimentally observed. It was only in 1920, when Pauli introduced nuclear spin and the associated nuclear magnetic moment, that the hyperfine structure in atomic spectra could be accounted for as due to the interaction of the atomic electrons with this nuclear moment. To this magnetic interaction was soon the quadrupole interaction added, between the nuclear quadrupole moment and the electronic electric field gradient.

It lasted until halfway the twentieth century until this hyperfine interaction was also observed in nuclear radiation. The nuclear hyperfine splitting could only be directly observed after the discovery of Rudolf Mössbauer in 1958 of the recoil-free emission and absorption of gamma-radiation by nuclei. Indirectly the presence of hyperfine splitting of nuclear levels could be observed in various phenomena influencing the nuclear radiation emitted by these nuclei. This led to a number of experimental techniques like perturbed angular correlation spectroscopy, low-temperature nuclear orientation, and ß-NMR. We will briefly discuss two of these techniques and illustrate them with an example, since they were also used to gather information on defects in semiconductors, the topic of this book.

For a description of the underlying theory of hyperfine interactions many textbooks are available. We especially refer to *Nuclear Condensed Matter Physics: Nuclear Methods and Applications* [1], meant as a textbook for university students entering the field of hyperfine interactions, and to *Hyperfine Interaction of defects in semiconductors* [2], which is specifically devoted to applications in semiconductor physics.

8.1.2 Mössbauer Spectroscopy

In 1958 Rudolf Mössbauer discovered the recoil-free emission and absorption of gamma radiation in nuclear transitions [3]. It earned him the Nobel Prize in 1961.

As illustrated in Fig. 8.1 the emission and absorption of gamma-radiation by nuclei leads to a shift and a broadening of the transition energy, due to the recoil of the atom and to lattice vibrations in the solid in which the atom is embedded. The existence of a zero-phonon transition, as discovered by R. Mössbauer, makes the resonant absorption in an absorber of the radiation emitted from a radioactive source possible. It can easily be made visible by Doppler-shifting the energy of the nuclear transition with moderate velocities (of the order of mm/s) as illustrated

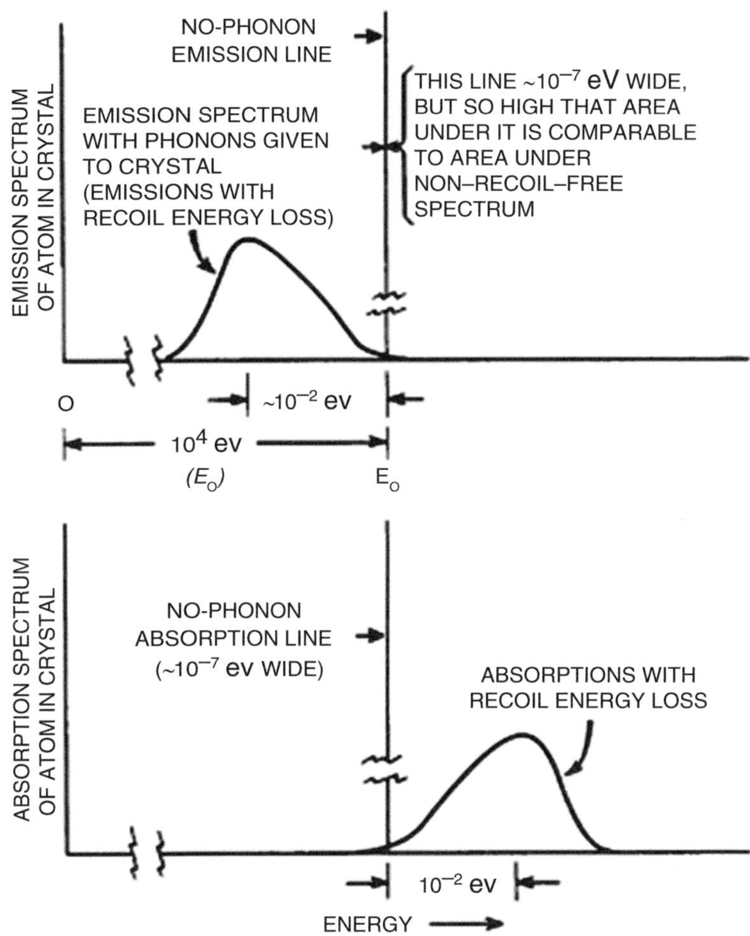

Fig. 8.1 Recoilless emission and absorption of gamma radiation (From Ref. [4])

in Fig. 8.2, and recording a so-called Mössbauer spectrum, in which the hyperfine splitting is made directly visible.

There are four main observables in a Mössbauer spectrum:

1. The *isomer shift*, which is the velocity position of the Mössbauer resonance. It indicates how much the transition energy in the absorber is different from the transition energy in the source, due to a different electron density at the nucleus. For practical use, the isomer shift is always expressed as an absorber isomer shift with respect to a certain standard material (like α-iron in the case of ^{57}Fe). Since the value of the isomer shift is determined by the electron density at the nucleus it is an excellent indication of the charge state of the probe atom (e.g. Fe^{2+} or Fe^{3+}).

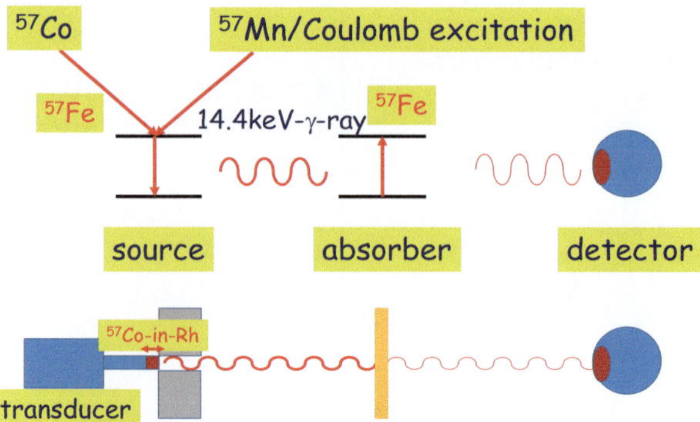

Fig. 8.2 Principle of emission Mössbauer spectroscopy and a configuration for absorption experiments

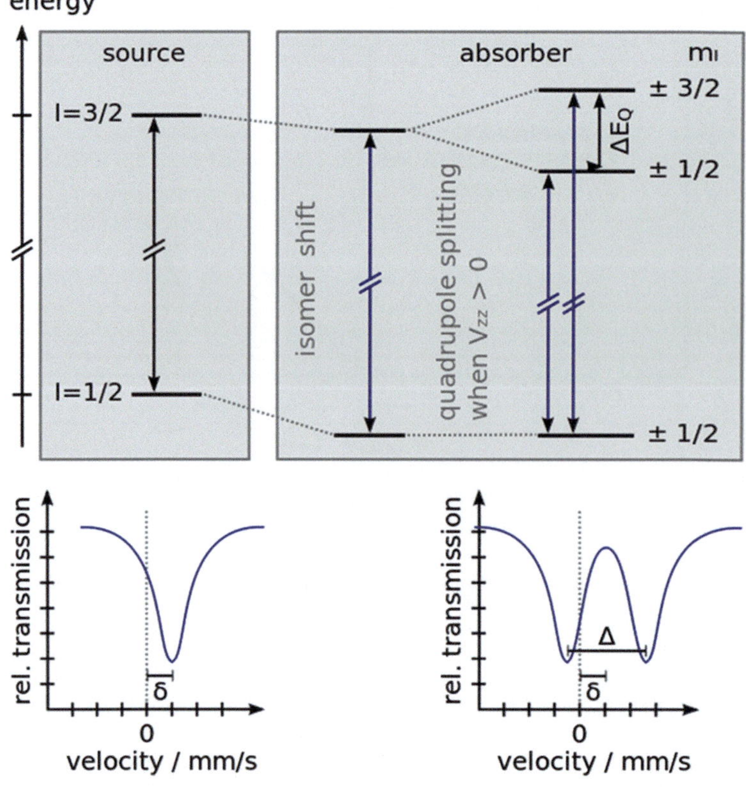

Fig. 8.3 Isomer shift and quadrupole splitting in a Mössbauer experiment (From Ref. [5])

2. The *hyperfine splitting*, which is the splitting of the Mössbauer resonance. Figure 8.3 shows the hyperfine splitting when an electric field gradient is present

at the site of the nucleus and interacts with the quadrupole moment of the excited nuclear state. In this simple case of a $3/2 \rightarrow 1/2$ transition, the resulting spectrum is a quadrupole doublet, the splitting of which is proportional to the electric field gradient at the nuclear site. The electric field gradient reflects in the first place the symmetry of the surrounding charge distribution and is e.g. zero when this symmetry is cubic. A probe atom occupying a substitutional or regular interstitial position without nearby lattice defects in a Si lattice will also have a zero quadrupole splitting. Mössbauer spectroscopy had an overwhelming use in magnetism studies. The magnetic splitting of a $3/2 \rightarrow 1/2$ transition gives rise to a six-line Mössbauer spectrum.

3. The *recoil-free fraction*, which is directly proportional to the intensity of the Mössbauer resonance. It measures the probability for recoilless emission and absorption of the gamma-radiation, and it is not surprising that it is strongly temperature dependent. The higher the transition energy, the lower the recoil-free fraction. Few Mössbauer transitions possess a substantial recoil-free fraction above room temperature, ^{57}Fe being a happy exception. The recoil-free fraction probes the phonon density of states at the site of the probe atom and is therefore different for probe atoms occupying different lattice sites or having defects in their surroundings. Since the recoil-free fraction has a well-understood temperature dependence, the recording of Mössbauer spectra as a function of temperature often allows disentangling complicated Mössbauer spectra.
4. The *linewidth* can also contain information on the solid in which the probe atom is embedded. The natural linewidth of a Mössbauer resonance is solely determined by the lifetime of the excited nuclear state, through Heisenberg's uncertainty principle. Apart from instrumental broadening of the Mössbauer level and broadening due to defects affecting the electron density, there is another potential source of broadening, which is temperature dependent. This occurs if the probe atom starts to move during the lifetime of the excited nuclear state (100 ns in the case of ^{57}Fe). As explained e.g. in *Mössbauer Spectroscopy: Tutorial Book* [6] diffusion of the probe atom gives rise to line broadening as well as particular phenomena like cage motion of the probe atom.

A Mössbauer splitting can occur as well in the absorber as in the source used to record the Mössbauer spectrum, or even in both. For the purpose of this book on defects in Si, it is important to dwell somewhat on this topic.

The terminology *absorption Mössbauer Spectroscopy* is used to designate experiments where the topic of investigation are Mössbauer probe atoms incorporated in the absorber. This accounts for more than 90 % of all Mössbauer studies. The major advantage is that it does not require handling of radioactivity. One uses a standard radioactive source which does not show any hyperfine splitting, a so-called "single line source" and the material of interest is the absorber. It does require a sufficient amount of Mössbauer probe atoms to be present in the absorber and a radioactive source of sufficient strength in order to be able to measure a Mössbauer spectrum within a reasonable amount of time. For impurities in semiconductors, which are the topic of this book, this is at the extreme of what is possible.

The terminology *emission Mössbauer spectroscopy*, on the other hand, is used for experiments in which the topics of investigation are Mössbauer probe atoms incorporated in the source. This requires the preparation of compounds containing radioactive material and hence a hot chemistry laboratory and adequate security. A subset of emission Mössbauer spectroscopy experiments is making use of implanted radioactive probes. A whole chapter in the recent book "*Mössbauer Spectroscopy: Tutorial Book*" [6] is devoted to this subject. The major advantage of this preparation method is the very low amount of probe atoms needed. Mössbauer probe atoms, under these circumstances can truly be characterized as observers or even "spies" in the solid, as is shown in Fig. 8.4.

The study of Fe in Si has challenged Mössbauer physicists for over 50 years. The complexity of the behaviour of Fe in Si was clearly underestimated in earlier studies where the role of precipitate formation during diffusion of ^{57}Fe (or its parent ^{57}Co) was insufficiently realized. Ion implantation was believed to avoid precipitate formation but also gave rise to complex spectra and interpretation problems.

After ion implantation the Mössbauer spectra of ^{57}Fe in Si in emission Mössbauer spectroscopy after implanting radioactive ^{57}Co [7] was dominated by two single lines. Also in early in-beam experiments by the Stanford group, starting from the ^{57m}Fe parent state, two lines were observed [8, 9] with asymmetric intensities. It was thought that the two single lines might be representative of two implantation sites, presumed to be substitutional and interstitial Fe in Si, but the observed dose dependence could not be accounted for. By applying an external magnetic field it could unambiguously be shown [10] that the high fluence ^{57}Fe (Si) spectrum was a quadrupole interaction doublet, with parameters very similar to those of amorphous Fe_xSi_{1-x} films [11] and hence associated with Fe atoms in an amorphized surrounding.

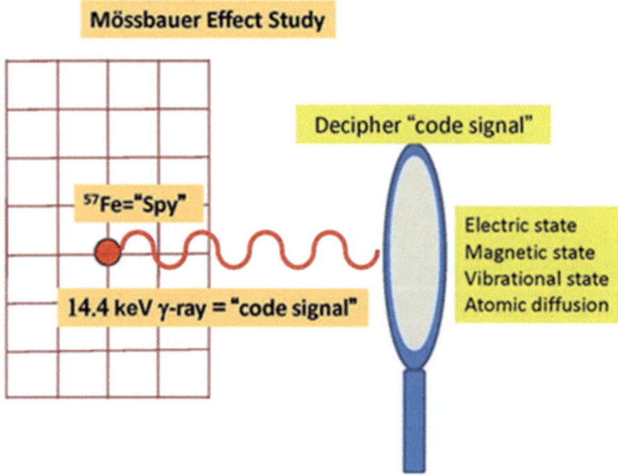

Fig. 8.4 Mössbauer probe acting as a "spy" in a solid (From Ref. [6])

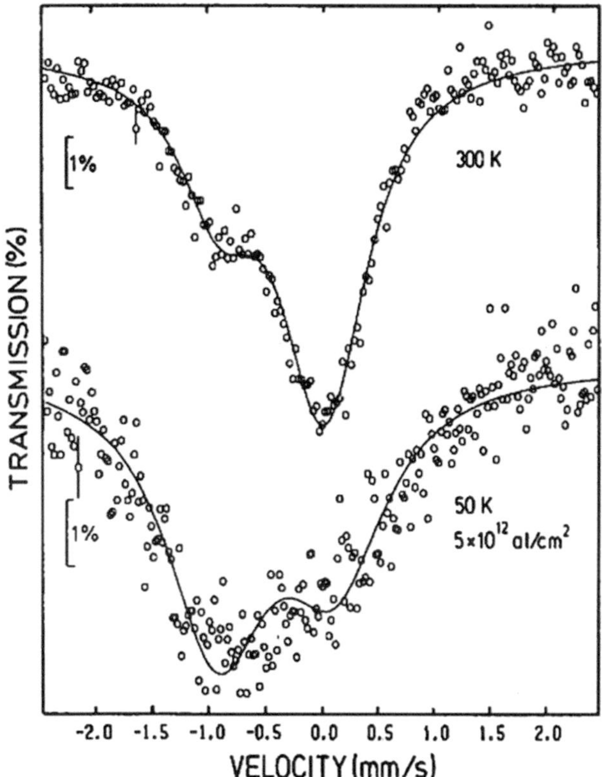

Fig. 8.5 ^{57}Fe spectra after ion implantation of ^{57}Co at 50 K and after subsequent annealing (From Ref. [12])

The origin of the asymmetry at lower fluencies was first accounted for by new emission Mössbauer experiments whereby ^{57}Co was implanted into Si held at the temperature of 50 K [12]. This cold implantation completely reversed the asymmetry (Fig. 8.5) in the spectrum compared to room temperature implanted samples. The model put forward was that at low temperatures, where vacancies are not mobile, ^{57}Co/^{57}Fe atoms land up in interstitial sites in Si, a site preferred by transition metals in Si. Upon increasing the temperature to room temperature, vacancies become mobile at about 100 K and are trapped by the interstitial Co atoms which become substitutional.

This model is basically still valid today. Many sophisticated experiments have been set up and fine details have been revealed about the behaviour of Fe in Si [5].

One such experiment is a Coulomb excitation recoil implantation study [13] at the Hahn-Meitner Institute in Berlin which led to an unambiguous identification of interstitial Fe in Si. The spectra, shown in Fig. 8.6, were interpreted as composed of the "amorphous" doublet and a superimposed single line (with isomer shift $\delta = +0.84(1)$ mm/s with respect to α-Fe).

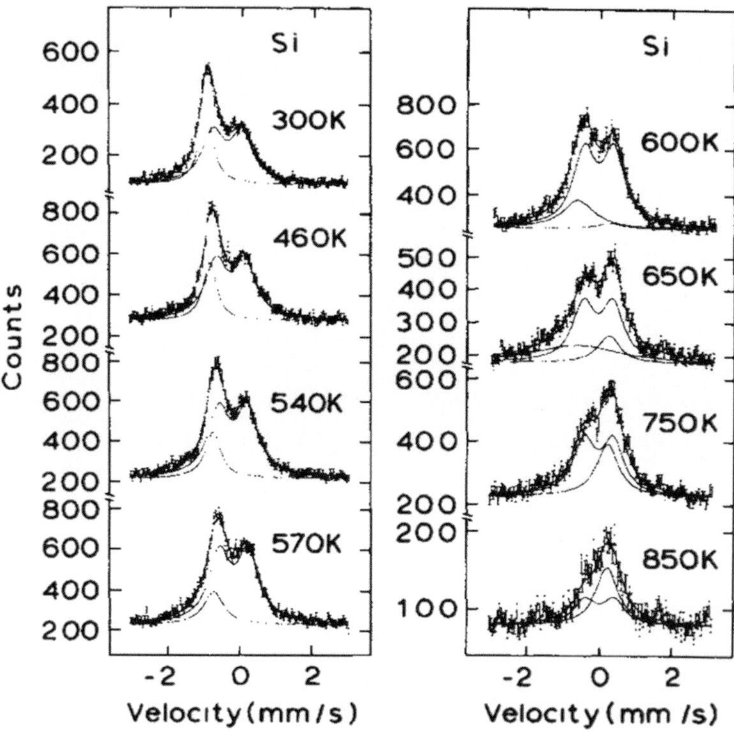

Fig. 8.6 After Coulomb-excitation and recoil- implantation of ^{57m}Fe in Si in-beam Mössbauer spectra were measured at temperatures between 300 and 850 K (From Ref. [13])

Due to the excellent statistics, an increase in the linewidth of this single line could be observed for higher target temperatures. This line broadening was analysed in terms of a diffusional motion of the iron atoms, which are able to make atomic jumps within the lifetime of the excited nuclear state. Diffusion coefficients were derived from the broadening of the Mössbauer lines, and these were found to be remarkably consistent with the known high- and low-temperature diffusion coefficients of Fe in Si.

Even more detailed information on Fe diffusion in Si could be obtained in on-line experiments at the ISOLDE facility at CERN where the very short-lived ($T_{1/2} = 1.5$ min) parent ^{57}Mn isotope, decaying to the ^{57}Fe Mössbauer transition, was implanted in Si samples with different doping levels: extremely p-type (p^{++}) and highly n-type (n+) [14]. As shown in Fig. 8.7, from the line broadening of the interstitial Mössbauer line charge state dependent diffusivity could be inferred.

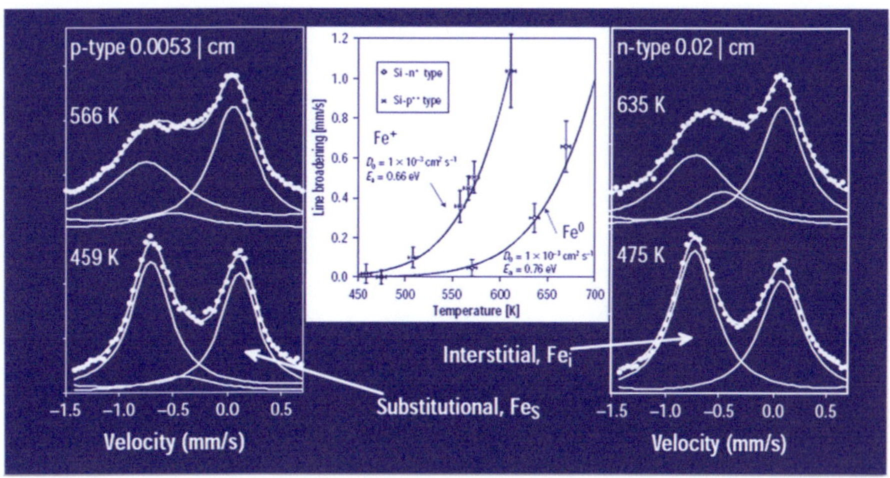

Fig. 8.7 Mössbauer spectra of ^{57}Mn/^{57}Fe in p-type and n-type Si wafers. The diffusivities were estimated from the line broadenings observed above 500 K (From Ref. [14])

8.1.3 *Perturbed Angular Correlations*

The hyperfine interaction can also be studied in an experiment where two subsequent gamma-rays, which are emitted in a decay scheme of a radioactive nucleus, are detected in two detectors as a function of time between the arrivals of the two gamma quanta. Since nuclei do not emit their radiation isotropically with respect to their spin direction, detecting the first of the two gammas by a fixed detector selects an ensemble of nuclei with their spin oriented preferentially with respect to the observation direction. During the lifetime of the intermediate level, the population of the hyperfine-split nuclear Zeeman levels is modified due to the reorientation of the nuclear spin, an effect which in a semi-classical picture is called the Larmor-precession of the nuclear spin around the direction of the hyperfine field. Hence, the frequency of this rotation can be monitored by detection of the second γ-ray by a second detector, making a fixed angle with the first one.

When the coincidence rate between the detection of γ_1 and γ_2 is measured as a function of time, then one observes a perturbation curve superimposed on the exponential nuclear lifetime decay curve, which clearly shows the spin precession (Fig. 8.8).

The amount of radioactive nuclei which is needed is orders of magnitude lower than in Mössbauer absorption spectroscopy and of the same magnitude as in Mössbauer emission spectroscopy. An absolute amount of 10^{11}–10^{12} radioactive probes is sufficient. While ^{57}Fe is the workhorse of Mössbauer Spectroscopy, because of its favourable parameters (high recoilless fraction, small natural linewidth), the most common Perturbed Angular Correlation probe is ^{111}Cd [parent ^{111}In].

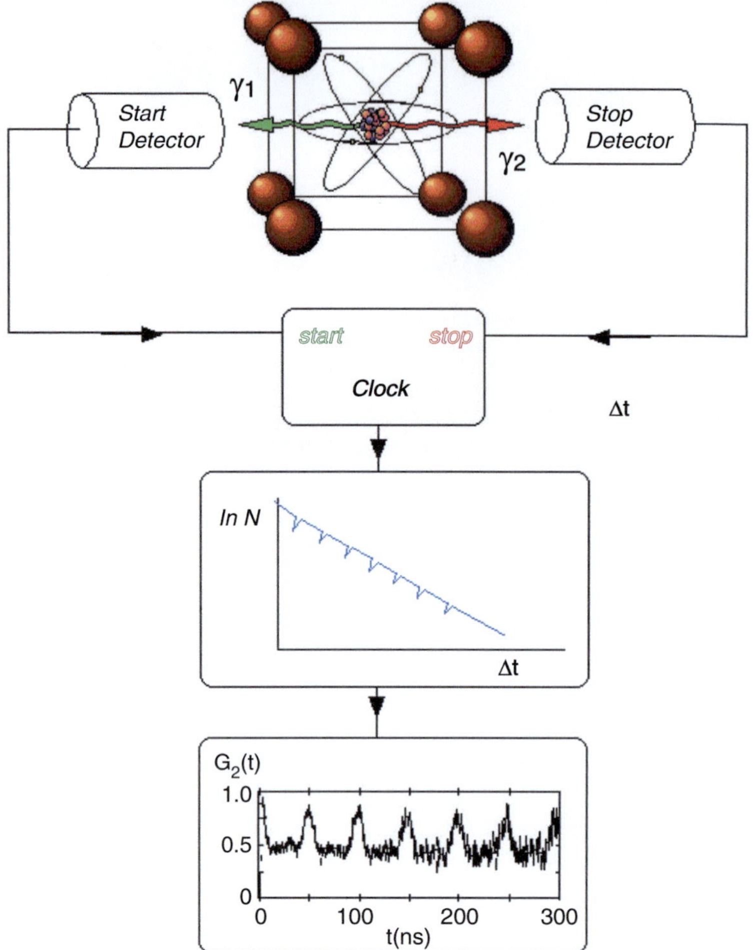

Fig. 8.8 Measurement scheme for Perturbed Angular Correlations (From Ref. [15])

The main observable in Perturbed Angular Correlation spectroscopy is the hyperfine splitting. Compared to Mössbauer spectroscopy there is no observable like the isomer shift or the recoil-free fraction. The latter is often considered an advantage in comparison to Mössbauer spectroscopy since it makes Perturbed Angular Correlation technique insensitive to temperature. Whereas the recoilless fraction is strongly temperature dependent and prevents using Mössbauer Spectroscopy at too high temperatures, Perturbed Angular Correlation spectroscopy is better suited to follow phenomena as a function of temperature.

The sensitivity on the hyperfine interactions strength is different. If interactions become too strong, the oscillations in the perturbation curve become too fast to be detected. Therefore, there is an upper limit to the interaction strength detectable by Perturbed Angular Correlations, which is not present for Mössbauer Spectroscopy. Both methods have a lower detectable limit in hyperfine interaction strength. If the interaction is too weak, a single oscillation in the perturbation curve takes longer than the time interval where enough statistics can be gathered. This determines the lower limit of the detectable interaction strengths. In Mössbauer Spectroscopy the hyperfine splitting can become so small that it is only present as a line broadening, for which there might be other competing origins. In practice this will depend on the isotope used.

The perturbation function sometimes shows a damping effect. This happens when a distribution of interaction frequencies is present. It can also be a sign of motion of the probe atom, accompanied by a reorientation of the spin taking place, giving rise to the loss of coherence over time. It allows studying diffusion phenomena also in Perturbed Angular Correlation spectroscopy.

A typical example relevant for this book on defects in Si is the study of pair formation in Si, studied by Perturbed Angular Correlation Spectroscopy [16]. Upon introduction of ^{111}In in As-doped Si, three different InAs pairs were found, dependent on the As concentration in the annealing temperature (Fig. 8.9).

8.1.4 ß-NMR

The ß-NMR is a technique to study hyperfine interactions that was developed around 1965 [17]. It is an extremely sensitive on-line technique using short-lived radioactive isotopes that decay by ß-emission. An ensemble of polarized nuclei is created, normally by selecting nuclei that recoil in a selected direction in the reaction plane. The target is placed in an external magnetic field and the degree of polarization of these ß-emitters is measured by a pair of plastic scintillators and attached detectors. An NMR signal from a small rf-coil is applied to destroy the polarization, and the frequency at which this happens is directly proportional to the hyperfine interaction strength. The degree of destruction is proportional to the number of nuclei that experience that particular hyperfine interaction strength. In the case of a nuclear quadrupole interaction several rf-frequencies, corresponding to the hyperfine transitions involved, are applied simultaneously to destroy the polarization. Figure 8.10 shows a diagram of the experimental set-up at ISOLDE.

The ß-NMR technique has been applied to study implanted boron in silicon [19, 20]. An increasing amount of substitutional boron was detected as function of target temperature as shown in Fig. 8.11 [20]. A well-defined electric field gradient could be assigned to B atoms in a non-substitutional configuration with <111> symmetry, but the exact configuration of this site could not be determined yet.

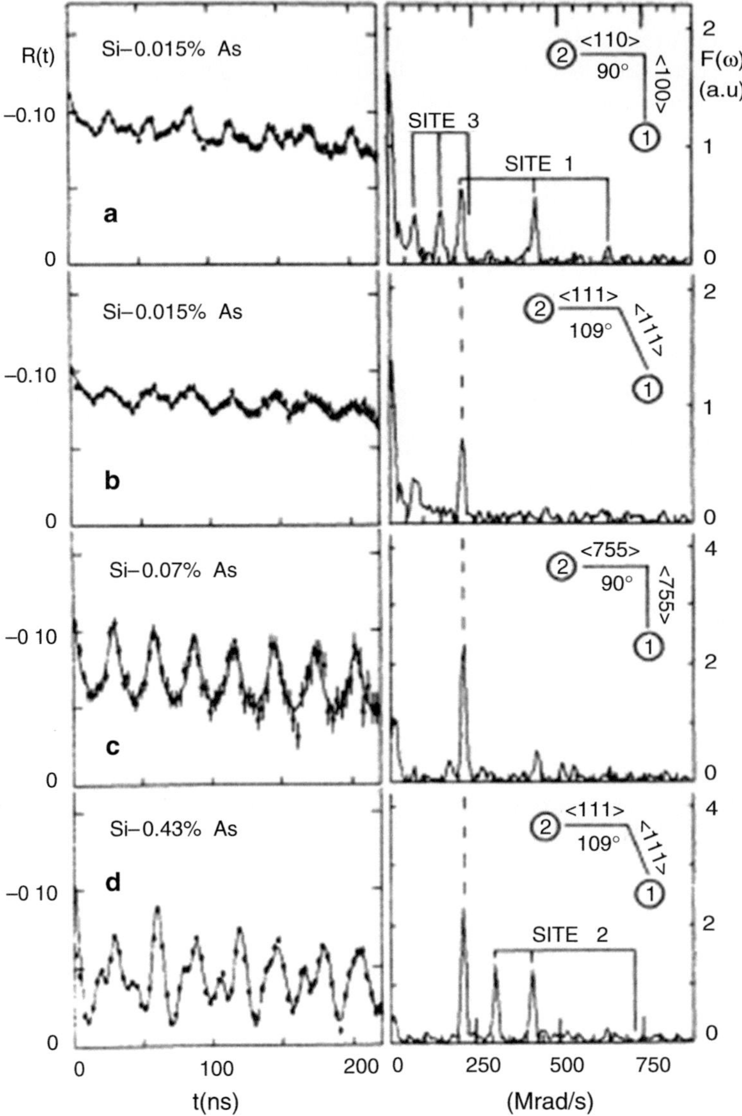

Fig. 8.9 Observation of three different InAs pairs in As-doped Si (From Ref. [16])

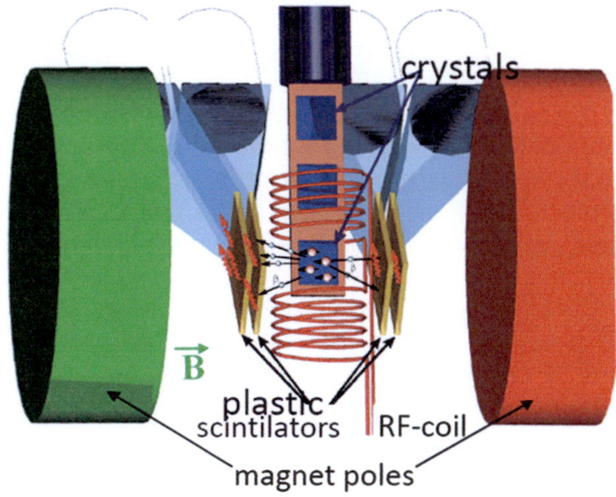

Fig. 8.10 Sketch of ß-NMR set-up at ISOLDE (From Ref. [18])

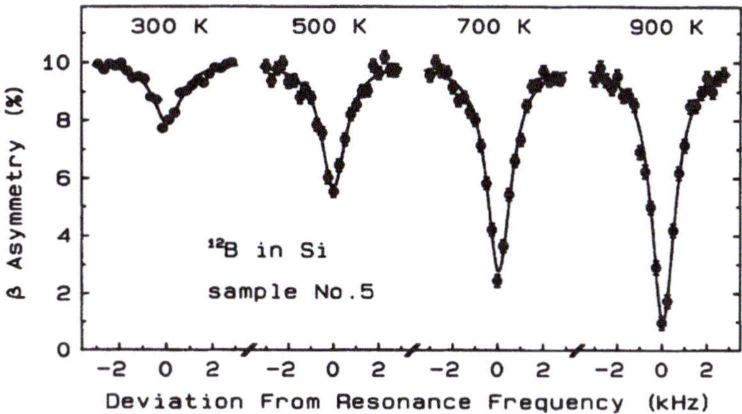

Fig. 8.11 ß-NMR signal of increasing amount of substitutional B atoms implanted into Si as a function of implantation temperature (From Ref. [20])

8.1.5 Non-hyperfine Interaction Techniques Involving Short-Lived Radioactive Isotopes to Study Defects in Semiconductors

Short-lived radioactive nuclei produced at high energy accelerators are also potential probes for other techniques than hyperfine interaction techniques. Among such techniques that have been used to study defects in semiconductors are emission channelling, tracer diffusion, deep level transient spectroscopy (DLTS), photoluminescence (PL) and Hall effect (HE). Tracer diffusion involving radioactive isotopes is a standard technique to measure diffusion rates. Not too short lifetimes are often required. The wide choice of available radioactive isotopes at facilities like ISOLDE offers ample possibilities to study diffusion of impurities in semiconductors [21].

Radioactive DLTS [22], PL [23] and HE [24] techniques are essentially the same as their non-radioactive counterparts. After the radioactive probe atom has been introduced into the semiconductor, measurements are performed as a function of time after the introduction of the probe atom. Parts of the DLTS, PL or HE signals which show a decrease in intensity following the nuclear lifetime of the radioactive probe atom can then with certainty be assigned to this probe atom. Emission channelling is the most intensively used method of these non-hyperfine interactions techniques and deserves a separate treatment since it differs substantially from classical channelling experiments.

8.1.5.1 Emission Channelling

The emission channelling technique (Fig. 8.12) differs on several aspects from conventional external channelling with α-particles. The short-lived radioactive probe atoms that are used in emission channelling experiments are in general electron or positron emitters. While positrons will travel along open channels much like α-particles, this is not the case for electrons. Since the emitted particles are guided by the potential of crystal axes or planes, positively charged particles will be guided through channels between crystal nuclei, the negatively charges particles will be channelled along a row or plane of nuclei, as shown in Fig. 8.13.

Essential in the emission channelling experiment is the two-dimensional position-sensitive detector [26] with the appropriate properties for detection of electrons or positrons in the decay of the short-lived radioactive probe. The result of a typical emission channelling experiment is shown in Fig. 8.14.

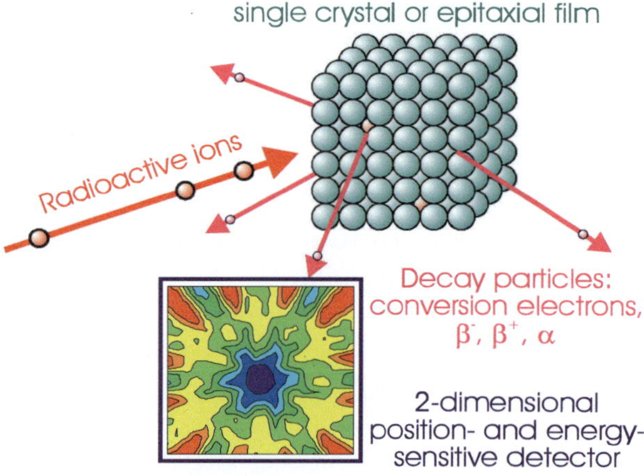

Fig. 8.12 Emission channelling scheme (From Ref. [25])

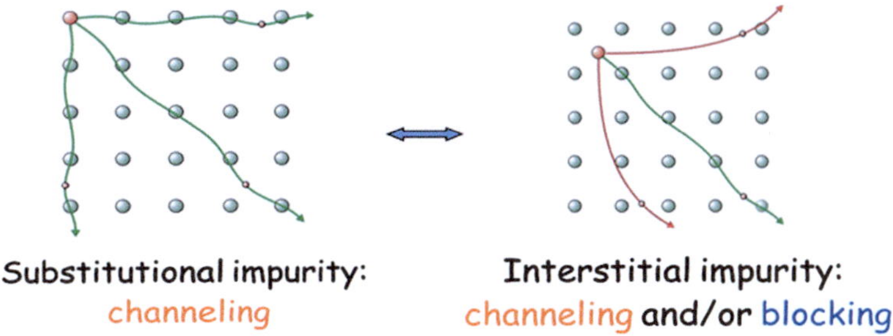

Fig. 8.13 Emission channelling of electrons (From Ref. [25])

From the analysis of the results, based on a comparison of the measured channelling patterns and simulated ones, one has to conclude that the implanted Er atoms reside preferentially on the regular tetrahedral interstitial position in the Si lattice.

8.1.5.2 Emission Channelling Study of Fe in Si

An interesting result with respect to the discussion later in this chapter, devoted to Fe in Si, is the emission channelling behaviour of implanted Fe in Si (Fig. 8.15). The implanted isotope is ^{59}Mn which is radioactive ($T_{1/2} = 4.6$ s) and decays with an average recoil of 200 eV to ^{59}Fe. The emission channelling experiment uses ß-radiation from ^{59}Fe which has a lifetime $T_{1/1} = 44.6$ days. The temperature dependence of the fractions observed in de emission channelling experiments shown in Fig. 8.15 are plotted in Fig. 8.16. No difference is observed between the different types and amounts of Si-doping used in these experiments. This is consistent with the fact that defect concentration created during the implantation process is about three orders of magnitude higher than the dopant concentration.

A more quantitative statement can be made about the lattice sites. The "mainly ideal substitutional site" means that the deviation from the equilibrium point is at most 0.079 Å. The value of 0.079 Å is a best guess assuming that Fe acts as a simple mass defect in the Si lattice.

The "displaced substitutional sites" are sites exhibiting static displacements around 0.4–0.7 Å from the ideal *S* position, and the "displaced tetrahedral interstitial sites" are located 0.3–0.8 Å from the ideal *T* position. Both in case of the near-*S* and near-*T* sites all possible displacements from the ideal *S* and *T* sites along <111> and <100> directions were tried in the fits. However, similar chi square values were obtained and the direction of the static displacement could therefore not be pinpointed.

It looks surprising that no Fe atoms were observed on the ideal tetrahedral site. This might be due to the fact that the emission channelling experiments were performed weeks after the implantation and the abundant vacancies had plenty

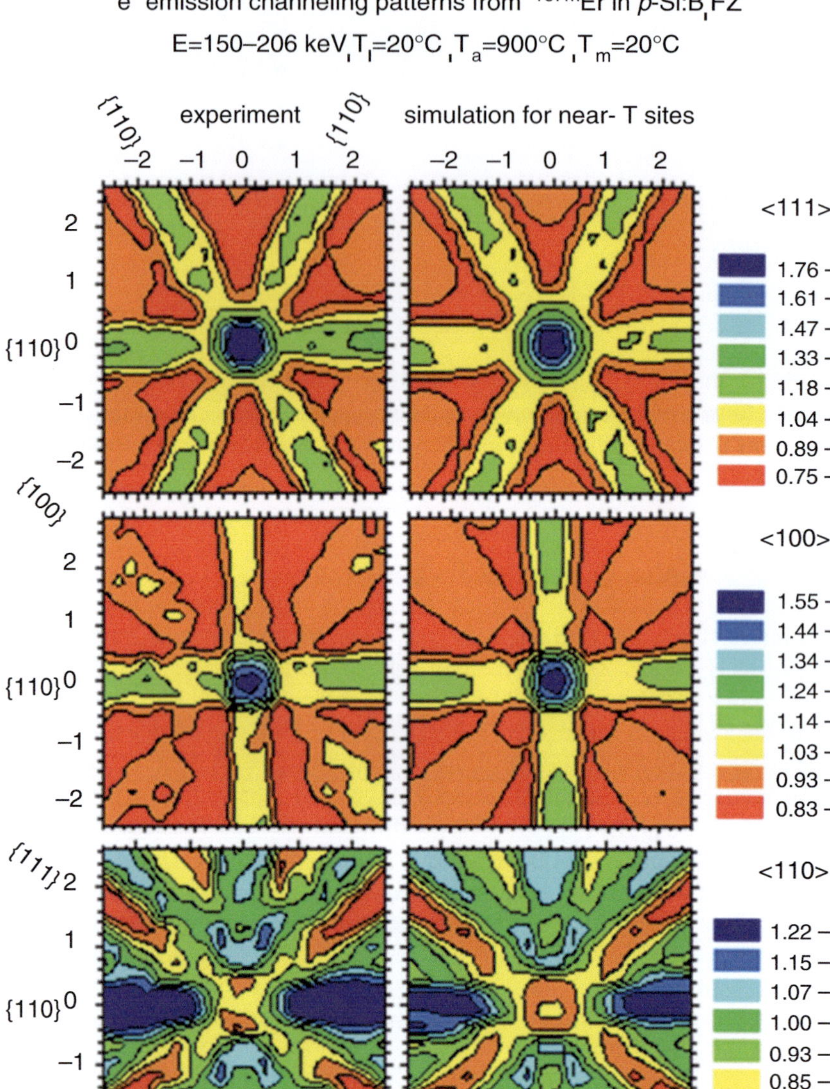

Fig. 8.14 Emission channelling of Er in Si (From Ref. [27])

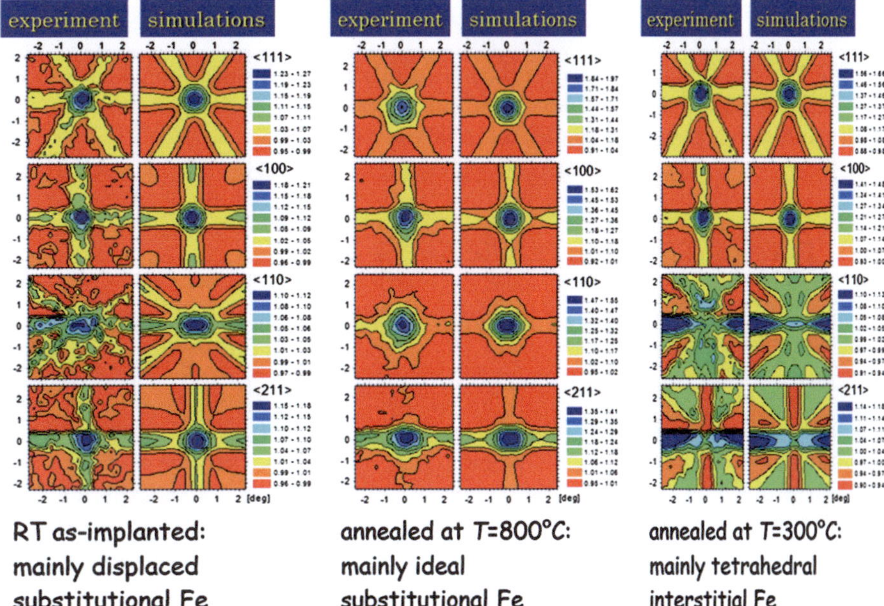

Fig. 8.15 Emission channelling of ^{59}Fe in Si (From Ref. [28])

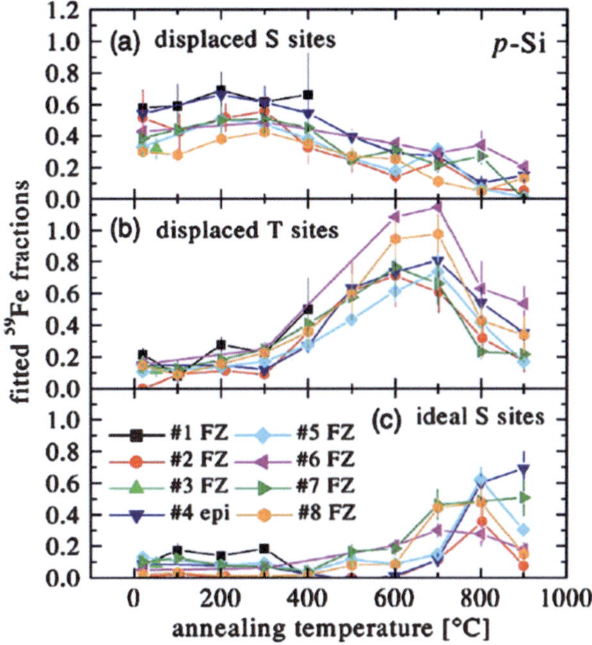

Fig. 8.16 Temperature dependence of the sites shown in Fig. 8.15 (From Ref. [28])

of time to associate with Fe_i. The “displaced tetrahedral interstitial sites” might therefore well correspond to Fe_iV_{Si} pairs.

The emission channelling observation of Fe on ideal substitutional sites is the most direct experimental evidence for substitutional Fe in Si, in nice agreement with the findings from Mössbauer spectroscopy.

8.2 Recent Mössbauer Studies on Fe in Si

8.2.1 Introduction to the Problems on Fe Impurities in Si

In this section we concentrate into the problems of Fe impurities in Si crystals. Metallic impurities in silicon materials have been studied for more than 70 years as one of the central research themes [29, 30]. There remain, however, still important questions unsolved. Silicon material industries are often suffering from contaminations during the device processes. Si solar cells, for instance, contain Fe impurities with a concentration from 10^{11} to 10^{17} Fe/cm^3, which strongly degrade the energy conversion efficiency. Fe atoms are considered to be “strong trapping centres” for carriers in the Si matrix, occupying on an interstitial site, as shown in Fig. 8.17, with a rather high diffusivity of 10^{-14}cm^2/s even at room temperature. Their distribution depends strongly on the defects such as vacancies, dislocations and grain boundaries in the crystals.

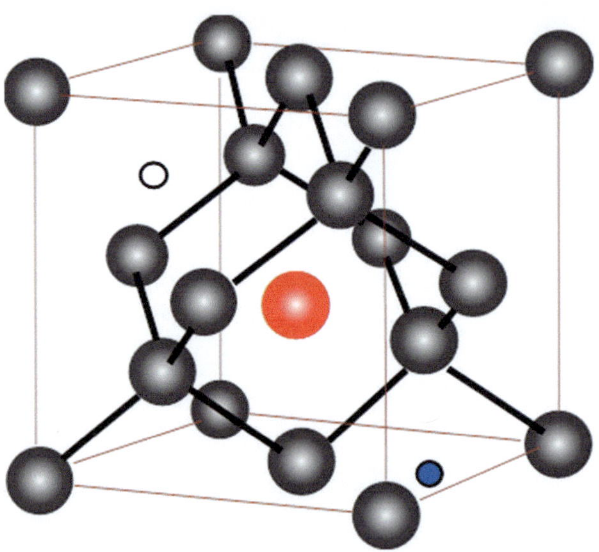

Fig. 8.17 Fe interstitial in Si matrix

An "interstitial" iron mapping was considered to be possible so far in B-doped Si using Photo-Luminescence (PL) [31] and μW-PCD [32]. The analysis was based on a Zoth and Bergholz model [33], taking into account the measurements on the carrier life times before and after breaking Fe-B pairs. This mapping technique can only be applied for electrically active Fe impurities, but not for inactive Fe impurities. Accordingly, a direct observation of all Fe impurities including interstitial Fe and Fe-B pairs appears to finally be necessary *on an atomistic scale*, in order to achieve not only the picture for Fe impurities in Si materials, but also Si wafer and device fully controlled processes.

Mössbauer spectroscopy appears to be ideal to characterize Fe impurities in Si crystal [7–14, 34–36]. Recently, a series of the experimental investigations [36–60] has been performed using Mössbauer spectroscopy on Fe impurities in p-type and n-type Si materials such as single crystal Si, multi-crystalline (mc) Si wafers, and even mc-Si solar cells. These results lead to the hypothesis that we need a new picture for Fe impurities in Si materials: the Fe impurities in the Si matrix exist not only on the interstitial sites with ${Fe_{int}}^{0/+1}$ and ${Fe_{int}}^{+}$-B^{-} pair, but also on a higher charge state of ${Fe_{int}}^{2+}$ associated with defects as well as on the substitutional sites with ${Fe_{sub}}^{0}$ and ${Fe_{sub}}^{-}$. Furthermore, these components are found to transform to each other in the Mössbauer spectra by changing experimental conditions such as temperature, external voltage, electron or light irradiation and external stress as well as by changing the Fermi level, carriers and their concentrations, and the device structures in the vicinity of ^{57}Fe probes.

In the following sections, first of all, recent results from Mössbauer spectroscopy will be discussed, summarizing the isomer shifts for different Fe components. Secondly will be presented: Mössbauer absorber spectra after annealing at 1273 K, measurements at high temperatures from 1273 K down to room temperature, Mössbauer absorption spectra after 100 keV ^{57}Fe implantation, and finally spectra under external stress. Furthermore, in order to discuss isolated Fe in Si materials, on-line emission spectra after GeV ^{57}Mn implantation in Si wafers will also be shown. Based on this complete series of experimental results a consistent interpretation will be presented for the components appearing under different experimental conditions.

8.2.2 *Isomer Shift to Distinguish Fe Impurities in Si*

Mössbauer spectroscopy resolves the hyperfine interaction between electrons and a nuclear probe, i.e. ^{57}Fe in the present case, observing directly the shifts and/or the splitting of the nuclear levels through recoil-free γ-ray resonant absorption or emission, which is called Mössbauer effect and is described in Sect. 8.1.2 of this chapter. The typical experimental set-up is shown in Fig. 8.3. Since the magnitudes of the shifts and the splittings are typically of the order of 10^{-7}–10^{-9} eV, we can use a Doppler energy shift of the γ-rays to scan the whole spectrum, moving the absorber with respect to the source with moderate velocities of the order of mm/s. Therefore, the spectrum is usually presented as a function of the Doppler velocity in mm/s for the case of ^{57}Fe spectrum. The Si sample containing ^{57}Fe impurities is either a source in accelerator facilities, or, more conventionally, an absorber in laboratory set-ups.

We can distinguish three major kinds of hyperfine interactions. The quadrupole interaction and the magnetic interaction give rise to a hyperfine splitting, as described in Sect. 8.1.2 of this chapter, which can be studied by other hyperfine interactions techniques as well. Mössbauer spectroscopy is unique in that it is sensitive also to the electric monopole interaction, which is the Coulomb interaction between the protons of the nucleus and the electrons (dominantly s-electrons) penetrating into the nucleus. This interaction causes a shift of the Mössbauer spectrum, which is called the "isomer shift δ [mm/s]" (see also Sect. 8.1.2 in this chapter). Isomer shift values depends on the charge state, high or low spin state, lattice site and bonding properties such as covalency and electronegativity. The isomer shift is influenced by defects in the surrounding Si matrix also.

8.2.3 Fe Components Observed in Mössbauer Spectra

Table 8.1 summarizes the isomer shifts for different Fe components observed at room temperature in Si materials. The same colouring for each component will be used throughout the rest of the chapter, when Mössbauer spectra are presented with the different Fe components. The isomer shift, which is proportional to the electron density at the nucleus, corresponds to the energy shift of the nuclear level for a Mössbauer component on a Doppler velocity scale of mm/s against the spectral centre of α-Fe at room temperature. The data have been obtained in different Mössbauer experiments on ^{57}Fe impurities, which were differently introduced into Si samples, and subsequently measured at different temperatures: (1) deposition at room temperature, and diffusion and measurements at high temperatures up to 1273 K [37–42], (2) highly energetic implantation of the mother isotope of ^{57}Mn/^{57}Fe into Si [36, 43–47] and (3) deposition and diffusion, and measurements at room temperature [48–52]. All the results can be analysed by the new model based

Table 8.1 Summary of Mössbauer components observed so far in Si

Component	Fe_{sub}	Fe_{sub}	Fe_{sub}	Fe_{sub} - defect	Fe_{int}	Fe_{int}-B	Fe_{int}	Fe_{int} - defect
Charge state of Fe	-1	0	+1	?	0	0	+1	+2
Isomer shift (mm/s)	-0.40 (5)	-0.17 (1)	+0.04 (2)	+0.15 (5)	+0.40 (1)	+0.42 (2)	+0.80 (2)	+0.80 ~ +1.20

on the interstitial Fe atoms with neutral, +1, and +2 states (defect associated), in addition to substitutional Fe with neutral, possibly +1 and −1 states. We also identify Fe-B pairs in a highly B-doped sample, which is very close to the isomer shift value for interstitial Fe with neutral charge state [51]. In Table 8.2, we summarize dominant Mössbauer experiments using completely different methodologies in different groups. One should notice that the different experiments use different nuclear decays, yielding different excess carrier concentration at the ^{57m}Fe nucleus, and consequently producing different charge states of ^{57m}Fe probes in Si materials.

Moreover, the assignments of the Fe components in Table 8.1 were originally based on a theoretical calculation for neutral interstitial Fe_{int} and substitutional Fe_{sub} [61]. There are, however, still rather large discrepancies between theoretical values and our experimental values, which are presented in Table 8.1. Nowadays, one could use much better theoretical tools [62], but the isomer shifts deduced for the charged interstitial and substitutional Fe atoms do not agree with the experimental values. Our assignments for the charged Fe states are, therefore, proposed by taking into account the experimental conditions at the ^{57}Fe nuclear probes in Si materials [46, 47, 51, 52] referring to the theoretical predictions for the Fe deep levels in Si [63, 64]. It should be mentioned that none of the spectra contained any Fe-silicides or Fe-oxides whose isomer shifts are well known. A peculiarity of Mössbauer spectroscopy is that it provides information on all Fe components present in the Si matrix, even if they are not electrically active, while all other evaluation techniques are only sensitive to electrically active components.

8.2.4 Unique Features of Mössbauer Spectroscopy

To control defects, light elements and metallic impurities in Si crystals, one intends to use nowadays an evaluation technique, which provides atomistic information. However, there is still no well-established technique accessible for single vacancy and interstitial in Si crystals, while dislocations and extended defects which might be decorated with metallic impurities such as Fe atoms can be directly observed, for instance, by High-Resolution Electron Microscope (HREM). Chapter 7 "Electron Beam Induced Current (EBIC) and Cathode Luminescence (CL)" provides such images of electrically active dislocations and extended defects. In addition, Chapter 3 describes "Deep Level Transient Spectroscopy (DLTS) and Laplace DLTS", yielding the deep levels of impurities in Si devices. On the other hand, light elements such as oxygen and nitrogen can be observed via Fourier Transform Infra-Red (FTIR) spectroscopy, combining with simulation techniques for the atomic vibrations of such light elements (Chap. 4). In the following sections, we demonstrate the results of Mössbauer spectroscopy, which appears to be ideal to study the basic nature of Fe impurities in Si materials. The experiments are always challenging because of the detection limit, and therefore, the spectroscopy can be applied mainly for Si materials which are intentionally contaminated by the stable

Table 8.2 Summary of experimental conditions and observed spectral components

	Diffused or implanted element and temperature	Total number of ^{57}Fe to obtain one spectrum	^{57}Fe depth profile and concentration	Time after diffusion or implantation	Radiation fluence	Charge states of interstitial ^{57m}Fe	Charge states of substitutional ^{57m}Fe	Reference
1.	100 keV ^{57}Fe implanted at RT&1073 K	$2 \times 10^{15}/cm^2$	$R = 87$ nm $\Delta \sim 32$ nm $2 \times 10^{20}/cm^3$	$\Delta t \sim$ days $\tau_{meas} \sim 100$ ns	^{57}Co source/3.7 GBq 5×10^6 γ/cm^2s	0	0	[53]
2.	^{57}Fe diffused and measured at RT-1273 K	$1 \times 10^{16}/cm^2$	$D = 1$–500 μm $\Delta 0.1$–200 μm $10^{17-21}/cm^3$	$\Delta t \sim$ weeks $\tau_{meas} \sim 100$ ns	^{57}Co source /3.7GBq 10^{4-6} γ/cm^2s	0/+1/+2	−1/0/+1?	[37–42]
3.	10 MeV ^{57m}Fe implanted at 300–850 K	$5 \times 10^{10}/cm^2$	$R = 10$ μm $\Delta \sim 5$ μm $1 \times 10^{16}/cm^3$	$\Delta t \sim 2$ ns $\tau_{meas} \sim 100$ ns	Coulomb excitation of ^{57}Fe 10^4 $^{57m}Fe/cm^2s$	+1	0	[13]
4.	^{57}Co diffused at 873–1473 K for 5–30 min	$\sim 5 \times 10^{13}/cm^2$	$D = 30$ nm $\Delta \sim 10$ nm $4 \times 10^{15}/cm^3$	$\Delta t \sim 30$ min $\tau_{meas} \sim 100$ ns	EC-decay of ^{57}Co to ^{57m}Fe	0/+1*	0	[35] *re-analysis by [60]
5.	30 keV ^{57}Co implanted at 50–300 K	2.5×10^{12} - 1 $\times 10^{13}/cm^2$	$R = 30$ nm $\Delta \sim 12$ nm 5×10^{19} - 2 $\times 10^{20}/cm^3$	$\Delta t \sim$ days $\tau_{meas} \sim 100$ ns	EC-decay of ^{57}Co to ^{57m}Fe	+1	0	[12]
6.	60 keV ^{57}Mn implanted at RT-800 K	$\sim 5 \times 10^{11}/cm^2$	$R = 57$ nm $\Delta \sim 22$ nm $2 \times 10^{17}/cm^3$	$\Delta t \sim 2$ min $\tau_{meas} \sim 100$ ns	β-decay of $^{57}Mn^+$ to ^{57m}Fe 10^8 $^{57}Mn^+/cm^2s$	+1	0	[14]
7	~ 500 MeV ^{57}Mn implanted at RT-1000 K	$\sim 2 \times 10^{11}/cm^2$	$R = 100$ μm $\Delta \sim 50$ μm $2 \times 10^{13}/cm^3$	$\Delta t \sim 2$ min $\tau_{meas} \sim 100$ ns	β-decay of $^{57}Mn^{25+}$ to ^{57m}Fe 10^7 $^{57}Mn^{25+}/cm^2s$	+1 (wafer) 0/+1/+2 (pn-junc.)	0 (wafer) −1/0 (pn-junc.)	[43–47]

isotope ^{57}Fe and as well as by mother isotopes of ^{57}Fe, i.e., either ^{57}Co or ^{57}Mn nuclear probes.

In the following we sum up the unique features of ^{57}Fe Mössbauer spectroscopy when it is applied for the study on Fe impurities in Si crystals.

1. The most important is that the 14.4 keV γ-rays are "recoil-free" absorbed or emitted without disturbing the electronic states nor the atomic vibrations of both Fe and Si atoms, i.e., Mössbauer effect. This is in contrast to other evaluation techniques such as PL, CL and EBIC where strong laser or electron beam are used to create the electro-active Fe states, and subsequently to measure emitted photons and induced current, respectively. The γ-ray intensity of 10^{5-8} photons/s is typically used in Mössbauer experiments, while PL, CL and EBIC techniques use many orders of magnitude higher intensities for excitation, i.e., 10^{18}–10^{20} photons and electrons/s, respectively.
2. Mössbauer spectroscopy provides information corresponding to all dominant Fe components existing in Si matrix, even if they are not electrically active, while all other evaluation techniques are only sensitive to electrically active components. Furthermore, different lattice sites and charge states of solid solution of Fe in Si can be distinguished in Mössbauer spectroscopy and have Mössbauer parameters which are different from Fe silicides and oxides.
3. The concentration of ^{57}Fe is proportional to the resonance area. The detection limit is different for source and absorber experiments, and close to 10^{11} and 10^{15} ^{57}Fe atoms/cm^3, respectively.
4. In Mössbauer spectroscopy, there are different measuring time-scales involved in the different experimental methods, depending on how ^{57}Fe probes were introduced into the Si matrix and subsequently how the measurements are triggered by the nuclear decay from the excited state of the 14.4 keV ^{57m}Fe nuclear level. Accordingly the atomistic information of Fe impurities is obtained through the hyperfine interactions during the nuclear life time, i.e., 98 ns of this level.
5. Not only the ^{57}Fe concentration, but also the carrier and their concentrations at around the ^{57}Fe probes are rather different in the following experiments which have been performed since more than 50 years: (1) ^{57}Fe implantation [6, 53], (2) ^{57}Fe diffusion at SIST [6, 37–42, 48–53], (3) ^{57m}Fe Coulomb-excitation and recoil-implantation at HMI [13], (4) ^{57}Co diffusion [35], (5) ^{57}Co implantation at Leuven [7, 12], (6) ^{57}Mn on-line separation and low-energy implantation at CERN [14] and (7) ^{57}Mn on-line separation and high-energy implantation at RIKEN [36, 43–47].
6. After ^{57}Co ($T_{1/2} = 270$d) [7, 12] and ^{57}Mn ($T_{1/2} = 1.45$ m) [14, 44–46] implantation into Si wafers the spectra consisted mainly of $Fe_{sub}{}^0$ and $Fe_{int}{}^+$, as presented in Table 8.2.

8.2.5 Mössbauer Spectra of ^{57}Fe Deposited on n-type Si Wafers

We first start looking at the Mössbauer absorber experiment on n-type Si wafers with a thickness of 540 μm containing 6×10^{18} Sb cm^{-3} and also 1×10^{12} P cm^{-3}, respectively [50]. The wafers were cut into square plates of 20×20 mm^2. After removing a surface oxide layer by a 20 % HF solution, ^{57}Fe stable isotopes were deposited onto the Si wafers with a thickness of 3.3 nm, and finally, the wafers were annealed at 1273 K for 1 week. We used a set-up enabling us to measure an in-situ Mössbauer spectrum under light illumination at room temperature. The Si sample was fixed in an acryl-holder with an angle of 45° against the 14.4 keV γ-ray direction. Accordingly, *the sample was kept to be electrically isolated from the ground during UV illumination.* The UV light source was a xenon lamp, and a mirror module was used to select an UV energy range between 3.5 and 5.5 eV. We measured each Mössbauer spectrum for 1 week with a 1.85GBq ^{57}Co-in-Rh source, yielding a γ-intensity of the order of 10^8 s^{-1} on the sample, which was ten orders of magnitude lower than the light photon intensity, i.e., 10^{18} s^{-1}. This means that the influence on the carrier production in the sample from the 14.4 keV and other higher energetic γ-rays emitted from the Mössbauer source are considered to be negligible.

Mössbauer spectra of ^{57}Fe in n-type Si (6×10^{18} Sb/cm^3) are shown in Fig. 8.18a. The uppermost spectrum was measured in the dark before UV illumination, the middle spectrum under UV illumination, and the lowest spectrum in the dark after UV illumination. The spectrum before illumination consists of two singlets. The spectrum under UV irradiation, however, changes from the spectrum in the dark: a broad singlet, C, appears, while the areal fractions of A and B decrease. Finally, the spectrum after illumination returns back to the initial spectrum in the dark, consisting only of A and B again. In Fig. 8.18b, furthermore, Mössbauer spectra of ^{57}Fe in n-type Si (10^{12} P/cm^3) are presented: in the dark (before), under UV illumination, and in the dark (after), respectively. Both the spectra before and under UV illumination consist of two singlets A, B. In the case of the spectrum after illumination, on the other hand, the spectrum consists of the components, A, B, and C and D in addition. In the case of the Sb-highly doped Si, the new component C appears only under illumination, but in the case of low P-doped Si the spectrum does not change, but the linewidth of singlet B gets slightly narrower. After illumination, however, the spectrum changes considerably: two singlets, C and D, are emerging at the expense of the area fractions of the singlets, A and B.

Before we interpret the components A, B, C and D in Fig. 8.18a, b, we should mention that we did not include potential Fe-silicides in the analysis of our spectra. This can be concluded by investigating a Mössbauer spectrum of poly-crystalline β-$FeSi_2$ which was measured at room temperature [59]: The β-$FeSi_2$ spectrum in Fig. 8.19 appears to be a broad doublet with resonances at around −0.2 and +0.3 mm/s. The former exists close to the component A in Fig. 8.18, while the latter between A and C. A fitting analysis of our spectra reveals that there is certainly no strong component of β-$FeSi_2$ present in our spectra, as one would expect from the reported Fe solubility [29].

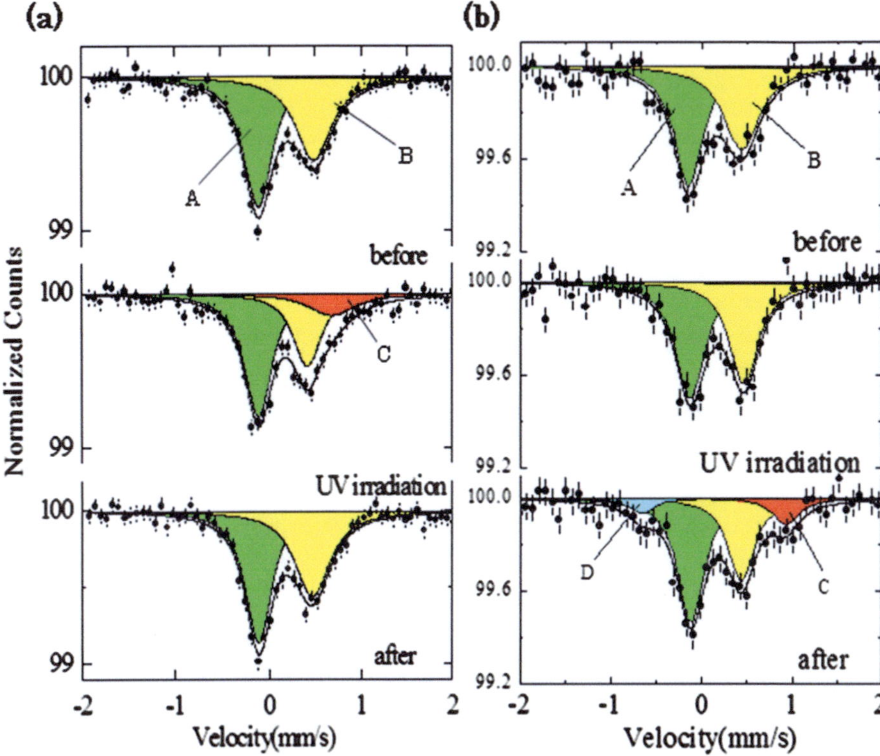

Fig. 8.18 Mössbauer spectra of ^{57}Fe in (**a**) n-type Si (6×10^{18} Sb/cm^3), and (**b**) n-type Si (1×10^{12} P/cm^3) (From Ref. [50])

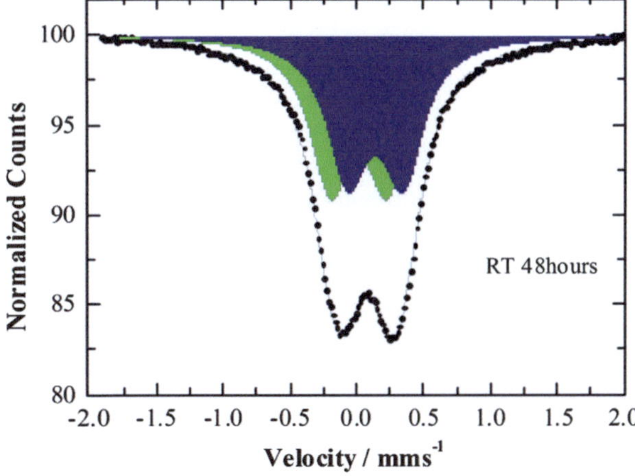

Fig. 8.19 Mössbauer spectra of $\beta FeSi_2$ at RT (From Ref. [58])

We have measured the spectra of ^{57}Fe doped n-type Si samples before, under and after UV illumination with a photon intensity of 10^{18} s^{-1}. The singlets A and B shown in Fig. 8.18 are the dominant components, which can be assigned to substitutional $Fe_{sub}{}^{0}$ and interstitial $Fe_{int}{}^{0}$, respectively, on the basis of theoretical calculations of the isomer shifts [61, 62] in addition to our previous experimental work described in this Chapter. In n-type Si "interstitial Fe" is considered to form a "donor" level at 0.40 eV from the valence band, and therefore, the charge state of interstitial Fe is expected to be neutral in both samples. On the other hand, the components C and D are observed as small satellites which must be produced only by the carrier injections through UV illumination and subsequently by the carrier trapping processes at ^{57}Fe probes. Accordingly, the singlet C appears to correspond to interstitial $Fe_{int}{}^{+}$, which must yield a lower electron density than that of $Fe_{int}{}^{0}$, i.e., more to the right on the Doppler velocity scale, while the singlet D is tentatively assigned to substitutional $Fe_{sub}{}^{-}$, related to an accepter state of substitutional Fe, the existence of which was predicted in a first principle calculation [63]. The isomer shift of the most left hand side resonance corresponds to the highest electron density in comparison with those of the other singlets, which appears to be consistent with the accepter state of $Fe_{sub}{}^{-}$. The same components as A, B, C, and D were observed immediately after highly energetic implantation of ^{57}Mn/^{57}Fe, where the formations of Fe clusters and silicides can be absolutely excluded because of the experimental conditions, as will be explained in the Sects. 8.2.11 and 8.2.12.

The heat treatment at 1273 K would provide a Gaussian distribution of ^{57}Fe substitutional atoms from the surface down to about 1 μm, and in addition, a rather homogenous distribution of ^{57}Fe interstitials at least down to 10 μm in the Si wafer, as was measured using SIMS [39]. This was confirmed by etching the surface layer and by the subsequent measurement of a spectrum, in which the substitutional component nearly disappeared, while the interstitial one remained. Accordingly, the substitutional Fe atoms are supposed to be illuminated directly from UV light. The interstitial Fe atoms are, however, expected to be distributed homogeneously, which means that only a small part of the interstitial Fe atoms are directly illuminated. The UV illumination may cause a transformation either from the substitutional Fe atom to the interstitial Fe atom or from the neutral substitutional Fe and the interstitial Fe to their charged states. Under illumination: in the low P-doped sample, the excess carriers react with the charged states of the substitutional and the interstitial Fe atoms effectively, so that only the neutral states can be seen, while in the highly Sb-doped sample, the transformation from the charged state to the neutral state of the interstitial Fe atoms seems to be partly suppressed due to the majority carriers and the Sb dopants. This results in the singlet of $Fe_{int}{}^{+}$. After illumination: in the case of low P-doped sample with a very low electron concentration, the excess carriers are supposed to diffuse and subsequently to react with neutral substitutional and interstitial Fe atoms, leading to the $Fe_{int}{}^{+}$ and $Fe_{sub}{}^{-}$ components which emerge in the dark after UV illumination, on one hand. In the case of highly Sb-doped sample, on the other hand, the excess carriers are compensated and therefore, the same spectrum is obtained as before illumination.

8.2.6 Low Energy ^{57}Fe Implantations into Si

As the second example, we discuss the Mössbauer absorber experiment after implantation of ^{57}Fe stable isotopes into Si wafers [53]. The ion implantation technique enables us to introduce a well-defined concentration profile of atoms such as boron and phosphorus dopants into semiconductors such as Si crystals. This technique is, therefore, widely used for electronic device processing, although the implantation produces also cascade defects consisting of vacancies and interstitials in the matrix due to energetic collisions of implanted atoms with host atoms during the slowing-down process. We apply this technique to study Fe impurities in Si matrix. We introduce ^{57}Fe probes into the Si matrix, and subsequently measure the transmission Mössbauer spectra, as shown in Fig. 8.20 (left).

$^{57}Fe^{+}$ ions were implanted into p-type CZ-Si (10^{19}B/cm^3) and FZ-Si (10^{15}B/cm^3) wafers at room temperature and also at 1073 K with an energy of 100 keV [53]. The total dose was 2×10^{15} ^{57}Fe /cm^2.Subsequently, Mössbauer absorption spectra were measured for 1 week at room temperature using a 1.85-GBq-^{57}Co-in-Rh source. The spectra are shown in Fig. 8.20 (left): the two upper spectra correspond to ^{57}Fe in CZ-Si (10^{19}B/cm^3), and the two lower to ^{57}Fe in FZ-Si (10^{15}B/cm^3), respectively. The

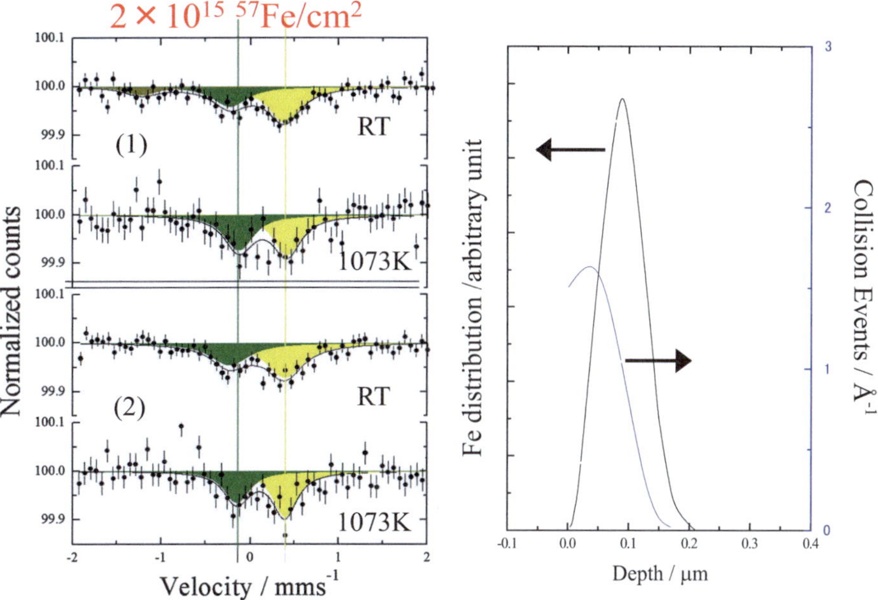

Fig. 8.20 (*Left*) Mössbauer spectra of ^{57}Fe in (1) CZ- and (2) FZ-Si measured at room temperature (From Ref. [53]). The implantations of $^{57}Fe^{+}$ ions with an energy of 100 keV were performed up to a total dose of 2×10^{15}^{57}Fe/cm^2 at room temperature and also at 1073 K, respectively (*right*). The depth profile of ^{57}Fe and defects distributions are simulated for room temperature without diffusion effect by SRIM [65]

depth profile of ^{57}Fe impurities and the numbers of collision events are shown in Fig. 8.20 (right) as a function of depth, which were estimated by a SRIM simulation [65] for the case of 100 keV-$^{57}Fe^{+}$-implantation at room temperature. The range and the straggling are 87 and 32 nm, respectively. As discussed in Sect. 8.1.2. earlier Mössbauer experiments [10] showed that 70 kV Fe implantation in Si at high fluence (10^{16} cm^{-2}) resulted in the formation of an amorphous region at the end of slowing-down process due to overlapping cascade damages. The present Mössbauer spectra were recorded with a somewhat higher implantation energy and lower fluence and the spectrum shape is similar for the RT implantation and for the implantations at 1073 K where amorphization should not be present. We therefore feel justified in analysing our spectra with two sharp absorption lines, suggesting that ^{57}Fe probes are stopping at lattice positions not in an amorphous region, but at well-defined lattice sites, as will be discussed in the following.

All the spectra mainly consist of substitutional $Fe_{sub}{}^{0}$ and interstitial $Fe_{int}{}^{0}$ components at the left (green) and the right hand sides (yellow), respectively. The area fraction of substitutional $Fe_{sub}{}^{0}$ is slightly higher after the implantation at 1073 K than after that at room temperature, suggesting that interstitial Fe atoms encounter vacancies produced by the implantation, forming substitutional Fe during implantation at 1073 K, because of the higher diffusivities of Fe interstitial and vacancy. The spectra indicate that the charge state of interstitial $Fe_{int}{}^{0}$ corresponds to the neutral state, although one would expect the charged state $Fe_{int}{}^{+}$ from the dopant concentrations. 100 keV-Fe^{+} ions will lose their energy mainly through elastic collisions during the slowing down processes. Accordingly, the following reactions are considered to occur possibly: $Fe_{int}{}^{+} + V^{-} \rightarrow Fe_{sub}{}^{0}$; $Fe_{int}{}^{+} + e^{-} \rightarrow Fe_{int}{}^{0}$. The vacancies V^{-} are considered to be distributed over the Fe profile, as is shown in Fig. 8.20 (right).

We have shown the absorber Mössbauer spectra, which were obtained either by the annealing at 1273 K for 1 week in Sect. 8.2.5, or by implantation at room temperature and also at 1073 K in the present section. The spectra of n-type Si wafers consist of interstitial $Fe_{int}{}^{0}$ and substitutional $Fe_{sub}{}^{0}$ as the dominant components, and additionally $Fe_{int}{}^{+}$ and $Fe_{sub}{}^{-}$ after trapping excess carriers injected by UV-illumination during the measurements. In p-type Si wafers, on the other hand, only the components of interstitial $Fe_{int}{}^{0}$ and substitutional $Fe_{sub}{}^{0}$ are observed, which is contradictory to what one would expect by taking into account the position of both the Fermi levels and the deep level for Fe interstitial, i.e., 0.38 eV above the Si valence band [29]. One should notice that all spectra clearly contain the substitutional component $Fe_{sub}{}^{0}$ and this is completely different from the model which has been accepted for Fe impurities in Si materials in the Si community [29, 30]. In order to clarify the nature of interstitial $Fe_{int}{}^{0}$, $Fe_{int}{}^{+}$ and $Fe_{int}{}^{2+}$, and substitutional $Fe_{sub}{}^{0}$ and $Fe_{sub}{}^{-}$, we investigated further ^{57}Fe deposited Si samples at high temperature, where ^{57}Fe probes are thought to be distributed over the whole sample depth.

8.2.7 High Temperature Absorber Experiments

The spectra of n-type FZ-Si with 1.9×10^{19} As/cm^3 [42, 53] are shown in Fig. 8.21 (left). The spectra were measured (1) at room temperature immediately after the deposition of ^{57}Fe with a thickness of 1.4 nm, (2) at 1273 K for about 10 days, and finally (3) at room temperature again after the measurement at 1273 K. The spectrum (1) consists of substitutional $Fe_{sub}{}^0$ and interstitial $Fe_{int}{}^0$ and $Fe_{int}{}^+$, respectively. The spectrum (2) at 1273 K can be fitted with three singlets, $Fe_{sub}{}^0$, $Fe_{int}{}^0$ and $Fe_{int}{}^{2+}$, the third component which is temporally assigned will be discussed later. The isomer shift of the $Fe_{int}{}^{2+}$ component corresponds to +1.2 mm/s, which could be due to an interstitial Fe in the vicinity of a defect. After the high temperature measurements, the spectrum (3) at room temperature can be fitted with $Fe_{sub}{}^0$ and $Fe_{int}{}^0$.

Furthermore, the spectra shown in Fig. 8.21 (right) of ^{57}Fe in p-type FZ-Si with 3.6×10^{15} B/cm^3 were measured in the same way as in Fig. 8.21 (left). The spectrum (1) consists of the same components of $Fe_{sub}{}^0$, $Fe_{int}{}^0$ and $Fe_{int}{}^+$, as can be seen in Fig. 8.21 (left), but additionally another component appears on the left hand site of $Fe_{sub}{}^0$, which appears to be due to $Fe_{sub}{}^-$ related to an accepter level predicted by theoretical work [63]. The spectrum (2) at 1273 K can be fitted with three singlets,

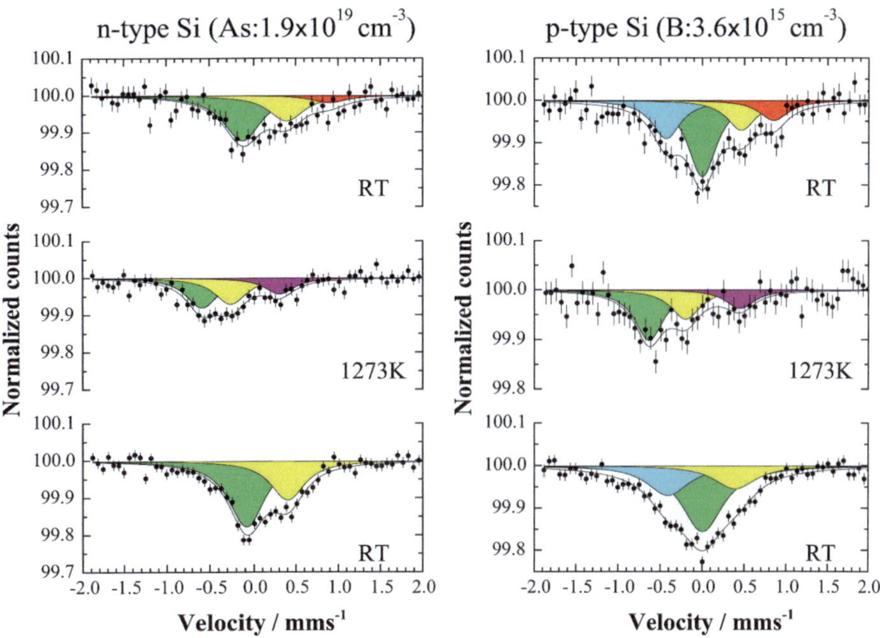

Fig. 8.21 (**a**) Mössbauer spectra of ^{57}Fe in n-type Si (As = 1.9×10^{19} cm^{-3}), and (**b**) Mössbauer spectra of ^{57}Fe in p-type Si (B = 3.6×10^{15} cm^{-3}), measured (1) at room temperature after ^{57}Fe deposition, (2) at 1273 K, and (3) at room temperature after high temperature measurement

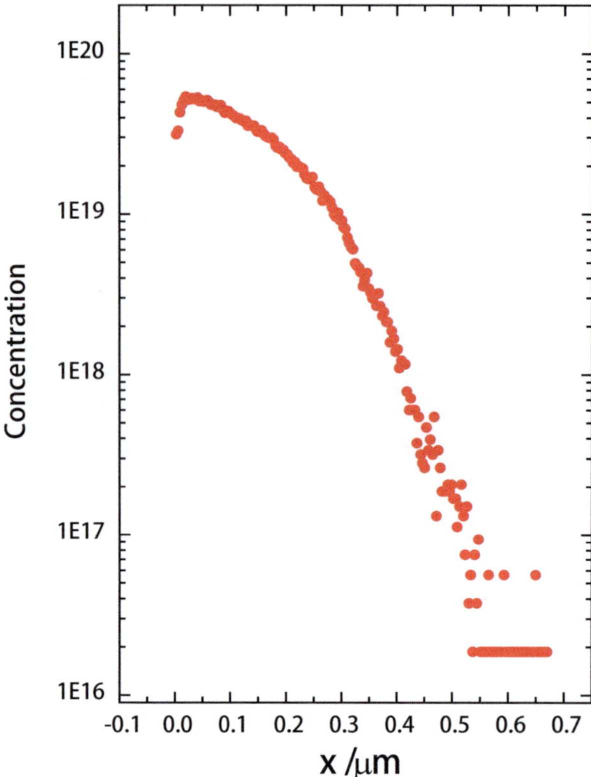

Fig. 8.22 Depth profile of ^{57}Fe after high temperature measurements at 1273 K

$Fe_{sub}{}^{0}$, $Fe_{int}{}^{0}$ and $Fe_{int}{}^{2+}$. The spectrum (3) at room temperature consists of three singlets, $Fe_{sub}{}^{0}$ and $Fe_{int}{}^{0}$ and $Fe_{sub}{}^{-}$. One of the samples was investigated by SIMS to measure the diffusion profile after the high temperature measurement, yielding a Gaussian profile due to Fe diffusion at 1273 K (Fig. 8.22). The diffusivity is rather low, of the same order as self-diffusion of Si in Si. The substitutional Fe atoms, therefore, are distributed near the surface down to several 100 nm, while the interstitial Fe components are supposed to be distributed over the whole range of the wafer due to the much higher diffusivity of interstitial Fe atoms in Si. The concentration of the interstitial Fe atoms is expected to be less than 10^{16} Fe/cm^3 which is close to the detection limit of SIMS.

In Fig. 8.23 we show the temperature dependence of the spectra of the ^{57}Fe doped p-type Si which were measured from 1273 K down to room temperature [42]. The spectra can be fitted by the same components as used in Fig. 8.21. As a reference, the normal temperature dependence of the centre shift for a Mössbauer component, i.e., the second order Doppler-shift, is shown with a dotted line in Fig. 8.23. The centre shift of $Fe_{sub}{}^{0}$ and $Fe_{int}{}^{0}$ components follows the second order Doppler shift,

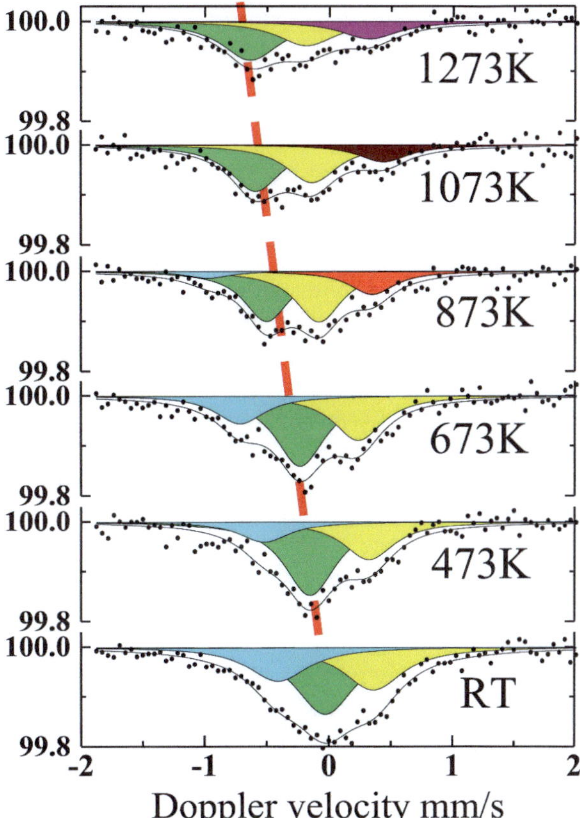

Fig. 8.23 Temperature dependence of Mössbauer spectrum of FZ-Si (1×10^{15} B/cm^3)

suggesting that the charge state is the same for all the temperatures. At around 800 K, the $Fe_{int}{}^{2+}$ component might transform to other components of $Fe_{sub}{}^{0}$ and $Fe_{sub}{}^{-}$, accompanied by anomalous centre shifts. This would be are related to the temperature dependence of the Fermi level. At around 800 K, the Fermi level goes to the middle position of the Si band gap. As lowering temperature, this energy level seems to be occupied, leading to this anomalous temperature dependence of the component. No Fe-silicide nor oxides are appearing in the whole temperature range.

In the next section we will investigate mainly the interstitial components of $Fe_{int}{}^{0}$, $Fe_{int}{}^{+}$ and $Fe_{int}{}^{2+}$ by changing the Fermi level either through applying the external voltage on a Schottky diode, or through doping differently.

8.2.8 *Fermi Level Dependences of the Mössbauer Spectrum of ^{57}Fe Deposited on Si Wafers at Room Temperature*

The Mössbauer spectra of ^{57}Fe deposited on Si wafers were measured either with external voltages, or with different Fermi-levels due to different dopant concentrations. Since the sample was not annealed at all and was also measured at room temperature, all the ^{57}Fe atoms are expected to be distributed near the surface. All the spectra indicate, however, that the dominant part of the deposited ^{57}Fe atoms appears to exist as a solid solution, such as interstitial and substitutional Fe atoms. It should be mentioned that none of the spectra contains Fe silicide.

Figure 8.24a shows the in-situ Mössbauer spectra of a Schottky diode measured under external voltages of −1.5, 0, and +1.5 V, respectively [52]. It is clear that external voltage influences strongly the in-situ Mössbauer spectra of the Al/^{57}Fe on p-n Si/Al sample. These spectra can be fitted by three superimposed singlets. The upper spectrum was obtained by applying a voltage of −1.5 V, indicating an absorption at around 0.80 mm/s larger than the middle spectrum without external bias. When a voltage of +1.5 V was applied, the spectral shape changed drastically both at around 0.40 mm/s and at 1.0 mm/s. Under external voltage of -1.5 and 0 V,

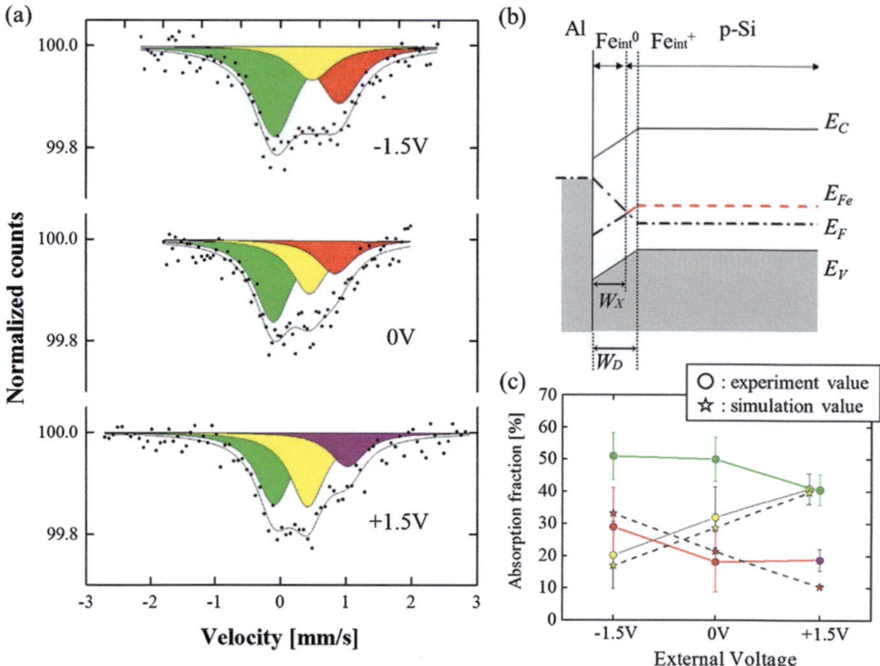

Fig. 8.24 (**a**) In-situ Mössbauer spectra of the Al/^{57}Fe on p-n Si/Al sample under external voltage (-1.5, 0, +1.5 V) (From Ref. [52]). (**b**) The band diagram in the vicinity of iron diffusion region. (**c**) The absorption fraction of each component in Fig. 8.24a

the spectra consist of $Fe_{sub}{}^0$, $Fe_{int}{}^0$, and $Fe_{int}{}^+$. Under external voltage of +1.5 V, on the other hand, the spectrum consist of $Fe_{sub}{}^0$, $Fe_{int}{}^0$, and $Fe_{int}{}^{2+}$. Figure 8.24c shows the area fraction of each component as function of the external voltage. As the external voltage is increased from −1.5 to 1.5 V, the fraction of $Fe_{int}{}^0$ increases, while the fraction of $Fe_{int}{}^+$ decreases. Furthermore, when applying the voltage of +1.5 V, the isomer shift of $Fe_{int}{}^+$ moves to that of $Fe_{int}{}^{2+}$, which is accompanied by the decrease of the fraction of $Fe_{sub}{}^0$.

Applying an external voltage to the Schottky diode, the ^{57}Fe atoms are distributed dominantly in a depletion region. In Fig. 8.24b, the band diagram is schematically shown in the vicinity of the ^{57}Fe diffused region. The barrier height between Al/p-Si is 0.92 eV based on the simple electron affinity model. There are other barriers at the interface between p-Si/n-Si and n-Si/Al, however, which will not influence the following discussion. By applying an external voltage to the sample, the width, W_D, of the depletion region is changed, leading to a change in the crossing point W_X between the Fermi level E_F and the Fe impurity level E_{Fe}. A similar situation has been often used in Deep Level Transient Spectroscopy (DTLS).

When half of deposited Fe atoms diffuse into the Si matrix as interstitial Fe_{int}, the diffusion length reaches about 0.5 μm after 1 week at room temperature. At the shallow region of the diffusion profile, as shown in Fig. 8.24b, the impurity level Fe_{int} is below the Fermi level E_F, resulting in that interstitial Fe_{int} will become neutral. On the other hand, at the deep region of the diffusion profile, the Fe_{int} level exists above the Fermi level E_F, and therefore, Fe_{int} will be positively charged. Based on the calculation of the diffusion profile of Fe_{int} and the crossing point derived from a simple linear band bending, the fractions of $Fe_{int}{}^0$ and $Fe_{int}{}^+$ are obtained for each of the external voltage condition. Figure 8.24c shows these simulation values in addition to the absorption fraction of each spectral component. From the comparison between experiment and simulation values, this model of the crossing point and the diffusion profile can explain the changes of the ratio between $Fe_{int}{}^0$ and $Fe_{int}{}^+$. When an external voltage of +1.5 V was applied, however, the experimental result shows a somewhat different tendency: The fraction of $Fe_{int}{}^0$ increases in fact, but the fraction of $Fe_{sub}{}^0$ deceases, and in addition, the charge state of $Fe_{int}{}^+$ seems to change to $Fe_{int}{}^{2+}$. The carrier flow may affect the charge state of Fe, but the reason is still not clear.

Figure 8.25 (left) shows the spectra of a CZ-Si wafer (1×10^{16} B/cm^3), which were measured in air at room temperature under a dark condition by two different set-ups (A) and (B) [52]: For the set-up (A) the γ-ray fluence was 3×10^4 γ/cm^{2}s by 0.5GBq-^{57}Co source, while for the set-up (B), the fluence was 5×10^6 γ/cm^{2}s by 3.7 GBq-^{57}Co source. For comparison, the spectrum measured by set-up (A) is shown in the upper most part of Fig. 8.25 (left). Surprisingly, the middle spectrum, measured by set-up (B), is quite different from the upper one by set-up (A). The middle spectrum by set-up (B) does not contain the component $Fe_{int}{}^0$ with 0.40 mm/s and consists only of the two components of $Fe_{sub}{}^0$ and $Fe_{int}{}^+$ with isomer shifts −0.10 and 0.80 mm/s respectively. The difference between set-up (A) and (B) is essentially the γ-ray fluence, which is two orders of magnitude different each

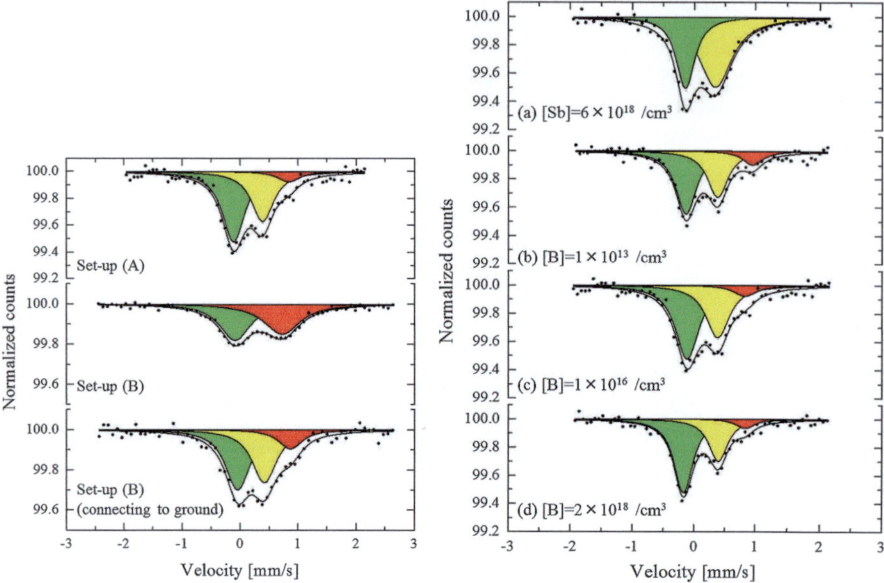

Fig. 8.25 (*Left*) Mössbauer spectra of p-type Si (1×10^{16} B/cm^3) measured by set-ups (*A*) and (*B*). (*Right*) Mössbauer spectra measured by set-up (*A*) for the ^{57}Fe deposited on Si wafers: (*a*) n-Si [Sb] = 6×10^{18}/cm^3, and p-Si (*b*) [B] = 1×10^{13}/cm^3, (*c*) [B] = 1×10^{16}/cm^3, and (*d*) [B] = 2×10^{18}/cm^3, respectively (From Ref. [52])

other. The total absorption area of the middle spectrum is smaller than those of other spectra, which appears to be due to a large dead-time of the detector. The lowest spectrum in Fig. 8.25 (left) is the spectrum when the sample surface was electrically connected to ground level. The component of 0.40 mm/s appeared again in the spectrum, and the spectrum was composed of three sharp lines of −0.10, 0.40, and 0.80 mm/s, as with the spectrum measured by the set-up (A).

Figure 8.25 (left) indicates that the different γ-ray fluence provides the changes in the spectra of the same sample [52]. Since we measure the Mössbauer effect, the charge state of ^{57}Fe atoms, which are observed in the spectrum, must not be affected by absorbing the 14.4 keV γ-rays. The excess carriers generated in the Si matrix by the γ-ray irradiation, i.e., both 136 and 122 keV γ-rays from the ^{57}Co source, may shift the quasi-Fermi level, leading to a change in the charge state of ^{57}Fe. When the sample surface was connected to the ground, the spectrum changed drastically. This suggests that the quasi-Fermi level turns back to the lower value corresponding to the set-up (A), leading to the spectrum consisting of three sharp lines of −0.10, 0.40, and 0.80 mm/s, as with the uppermost spectrum measured by the set-up (A).

In a new series of experiments, further, we study the dopant dependence of the spectrum [52]. Here, we discuss the fraction changes between Fe_{int}^0 and Fe_{int}^+.

Based on the Shockley-Read-Hall model in a steady state condition, the probability f_T of electron occupation on the trap level E_{Fe} is:

$$f_T = \frac{1}{1 + \exp\left[(E_{Fe} - E_F)/kT\right]} \tag{8.1}$$

where k is Boltzmann's constant, and T temperature. However, the spectra in Fig. 8.25 (right) (b–d) cannot be explained by Eq. 8.1 using the Fermi level calculated from the dopant concentration. In order to explain the fractions of $Fe_{int}{}^0$ and $Fe_{int}{}^+$ components, it appears to be necessary to consider an additional influence from interstitial Fe_{int} itself as a donor. In the region of $[Fe] > [B]$, holes due to B acceptor are compensated by electrons due to Fe donor as a semi-insulating silicon. The electrical neutrality condition is $[B] = [Fe]\ (1\text{-}f_T)$, therefore, $f_T \sim 1\text{-}[B]/[Fe]$. In the region of $[Fe] \gg [B]$, $f_T = 1$, and then the states of Fe_{int} are occupied by electrons and become neutral. If, on the other hand, Eq. 8.1 is applied in the region of $[Fe] < [B]$, then the charge state of Fe_{int} changes to positive depending on [B]. In this experiment, Fe atoms diffused into Si within the region of $[Fe] > [B]$. Therefore, the Fe_{int} component of the spectra in Fig. 8.25 (right) (b–d) should mainly consist of $Fe_{int}{}^0$ independent of [B]. Furthermore, since Fe atoms are expected to form Fe-B pairs, they will consume all B atoms within the range of diffusion, and the [Fe-B]/[Fe] is 2.3 % even for (d) $[B] = 2 \times 10^{18}/cm^3$.

We have investigated the Mössbauer spectra of ^{57}Fe-deposited Si wafers with different Fermi levels. Clearly, the external voltage applied to a Schottky junction causes a change in the fractions of $Fe_{int}{}^0$ and $Fe_{int}{}^+$ components. In the case of dopant dependence, on the other hand, the spectra are not influenced so much, as is expected from the simple Fermi level dependence due to different boron concentration. This was discussed by taking into account of the influence of Fe_{int} itself as a donor within the region of $[Fe] > [B]$. Also, we observe clearly that both the γ-ray fluence and the electrical connection of the Si wafer to the ground produce an excess carriers, resulting in a big change in the Mössbauer spectrum. A similar change was also observed in a spectrum of a multi-crystalline Si wafer with $[B] = 1 \times 10^{16}\ /cm^3$, which consisted of $Fe_{sub}{}^0$ and $Fe_{int}{}^+$.

In the next section we will report on a challenge to search for Fe-B pairs in the Mössbauer spectrum, a method which has been used to estimate the Fe interstitial distribution in Si materials widely [31–33].

8.2.9 Search for Fe-B Pairs in a Highly B-doped Si Wafer

A highly B-doped Si wafer was deposited with ^{57}Fe impurities, and subsequently Mössbauer spectra were measured at room temperature before and after PL mapping, in order to observe the Fe-B pairs directly [51]. In addition, we investigated the Mössbauer spectra under visible light illumination to study the dissociation processes of the Fe-B pairs. We found experimentally the Mössbauer components, which could be assigned to the Fe-B pairs, immediately after the ^{57}Fe deposition as well as after PL mapping. A highly doped single-crystal Si wafer (10^{19} B/cm^3), which was a square of 10×10 mm^2 with a thickness of 520 μm, was chemically etched by 20 % HF for a few seconds. A 2.3 nm-thick ^{57}Fe layer (95 % enriched) was deposited under a vacuum of 10^{-5} Pa by electron-beam evaporation on the surface. Subsequently, PL mapping was performed using the 532 nm line of a diode pumped solid state laser as an excitation source. Mössbauer transmission spectra were measured at room temperature with and without visible light illumination for about 7 days.

Figure 8.26 shows the Mössbauer spectra; (a) immediately after the ^{57}Fe deposition, (b) after PL mapping, (c) under visible light illumination, and (d) in dark after light illumination. The spectrum (a) can be fitted by a superposition of four Lorentzians; two major components are assigned to the neutral states of substitutional $Fe_{sub}{}^0$ and interstitial $Fe_{int}{}^0$ with the isomer shifts of −0.15 and 0.40 mm/s, respectively. In addition, there are two other minor components; one exists at the right-hand side of the $Fe_{int}{}^0$ component, and is related to "a charge fluctuating component" between $Fe_{int}{}^{1+}$ and $Fe_{int}{}^{2+}$ ions with an isomer shift of 0.89 mm/s, and the other is preliminary assigned to an "Fe-defect cluster" with an isomer shift of 0.15 mm/s. The Fe-defect cluster component was observed in p-type wafers immediately after ^{57}Fe deposition and the fraction is different for different venders of wafers. Since this Fe-defect component has been observed after the deposition at room temperature, this Si wafer seems to contain a rather high defect density on the surface.

After PL mapping, the spectrum in Fig. 8.26b consists of two components; one component emerges at the expense both of the $Fe_{sub}{}^0$ and "Fe-defect cluster" with an isomer shift of 0.04 mm/s on the left-hand side, and the other with that of 0.42 mm/s on the right-hand side. The former component is newly observed, and is tentatively assigned to "$Fe_{sub}{}^{1+}$ ions" adjacent to a vacancy. On the other hand, the latter component is the $Fe_{int}{}^0$ atoms and/or the "$Fe_{int}{}^+$-B^- pairs", both of which must yield the similar isomer shift i.e., the charge density at the ^{57}Fe nucleus. The laser irradiation for the PL mapping causes drastic changes on the Fe impurities which are diffused into the Si matrix at room temperature and are subsequently distributed near the surface. This process leads to different kinetic reactions with the lattice defects and the dopants, finally forming the components observed in the Mössbauer spectra. In the present study, the ^{57}Fe concentration near the surface is estimated to be 5×10^{20}/cm^3 by electron probe micro-analysis (EPMA) with different accelerating voltage, and is considered to be of almost the same order as the B concentration of 10^{19} /cm^3. So that the laser irradiation is expected to induce the formation of $Fe_{int}{}^+$-B^- pairs as the result from a kinetic reaction between $Fe_{int}{}^+$ and B^- dopants.

Under visible light illumination, the spectrum in Fig. 8.26c consists of the $Fe_{sub}{}^0$ and $Fe_{int}{}^0$ components with the area fraction of 66 % and 34 %, respectively. The $Fe_{sub}{}^+$ component on the left-hand side in (b) changes to the stable $Fe_{sub}{}^0$, while "$Fe_{int}{}^+$-B^- pairs" on the right-hand side in (b) dissociate to the $Fe_{int}{}^0$ atoms and B^-

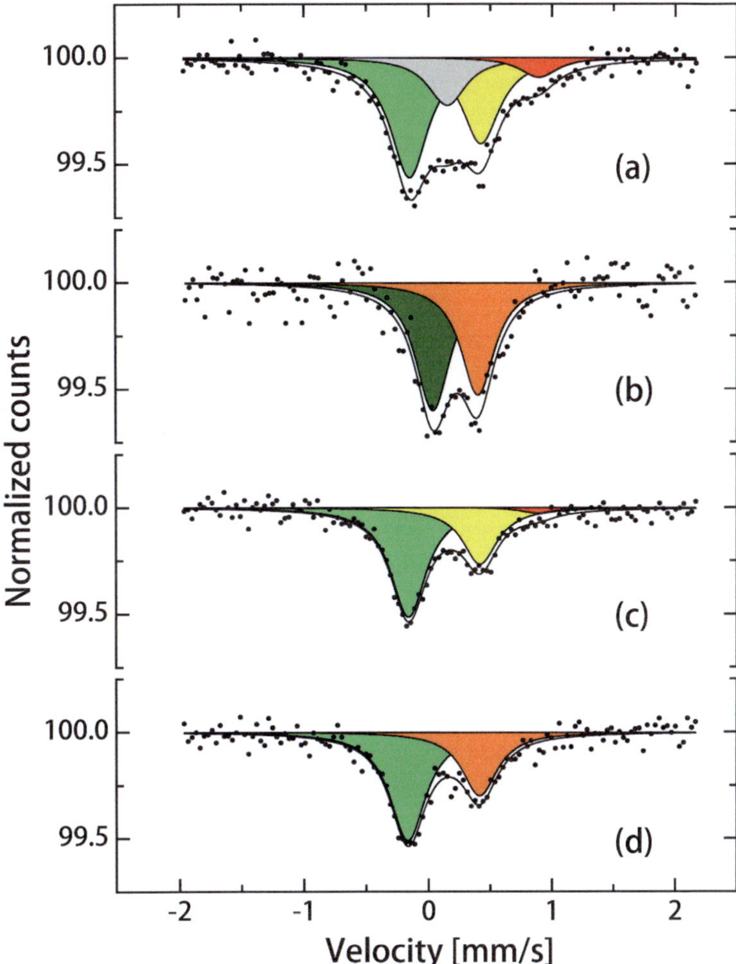

Fig. 8.26 Mössbauer spectra of a highly B-doped single-crystal Si wafer: (*a*) immediately after ^{57}Fe deposition, (*b*) after PL mapping, (*c*) under visible light illumination, and (*d*) after light illumination. The spectra of (*a*), (*b*), and (*d*) were measured in dark [51]

dopant by visible light illumination. The increase of the area fraction of the $Fe_{sub}{}^{0}$ component means that some $Fe_{int}{}^{0}$ changes to $Fe_{sub}{}^{0}$ during the diffusion process from the surface to the bulk matrix.

The charge state of the Fe interstitial component in Fig. 8.26c is neutral under visible light illumination. By the light illumination, however, there might emerge an additional interstitial component at the most right hand site of the spectrum same as in (a), which seems to be originating from the charge fluctuating component between $Fe_{int}{}^{1+}$ and $Fe_{int}{}^{2+}$ ions. This component must be associated with defects, leading to the degradation of the energy conversion efficiency for Si solar cells, because both Fe_{int}^{1+} and Fe_{int}^{2+} ions must trap strongly the excess carriers produced by light illumination [50]. Under dark condition after light illumination, spectrum (d) is similar to spectrum (c), again containing Fe_{int}^{+}-B^{-} pairs.

8.2.10 Direct Observation of a Stress Induced Diffusion of ^{57}Fe in Si Wafer

We have developed an Instron-type tensile testing machine specially designed for Mössbauer spectroscopy, which is shown in Fig. 8.27 [48, 49], and have applied external stress to a ^{57}Fe deposited n-type Si wafer. Mössbauer absorber experiments were performed at room temperature in a vacuum under applying external bending stress (Fig. 8.27). A n-type silicon wafer ($P \sim 10^{12}$ cm^{-3}) was cut into two square disks of $20 \times 10 \times 0.53$ mm, and then ^{57}Fe impurities were deposited on the Si substrates with a thickness of 3 nm in a vacuum of 10^{-6}Pa. The Si samples were subsequently annealed at 1273 K for 1 week. This Si sample was mounted in the Mössbauer set-up. Since two strain gauges were fixed on the both sample surfaces as shown in the inset of Fig. 8.27, a strain on each side of the sample could be measured separately as a function of the external stress. Figure 8.27 (right) shows the stress–strain curves when a compressive stress was applied to the Si sample along [011] direction. A positive and negative strain represents a tensile and a compressive strain, respectively, indicating that the Si sample was bent under the external stresses. As indicated in Fig. 8.27, the tensile strain or the compressive strain can be applied selectively either on the ^{57}Fe-deposited side or on the non-

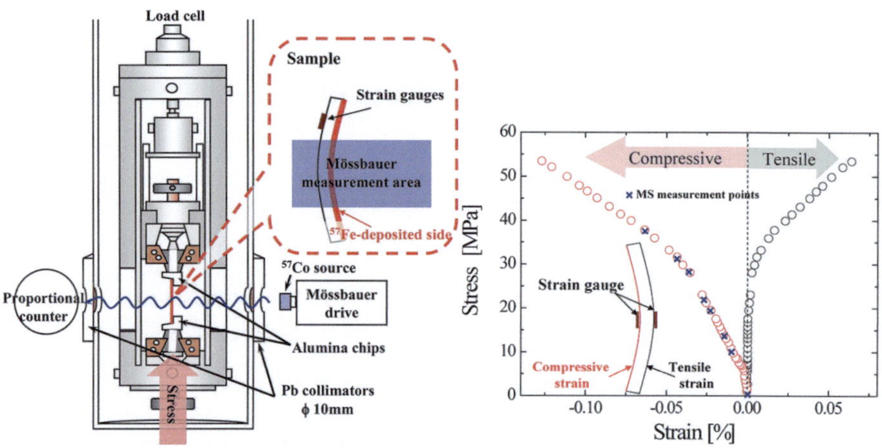

Fig. 8.27 (*Left*) An experimental set-up for in-situ Mössbauer measurements under external bending stress applied by a tensile testing machine. (*Right*) A stress strain curve observed (From Refs. [48, 49])

deposited side. A 1.85 GBq ^{57}Co-in-Rh standard source was used for transmission Mössbauer experiments at room temperature. The sample was mounted between alumina blocks in a vacuum chamber. Mössbauer absorption spectra of the ^{57}Fe doped Si sample were measured as a function of the external stress up to 37 MPa.

Figure 8.28 shows typical Mössbauer spectra of ^{57}Fe-doped Si with (a) tensile stress and with (b) compressive stress on the ^{57}Fe-deposited side. After the annealing at 1273 K for 1 week, the spectra are well fitted with three or two sets of singlets $Fe_{sub}{}^0$, $Fe_{int}{}^0$, and $Fe_{int}{}^+$. The isomer shifts for the interstitial $Fe_{int}{}^0$ and the substitutional Fe $_{sub}{}^0$ do not change under the external stress. However, the linewidths (full width at half maximum) of the interstitial Fe component and the areal fractions are found to depend strongly on the external stress, as can be clearly seen in the Mössbauer spectra (Fig. 8.28). At around 28 MPa, both the linewidth and the areal fraction seem to return toward the initial values before applying the external stress. According to the stress–strain curves as shown in Fig. 8.27 (right), the slope of each curve changed suddenly at 28 MPa, suggesting a buckling of the sample. Thus the stress field might be localized at near the centre of the sample, while it is relaxed in the most of the sample area for the Mössbauer measurements. Accordingly, a decrease of the interstitial linewidth and an increase of the substitutional areal fraction might be observed at 28 MPa.

When the external stress increases, the line width of interstitial $Fe_{int}{}^0$ component increases and then decreases again at around 28 MPa. The line broadening could

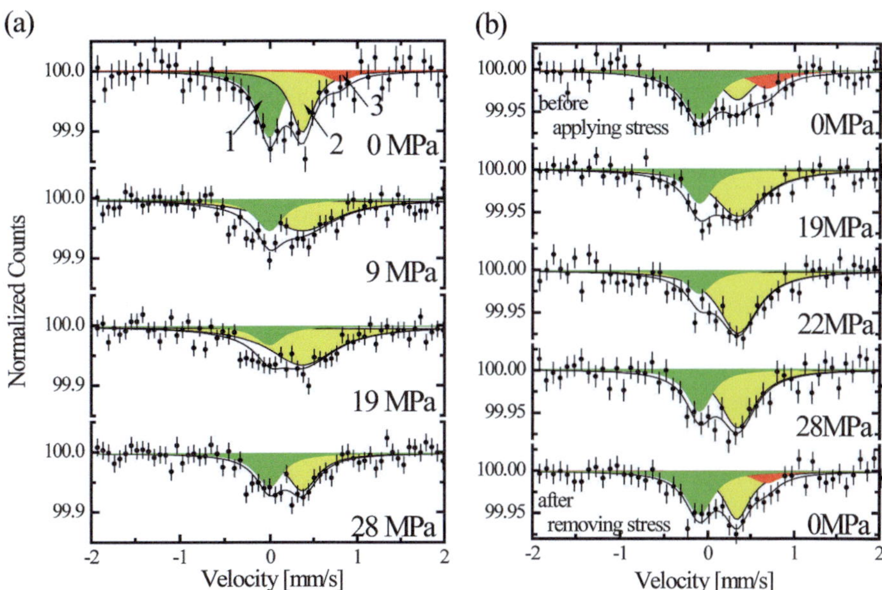

Fig. 8.28 Mössbauer spectra of ^{57}Fe-doped Si sample with (**a**) tensile (**b**) compressive stress on the ^{57}Fe deposited side (From Refs. [48, 49]). The components 1, 2, 3 are interpreted as substitutional $Fe_{sub}{}^0$, interstitial $Fe_{int}{}^0$, interstitial $Fe_{int}{}^+$, respectively

be explained either by the stress-induced fast jumps of ^{57}Fe probes within the life time of 14.4 keV Mössbauer state, or by a distribution of the isomer shift due to a change in the electron density. The latter possibility, however, can be excluded, because the isomer shift of the interstitial $Fe_{int}{}^{0}$ component is not expected to depend on the applied stress, although the Si lattice would be dilated or contracted around ^{57}Fe nuclei. Therefore, the line broadenings of Fe component must arise from the fast jumps of interstitial $Fe_{int}{}^{0}$ atoms when the bending external stress is applied. From the value of the line broadening, the diffusion coefficient is estimated to be of the order of 10^{-9} cm^2/s at room temperature, leading to an activation enthalpy of 0.35 eV for interstitial Fe jumps. This value is much smaller than that of 0.67 eV, which was reported [29]. The area fraction of $Fe_{sub}{}^{0}$ decreases by 30 % up to 19 MPa, and then increases again at 28 MPa. In Mössbauer spectroscopy, the absorption area is proportional to the number of ^{57}Fe probes, suggesting that the interstitial Fe atoms are newly formed from substitutional Fe atoms by applying the external stress. Therefore, the external bending stress appears to enhance the formation process from the substitutional to the interstitial Fe due to a change in the diffusion barrier.

We have discussed so far the results from the Mössbauer absorber experiments, where ^{57}Fe stable isotopes were deposited and the spectra of different Si wafers were measured at different conditions, such as under different temperatures, light illuminations, external voltages, and also external stress. Although all the spectra could be analysed by a superposition of singlet components, which consist of substitutional $Fe_{sub}{}^{0}$ and interstitial $Fe_{int}{}^{0}$ as dominant components, and further small components such as $Fe_{sub}{}^{-}$, $Fe_{int}{}^{+}$ and $Fe_{int}{}^{2+}$. However, our experiments might be suffering from clustering and precipitation of ^{57}Fe probes in the Si matrix, because we had to use a considerable high concentration of ^{57}Fe and therefore, someone might claim that Mössbauer spectroscopy does not observe isolated ^{57}Fe atoms. Accordingly, we need an experimental methodology which is by principle completely free from Fe clustering and/or precipitations. In the following we will show that such ideal experimental conditions exist when using an on-line nuclear fragment separator for Mössbauer experiment on $^{57}Mn/^{57}Fe$ in Si crystals and solar cells.

8.2.11 High Energy ^{57}Mn Implantation into Si

We have developed an on-line set-up for $^{57}Mn/^{57}Fe$ Mössbauer spectroscopy at RIKEN [43–47], where one can implant ^{57}Mn, i.e. a mother isotope of ^{57m}Fe, into Si wafers with an energy of several 100 MeV. The set-up is schematically presented in Fig. 8.29. Using the Ring cyclotron (K = 540 MeV) combined with the LINAC (16 MV) at RIKEN, ^{58}Fe nuclei are accelerated up to the energy of 62 MeV/u. The particles are impinged on a Be target, consequently producing projectile nuclear fragments. ^{57}Mn nuclei with a half-life of 1.45 m are separated from the fragments

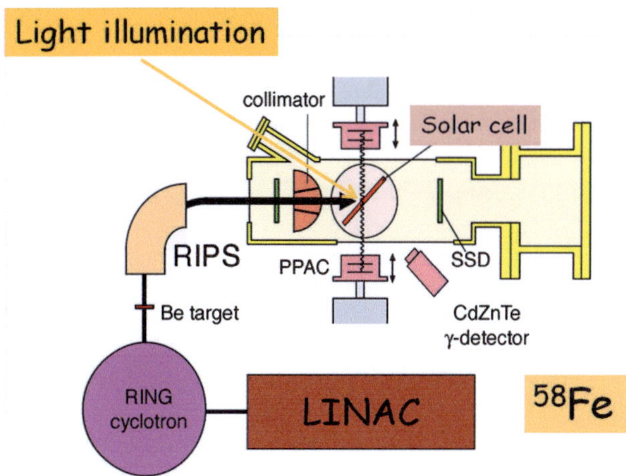

Fig. 8.29 Experimental set-up for ^{57}Mn implantations at RIKEN

using RIPS, and are finally implanted into a specimen through an energy degrader of aluminium foil with a thickness of 200 μm. The final implantation energy after the energy degrader is about 17 MeV/u, leading to an implantation range of about 250 μm and a straggling of about 100 μm. The ^{57}Mn beam direction is 45° to the Si surface and the spot size of the beam is 20 mm in diameter. The beam intensity is typically 2×10^8 ^{57}Mn/s at the measuring chamber. ^{57}Mn nuclei experience β^-decay directly to the 14.4 keV excited state of ^{57}Fe, and the 14.4 keV γ-rays are used for the Mössbauer experiments.

The energy loss of highly energetic ^{57}Mn is dominated by electronic stopping power, creating electrons and holes along the nuclear tracks. Finally, ^{57}Mn probes come to rest at the end of nuclear tracks in the Si matrix, and are distributed in a region in the wafer of about 100 μm straggling width. Subsequently, the ^{57}Mn nuclei undergo β^- decay to ^{57m}Fe$^+$, leaving from their lattice positions (i.e., an "internal implantation" due to the β-decay recoil-energy of 60 keV), and come to their final lattice positions on substitutional and interstitial sites. We measure ^{57m}Fe Mössbauer spectra during the life time of 98 ns, in which the capture reactions of the ^{57m}Fe atom would occur with the excess carriers. Consequently, the Fe components $Fe_{sub}{}^0$ and $Fe_{int}{}^+$ will be formed and observed in the spectrum, as is shown in Fig. 8.30. Furthermore, these capture processes are different between a Si wafer and a Si device, for instance a Si-solar cell, due to the different carrier flows as well as reactions. Indeed, during a solar cell operation, i.e., under light illumination, it was successfully observed that the components of $Fe_{sub}{}^0$ and $Fe_{int}{}^+$ transform to those of $Fe_{sub}{}^-$, $Fe_{int}{}^0$ and $Fe_{int}{}^{2+}$ by captures due to a directional excess carrier flow [47]. The details will be discussed in the following section.

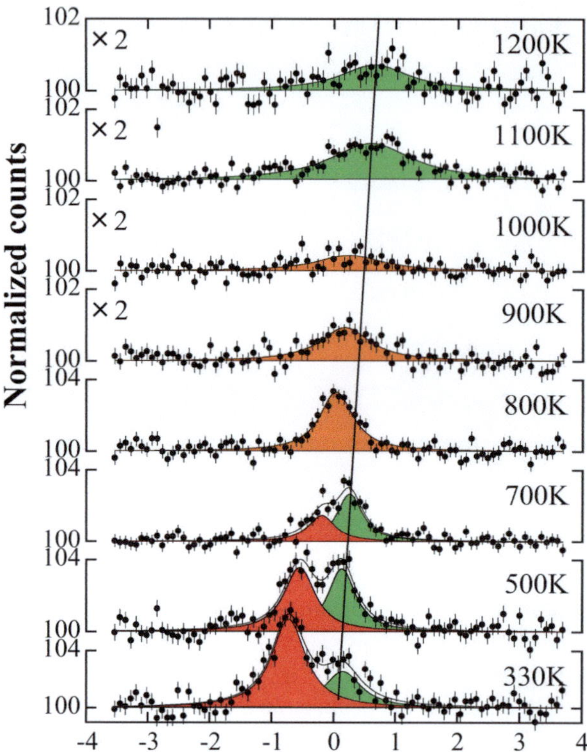

Fig. 8.30 Mössbauer spectra of ^{57}Fe in n-type FZ-Si measured between 330 and 1200 K. Notice that the vertical scale for the region above 900 K is blown up with a factor of two, in order to show the small and broad resonance effects more clearly (From Ref. [46])

As a typical example obtained from this newly developed on-line implantation technique, Fig. 8.30 shows Mössbauer spectra of ^{57}Fe in n-type FZ-Si in the temperature region between 330 and 1200 K. The spectra from 330 to 700 K can be fitted by two Lorentzian singlets, while the spectra from 800 K up to 1200 K can be analysed only by a broad singlet. The two singlets at 330 K at the left and the right hand site have been assigned to $Fe_{int}{}^{+}$ and $Fe_{sub}{}^{0}$ atoms on interstitial and substitutional sites in the Si matrix, respectively. This assignment is based on a theoretical calculation of the isomer shifts [61], which relate with the charge density at ^{57}Fe nucleus. The spectra and their fitting parameters are changing anomalously with elevating temperature. The resonance area of the interstitial ^{57}Fe component at the left hand side decreases strongly above 500 K. A "peak position of a component" in the Doppler velocity scale provides us with information both on the different lattice sites of ^{57}Fe atoms and their charge states. This shift is called "centre shift", depending on the isomer shift and the second-order Doppler (SOD) shift. As a reference, the line in Fig. 8.30 shows the usual temperature dependence of a spectrum component due to the SOD shift. In this case both the interstitial

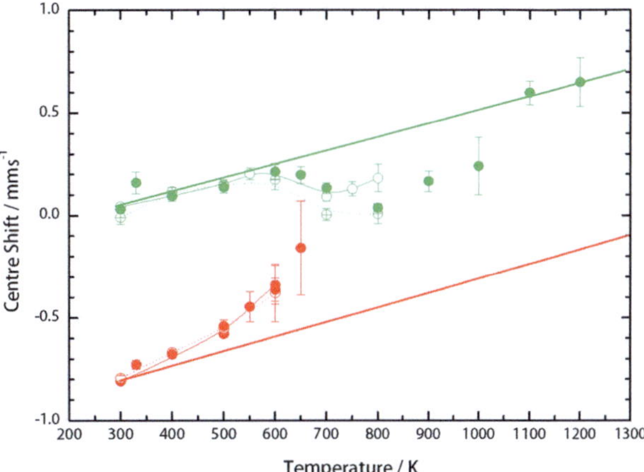

Fig. 8.31 Centre shifts of Fe_{int}^{+} (*red*) and Fe_{sub}^{0} (*green*) component are plotted as function of temperature, which are compared with the lines expected from the second order Doppler shift (From Ref. [46])

and the substitutional components, however, do not simply follow the SOD shift. The centre shifts of each component are plotted as function of temperature in Fig. 8.31, in which other data points obtained from p-type Si wafers are also shown. The centre shift of the interstitial component deviates from the SOD shift (shown as red and green lines), moving continuously to the substitutional position with increasing temperature. This anomaly is accompanied with a strong decrease of the resonance area. Furthermore, the centre shifts of the singlet between 800 and 1000 K are different from the extrapolation of the centre shifts of both components, but above 1100 K the shift goes back to the position of the substitutional site again (Fig. 8.31).

The relaxation behaviour observed in this experiment can be interpreted in terms of a diffusion–reaction process of interstitial Fe atoms with vacancies within the time scale of 100 ns, leading to the formation of substitutional Fe atoms in the Si matrix. This model is shown in Fig. 8.32. Instead of the above simple fitting analysis, we now try to reproduce the whole changes observed in the spectra up to 1200 K by using a set of the kinetic equations of the interstitial $C_I{}^{Fe}$ and the substitutional ^{57}Fe $C_S{}^{Fe} = (1 - C_I{}^{Fe})$ reacting with vacancies C_V and self-interstitials C_I which arises from the above mentioned "diffusion–reaction [46]". Here, K1, K2, K3 and K4 are the reaction coefficients.

$$\frac{dC_I^{Fe}}{dt} = K1\left(1 - C_I^{Fe}\right) - K2C_I^{Fe}C_V$$

$$\frac{dC_V}{dt} = K1\left(1 - C_I^{Fe}\right) - K2C_I^{Fe}C_V + K3\left(1 - C_V\right) - K4C_VC_I$$

$$\frac{dC_I}{dt} = K3\left(1 - C_V\right) - K4C_VC_I$$

$$\Phi(\omega) = \frac{const}{\Gamma + \Delta\Gamma_S + i\left(\omega - \varpi - \alpha\cdot\Delta\omega\right) + \frac{(\Delta\omega)^2\left(1-\alpha^2\right)}{\Gamma+\Delta\Gamma_I+i(\omega-\varpi+\alpha\cdot\Delta\omega)+\omega_R}}$$

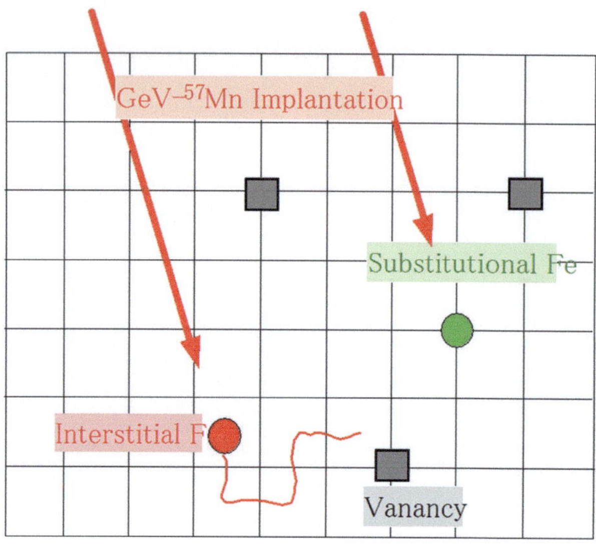

Fig. 8.32 A model describes the relaxation behavior

In order to take account of the relaxation behaviours, we use a theoretical model proposed by Dattagupta [66] which was originally applied for a system with a charge fluctuation. In addition, we have to incorporate the effects, $\Delta\Gamma_S$ and $\Delta\Gamma_I$, from the diffusion broadenings of substitutional and interstitial Fe to the spectra $\Phi(\omega)$, respectively. Here ω_R is a relaxation rate, $\alpha = p_s - p_i$ where p_s and p_i are the substitutional and the interstitial fractions, respectively. $\varpi = 1/2\ (\delta_s - \delta_I)$ and $\Delta\omega = 1/2\ (\delta_s + \delta_I)$ where δ_s and δ_I are the isomer shifts for the substitutional and the interstitial components, respectively. $\Delta\Gamma_S$ and $\Delta\Gamma_I$ are the line broadenings due to diffusion. The values of p_s and p_i are deduced from the numerical solutions which are obtained from the kinetic equations for the diffusion–reaction mainly due to the interstitial jumps within 100 ns [46]: In this model an interstitial Fe atom is jumping with an activation enthalpy of 0.67 eV [29] within the lifetime of ^{57}Fe, 100 ns, and subsequently finding a vacancy, which results in the formation of the substitutional Fe atoms. The number of vacancies related with one interstitial Fe is a parameter. A result obtained from the simulation is shown in Fig. 8.33. The general tendency of the temperature dependence of the Mössbauer spectra, which is shown in Fig. 8.30, appears to be reproducible by the simulation.

After GeV-^{57}Mn/^{57}Fe implantation into Si wafers, ^{57}Fe atoms are found to occupy interstitial and substitutional sites. Interstitial Fe atoms start jumping already above 300 K within the lifetime of ^{57}Fe nuclei, and the simultaneous relaxation effects on the centre shift suggest that the charge state of interstitial Fe atoms changes with increasing temperature. Substitutional Fe atoms are formed via a kinetic reaction between interstitial Fe and excess vacancies above 600 K. The process must be related to the recovery processes to the equilibrium states of the Si lattice around the nuclear probes. Above 1000 K substitutional Fe atoms start migrating within the life time of ^{57m}Fe, leading to long-range diffusion via vacancies in the Si matrix.

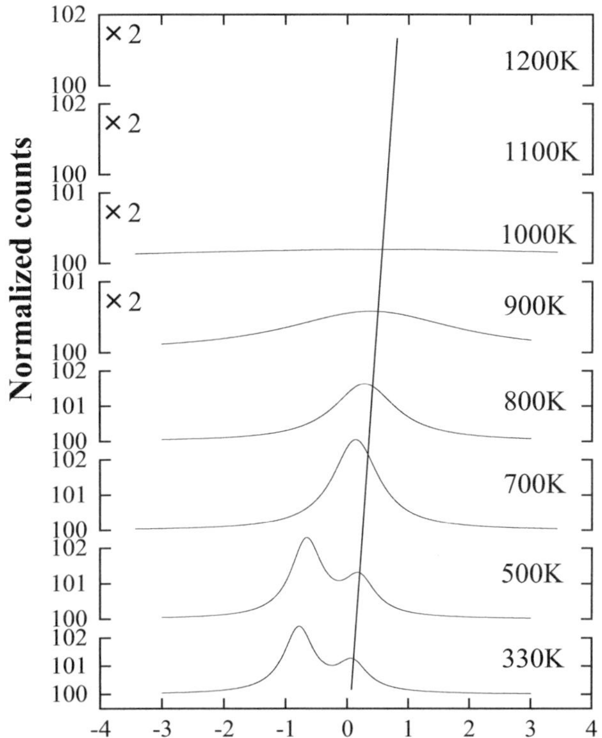

Fig. 8.33 The model reproduces the overall temperature dependence of the Mössbauer spectra of ^{57}Fe in n-type FZ-Si measured between 330 and 1000 K (From Ref. [46])

8.2.12 High Energy ^{57}Mn Implantation into mc-Si Solar Cell

In order to study Fe impurities inside a Si solar cell (Fig. 8.34), the cross section of which is schematically drown in Fig. 8.35, we have to introduce, first of all, isolated ^{57}Fe probes into a region where the carrier trapping processes are occurring during solar cell operation, i.e. light illumination. At the RI-beam facility in RIKEN we used the same technique as described above, i.e., the projectile fragmentation combined with ^{57}Mn/^{57}Fe implantation Mössbauer spectroscopy. The nuclear probes are deeply implanted into the whole p-layer of a mc-Si solar cell through the n-layer, and subsequently Mössbauer spectra can be measured under solar cell operation immediately after each implantation of ^{57}Mn within 2 min. This exotic measuring technique guarantees that all Mössbauer components in the spectrum are originating from isolated ^{57}Fe probes, but not from Fe clusters and Fe precipitates. The spectra of ^{57}Fe both in p-type mc-Si and in the p-region of the p-n junction solar cell which were measured at 400 K with/without light illumination, are shown in Fig. 8.36 [47].

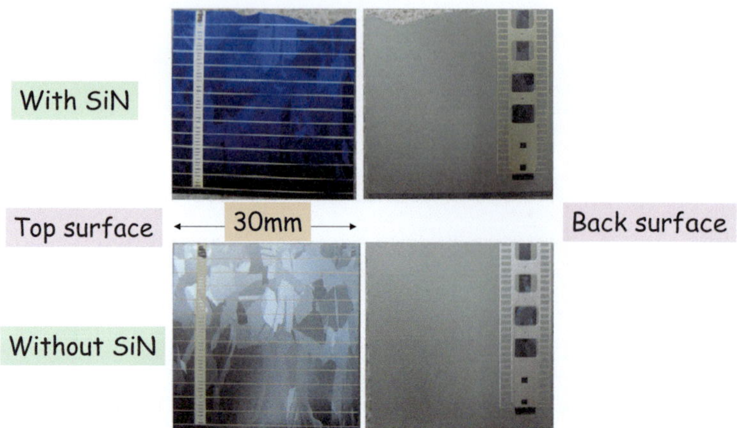

Fig. 8.34 Samples of mc-Si solar cell with and without Si-N layer

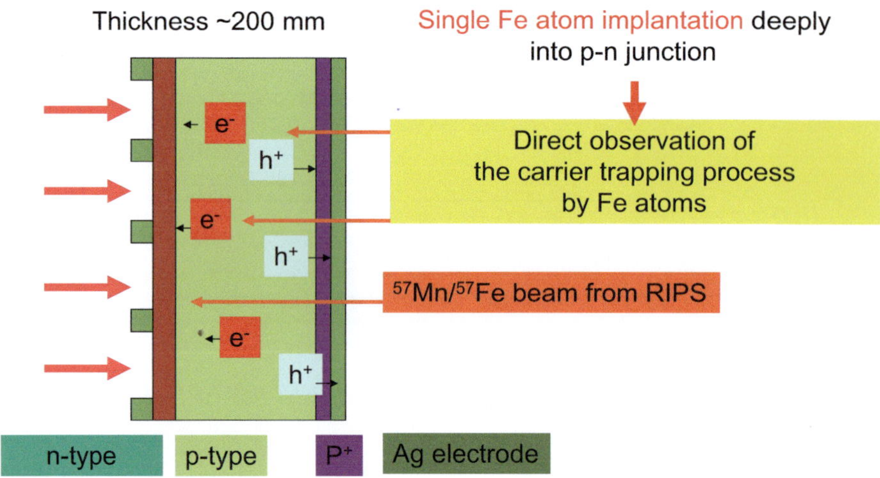

Fig. 8.35 Carrier trapping on implanted ^{57m}Fe after β-decay from ^{57}Mn

In Fig. 8.37, the spectra of a mc-Si solar cell measured at 300 K are presented [50]: (a) before, (b) under and (c) after Xe-lamp illumination. Additionally, the differences between (a) and (b), (b) and (c), and (a) and (c) are plotted to show the changes in the spectra. To guide the eye, a dotted line is also plotted at the isomer shift corresponding to the $Fe_{int}{}^{2+}$ state.

Please notice that the Doppler velocity scale, i.e. the energy scale of the spectrum is converted in this source experiment so that it can be compared to more common absorber experiments. No considerable change can be seen in the difference between (a) and (c), indicating no defect accumulations around the ^{57}Mn/^{57}Fe probes during the measurements, although the implantation energy reaches up to GeV. Under light illumination, the interstitial $Fe_{int}{}^{2+}$ and $Fe_{int}{}^{1+}$ partly disappear, accompanied by an

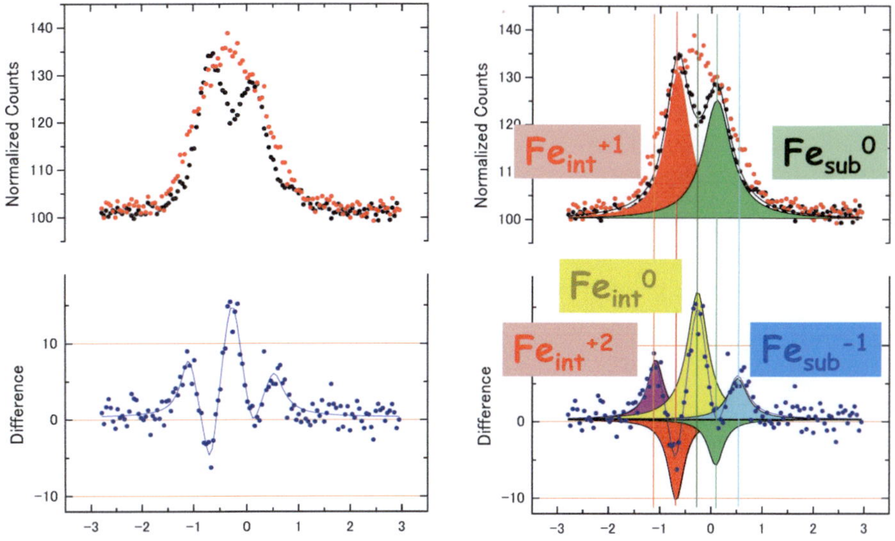

Fig. 8.36 Mössbauer spectra under light illumination (From Ref. [47])

increase of other components, manifesting themselves in the excess carrier trappings on the interstitial Fe atoms. When the light is switched off, the spectrum appears as before.

The interstitial Fe_{int}^{2+} can be observed by this highly energetic implantation technique only in a sample such as a mc-Si wafer, which must contain a considerable amount of lattice defects. Accordingly, this higher charged state must be related with Fe interstitials adjacent to the defects in the solar cell already existing before the measurements. This is, in fact, the first in-situ observation of the carrier trapping processes at Fe impurities in mc-Si solar cell, which degrades the energy conversion efficiency.

8.2.13 ^{57}Fe Mapping by a Mössbauer Microscope

We have developed a Mössbauer spectroscopic microscope [54–58] which measures conversion and Auger electrons emitted after "Mössbauer effect", i.e., recoil-free nuclear resonance absorption of focused 14.4 keV γ-rays (Figs. 8.38 and 8.39). This microscope is, therefore, sensitive only to ^{57}Fe, and the resolution reaches down to 50 μm so far. In the mc-Si matrix, the distribution of Fe impurities must be strongly correlated with the defects such as vacancies, dislocations and grain boundaries. The mapping intensity of the MS image depends not only on the number of ^{57}Fe, but also on the Mössbauer spectral components. Accordingly, by adjusting the Mössbauer resonance condition to each spectral component, we are able to measure

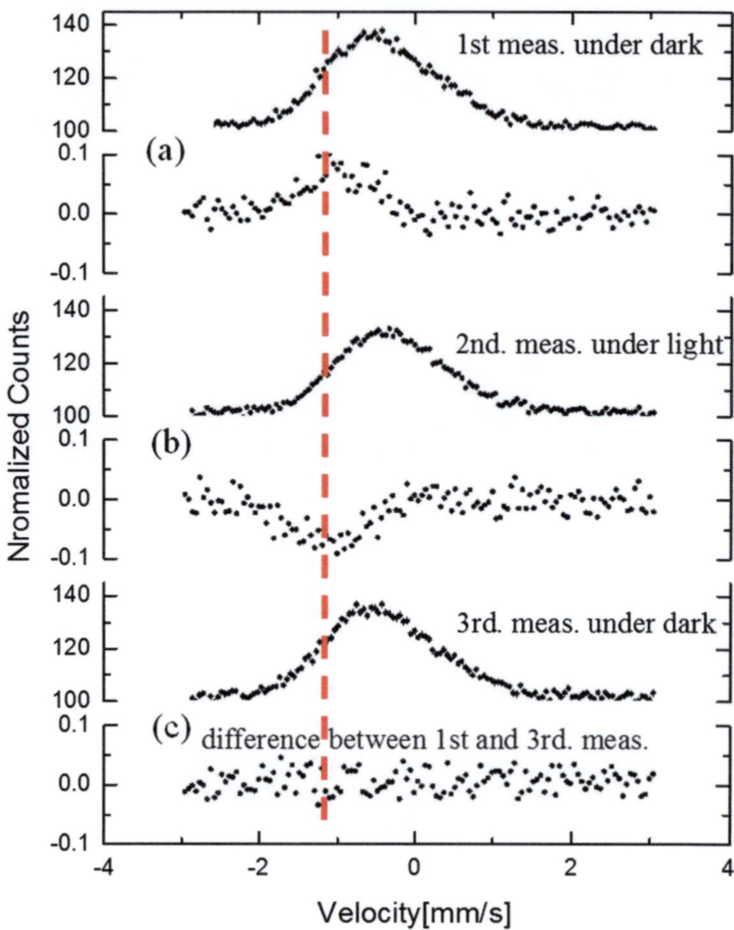

Fig. 8.37 Mössbauer spectra of mc-Si solar cell: (*a*) before and (*c*) after measurements and (*b*) under light illumination, and the differences between (*a*) and (*b*), (*b*) and (*c*), and (*a*) and (*c*), respectively (From Ref. [50])

separately the mapping images for different components. We have applied this new technique to study a mc-Si solar cell without SiN layer, in which ^{57}Fe impurities were intentionally deposited with a thickness of 1.7 nm, and subsequently, were diffused into the n-region with electrodes at room temperature.

Figure 8.40 shows the Fe mapping images of the ^{57}Fe-diffused mc-Si wafer corresponding to (a) substitutional $Fe_{sub}{}^0$, (b) interstitial $Fe_{int}{}^0$, (c) interstitial $Fe_{int}{}^+$, and (d) the CCD picture of measured region of the mc-Si solar cell, and (e) the Mössbauer spectrum, respectively. In the picture (d) two electrodes can be seen. Roughly speaking, the mapping intensities of (a), (b) and (c) are proportional to the area intensities corresponding to $Fe_{sub}{}^0$, $Fe_{int}{}^0$ and $Fe_{int}{}^+$, respectively. It should be

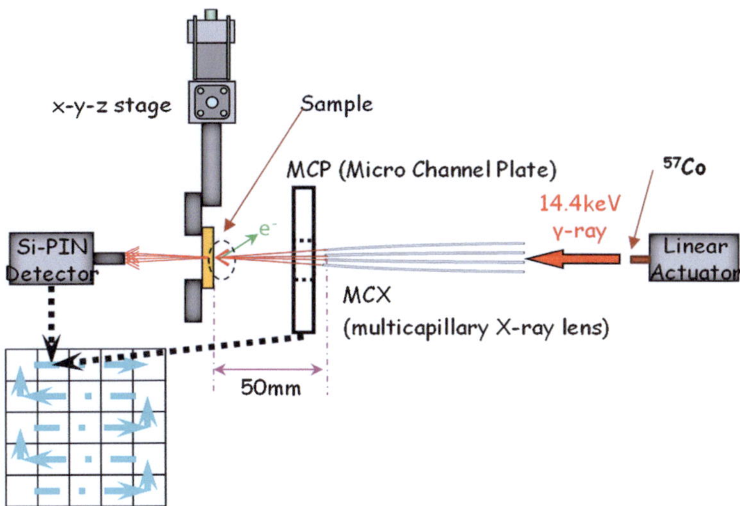

Fig. 8.38 Principle of Mössbauer microscope by focusing 14.4 keV-γ-rays using MCX lens combined with XYZ stage and with MCP-electrons and Si-PIN γ-rays detection systems

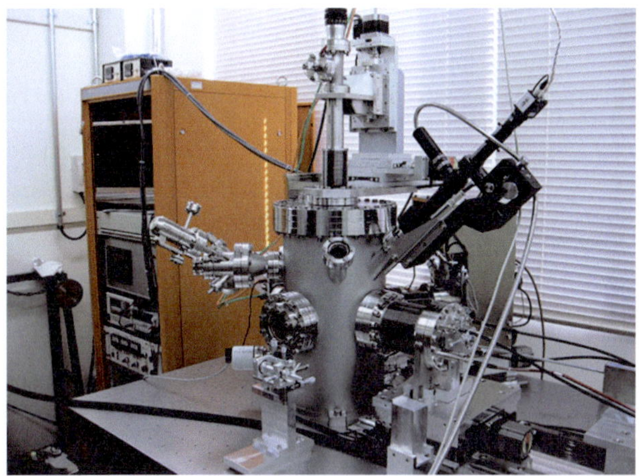

Fig. 8.39 Photograph of Mössbauer microscope

noted that the Fe impurities distribute not only along the grain boundaries, but also in the grain matrix.

Finally, Fig. 8.41 shows another typical example of a mapping image observed in mc-Si solar cell with silver electrodes as-received [56], which was not intentionally contaminated with ^{57}Fe impurities. The sample was specially made without Si–N surface layer for our investigations. Figure 8.41 represent (a) a conventional photo of the mc-Si solar cell, (b) a mapping image with vibrating ^{57}Co source, (c) without vibrating ^{57}Co source and (d) a subtraction (b–c). The mapping images are supposed

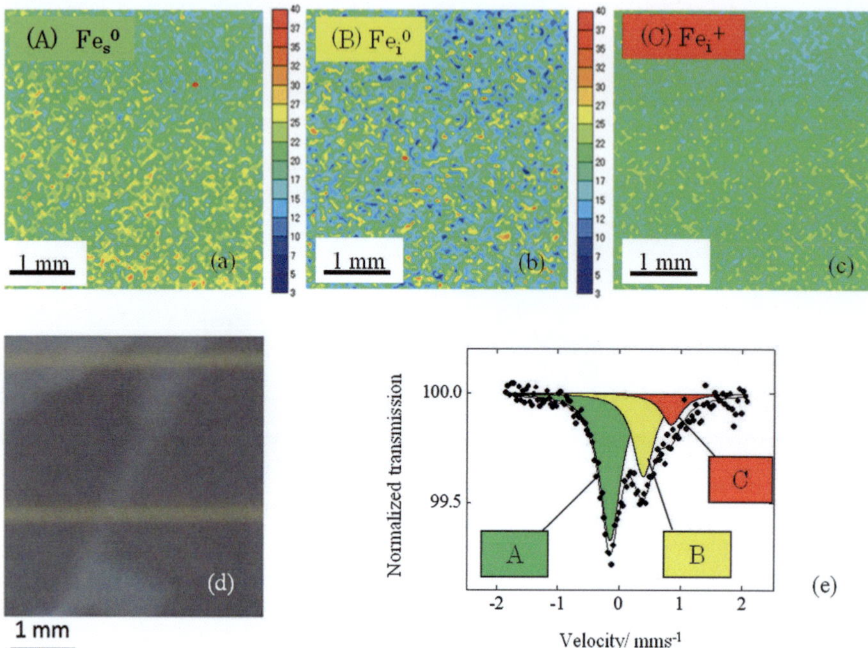

Fig. 8.40 Mössbauer mapping images: (**a**) substitutional $Fe_{sub}{}^{0}$, (**b**) interstitial $Fe_{i}{}^{0}$, (**c**) interstitial $Fe_{int}{}^{+}$, (**d**) measured region of mc-Si solar cell with electrodes, and (**e**) the transmission Mössbauer spectrum. The color scale for the mappings of (**a**), (**b**) and (**c**) is the total electron counts at each mapping point (From Ref. [50])

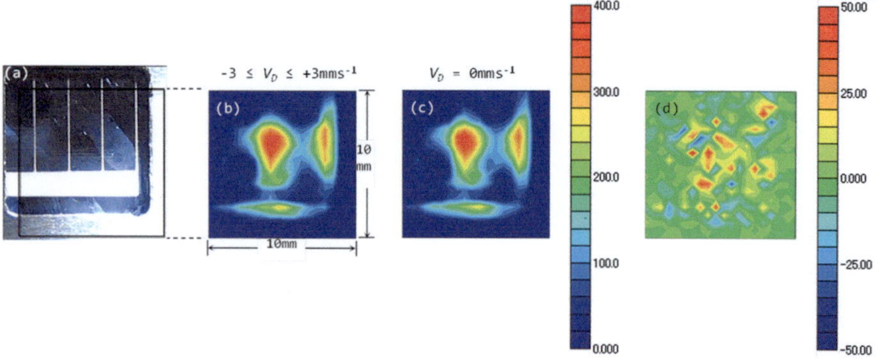

Fig. 8.41 ^{57}Fe mapping images of a mc-Si solar cell as-received (**a**) without intentionally ^{57}Fe deposition corresponding to: (**b**) whole ^{57}Fe impurities, (**c**) substitutional $^{57}Fe_{sub}{}^{0}$, and (**d**) interstitial $^{57}Fe_{int}{}^{0}$ and $^{57}Fe_{int}{}^{+}$ components (From Ref. [56])

to correspond to (b) the whole ^{57}Fe impurities, (c) substitutional $^{57}Fe_{sub}{}^{0}$, and (d) interstitial $^{57}Fe_{int}{}^{0}$ and $^{57}Fe_{int}{}^{+}$ components, respectively. These Fe contamination appears to be due to Ag electrode processing, and therefore, the microscope must play an important role to solve the serious problem from a PV cell degradation.

8.3 Summary

The application of nuclear methods for defects and impurities in Si are reviewed in this chapter. Clearly shown are that the nuclear probes provide us “excellent eyes” on an atomistic scale inside the Si materials, which clarify the lattice positions, the charge states, as well as their dynamic behaviors such as diffusion and carrier trapping processes. It should be stressed, however, that one technique using one nuclear probe has a unique time scale for the measurement not only due to the nuclear lifetime, but also due to the experimental method how the nuclear probes are introduced under different experimental conditions. Therefore, there is an experimental limitation to study the space correlations between the defects, the carriers and the nuclear probes as the impurities in Si materials. Furthermore, the solubility of the nuclear probe is often extremely low in the Si matrix, providing us a challenge against the detection limit to achieve a clear-cut understanding of the physics involved. Accordingly, one should not only use one nuclear method, but also other evaluation techniques available in the research field on Si materials. Especially desirable are theoretical calculations of the hyperfine parameters, which have vital importance to assign the components observed in a spectrum. In our long history of hyperfine interaction studies, it has been difficult to reproduce experimental findings because of a limitation in the computer capacity. We hope very much that we are already coming into the time where all the experimental findings from the nuclear methods can be understood on the basis of the theoretical predictions.

As one of the most challenging research themes we have chosen is the problem of Fe impurities in Si materials: (1) Different Mössbauer experiments on ^{57}Fe impurities into Si samples have been performed after deposition at room temperature, and diffusion and measurements at high temperatures up to 1273 K, ^{57}Fe implantation into Si, and deposition and diffusion, and subsequent measurements at room temperature. The spectra are found to consist of not only interstitial ^{57}Fe, but also of substitutional ^{57}Fe. Both the lattice sites and the charge states of Fe impurities in Si materials change depending on the experimental conditions such as temperature, light illumination, external stress, and electrical connection to the ground as well as external voltage.

(2) We have investigated the Mössbauer spectra of ^{57}Fe-deposited and also ^{57}Fe -implanted Si wafers with different Fermi levels. Clearly, the external voltage applied to a Schottky junction causes a change in the fractions of Fe_{int}^{0} and Fe_{int}^{+} components. In the case of dopant dependence, on the other hand, the spectra are not influenced so much, as is expected from the simple Fermi level dependence due to different boron concentration. This was discussed by taking into account the influence of Fe_{int} itself as a donor within the region of [Fe] > [B]. Also, we observe clearly that both the γ-ray fluence and the electrical connection of the Si wafer to

the ground produce an excess carriers, resulting in a big change in the Mössbauer spectrum. A similar change was also observed in a spectrum of a multi-crystalline Si wafer with $[B] = 1 \times 10^{16}/cm^3$, which consisted of $Fe_{sub}{}^0$ and $Fe_{int}{}^+$.

(3) The isolated Fe atoms can be experimentally obtained in n-type Si wafers from 330 to 1200 K by a newly developed on-line Mössbauer experimental set-up using a RIKEN on-line nuclear fragment separator, RIPS. Combining this facility with a high-energy implantation technique provides us with a unique experimental condition which enables us to follow the dynamical behavior of nuclear probes such as $^{57}Mn/^{57}Fe$ in any material even in a system where no practical solubility has been reported.

(4) In addition, the carrier trapping processes have been studied by highly energetic implantation, or by diffusion processes. The different Fe charge states on different lattice sites could be detected directly during implantation and/or light illumination. Clearly observed in the on-line and the conventional experiments is that the interstitial components of $Fe_{int}{}^0$, $Fe_{int}{}^+$ and $Fe_{int}{}^{2+}$ change their charge states under light illumination through the carrier trapping. The Fe mapping images show that the Fe impurities distribute differently in different grains.

(5) Based on the experimental results from the series of Mössbauer spectroscopy, we have proposed a possible model of Fe impurities in Si materials, which were detected and interpreted mainly by taking account of the experimental conditions. It is clear that Fe impurities exist not only as interstitial $Fe_{int}{}^{0/+}$, $Fe\ _{int}{}^+\text{-}B^-$, but also as $Fe_{sub}{}^{-/0}$ and defect associated $Fe_{int}{}^{2+}$ which could be transformed into each other depending on external conditions. In a next step, it is essentially important to investigate the spatial correlations between the above mentioned Fe components and lattice defects by the 3D Mössbauer Spectroscopic Microscope under development.

References

1. Schatz, G., Weidinger, A.: Nuclear Condensed Matter Physics: Nuclear Methods and Applications. Wiley, New York (1995)
2. Langouche, G. (ed.): Hyperfine Interaction of Defects in Semiconductors. Elsevier, Amsterdam (1992)
3. Mössbauer, R.L.: Gammastrahlung in Ir191. Zeitschrift für Physik A **151**, 124–143 (1958)
4. Cohen, R.L. (ed.): Applications of Mössbauer Spectroscopy: Volume 1. Academic Press, New York (1976)
5. Wikipedia. Mössbauer Spectroscopy
6. Yoshida, Y., Langouche, G. (eds.): Mössbauer Spectroscopy: Tutorial Book. Springer, Berlin (2013)
7. Langouche, G., de Potter, M., Dézsi, I., Van Rossum, M.: Mössbauer study of the microscopic surrounding of Co atoms implanted in Si and Ge below the full amorphization limit. Radiat. Eff. Lett. **67**, 101–106 (1982)
8. Latshaw, G.L.: A Mössbauer effect study of iron implanted into silicon, diamond and germanium. PhD thesis, Stanford University (1971)
9. Latshaw, G.L., Russell, P.B., Hanna, S.S.: In-beam implantation of iron into germanium, silicon and diamond studied by the Mössbauer effect. Hyperfine Interact. **8**, 105–127 (1980)

10. Sawicka, B.D., Sawicki, J.A.: Evidence of the electric quadrupole coupling of ^{57}Fe implanted in silicon. Phys. Lett. **A64**, 311–312 (1977)
11. Sawicki, J.A., Sawicka, B.D.: Correlation of ^{57}Fe isomer shift, quadrupole coupling and interatomic distance in group-IV elements. Phys. Stat. Sol. B **86**, K159–K161 (1978)
12. Langouche, G., de Potter, M.: Identification of substitutional and interstitial Co implanted in Si. Nucl. Instrum. Methods Phys. Res. B **19**(20), 322–324 (1987)
13. Schwalbach, P., Laubach, S., Hartick, M., Kankeleit, E., Keck, B., Menningen, M., Sielemann, R.: Diffusion and isomer shift of interstitial iron in silicon observed via in-beam Mössbauer spectroscopy. Phys. Rev. Lett. **64**, 1274–1277 (1990)
14. Gunnlaugsson, H.P., Weyer, G., Dietrich, M., Fanciulli, M., Bharuth-Ram, K., Sielemann, R., and the ISOLDE Collaboration: Charge state dependence of the diffusivity of interstitial Fe in silicon detected by Mössbauer spectroscopy. Appl. Phys. Lett. **80**, 2657–2659 (2002)
15. http://defects.physics.wsu.edu/pac_setup.html
16. Wichert, T., Swanson, M.L., Queneville, A.F.: Formation of In-As complexes in silicon observed by the perturbed angular correlation technique. Phys. Rev. Lett. **57**, 1757–1760 (1986)
17. Sugimoto, K., Mizobuchi, A., Nakai, K., Matuda, K.: Nuclear magnetic resonance of polarized ^{17}F formed through the ^{16}O(d, n)^{17}F reaction. Phys. Lett. **18**, 38–39 (1965)
18. http://isolde.web.cern.ch *courtesy of M. Stachura*
19. Minamisono, T., Nojiri, Y., Deutch B.I., Asahi K.: Hyperfine interactions of polarized β-emitting ^{12}B and ^{12}N in Si, Ge and GaP. Hyperfine Interact. **15**(16), 543 (1983)
20. Metzner, H., Sulzer, G., Seelinger, W., Ittermann, B., Frank, H.-P., Fischer, B., Ergezinger, K.-H., Dippel, R., Diehl, E., Stockmann, H.-J., Ackermann, H.: Bulk-doping-controlled implantation site of boron in silicon. Phys. Rev. B **42**, 11419–11422 (1990)
21. Wolf, H., Wagner, F., Wichert, T.: Anomalous diffusion profiles of Ag in CdTe due to chemical self-diffusion. Phys. Rev. Lett. **94**, 125901 (2005)
22. Achtziger, N., Forkel-Wirth, D., Grillenberger, J., Licht, T., Witthuhn, W.: Identification of deep bandgap states in 4H- and 6H-SiC by radio-tracer DLTS and PAC-spectroscopy. Nucl. Instr. Meth. Phys. Res. B. **136**, 756–762 (1998)
23. Johnston, K., Henry, M.O., McCabe, D., McGlynn, E., Dietrich, M., Alves, E., Xia, M.: Identification of donor-related impurities in ZnO using photoluminescence and radiotracer techniques. Phys. Rev. B **73**, 165212 (2006)
24. Von Nathusius, C., Vianden, R.: Hall effect measurements on transmutation doped semiconductors. Hyperfine Interact. **129**, 391–400 (2000)
25. http://isolde.web.cern.ch *courtesy of U. Wahl*
26. Wahl, U., Correia, J.G., Czermak, A., Jahn, S.G., Jalocha, P., Marques, J.G., Rudge, A., Schopper, F., Soares, J.C., Vantomme, A., Weilhammer, P., and the ISOLDE Collaboration: Position-sensitive Si pad detectors for electron emission channeling experiments. Nucl. Instr. Meth. Phys. Res. A. **524**, 245–256 (2004)
27. Wahl, U., Vantomme, A., Langouche, G., Marques, J., Correia, J.G.: Direct evidence for tetrahedral interstitial Er in Si. Phys. Rev. Lett. **79**, 2069–2072 (1997)
28. Wahl, U., Correia, J.G., Rita, E., Araujo, J.P., Soares, J.C.: Lattice sites of implanted Fe in Si. Phys. Rev. B **72**, 014115 (2005)
29. Istratov, A.A., Hieslmair, H., Weber, E.R.: Iron and its complexes in silicon. Appl. Phys. A **69**, 13–44 (1999)
30. Istratov, A.A., Hieslmair, H., Weber, E.R.: Iron contamination in silicon technology. Appl. Phys. A **70**, 489–534 (2000)
31. Macdonald, D., Tan, J., Trupke, T.: Imaging interstitial iron concentrations in boron-doped crystalline silicon using photoluminescence. J. Appl. Phys **103**, 073710–073717 (2008)
32. Schubert, M.C., Kerler, M.J., Warta, W.: Influence of heterogeneous profiles in carrier density measurements with respect to iron concentration measurements in silicon. J. Appl. Phys **105**, 114903–114906 (2009)
33. Zoth, G., Bergholz, W.: A fast, preparation-free method to detect iron in silicon. J. Appl. Phys. **67**, 6764–6771 (1990)

34. Langouche, G.: Characterization of semiconductors by Mössbauer spectroscopy. In: Long, G.L., Grandjean, F. (eds.) Mössbauer Spectroscopy Applied to Inorganic Chemistry, vol. 3, pp. 445–512. Plenum Press, New York (1989)
35. Gilles, D., Schroter, W., Bergholz, W.: Impact of the electronic structure on the solubility and diffusion of 3d transition elements in silicon. Phys. Rev. B **41**, 5770–5782 (1990)
36. Langouche, G., Yoshida, Y.: Ion implantation. In: Yoshida, Y., Langouche, G. (eds.) Mössbauer Spectroscopy – Tutorial Book, pp. 267–303. Springer, Berlin (2013)
37. Yoshida, Y., Shimura, F.: In-situ observation of diffusion and segregation of Fe atoms in Si crystals at high temperature by Mössbauer spectroscopy. In: Huff, H.R., Gösele, U., Tsuya, H. (eds.) Proceedings of 8th International Symposium on Silicon Materials Science and Technology, vol. 98-1, pp. 984–996. ECS, San Diego (1998)
38. Yoshida, Y.: Direct observation of substitutional and interstitial Fe atoms in Si by high-temperature and in-beam Mössbauer spectroscopy. In: *PV2003-03 of ECS, ALTECH 2003* Analytical and Diagnostic Techniques for Semiconductor Materials and Processes, 479–482. Salt Lake City (2003)
39. Yoshida, Y., Ogawa, S., Arikawa, K.: Direct observation of substitutional Fe atoms in Si and SOI wafers at 1273 K. Physica B **340–342**, 605–608 (2003)
40. Yoshida, Y., Horie, S., Niira, K., Fukui, K., Shirasawa, K.: In-situ observation of iron atoms in multicrystalline silicon at 1273 K and 300 K by Mössbauer spectroscopy. Physica B **376–377**, 226–230 (2006)
41. Yoshida, Y., Aoki, S., Sakata, K., Suzuki, Y., Adachi, M., Suzuki, K.: Iron impurities in multicrystalline silicon studied by Mössbauer spectroscopy. Physica B **401–402**, 119–122 (2007)
42. Yoshida, Y., Suzuki, Y., Matsushita, A., Suzuki, K., Sakata, K.: Fermi level dependence of Mössbauer spectroscopic components corresponding to iron interstitials and their clusters in silicon. Physica B **401–402**, 167–170 (2007)
43. Yoshida, Y., Kobayashi, Y., Hayakawa, K., Yukihira, K., Shimura, F., Yoshida, A., Diao, X., Ogawa, H., Yano, Y., Ambe, F.: In-beam Mössbauer study on interstitial and substitutional ^{57}Mn/^{57}Fe jumps in Si. Defect Diffus. Forum **194–199**, 611 (2001)
44. Yoshida, Y., Kobayashi, Y., Yoshida, A., Diao, X., Ogawa, S., Hayakawa, K., Yukihira, K., Shimura, F., Ambe, F.: In-beam Mössbauer spectroscopy after GeV-Ion implantation at an on-line projectile-fragments separator. Hyperfine Interact. **141–142**, 157–162 (2002)
45. Yoshida, Y., Kobayashi, Y., Hayakawa, K., Yukihira, K., Yoshida, A., Ueno, H., Shimura, F., Ambe, F.: In-situ observation of substitutional and interstitial Fe atoms in Si after GeV-implantation: an in-beam Mossbauer study. Physica B **376–377**, 69–72 (2006)
46. Yoshida, Y., Kobayashi, Y., Yukihira, K., Hayakawa, K., Suzuki, K., Yoshida, A., Ueno, H., Yoshimi, A., Shimada, K., Nagae, D., Asahi, K., Langouche, G.: ^{57}Fe diffusion in n-type Si after GeV implantation of ^{57}Mn. Physica B **401–402**, 101–104 (2007)
47. Yoshida, Y., Suzuki, K., Kobayashi, Y., Nagatomo, T., Akiyama, Y., Yukihira, K., Hayakawa, K., Ueno, H., Yoshimi, A., Nagae, D., Asahi, K., Langouche, G.: ^{57}Fe charge state in mc-Si solar cells under light illumination after GeV- implantation of ^{57}Mn. Hyperfine Interact. **204**, 133–137 (2011)
48. Suzuki, K., Yoshida, Y., Kamimura, T., Ichino, M., Asahi, K.: Iron diffusion in silicon under external stress. Physica B **404**, 4678–4680 (2009)
49. Suzuki, K., Yoshida, Y., Hayakawa, K., Yukihira, K., Ichino, M., Asahi, K.: Observation of iron impurity diffusion in silicon under bending stress by Mössbauer spectroscopy. Hyperfine Interact. **197**, 213–217 (2010)
50. Yoshida, Y., Tsukamoto, Y., Ichino, M., Tanaka, K.: Direct observation of carrier trapping processes on Fe impurities in mc-Si solar cells. Solid State Phenom. **205–206**, 40–46 (2014)
51. Tanaka, K., Watanabe, T., Uenoyama, T., Ino, Y., Yoshida, Y.: Search for FeB pairs in B-highly doped Si wafers by Mössbauer spectroscopy. In: Proceedings of the 7th Forum on the Science and Technology of Silicon Materials, 107–111 (2014)
52. Ino, Y., Tanaka, K., Yoshida, Y.: Direct observations of Fe impurities in Si with different Fermi levels by Mössbauer spectroscopy. Solid State Phenom. **242**, 205–210 (2016)

53. Yoshida, Y., Ino, Y., Tanaka, K.: Mössbauer spectroscopy on Fe impurities in Si materials. Solid State Phenom. **242**, 211–217 (2016)
54. Yoshida, Y., Suzuki, K., Hayakawa, K., Yukihira, K., Soejima, H.: Mössbauer spectroscopic microscope. Hyperfine Interact. **188**, 121–126 (2009)
55. Yoshida, Y., Kamimura, T., Ichino, M., Hayakawa, K., Yukihira, K., Soejima, H.: Mössbauer spectroscopic microscope. J. Phys. Conf. Ser. **217**, 012003–012004 (2010)
56. Yoshida, Y., Hayakawa, K., Yukihira, K., Ichino, M., Akiyama, Y., Kumabe, H., Soejima, H.: Development and applications of "Mössbauer cameras". Hyperfine Interact. **198**, 23–29 (2010)
57. Hayakawa, K., Tsukamoto, Y., Akiyama, Y., Kurata, M., Yukihira, K., Soejima, H., Yoshida, Y.: Deployment of system and technology for Mössbauer spectroscopic microscope. Hyperfine Interact. **206**, 79–82 (2012)
58. Tanaka, K., Akiyama, Y., Hayakawa, K., Yukihira, K., Yoshida, Y.: Mapping analyses of Fe-diffused mc-Si using Mössbauer microscope and photo- luminescence. Hyperfine Interact. **206**, 75–78 (2012)
59. Kamimura, T., Udono, H., Yoshida, Y.: not published (2008); Kaminura, T.: Master thesis, Shizuoka Institute of Science and Technology, Japan (2009)
60. Yoshida, Y., Ino, Y.: To be published (2015)
61. Kübler, J., Kumm, A.E., Overhof, H., Schwalbach, P., Hartick, M., Kankeleit, E., Keck, B., Wende, L., Sielemann, R.: Isomer-shift of interstitial and substitutional iron in silicon and germanium. Z. Phys. B **92**, 155–162 (1993)
62. Coutinho, J.: Private communication (2015)
63. Estreicher, S.K., Sanati, M., Gonzalez Szwacki, N.: Iron in silicon: interactions with radiation defects, carbon, and oxygen. Phys. Rev. B **77**, 125214–125219 (2008)
64. Backlund, D.J., Estreicher, S.K.: Ti, Fe and Ni in Si and their interactions with the vacancy and the A center: a theoretical study. Phys. Rev. B **81**, 235213–235218 (2010)
65. Ziegler, J.F., Biersack, J.P.: In *SRIM* – The stopping and range of ions in solids, http://www.srim.org/ (2009)
66. Dattagupta, S.: Time-dependent effects and relaxation in Mössbauer spectroscopy. In: Dickson, D.P.E., Berry, F. (eds.) Mössbauer Spectroscopy, pp. 198–218. Cambridge University Press, London (1986)

Chapter 9
Defect Engineering in Silicon Materials

Werner Bergholz

Abstract Point Defects (i.e. vacancies and selfinterstials are zero-dimensional defects) in silicon. In solids in general, such defects are unavoidable, their presence follows from the Second Law of Thermodynamics; for a given temperature and for each species of intrinsic point defects (vacancy, self-interstitial). In contrast, impurities for which there is an equilibrium solubility, can be avoided by suitable countermeasures. Likewise, one-dimensional defects (dislocations), two-dimensional defects (e.g. stacking faults) and three-dimensional defects (e.g. voids or precipitates of impurities) can be avoided by defect engineering; they are not inevitable by the laws of thermodynamics. To turn silicon into practical devices (e.g. integrated circuits, photovoltaic cells), defect engineering is part and parcel of the overall process and production technology. First, the fundamentals will be explained, e.g. the analogy between the chemistry of ions in water as a "substrate" and the "chemistry" of point defects in silicon, the most perfect, purest solid material available. Secondly, the application of these fundamental defect engineering principles will be described for various technology areas. The focus will be on microelectronics, photovoltaics will also be mentioned. Topics in microelectronics will include the impact of oxidation, ion implantation, reactive ion etching and thermal processes for diffusion of dopant atoms. Gettering of metal contamination will also be considered. It will be elucidated that there is a close interaction of metal impurities and extended defects, and that an uncontrolled metal contamination is a source of unacceptable quality risks in the production of microelectronic devices and photovoltaic cells.

Keywords Point defects • Dislocations • Stacking faults • Metal contamination • Device failure • Quality engineering • Key control characteristic

---

W. Bergholz (✉)
Jacobs University Bremen, Bremen, Germany
e-mail: w.bergholz@jacobs-university.de

Y. Yoshida, G. Langouche (eds.), *Defects and Impurities in Silicon Materials*, Lecture Notes in Physics 916, DOI 10.1007/978-4-431-55800-2_9

9.1 Introduction

In this section the scope and the objectives of defect engineering, i.e. the application of this knowledge in the semiconductor industry, will be outlined against the backdrop of the remarkable perfection of electronic grade silicon material for the microelectronics industry. Also, the transfer of the material knowledge gained in the microelectronics industry to the photovoltaics (PV) industry which in terms of silicon consumption has overtaken microelectronics will be briefly dealt with in this chapter:

Before going into any detail, the way how a good connection between science (in the preceding chapters) and application has to be explained: The term "Engineering" implies that in this chapter the scope has to be shifted from an emphasis on science to generate new knowledge about materials to how to gain new insights with regards to the application of silicon materials science to a corporate semiconductor technology R&D environment or in a medium and large scale production environment. This has two implications:

(a) In terms of the analytical methods many of the scientifically oriented state-of-the-art detection and analytical methods have to be replaced by high throughput/short cycle time methods.
(b) In R&D it is possible to create clearcut "boundary conditions" for experiments and systematic studies. In a production environment one is faced with complexity imposed by the production process.

Also, new concepts beyond Si materials have to be included, such as Statistical Process Control (SPC) of semiconductor manufacturing processes, with implications. Finally, the description of complex "real-word defects in semiconductor and photovoltaic production scenarios" necessitates the use of engineering concepts which are less rigid than "pure" science concepts but enable the engineer to understand the rationale behind practical defect engineering and enable him/her to apply such concepts to new technologies and unprecedented defect formation mechanisms.

Important sources for this chapter are books and key papers on silicon and semiconductor technologies, training materials from internal training for engineers in Siemens Semiconductors and last but not least real and generally important defect scenarios that have occurred in 30 years of microelectronics and photovoltaics production. In the following we strive to keep a good balance between academic rigor and a pragmatic approach needed to cope with the complexity of semiconductor processes. We are led by the objective that the reader will be enabled to monitor and analyze defect formation phenomena to identify the root causes of defect formation and develop countermeasures by a profound understanding of the defect scenarios.

It is not an exaggeration to state that microelectronics-grade silicon is the most perfect and clean material ever made [1, 2]. The mainstream production of integrated circuits, such as central processing units (CPUs), dynamic random access memories (DRAMs), microprocessors, chip cards etc. is carried out on single-crystal silicon wafers of 150–300 mm diameter, with 450 mm diameter wafers at the R&D and sooner or later at the pilot production stage [3]. Advanced microelectronic devices would simply not work on Si wafers which are not a single-crystal and dislocation free [4], or which do not have the same mechanical (macroscopic shape) and structural (microscopic) perfection as wafers for integrated circuit production, or which are not as pure in volume and clean at the surface. This is not true for the photovoltaics industry, where relatively impure and unpolished multicrystalline silicon has the largest share in the production of PV cells and modules so far.

As the first step to a profound understanding of defect engineering, it is instructive to briefly review some of the most remarkable and relevant properties of silicon wafers [2]:

1. They are single crystals, the wafer surface is parallel to one of the {100} planes, a flat or a notch as the fiducial mark for alignment of the wafers indicates one of the <110> directions, as shown in Fig. 9.1a and b (recently, notchless wafers with a laser mark to indicate the crystal direction have been under discussion in the SEMI standards development organization, also see [5]). A single crystal means, that the atomic rows and planes are perfect from wafer surface to wafer surface in any direction, see also Fig. 9.3. For the sake of completeness it has to be mentioned that some power devices are also made on {111} oriented wafers.
2. The deviation of the local flatness of the wafer corresponds to about the thickness of a human hair (80 μm) on a football pitch [2]. Also the other dimensions, such as thickness, diameter, edge profile and the shape and size of the notch are well defined by tight geometric specifications.
3. The purity of the material with respect to metal contamination in the bulk is by far a smaller fraction than one person of the whole world population. The most abundant impurities are carbon and oxygen, and appropriate dopant atoms, which

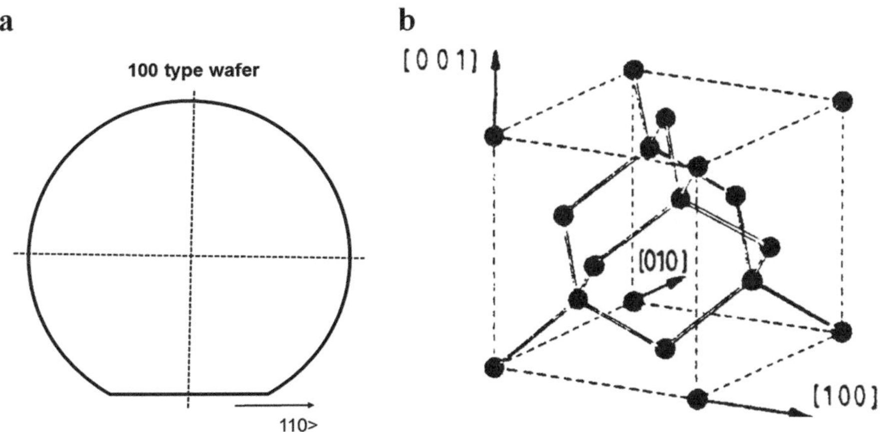

Fig. 9.1 (**a**) Sketch of a Si wafer with 100 surfaces. (**b**) Sketch of an elementary cell of the silicon lattice

are incorporated in well-defined and specified concentrations during the crystal growth process (to engineer the resilience of the wafer to thermal stress and to enable well defined intrinsic gettering, as will be described later). Although the concentrations of oxygen and carbon are only of the order of parts per million, they are critical for the robustness of the wafers against stress and for defect engineering, as will be described later in detail [7, 8].

A good source for the actual state of the art for manufacturing silicon wafers for the microelectronics industry can be found in the SEMI-standard M1 [2] and information about the future technical requirements are described in the International Roadmap for Semiconductors ITRS [3].

4. As briefly alluded to before, in the PV industry the wafers are by far not as clean and as perfect as Si wafers for microelectronics. The raw wafers used to produce cells are just sawn from a silicon ingot or brick, they are not edge rounded, and they are as thin as compatible with processes, currently around 150 μm, i.e. twice the thickness of a human hair (wafers for microelectronics are typically 625–925 μm) [9]. The reason for this drastically reduced technical sophistication is that the cost of the silicon wafers must be as low as possible, because ultimately the cost to produce electric energy by means of silicon PV cells must be low enough to compete with the standard ways to produce electric energy by burning fossil fuels or by using nuclear power [10]. Since the purification of silicon material is expensive (even under reduced purity requirements), the wafers are made "paper – thin" to use as little of the expensive purified material as possible and to avoid any technical specification which is not really needed for the performance and durability of the end product. All the same, also a PV wafer has to comply with the relevant technical specifications, which however as mentioned, are much less demanding.

So, the demanding properties according to the technical specification for microelectronics silicon wafers are necessary in most instances to warrant the full functionality and reliability/durability of the integrated circuits produced, in contrast to the situation in the PV industry, where the technical properties are much less demanding.

To apply the knowledge presented in the preceding chapters of this book to practical problems in technology development and to mass manufacturing, i.e. in defect engineering, needs the introduction of additional concepts which are rooted in general process and technology management and presents the challenge to apply the silicon science and technology presented so far in a comprehensive and concatenated manner, with many potential interactions of different phenomena.

So, the purpose of defect engineering is to establish and maintain the properties (1)–(4) as well as possible throughout the complete integrated circuit or PV-cell manufacturing processes, including the procurement of the wafers, since defects can be already "built-in" in the raw silicon wafer before any device process, likewise the robustness of the wafers to cope with e.g. mechanical stresses can be determined already by the silicon wafer manufacturing process as will be explained in later sections. Such defect engineering is indispensable in order to prevent the detrimental consequences of defects (such as leakage currents in pn-junctions, gate oxide shorts, drift of transistor parameters etc. for microelectronics and below par electrical conversion efficiency for PV cells), and other problems in the performance of the reliability/durability of these products [11–15]. In spite of the much less stringent requirements for PV silicon wafers, defects do play a role to detract the PV cell performance, and it can be stated that improvements of the conversion efficiency of PV cells that have been achieved over the last decades are both due to improvements in the manufacturing process stability in terms of reproducibly achieving the critical device parameters, improvement in the design and last but not least have been achieved by reducing defect densities. More detailed information for these facts can be found in Refs. [16, 17].

Remarkably, in addition to the prevention of defects in a number of instances defect engineering serves to intentionally introduce defects in a controlled fashion to support or impart certain device functions.

Thus the scope of defect engineering is the prevention of those defects which harm the integrated circuits and to create defects which support or enhance the functionality and/or reliability/durability of the device.

This chapter will mainly focus on defects that are relevant for device processing, it will not cover in detail defect formation during the manufacture of silicon ingots or wafers at the suppliers of silicon wafers, this is the subject of other chapters of this book.

In microelectronics, defect engineering includes both defects in the silicon material and defects in the technology layers above the silicon wafer (such as metal lines). The latter are mainly, but not exclusively due to particles that cannot be avoided entirely in spite of a cleanroom [18], and in spite of equipment that is low in particle shedding and materials which are ultraclean, both with respect to particles and impurities.

In this chapter we will only deal with the defects in the silicon wafer, not the defects which are defects in the structure of the technology layers (such as short or open circuits in metal lines in one of the metallization layers above the silicon).

9.2 What Does Defect Engineering Mean in Detail?

As mentioned in the introduction to this chapter, defect engineering means that the defect formation has to be prevented and/or promoted in a "controlled" fashion, in an industrial manufacturing environment. It is one of the desired learning outcomes of this chapter that the difference in "boundary conditions" between a research laboratory situation and an R&D pilot production line and/or a large volume manufacturing is clearly understood, with respect to the defect formation mechanisms and the control of defects, which includes assessment of the individual unit process performance and the integrated process performance in terms of defect types and densities. It will turn out that often, defect formation/prevention is not a question of optimizing a unit process but to understand and modify the integrated process so as to remove undesired detrimental interactions between processes.

To be able to implement state-of-the art defect engineering, the following three areas of competence are required:

(a) Scientific understanding and modelling of defect formation and prevention. The preceding chapters of this book have been instrumental to this purpose. Very often, defects are avoided by trial and error through variation of the process parameters. Time and again, experience has shown, that without a profound understanding of the defect formation mechanism, this is neither a safe nor an efficient method, and recurrence of defects when some other process parameters are changed, is frequently observed in such situations, due to interactions between different process steps that have been neglected or not understood. A significant part of this chapter will be dedicated to teaching by the introduction of concepts and examples how such interactions can be detected or even anticipated at the stage of process/equipment design.
(b) In manufacturing, there is a strict requirement for effective (i.e. high success rate, high yield and good quality [i.e. within the specified defect density targets]) and efficient production (achieve a high throughput with i.e. with as low cost for consumables [such as Si wafers and other materials] and equipment as possible). In the context of defect engineering this means that the methods to monitor the defect densities must be fast and cover a large enough area of the processed wafers not to interfere too much with the speed and sequences of the production process (i.e. the cycle time targets for the unit processes and time coupling requirements which exist between some processes should not be jeopardized) [15], and to cover a representative number of samples. Also, they must satisfy the needs of statistical process control (SPC), which is a pivotal quality management tool that is indispensable in microelectronics to ensure that processes are predictably under control [16, 19]. SPC is also becoming more and more relevant in the photovoltaics industry (although initially this was not the case see [16]). Therefore, the defect monitoring methods in manufacturing are frequently different from sophisticated but low throughput, time-consuming analytical methods in a science laboratory, which would be more frequently used to understand and model defect formation in the context of research. Figure 9.2 illustrates one "production-oriented" method to detect, e.g., contamination by Co, Ni and Cu, and with limitations also Fe (these are ubiquitous in a technical environment and extremely detrimental to devices) on a large wafer area in a short time [6, 15, 18, 20]. The method is called haze test, it consists of driving in any surface impurities by e.g. a 30s rapid anneal step at 1200 °C and delineating the defects that have formed in the process by a suitable defect etch, which converts invisible surface near precipitates of Ni and Cu (and other extended defects) into etch pits which can be visualized on the whole wafer in one photograph by dark field illumination, see also [6]. With this method it is possible to test a process (mechanical transport as in the illustrated example, or chemical or dry etch processes) in 15 min over the surface of a

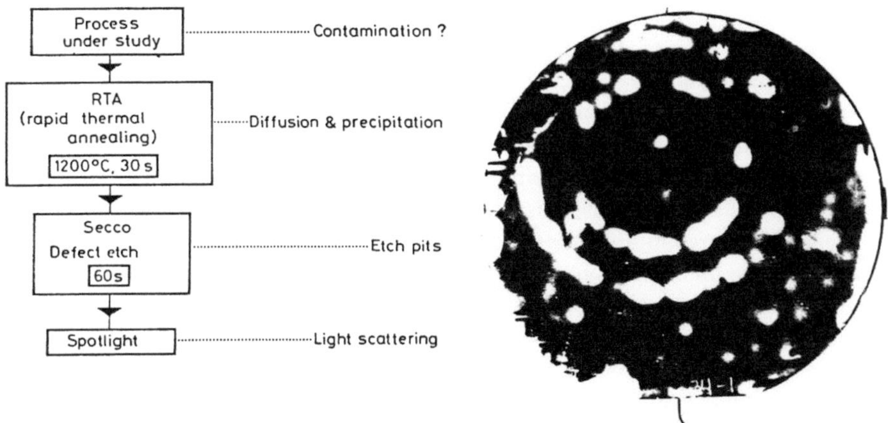

Fig. 9.2 Illustration of the haze test for fast diffusing metal impurities. *Left hand side*: Process steps for the rapid haze test to detect any Co, Ni or Co impurities (and with limitations, Fe) on or in the wafer (total duration as short as 15 min, this preparation step can also be used if the detection is electrically, e.g. by microwave PCD method mentioned in Chap. 3); *right hand side*: result of the test: Ni contaminated areas are seen as bright spots (light scattering from the etch pits), the sensitivity of the method better than 10^{11} cm^{-3}. Note that in this example the original contamination was on the wafer back surface, the haze picture has been taken of the front side. That implies that in 30s the Ni diffused right through the thickness of the silicon within 30s (!) and also spread laterally over several mm, as indicated by the circular bright areas (The wafer is a 100 mm diameter wafer) [6]

complete wafer. So the method is a large area fast detection method with the additional benefit that it shows up the contamination pattern on the wafer, which often is a direct indicator of the contamination source. If parallel processing of several test wafers is done, the time per wafer is reduced to below 5 min. Other methods for fast large-area detection of metal impurities and other defects exist and examples will be given at appropriate places in this chapter. Also, in a later section, an example for the fast large-area electrical detection of these impurities will be presented, the method is equivalent in sensitivity to the DLTS method presented in Chap. 3, and less specific, but has become the mainstream method for the detection of metal contamination both in microelectronics and PV technology. As will be explained in detail in Sect. 9.5 of this chapter, the prevention of metal contamination is one of the key factors for the suppression of the formation of extended defects. Already at this stage it is pointed out that for Fe, Co and Mn an interesting scientific method exists, namely Mössbauer spectroscopy. The method has the unique advantage that it can detect all Fe (in absorber experiments) and all Co/Mn (in source experiments, see Chap. 8 for details regarding source and absorber experiments), and at the same time it has spectroscopic properties, namely that different species (isolated impurities on different sites in the crystal, agglomerates, precipitates) can be separated. The method and its applications and the additional information obtained and resulting open questions are explained in Chap. 8 and have contributed significantly to the understanding of the behavior of transition metals in silicon, although it has to be conceded that to date no overall consistent model to explain all empirical observations has been arrived at so far.

(c) A profound knowledge of the critical parameters in unit processes which are relevant for defect formation or prevention, since as pointed out earlier a controlled defect density is a pre-requisite for consistent performance and reliability/durability of the final products [4, 12–15]. Obviously, the profound knowledge mentioned in a. is a pre-requisite for such endeavors. In the microelectronics industry, the automotive quality management system standard ISO TS 16949 (see http://www.iso.org/iso/catalogue_detail?csnumber=52844) is a "must" (this is because a significant part of microelectronic products is for automotive applications), and one requirement of the standard is to define and monitor parameters critical to quality. By standard, such a parameter is called "special characteristic". This is why these parameters are also called "key control characteristics" (KCCs). Since defects are obviously critical to quality (in the case of the defects introduced on purpose their absence can also affect the performance and/or the reliability/durability negatively!), all defect densities and/or the associated process parameters are by definition KCCs and should ideally be monitored by SPC.

In the following chapters all three aspects (a)–(c) will repeatedly be mentioned where appropriate for a clear logical connection of this abstract concept with manufacturing reality and to create a more profound understanding of how quality cannot tested into the end product but "manufactured into the product", so the manufacturing process must predictably create this quality. This is the basic philosophy behind defect engineering, as a foundation it needs the knowledge presented in the preceding chapters.

9.3 Classification of Defects

The starting point in defect engineering is modelling and understanding defects, i.e. the aspect a. from Sect. 9.2. As the first step, the crystal structure of silicon and how the crystal structure is "embedded" in the external shape of silicon wafers used in microelectronics manufacturing has to be described.

Silicon crystallizes in the diamond lattice, i.e. the each Si atom is bonded to four nearest neighbors (compare Fig. 9.1b), the bonds are directional covalent bonds in a tetrahedral arrangement. A Si wafer is a single crystal without any extended crystal defects, i.e. the elementary cell is reproduced faithfully throughout the wafer, if the lattice constant would be about 50 million times larger as it actually is, a <100> wafer would look like in Fig. 9.3.

As the next step towards a scientific and technical understanding for defects we introduce a commonly used classification of crystal defects. Crystal defects are

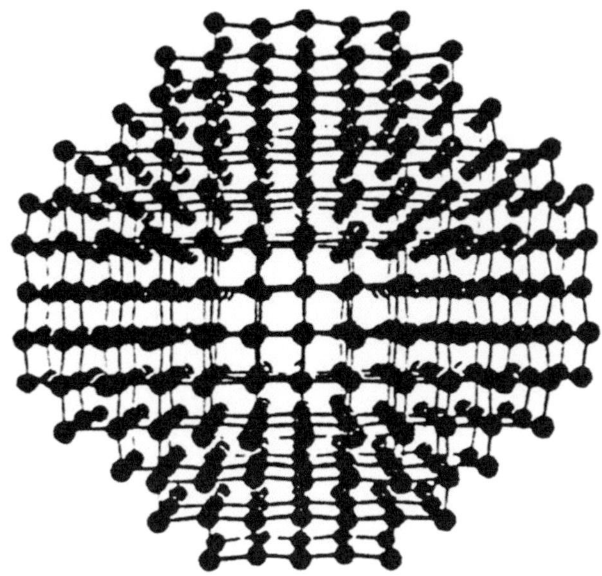

Fig. 9.3 View of a <100> wafer if the lattice constant would be 50 million times larger than it is in reality

any deviations from the ideal single crystal. The usual classification of defects is according to their dimensions.

9.3.1 Zero-Dimensional Defects

These defects are called point defects. They can be either the "wrong" atoms on a regular site of the crystal lattice, i.e. an impurity atom (larger or smaller than the host lattice atoms, with or without a different number of valence electrons), or they can be intrinsic defects. That means that there is either a missing silicon atom where there should be one or there is an extra Si atom not on a regular lattice site. The first type of defect is called a vacancy, the second one is termed self-interstitial, since an extra Si atom is "squeezed" in between regular lattice sites as there is no room on a regular lattice site. This is possible since the diamond lattice is a relatively open structure, compare Fig. 9.1b. Likewise, impurity atoms can be on such interstitial sites, and as will be explained in detail later, such impurity atoms can diffuse up to ten orders of magnitude faster than substitutional impurity atoms, the haze test explained above is an illustration of such fast diffusion.

As a rule, these defects introduce mechanical distortions to the crystal lattice in the immediate neighborhood as indicated in the 2D sketch of Fig. 9.4. It is to be noted that point defects can also form pairs or clusters of more than two point

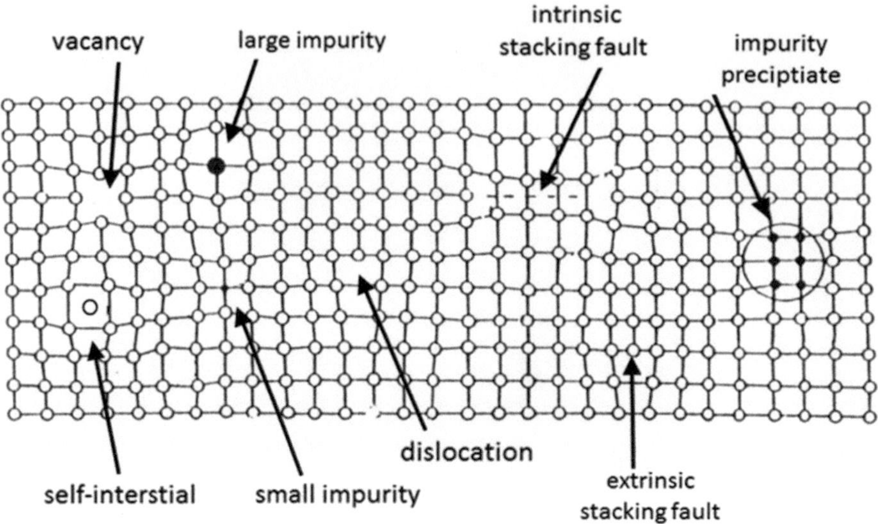

Fig. 9.4 Sketch of 0-D, 1-D, 2-D and 3-D crystal defects as a projection into a plane, using a *square* lattice

defects, consisting of equal or different point defects. This is also the case in metals, because the additional energy spent on the mechanical distortion can be lowered that way.

In silicon, there is second way to lower the energy of the system, which we pragmatically call a second "driving force" for pairing and/or clustering: In a semiconductor, the point defects can be neutral or singly or multiply charged, so there is an additional degree of freedom for an energy lowering of the system by coulomb interaction between the defects or hybridization of the electronic wave functions of the defects. Furthermore, it has to be realized that depending on the position of the Fermi level and on the donor/acceptor levels in the band gap, more than one charge state can exist at the same time, for a detailed presentation the reader is referred to the book by Tuck [21] or [18, 22]. The consequences of this fact will be followed up below and in a later section.

The most important and striking property of intrinsic point defects (vacancies, interstitials) is that they cannot be avoided. Point defects have a certain thermal equilibrium concentration which depends on the temperature following an Arrhenius (i.e. exponential) law. The reason for their existence is founded in the second law of thermodynamics since the free energy of a solid can be lowered by the entropy term -TS, and point defects can be generated in principle at any site of the solid, leading to a large entropy. Furthermore, the solubility can in addition depend on the doping level, due to the fact that other than neutral charge states can exist in addition to the neutral impurity or point defect, since this is an extra thermodynamic species, it adds to the solubility [21, 22]. This is a very significant factor for manufacturing of electronic devices and photovoltaic cells, especially in connection with gettering of unwanted metal impurities (i.e. the removal of detrimental metal impurities from active device regions to place in the wafer where they are not harmful).

Intrinsic point defects are always present in thermal equilibrium. Under normal circumstances the concentration of intrinsic point defects is so low that their effects on the product are negligible. However, under certain process conditions, vacancies and/or interstitial can agglomerate to form clusters or larger extended defects, and then they can become very detrimental to the device and their formation has to be prevented. If there is enough time given for a judicious thermal equilibration program, the agglomeration of intrinsic defects to extended defects can, in principle, be prevented (see Chaps. 4 and 5, and [1]). Also, the presence of impurities can be suppressed to a large degree by suitable counter measures, i.e. ultra clean materials and ultraclean processing, such strategies and technological countermeasures are essential constituents of the defect engineering toolbox [11–15].

So, remarkably, even a perfect silicon single crystal MUST contain a certain number of vacancies and self-interstitials at a given temperature, in thermal equilibrium. The thermal equilibrium concentrations depend strongly on temperature and other factors, such as the doping level. A point of particular importance is, as mentioned above, that point defect concentrations in excess of the thermal equilibrium concentration can coalesce into extended two-dimensional or three-dimensional defects, e.g., during crystal growth (compare Chaps. 4 and 5) or during any high temperature device processing steps, under some circumstances even at room temperatures, as detailed in later sections.

9.3.2 One-Dimensional Defects (1D Defects)

Contrary to point defects, the existence of extended 1D, 2D or 3D extended defects is not "mandated" by the principles of thermodynamics; they can be avoided altogether by suitable defect engineering strategies, i.e. the right process conditions and suitable precautions.

For the further understanding of geometry of extended defects, it is important to get some insight into the geometric structure of such extended defects in silicon, in particular to understand how extended defects are drawn often drawn in a 2D projected view in textbooks or scientific publications, whereas in reality they are three-dimensional objects in the silicon lattice.

The most important and frequent one-dimensional defects are dislocations, i.e. line – type defects. By definition, a dislocation is the edge of a crystal lattice plane that ends inside the crystal, as sketched for a 3D cubic lattice in Fig. 9.5.

The sketch in Fig. 9.5 is a simplified representation of a real geometry of dislocations in a diamond lattice. Figure 9.6 shows a ball and stick model of the most common dislocation type, the 60° dislocation. For most situations in defect engineering it is not necessary to consider the detailed geometry of a dislocation in the diamond lattice, it is however relevant to understand the orientation of the glide plane if slip in a wafer occurs, which is sketched in Fig. 9.7.

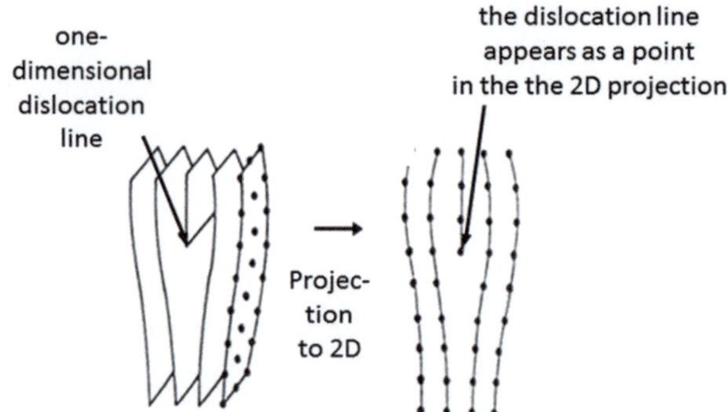

Fig. 9.5 Schematic drawing of a dislocation in 3D and how it is projected into 2D for easier drawing. The lattice planes are indicated as lines in the 2D projections, lattice points are only sketched for one of the planes in the 3D sketch. In the 2D projection the dislocation appears as a point, which is an "end-on" view of the dislocations. Note that dislocations can be and are as a rule curved rather than straight as in this illustration. Dislocations are also represented in Fig. 9.4

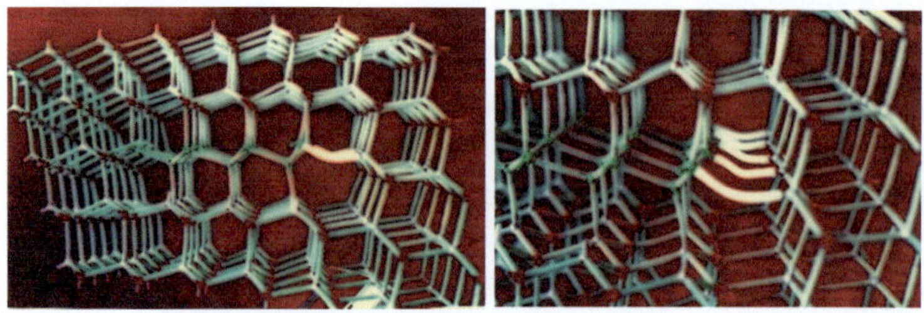

Fig. 9.6 Ball-and stick-model of a 60° dislocation in silicon. *Left* is an overview, *right* a close-up photograph of the core of the dislocation. The normal covalent bonds between the silicon atoms are shown in *green*, the "dangling" bonds at the end of the extra half plane are indicated by *white pipes* (which also point to the end of the extra half plane)

9.3.3 Two-Dimensional Defects

If an extra half plane is inserted into a crystal which nowhere intersects the surface of the wafer, as sketched in Fig. 9.8, (or if a plane is removed), a two-dimensional defect results. Such defects are called extrinsic or intrinsic stacking faults (SFs), respectively. It is obvious, that a stacking fault is bounded by a (partial) dislocation, which lies entirely inside the crystal. Both types of stacking faults are also represented in Fig. 9.4.

An important property of stacking faults and dislocations in connection with device processing is that they can absorb or emit intrinsic defects, so that they can grow or shrink (i.e. the dislocation around it "climbs", as opposed to glides if it is not an in-lattice-plane movement), so they can move into active device regions even if they were originally formed away from them. In addition, for defect engineering it is important to note that they can act as sources or sinks for intrinsic defects and

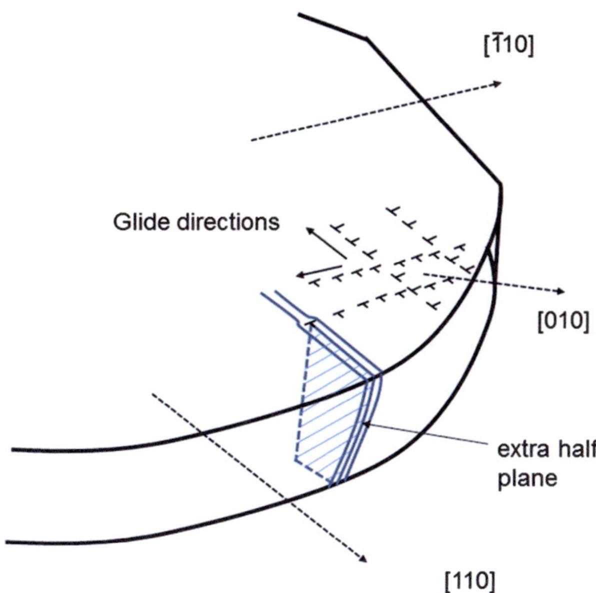

Fig. 9.7 Schematic drawing of slip via (111) glide planes in a (100) wafer by excessive thermal stress and the orientation of the extra half plane relative to the wafer external surfaces and also to the crystal directions in the wafer plane for a (100) wafer. As indicated in the figure, the dislocations arrays typically appear in lines, the <010> directions. Such dislocation formation is typical for too harsh thermal processing or unsuitable combinations of layers deposited on a silicon wafer before thermal processing (After [4])

impurities and can thus influence the concentration of intrinsic point defects. This can both be detrimental and useful, depending on the situation. Stacking fault growth is particularly significant in connection with e.g. oxidation (see later sections).

9.3.4 Three-Dimensional Defects

Three-dimensional defects in silicon can be introduced unintentionally (e.g. the aggregates of vacancies, also called crystal originated pits (COPs), see also Chaps. 4 and 5, or precipitates of unwanted impurities), or on purpose, e.g. precipitates of oxygen, which have mostly the structure of amorphous silicon dioxide [7, 8]. Oxygen precipitates and defects that can be co-created (e.g. stacking faults) are instrumental to remove unwanted metal precipitates from the device layer near the surface of the wafer to the bulk, a process called intrinsic gettering (see a later subsection of this chapter and Chap. 6). As mentioned before, extended defects are not necessary but can form unintentionally under unsuitable process conditions for their prevention (or can be engineered to be formed by using a process which provokes their formation, relevant detailed information to be found in Chaps. 4, 5 and 6).

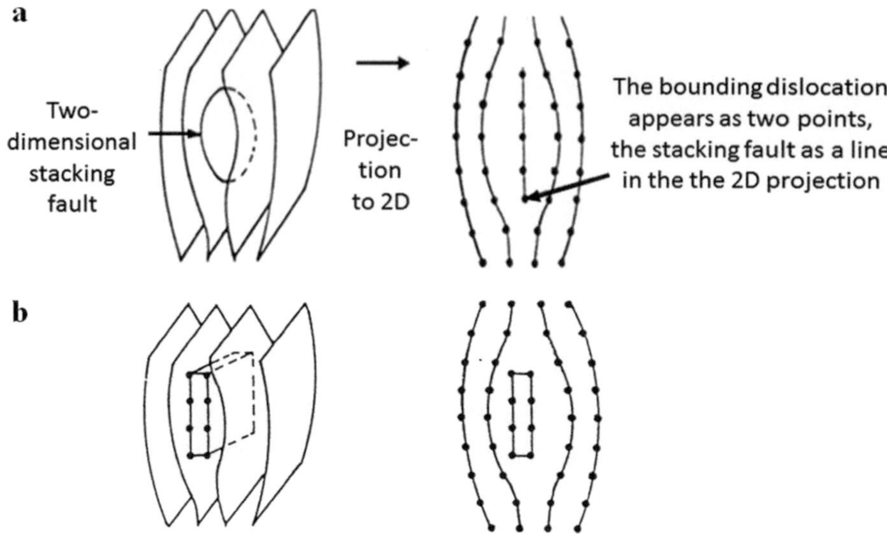

Fig. 9.8 (**a**) Schematic drawing of an extrinsic stacking fault in 3-D and in 2-D projection. (**b**) Schematic drawing of a precipitate in 3D (*left hand side*) and in the 2D projection (*right hand side*)

Once formed, all defects can interact and react with each other, which can give rise to quite complex defect reaction and interaction scenarios. This is at the core of defect engineering, and is ultimately the chief cause that metal impurities, if not controlled, constitute an unacceptable risk to the performance and the reliability/durability of microelectronic and photovoltaic devices and products [12–15, 23].

The task of defect engineering is to control such reactions and interactions. The interaction can be direct (e.g. by exchange, absorption or emission of point defects), but it can also be indirect by elastic interaction or, most important, by electrical interaction, since most defects can assume different charge states, depending on the Fermi level in the active device regions and in the rest of the wafer (in a photovoltaic cell or a power semiconductor, the entire wafer is active device region, in microelectronic products other than power semiconductors only a very thin surface layer of a few μm thickness is the active device area).

9.4 The Electrical Activity of Defects in Silicon and Some Consequences

In defect engineering, the fact that in silicon most defects are electrically active is of paramount importance, both in terms of direct electrical effects on the recombination and generation lifetimes (compare Chap. 3), and in terms of indirect effects by influencing the solubility and diffusivity of defects [21, 22] and reactions between defects. The effects are further complicated by the fact that such defects can act as single or multiple donors/acceptors or even both, i.e. an "amphoteric" behavior, a term used in chemistry for elements that can act as acids and as bases. In fact, it is insightful to note that there is an interesting analogy between the chemistry in water and the "defect chemistry" in silicon. This idea has been put forward by Hannay [24] and is a very useful concept for defect engineering, because well-known principles from wet chemistry that many engineers are familiar with, such as the law of mass action, can be applied in an analogue fashion, see also [21] for more detailed explanations.

In this analogy, silicon assumes the role as a host, as water does for liquid chemistry. Donor impurities such as phosphorus correspond to a base, since they add an electron to the conduction band, in analogy to the fact that ammonium hydroxide NH_4OH adds an OH^- to water. The two corresponding "chemical" reactions are:

$$\mathrm{P} \rightarrow \mathrm{P}^{+} + \mathrm{e}^{-}$$

$$\mathrm{NH_4OH} \rightarrow \mathrm{NH_4}^{+} + \mathrm{OH}^{-}$$

Likewise, an acceptor impurity such as boron corresponds to an acid.

The law of mass action holds for all cases, also for the absence of any acids in water, or the absence of significant impurities (in comparison to the intrinsic concentration of holes and electrons) in silicon. In such a special situation (normally on speaks of undoped intrinsic silicon) the product of the concentrations of the two types of ions/charge carriers electrons e and holes h (as denoted by square brackets) is constant for a given temperature, and the equilibrium constants K for the reaction depends exponentially on temperature, which reflects the energy needed to break the respective bonds, as illustrated in Fig. 9.9.

$$\left[\mathrm{H}^{+}\right]\,[\mathrm{OH}^{-}] = \mathrm{K}_{\mathrm{water}}$$

$$\left[\mathrm{e}^{+}\right]\,[\mathrm{h}^{-}] = \mathrm{K}_{\mathrm{silicon}}$$

In complete analogy to the fact that the solubility of, e.g., Fe^{+} ions in water depends on the pH-value, the solubility of impurities can be influenced significantly be changing the concentration of boron acceptor impurities in Si, the mathematics to quantify the dependence has to take into account the Fermi level (not relevant for water), but is in principle analogous:

Figure 9.10a shows experimental values from Hannay for the impurity Li, which is a shallow donor, similar to phosphorus:

The phenomenon, that the solubility of Li can be increased by doping can be qualitatively understood considering the reaction, and remembering the law of mass action from chemistry:

$$\mathrm{Li} \rightarrow \mathrm{Li}^{+} + \mathrm{e}^{-}$$

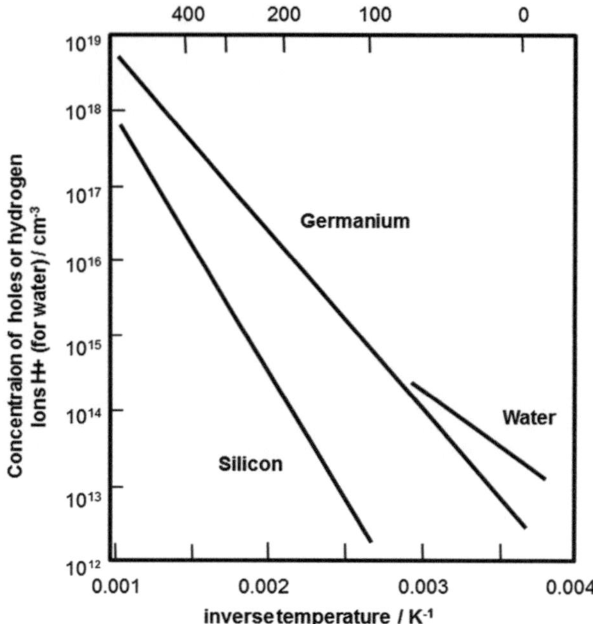

Fig. 9.9 Intrinsic carrier concentration in Si and Ge, and in water, and the concentration of OH− and H+ in neutral pH7 water (which corresponds to intrinsic silicon), after Hannay [24]

The "product" e^- of the ionization reaction of the Li impurity (which is a donor) is removed from the reaction by the addition of acceptors, which contribute holes h^+, which serve to neutralize (i.e. eliminate) the electrons e^-. Hence, by the law of mass action, the equation is driven to the right. This means, that in addition to the "normal" solubility of Li in intrinsic undoped silicon an additional solubility results from the additional thermodynamic entity Li^+. In technology, this defect engineering "tool" has been used in the production of Li drifted silicon nuclear detectors at times when it was not possible to manufacture silicon crystals pure enough to use intrinsic silicon for nuclear detectors (which is possible nowadays, that technology has been obsolete for more than 30 years) [25].

Similar phenomena have been observed by Gilles et al. [22] for the solubility of the transition metals Mn, Fe and Co in silicon as a function of doping level, as shown in Fig. 9.10b for Fe in Si at 700 °C:

The solubility increases by orders of magnitude for both heavy n-type and p-type doping. For n-type doping there are strong indications that this must be due to an immobile i.e. non-interstitial Fe species. This finding is especially important for Chap. 8 in which the results for Fe, Co and Mn in silicon from nuclear methods found strong evidence for substitutional Mn, Fe, Co, which is somewhat at variance with the other data on the interstitial and substitutional solubilities of these elements in silicon.

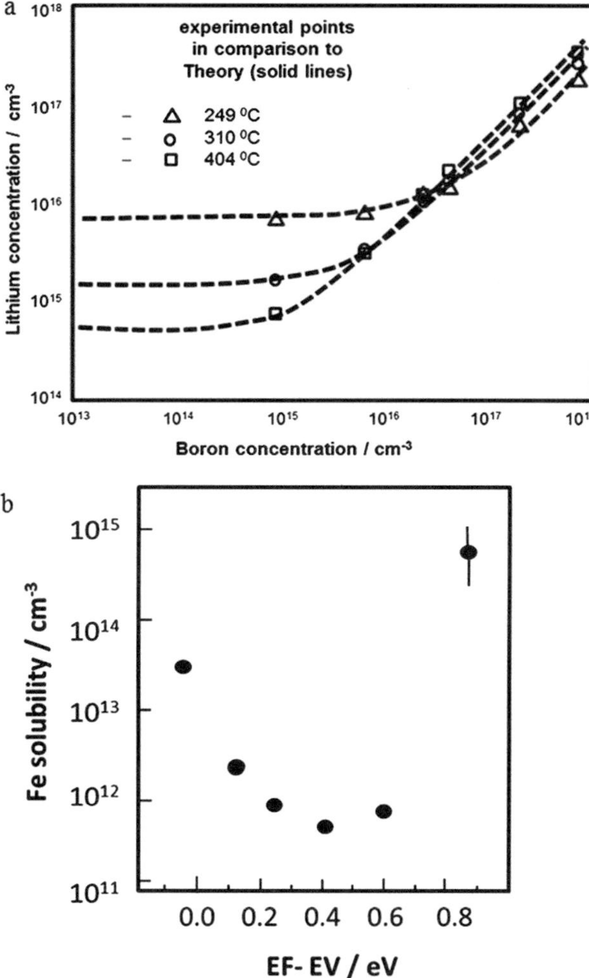

Fig. 9.10 (**a**) Li solubility as a function of the acceptor concentration for three different temperatures, data from [24] (**b**) The Fe solubility in silicon at 70 °C has been measured by radioactive tracers as function of the Fermi level for different doping levels, and is plotted as a function of the Fermi level at 700 °C, after Gilles et al. [22]

The strong doping dependence of transition metals is one component of the scientific basis for modelling phosphorus diffusion gettering and the larger resilience of heavily boron doped epitaxial wafers against metal contamination. Also, in photovoltaics the effect of the heavy aluminium doping by the Al paste on the back surface of standard photovoltaic cells has most likely led to a high resilience against metal contamination through the gettering action of the highly Al doped back surface layer, in combination with the phosphorus diffusion gettering during the n^+ emitter diffusion for the basic n^+p photovoltaic cell [17].

There are many more phenomena associated with the fact that defects can occur in different charge states in silicon, some of which will be mentioned in the following sections. In all defect engineering efforts, the potential impact of charge states on the solubility, the diffusion constant and potential interaction with other defects has to be kept in mind.

9.5 Formation Mechanisms of Extended Defects and How to Suppress Their Formation in Defect Engineering

As mentioned before, extended defects are not required by thermodynamics. Whether extended defects form or whether their formation can be prevented both during the crystal growth or during the processing in microelectronics/photovoltaic manufacturing depends on a kind of "competition" between two "forces" (not mechanical but proverbial forces). In strict science term it is the tendency of the system in question to lower its energy by the formation of defects for A, and the activation energy or kinetic barrier for defect formation for B, which help to understand different pathways for extended defect formation and the problem caused in electronic devices [6, 15]:

(A) Driving forces: These are the super-saturations of intrinsic defects and/or impurities and/or thermal/mechanical stresses due to temperature gradients or different thermal expansion coefficient of silicon, silicon dioxide, nitride etc. (circles on the left in Fig. 9.11).
(B) Impeding forces: These are the nucleation barrier (activation barrier, see Fig. 9.12) for the formation of extended defects and the fact the diffusion of point defects can be slow (if nothing moves, nothing can happen even though it would be energetically favorable), i.e. a kinetic barrier.

With respect to the principles (a)–(c) described in Sect. 9.2, Fig. 9.11 is a practical "map" or guideline how to prevent the formation of extended defects by defect engineering interventions. The defect engineering tools developed on this basis can be categorized according to two strategies, strategy I is to reduce the "driving forces", and strategy II is to strengthen the "impeding forces", usually both types of tools are applied simultaneously.

Strategy I Reduce the driving forces, such as thermal/ layer stress or the super saturation of intrinsic defects

In thermal processing the thermal stress can be done by suitably slow temperature ramps from the furnace stand-by temperature to the processing temperature and back after the end of the process, and slow enough move-in and move-out of the wafers into and out of the furnaces. Both too fast ramp-up and too fast move-in (or ramp down and move-out) lead to significantly higher temperatures of the outer

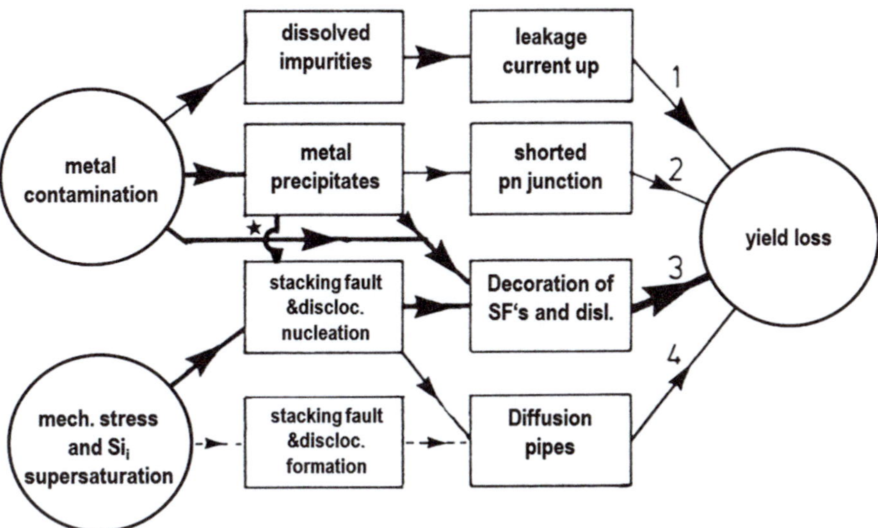

Fig. 9.11 Different pathways from driving forces to the formation of extended defects and or electrically active point defect impurities to electrically detrimental defects [6]

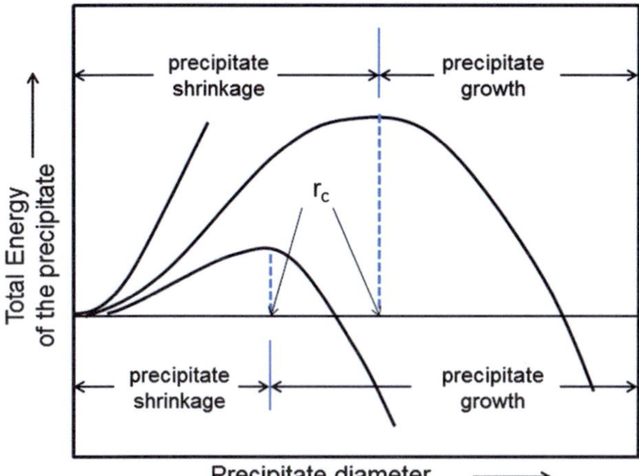

Fig. 9.12 Schematic representation of the energy barrier for the nucleation of extended defects. The chemical energy gained by reduction of the supersaturation of intrinsic or point defects is proportional to the volume (i.e. proportional to radius r cubed) whereas the additional energy needed to create the additional surface (proportional to the square of r). Therefore, for small radii, there is barrier which has to be surmounted (spending energy for the additional surface energy without much gain energy from reducing the supersaturation), which is an "impeding" force for the formation of extended defects

Fig. 9.13 Photograph of wafers moving into a furnace at a stand-by temperature of 800 °C. Note the close spacing of the wafers in the quartz wafer carrier which is needed to reach a reasonable productivity, to accommodate as many wafers as possible in the limited space of a furnace tube. The quartz has been manufactured from high purity quartz to reduce metal contamination

rim/center of the wafers (respectively) which are relatively closely spaced in a boat, see Fig. 9.13.

It is a general experience that a temperature difference of 100 K between the edge and the center of the wafers can occur during move-in and ramp-up which is close to the process window limit, beyond which plastic deformation will be induced since the critical shear stress for dislocation formation is exceeded [4, 12, 13]. Such dislocations result in wafer warpage to form a saddle or propeller-like shape (since the perimeter of the wafer is larger than the center) which can lead to problems with photolithography and which can cause excessive leakage current in devices via the dislocations generated. During move-out the opposite temperature gradient develops, experience (and simulation) shows that the dislocations will develop in the center of the wafers, which is larger than the perimeter, deformation will be like a bow. Figure 9.14 shows an example of a untypically strong formation of dislocation formation by excessive thermal stress. To prevent such defects by very slow ramps and move-in and move-out would be easy, however the resulting loss in productivity and the corresponding increase in cost would not be acceptable. So defect engineering has to identify a good compromise between the two requirements high productivity and the prevention of defects.

More specifically, defect engineering has to determine the process limits and a sufficient safety margin for the process parameters move-in/out and speeds and ramp rates so that the process is robust against small variations of the material

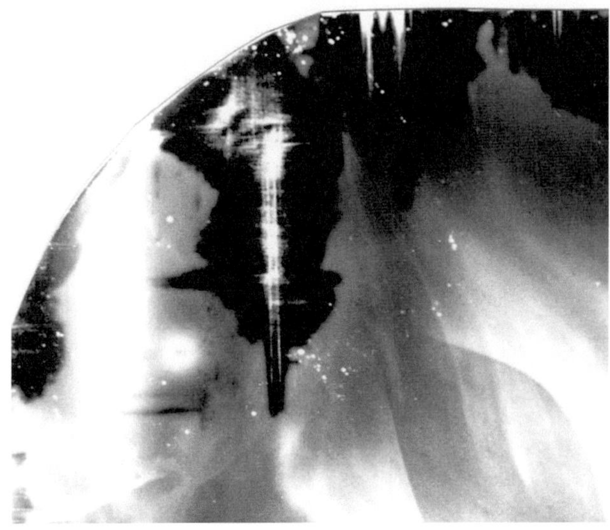

Fig. 9.14 Dark field image of part of a wafer which had undergone excessive temperature stress (during rapid thermal anneal as part of a haze test), where the dislocation arrays in the [100] direction are well visible via the etch pits of the dislocations (additionally visible: haze from metal contamination, which is not present in the dislocated areas due to gettering of the metal impurities by the dislocations). The wafer is has been defect etched to show develop etch pits where the dislocation penetrate the surface, the light scattering from the etch pits shows up the intersection of the glide planes with the surface, where a high density of etch pits shows up as lines in the <110> directions, also compare Fig. 9.7

(which depends, among other things on the oxygen content, doping level and thermal history before the thermal process in question [4]), but at the same time does not waste equipment capacity. It is also obvious from first principles in radiation physics that a larger distance between the wafers would decrease the tendency for dislocation formation by thermal stress; however this implies again lower productivity since furnace tubes have a limited length.

Likewise, the layer stress between silicon and a silicon nitride layer due to different thermal expansion coefficients can be reduced by adding a thin silicon dioxide layer between the silicon and the nitride [4]. The silicon dioxide gets viscous above a certain temperature and allows the glide of the nitride on top of the silicon, very similar to a lubricating layer between two plates.

In terms of the topic (b) for the requirements of defect engineering mentioned in Sect. 9.2 for manufacturing, an efficient and effective method to make sure that there is no formation of dislocations is needed. Such a fast low-cost method is the etching method for test wafers or for partially or fully processed production wafers. In order to satisfy the requirements of a quality management system, there must be a standard operating procedure in place for defect etching, in combination with a standard procedure to inspect the wafers by dark field illumination and evaluated the length of glide lines (if any), as shown in Fig. 9.14, and additional inspection

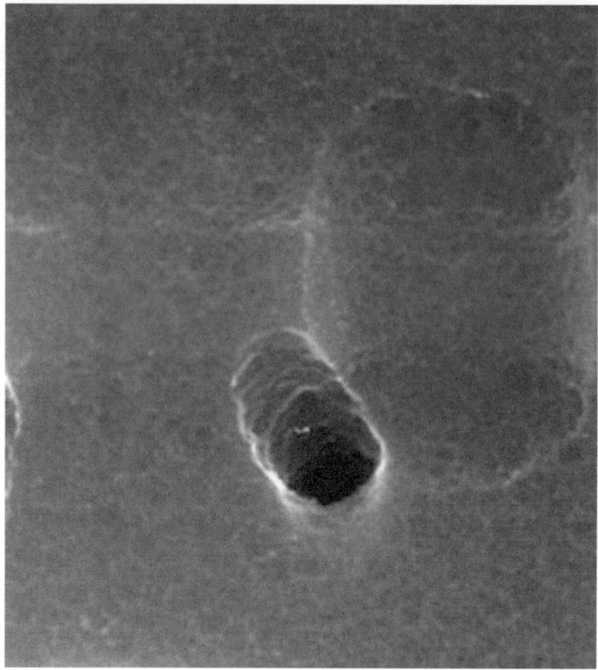

Fig. 9.15 Etch pit of a dislocation in the center of the photograph. The slight etch depression next to it is where the direct contact for a 4 Mbit DRAM was located (The etching depression is due to the residual damage from the implantation of the contact). The field of view in the micrograph is approximately 1.5 × 2 μm

by either optical microscopy according to a fixed inspection plan (i.e. the number of fields of view at which magnification) or by SEM with a similar inspection plan. Figure 9.15 shows the etch pit of a dislocation after Secco defect etching, the funnel-like etch pit is very characteristic, so an unambiguous attribution of that type of etch pit to a dislocation is possible (for details and reference on defect etching, see Chap. 6, and [26]).

Another very efficient way to check for dislocations is the use of X-ray topography [26], in which the dislocation arrays inside the whole volume of the wafer show up as contrast, an example of an X-ray topograph is shown in Fig. 9.16, which images approximately a quarter of a wafer of a 150 mm wafer (equipment which can image a whole wafer can be purchased to-day, but was not available at the time when the X-ray topograph was recorded). The dislocations on the glide planes in the <100> direction are clearly visible. In addition, additional dislocations have emerged near the wafer flat. The nucleation of these dislocations was most likely due to mechanical damage from the laser marking near the wafer edge.

In terms of topic (c) for defect engineering under production conditions, such regular process inspection results then have to be entered in to a run chart (process results plotted vs. the run number or date/time) and, what is better, developed into

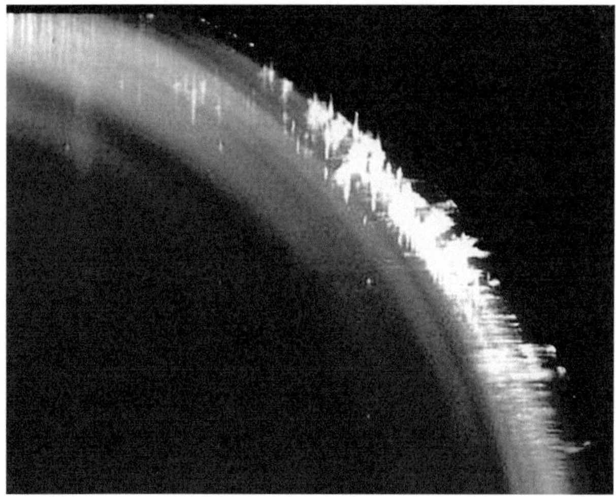

Fig. 9.16 X-ray topograph of part of a 150 mm wafer. The extensive dislocation formation in the <100> direction is clearly visible by a crisscross pattern of bright lines (the imaging condition is set such that the lattice distortion around the dislocation core gives rise to additional reflected X-ray intensity). In addition one can see many dislocations near the wafer flat and a ring-like contrast, the latter stems from the lattice distortion of oxygen precipitates and additional secondary bulk microdefects triggered by oxygen precipitation

an SPC (statistical process control) control chart [19]. If according to the definition of SPC (i.e. cpk-value larger than 1.67), the process is under control, there is a probability lower than 3.4 in 1 million that that process will generate dislocations in the next production runs, i.e. quality is "predictably" manufactured [19].

The second type of "driving force" frequently encountered in microelectronic and photovoltaic production is the super saturation of point defects and impurities.

The most frequent process for which this applies is oxidation. Oxidation leads to a super saturation of self-interstitials, which can result in the formation of oxidation induced stacking faults (OISFs) [4], if the super saturation is high enough, and/or if there are nuclei which significantly lower the nucleation barrier. The driving force can be reduced e.g. by reducing the oxidation rate, so the formation of OISFs can be suppressed, via avoiding exceeding the nucleation barrier. To kind of visualize the super saturation via the growth rate of extrinsic stacking faults, Fig. 9.17 shows the lengths of stacking faults which were intentionally nucleated as a function of temperature for dry oxidation for 3 h under conditions where the reduced nucleation barrier was easily surmounted. The different growth rates at different temperatures and crystal orientation demonstrate that the production rate and the resulting super saturation of self-interstitials during oxidation depend on the processing temperature and the oxidation rate (influenced via the oxygen concentration in the gas fed through the furnace).

It terms of defect engineering for production control (b) and (c), similar principles apply as for the routine checks for dislocations, namely defect etching

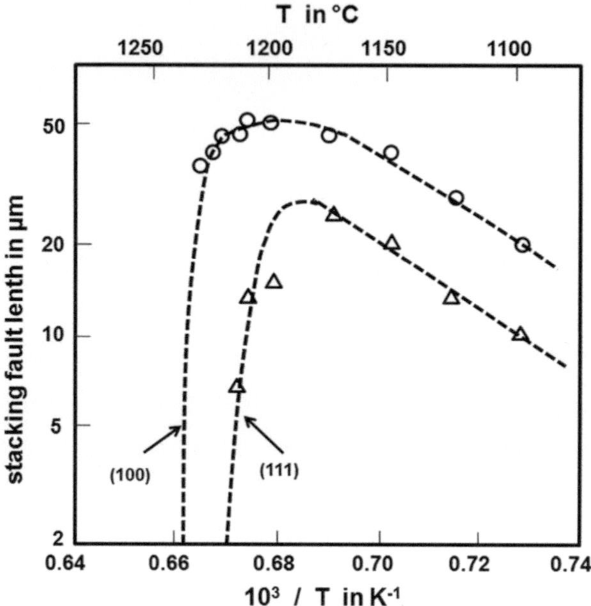

Fig. 9.17 Length of stacking faults, as measured by defect etching and optical microscopy. The length of the stacking fault reflects the super saturation with self-interstitials caused by oxidation at different temperatures and different crystal surfaces, after data from [4]

in combination with a test plan and a run chart of an SPC chart. Also, a good compromise between high productivity (= high oxidation rate) and robustness of the process against spurious formation of stacking faults has to be determined from process window experiments.

So, to reduce the driving forces most of the time does come at a cost (usually in terms of lost productivity, but it can also result in higher cost for better wafers or more expensive equipment). Therefore, it is clear that in parallel, the second defect prevention strategy has to be employed.

Strategy II Strengthen the "impeding forces"

Quite frequently, the cost for reducing the driving forces is simply too high, and/or the process is difficult to control, such as that the nucleation barrier for the formation of extended defects is lowered by the presence of metal impurities. Such metal impurities have an extremely low nucleation barrier themselves, and diffuse very fast at the same time, which is a most "unfortunate" combination, as will be shown later [15], the nucleation barrier of the formation of dislocations and stacking faults can be dramatically reduced by formation of metal impurity precipitates (which often have a small nucleation barrier and a negligible kinetic barrier, see several pathways in Fig. 9.11). Figure 9.18 is a striking example how the presence of metal impurities can dramatically enhance the formation of OISFs: For

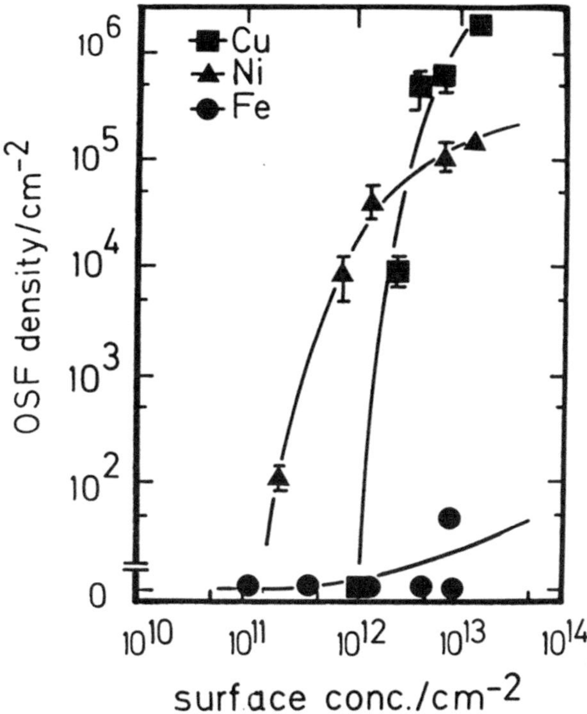

Fig. 9.18 Enhancement of the formation of oxidation induced stacking faults by surface metal impurities introduced intentionally before oxidation, after Hourai et al. [27]

a given process with negligible metal contamination the process does not generate oxidation-induced stacking faults. However, for Cu or Ni impurity concentrations which correspond to less than one person in the world's population (this means an impurity concentration of 10^{12} cm^{-3}, one cubic centimeter of Si contains 5×10^{22} atoms) the stacking fault density increases by up to a factor of 100,000! [27].

Therefore it is clear that a central requirement to strengthen the nucleation barrier is to prevent metal contamination as best as possible. As mentioned in Sect. 9.2, it is absolutely necessary to have fast detection methods so that the up to 1000 process steps in microelectronics manufacturing can be regularly screened for metal contamination. Figure 9.19 [28] gives an overview of typical metal contamination levels in microelectronic manufacturing. It is obvious that plasma etching process and implantation are particularly prone to metal contamination, and that also handling via wafer chucks made from stainless steel or other materials which contain dangerous metals can contribute significantly to metal contamination, the published results are congruent with the practical experience of the author in 20 years industry experience [11–15].

The permissible metal contamination levels in photovoltaic production are much higher [17], and therefore it would not be economical to dramatically reduce these

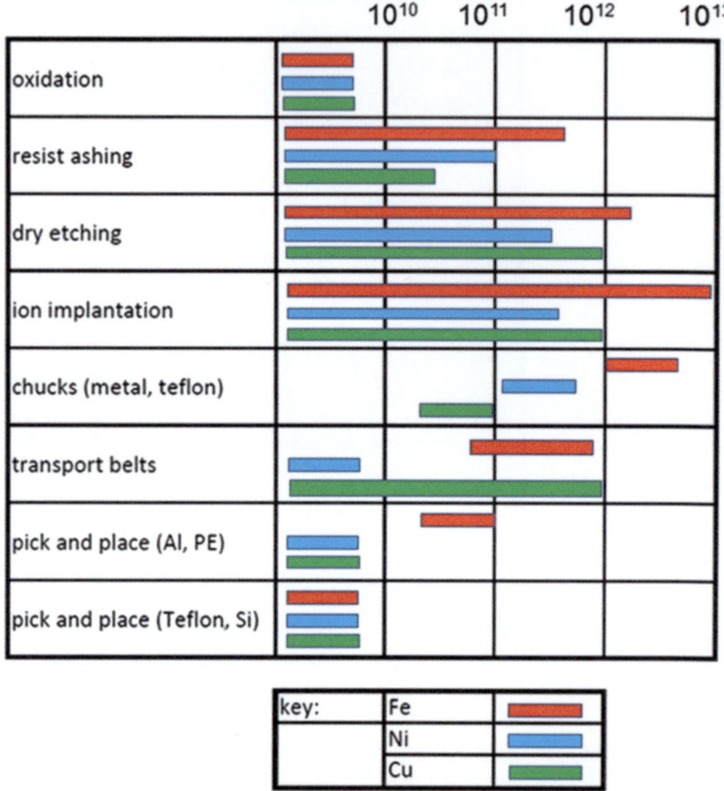

Fig. 9.19 Typical metal contamination levels in microelectronics production. Depending on the particular device process, the typical requirements may be even lower than the concentration limits indicated by the traffic light symbols shown here. Data according from [28], the units given on top of the diagram are cm^{-3}, i.e. volume concentrations

levels by using e.g. cleaner but significantly more expensive chemicals. However, it is predictable as designs of PV cells get more sophisticated, the allowable metal contamination levels will go down appreciably.

So, for effective and efficient process control of extended defects, control of metal contamination levels, according to (b) and (c) of defect engineering principles under mass production is absolute necessity, which implies the needs fast and efficient methods to detect metal contamination, such as the haze method described earlier, or the microwave PCD method described in Chap. 3. While most electrical and optical detection methods, such as DLTS, FDLTS, FTIR or electro and photoluminescence methods (see Chap. 3) are very good methods to identify and in-depth characterize the metal impurities, they are relatively slow and some of them very expensive methods with a sampling area of the order of mm^2. On the other hands, electrical methods based on measurement of the minority carrier lifetime (e.g. microwave PCD) are normally not impurity specific, but very fast

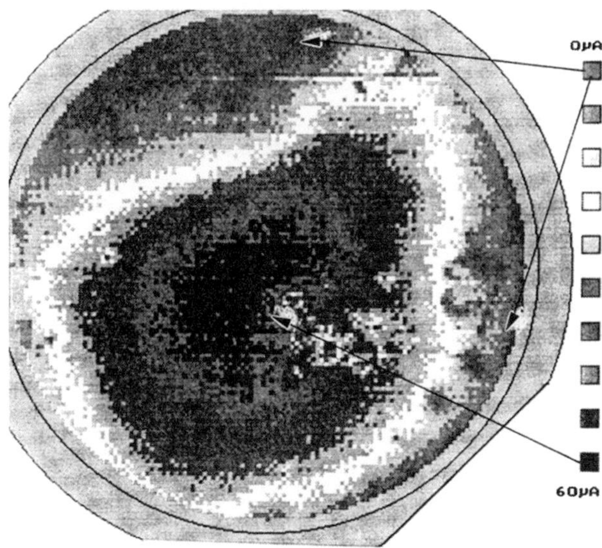

Fig. 9.20 Diffusion length wafer map of a reactive ion etched and rapid thermal annealed wafer to drive in any process-induced metal contamination into the wafer volume. The measure pattern is the magnitude of the photo induced current on the wafer back surface which results from a laser beam which is scanned over the wafer front surface. The technique has been named as Elymat technique. The whole volume is sampled because the minority carriers generated on the wafer front surface have to diffuse to the back surface where they are collected in a space charge region of the an electrolyte semiconductor contact. The contact with a low surface recombination velocity is implemented by a dilute HF electrolyte. The Elymat technique is not as popular as the microwave PCD method or the SPV method, due to the complications which arise from the safety issues that have to be controlled when using HF

and can sample large wafer areas (essentially the complete wafer surface), such as SPV (surface photo voltage, [15]) or they can even sample the complete wafer volume, if the diffusion current through the wafer is measured, as in the Elymat method (compare also Fig. 9.25, [29, 31]). Figure 9.20 shows the result of such a whole volume measurement of the spatially resolve photocurrent measurement of a dry etched test wafer. The measurement principle is explained in the captions of Figs. 9.20 and 9.25 and in Ref. [29]. Since a reference wafer would have a diffusion length at least ten times as high and almost homogenous over the wafer, it is demonstrated that the contamination is considerable and not homogenous over the wafer. With the help of such wafer maps it is possible to

- identify the localized sources of the metal contamination in the plasma etching equipment
- to test whether equipment and/or process modifications can lower the metal contamination level
- whether different cleaning steps can or cannot remove the impurities

A similar wafer map could have been generated with the microwave photoconductivity method explained in Chap. 3, with the proviso that only a surface layer of a few micrometer thickness is sampled. Commercial equipment to generate wafer maps of this kind is also available for the SPV method. In the case of SPV, the top 5–50 μm layer is sampled for metal contamination, depending on the illumination conditions, which can be chosen within certain limits [32].

Another important way to strengthen the "impeding forces" is obviously to go to lower temperatures in processing, since this slows down diffusion processes, also for some metals (if "nothing moves" or does not move fast enough, nucleation of defects with the help of metal contamination is prevented kinetically) This constitutes increasing the kinetic barrier to defect formation.

This will be illustrated in more detail in the next section, in which more complex defect formation scenarios will be described, which are typically encountered in microelectronic and/or photovoltaic production. A lower processing temperature may of course at the same time slow down the actual diffusion or annealing process needed in the device production, so here similar productivity issues may arise.

9.6 Selected Practically Important Examples for Defect Formation Mechanisms and Prevention Measures from Microelectronics Manufacturing

In this section the knowledge and understanding about fundamental principles of defect formation (as outlined in the previous sections) will be put into the context of common defect formation and defect prevention scenarios in microelectronics production. In this way, we will further explore the defect mechanism paths of Fig. 9.11 and in particular give examples for the electrical effects of defects on devices, i.e. up to when the defects affect not only the performance but also the functionality and/or the reliability of the electric devices. We will always start in one or both of the two circles representing the primary "driving forces" of defect formation on the left of the pathway map, namely super saturation of intrinsic defects, mechanical stress and metal contamination.

Due to the ubiquitous and widely varying metal contamination in microelectronic and PV production, which together with the "unfortunate" properties of these impurity atoms [15] makes metal contamination the highest-ranking risk factor for defect engineering and defect control for defects in silicon, we will first describe the general properties of the practically most important 3d-transition metal impurities in some detail. Subsequently, we will consider examples for more complex interaction of extended defects and metal impurities, and finally we will treat gettering by intentionally introduced defects in the bulk to remove metal impurities from active device areas, which is the beneficial rather than detrimental effect of the complex interaction of metal impurities and extended defects.

9.6.1 3d-Metal Impurity Diffusion and Precipitation

In Sect. 9.5 and in Chap. 3 the detrimental effect of metal contamination on the minority carrier diffusion length in silicon was already mentioned, and how this effect can be used to scan a wafer or a process in a fast and efficient manner for the presence of metal contamination. In this subsection, we will turn our primary attention to the most critical and detrimental effects associated with an uncontrolled metal contamination, the formation of metal precipitates, which can lead to a drastic increase in the formation of extended defects via lowering the nucleation barrier for dislocations and stacking faults, with potentially dramatic consequences for the manufactured product [4, 13, 23].

Many of the transition metal impurities (in particular Fe, Co, Ni, Cu and Pd) in silicon combine two rather "unfortunate" properties, which make it extremely easy for them to generate precipitates (i.e. three-dimensional defects), where Co, Ni and Cu are in particular dangerous because of a third "unfortunate" property.

The first property is that these elements diffuse extremely fast for solid state diffusion, up to ten orders of magnitude faster than doping elements like boron or phosphorus, as illustrated in Fig. 9.21. This means that at common diffusion temperatures these elements can distribute almost homogeneously throughout the wafer. Remarkably, the diffusion constants at those device processing temperatures

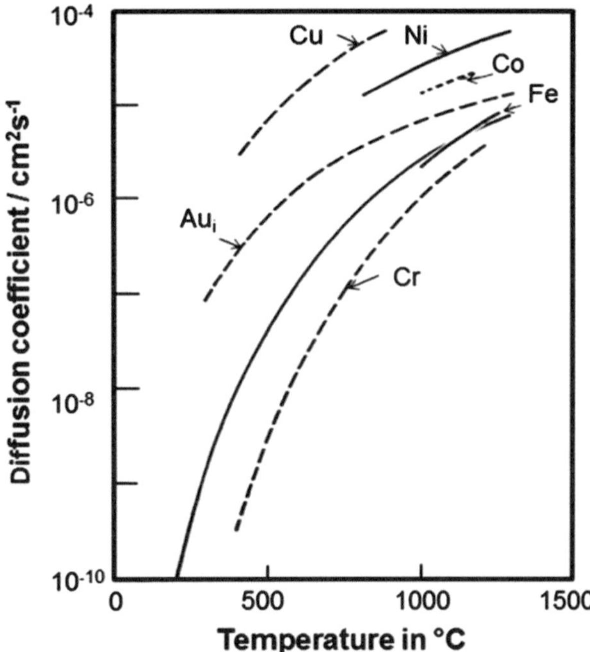

Fig. 9.21 Diffusion coefficient of some transition elements in silicon (After Graff [20])

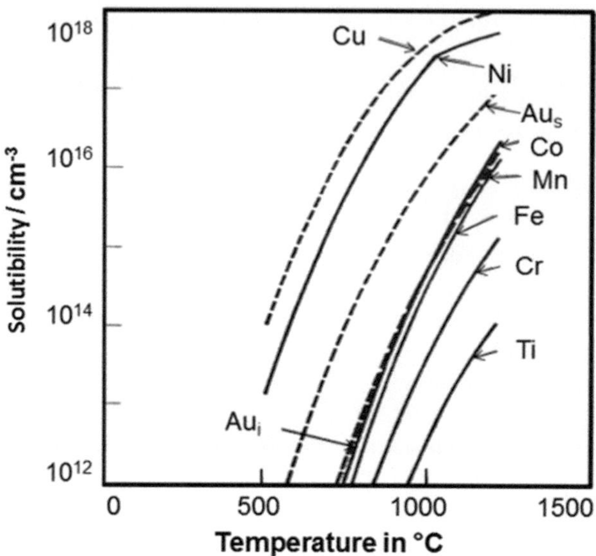

Fig. 9.22 Solubility of transition elements in silicon (After Graff [20])

are of the order of 10^{-6}–10^{-4} cm^2s^{-1}, which is a range typical for the diffusion of e.g. ink molecules in water.

The second "problematic" property is that the solubility is very low in absolute terms (a few parts per million at diffusion temperature) and that it drops rapidly with temperature (see Fig. 9.22), which means that during cooling a high super saturation develops, so a driving force for precipitation even for μg of contamination is extremely high. In addition, the precipitation process is further supported by the high diffusion coefficient which "unfortunately" falls very slowly with temperature, so the kinetic barrier is very low.

The third undesirable additional property of Co, Cu and Ni is that the initial stages of the precipitates (e.g. $NiSi_2$) are isomorphous to the silicon lattice with a very small lattice mismatch [15]. So the nucleation barrier is almost negligible, since the surface energy and lattice distortion energies are small, which are the main factors which determine the energy barrier for extended defect nucleation. So the energy barriers is unusually low as well.

As for dislocation and stacking fault detection, the "work horse" in defect engineering for the routine detection of metal precipitation is defect etching. As an example and illustration of this extremely fast nucleation Fig. 9.23 shows etch figures from Cu precipitates, which have formed after a haze test with intentional Cu contamination from the wafer back surface, metal precipitates can be found mainly at the two surfaces (and very few in the bulk, unless intrinsic gettering is active). Further details about defect etching, in particular how to replace the environmentally undesirable chromium containing defect etchants (mainly used in the examples) by alternative acid mixtures which have the same sensitivity can be found in Chap. 6.

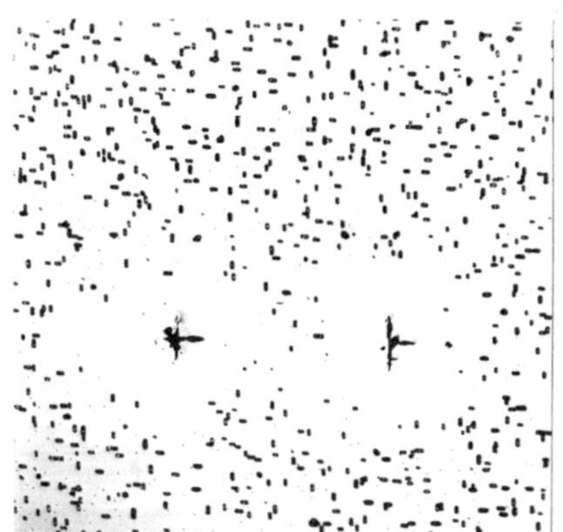

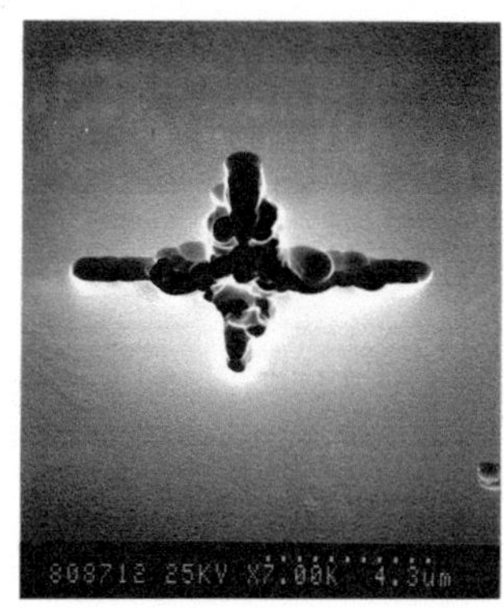

Fig. 9.23 Microscope image of the etch pits of Cu precipitates after a haze test, as described in Sect. 9.2. On the left hand side in the optical microscope image, there is a high density of etch pits of $CuSi_2$ precipitates, which consist of several layers of $CuSi_2$ on {111} crystallographic planes, therefore the etch pits resemble those of stacking faults. The two larger more complex etch pits in the center are secondary Cu precipitates which involve the formation of dislocations nucleated on the initial precipitates on {111} planes. A typical etch pit like the two complex pits in the center of the optical microscope image on the left is enlarged in an SEM image on the right hand side. Note that the large secondary precipitates have obviously emptied their vicinity from Cu in a very effective manner, since there are no $CuSi_2$ precipitates in an area around the larger precipitates. In other words, a re-dissolution of the original precipitates and a kind of gettering of the Cu to the secondary defects has happened. Quite remarkably, all this has occurred during the 10 s or so of cooling from 1200 °C to near room temperature in the rapid thermal annealing equipment used. So primary nucleation, re-dissolution, diffusion over a distance of approximately 10 μm and re-precipitation have all occurred during this time

As explained in the caption of Fig. 9.23, there is an "internal" competition between two types of Cu precipitates, the ones which resemble stacking faults are easier to nucleate (they are almost perfectly isomorphous to the silicon lattice), and the larger precipitates, which are colonies of small Cu precipitates and associated dislocation complexes. They compete for the Cu in super saturation, obviously the second more complex Cu precipitates win.

Similar to the competition between different Cu precipitates, there can be a competition and interaction and mutual enhancement of precipitation between different metals. Figure 9.24 demonstrates this interaction on a haze tested wafer which has been intentionally contaminated by slight scratching of different metal wires on the wafer back surface and a haze test described in Sect. 9.2.

Metal contamination can practically occur in any process, and in many instances there will be several of the hazardous metals present simultaneously, therefore the interaction demonstrated in this artificial contamination experiment is relevant for production. As already briefly mentioned in the preceding section, experience has shown that high temperature diffusion, dry etching of silicon surfaces and ion implantation all have a particularly high risk of contamination [14], and even

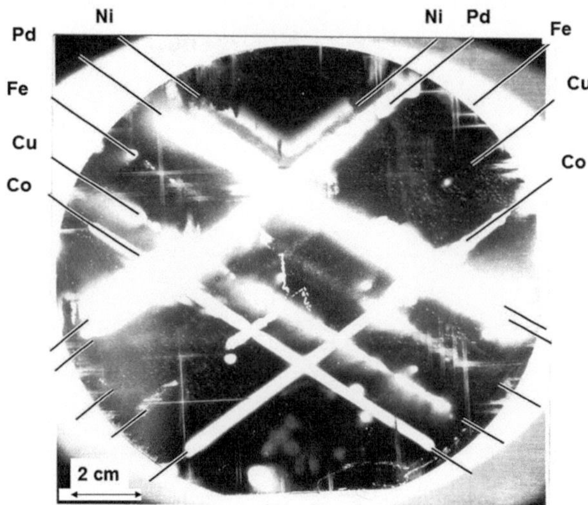

Fig. 9.24 Haze test of a wafer after intentional contamination by Ni, Pd, Fe, Cu and Co along the lines indicated. *Bright* areas represent regions of a high density of precipitates. Pd has the strongest tendency for precipitation. Fe displays the weakest precipitation. Note that there are obviously interactions between some of the metals at the crossing points of the lines. Also visible: the etch pits of dislocations on their respective *glide lines*, as *thin bright lines* in <110> direction, after [33]

cleaning (!) can, under certain circumstances lead to additional metal contamination (see also Fig. 9.27), [31].

Thus, in the sense of item (c) of Sect. 9.2, these are the processes to be primarily screened by defect engineering, because of the high and hazardous defect formation potential of the metals themselves, and because of the interaction with other extended lattice defects (see Fig. 9.11) which will be explained and illustrated further in Sect. 9.6.3 of this chapter. So, the task of defect engineering is to investigate the main factors which potentially increase metal contamination and counter measures to take to prevent it, and how to monitor the "state of health" of the suspicious process regarding metal contamination at an acceptable cycle time and for acceptable cost.

Such a screening concept is typically implemented by a combination of the methods to detect metal contamination via their effect on the minority carrier diffusion length (SPV, Elymat, Microwave photoconductivity) and defect etching/X-ray topography.

The least common of the electrical methods, the Elymat technique [29] pioneered was by Lehmann and Föll. It is at the same time the most "versatile" method (also see Fig. 9.25), since it is effectively a way to build an "instantaneous" test diode, both on the front and back sides of the wafer using a semiconductor electrolyte contact by HF. As pointed out before, the HF electrolytic contact ensures a low surface recombination, so that the effect on the minority carriers generated by a

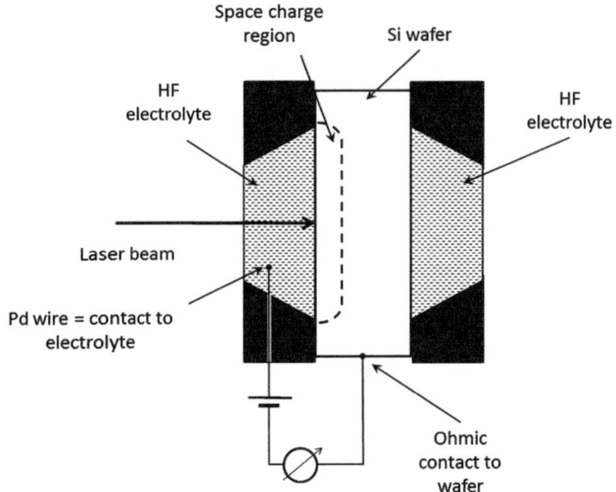

Fig. 9.25 Schematic cross section of the Elymat measurement set up, after [29]. The wafer is immersed from both sides by dilute HF, Pt wires in the electrolyte and tungsten wires on the edge of the wafer enable that an electrolyte-semiconductor junction can be formed, and at reverse bias, a space charge region can be made larger or smaller by an externally applied voltage, which will collect the minority carriers generated by a laser beam scanned across the front surface of the wafer. For routine monitoring it is usual not to use the wafer map itself, but the Key Control Characteristics are average values across the wafer and/or the standard deviation of the several hundred or thousand (depending on the chosen resolution) measured points across the wafer. Both the average value and the standard deviation in run charts or SPC charts are good indicators whether a process is in its normal state or whether untypically high contamination levels have occurred either over the whole wafer or locally (as detected by an increased standard deviation). In such a case, the pattern visible on the wafer map will be conducive to identifying the root cause of the contamination, almost like a fingerprint of the contamination source [29, 30]

laser beam scanned over the front surface can be studied (1) via the photocurrent from the front surface to the back surface (2) via the photocurrent collected at the front surface itself. The former is a measure of the diffusion length in the whole wafer volume, the latter is an indicator of surface near defects, since the diffusion distance is negligible, but the quality of the depleted zone and the junction to collect the carries is affected by defects, if these are present. In addition the dark leakage currents of both the back and front surface electrolytic contacts provide additional information on defects either caused by metals and/or by extended defects (however, these are naturally not spatially resolved but give a "fudged" quality indicator of the front or back surface).

In the routine monitoring of metal contamination, some metal contamination will almost always be found, so it is important to know for each process, what is

(i) the typical metal contamination level, and what is
(ii) the maximum contamination level which can be tolerated by the process

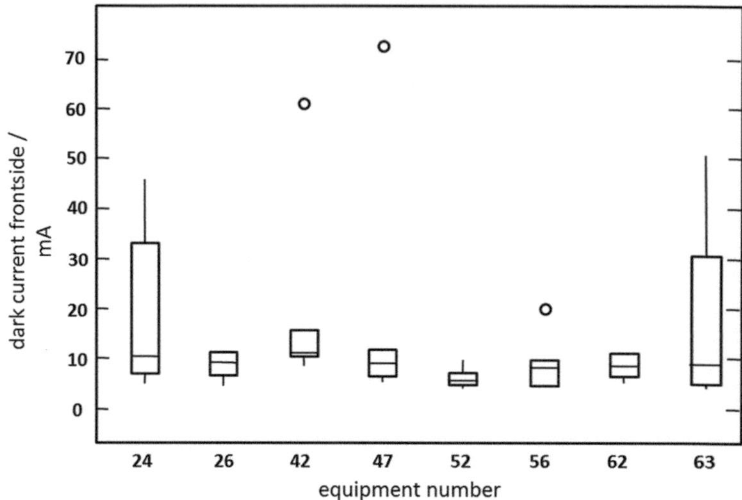

Fig. 9.26 Comparison of eight gate oxide furnaces in an Elymat cross sectional metal contamination study, the data are represented in box plot form. The quantity to indicate the metal contamination level is not the minority carrier lifetime, but the leakage current of the front site electrolyte-semiconductor contact, which in a way is a test device to study the effect of metal precipitates on a semiconductor space charge layer, in this case caused be a reverse biased electrolyte-semiconductor junction, so high is "bad", low is "good"

In order to get a representative data set for i., one can do a time series study ("longitudinal" study) on one equipment for a number of consecutive days or weeks, or a comparison of the equipment of one kind that are in use in that particular processing line ("cross sectional" study).

Figure 9.26 shows an example of a cross-sectional study and is, at the same time, an example of one of several ways how measurements using the Elymat method can be used. The equipment monitored were oxidation furnaces. The furnace tubes obviously differ considerably in their contamination levels, as indicated by the different levels of the electrolyte-silicon leakage currents measured without laser illumination (presumably due to different precipitate densities or sizes in the surface areas). By using in parallel a method to detect the type and concentration of impurities in terms of atoms/cm^{-3} one could in principle try to calibrate the leakage current vs. concentration. Since defect formation is a complex non-linear process, this would probably not work very well, but is not really needed: The important information for the defect engineer is already there: The furnace tubes 24 and 63 have an untypical high leakage current (i.e. contamination level) and need to be improved, the baseline level is represented by the other furnace tubes.

Figure 9.27 shows another example of a "cross sectional" monitoring study, in which different cleaning equipment has been characterized by the backside photo current, which samples the whole wafer volume. Contrary to the previous example, "good" is a high value of photocurrent, since defects decrease the photocurrent.

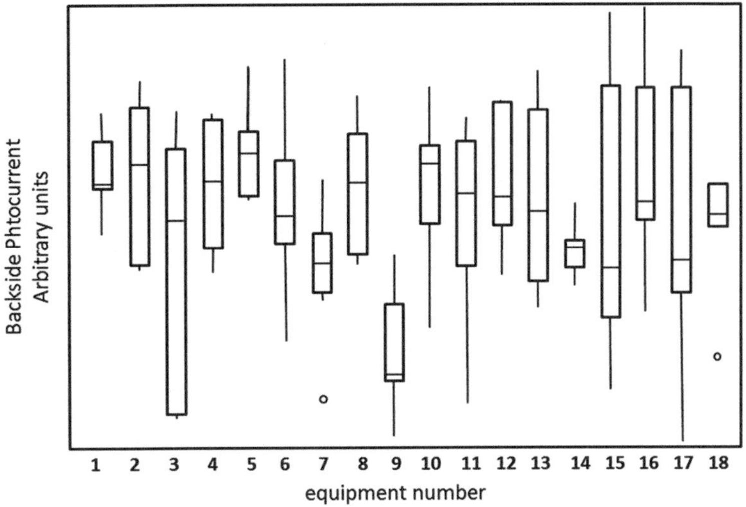

Fig. 9.27 Example for the weekly monitoring of contamination levels for wet cleaning equipment via the backside photocurrent represented in box plot form. By parallel investigations with DLTS it was established that the diffusion current in this case is mainly affected by the Fe contamination level. It is obvious that in equipment 3 and 9 the contamination is significantly higher in than in the other cleaners. However, the level was still within the permissible specification, but indicated a "special cause" (in the sense of statistical process control) for the contamination level which had to be eliminated by the responsible wet chemical process engineer before it could reach critical levels. Note that a higher level of contamination leads to lower current values, since it reduces the minority carrier diffusion length, so here in contrast to the previous graph high is "good" and "low" is bad

In the previous case, "good" was a low value, since defects cause higher leakage currents.

To demonstrate that the backside photo current measurement is indeed correlated to the Fe contamination level, a systematic correlation study on test wafers was performed, in which the BPC Elymat values were compared to the Fe concentration levels determined by the SPV technique on the same wafers. Figure 9.28 shows the results. There is a clear correlation and the expected dependence over several orders of magnitude. The scatter is due to the fact, that potentially other impurities also influence the BPC photocurrent marginally, and probably more significantly, that the two methods average over different areas, which causes some variability of the results, since the contamination level is not homogenous across the wafer, which is the case in such wafers.

To gain insight into question ii concerning the permissible contamination level, one needs to know HOW does a particular type of metal contamination cause leakage currents or other device failure, and to investigate the critical level of contamination level by process window or intentional contamination experiments [27].

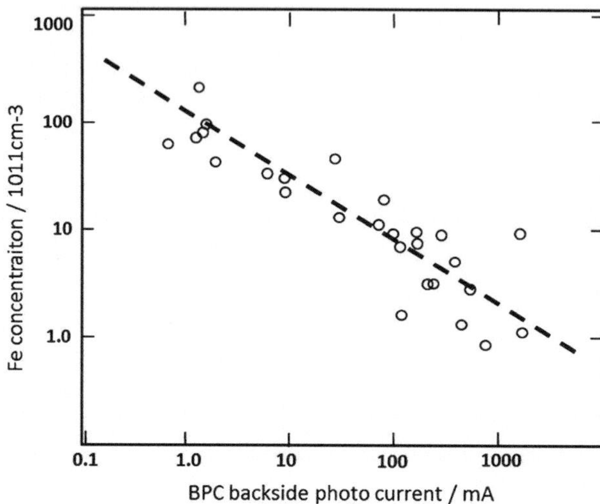

Fig. 9.28 Correlation between the Fe concentration in test wafers, as determined by SPV, and the Elymat photocurrent values, all values measured on the same test wafers

For Fe, which is typically the most important and at the same time, ubiquitous contamination element, the main degradation effect is the reduction of the minority carrier lifetime. By contrast, the main degradation mechanism of the Ni, Cu and Co is via their strong tendency for the formation of precipitates (as illustrated above), which can lead to several different device failure mechanisms, such as direct causing shorts in pn – junctions or thinning gate oxides [15, 34].

For Fe, an empirical correlation between the Fe concentration for gate-oxide test wafers and the yield in a gate oxide test method is shown in Fig. 9.29: For a threshold level of 5×10^{11} cm^{-2} surface concentration, the gate-oxide yield starts to degrade, and the variability of the results increases simultaneously [31]. Since empirical correlations have shown that the results of such tests correlate with the burn-in failure rate for DRAMs [23], this contamination level is absolutely critical for the production of CMOS and BiCMOS devices with a large gate-oxide area. In particular this is true for DRAM production since DRAMs have the largest gate-oxide area due to the high packing density of transistors in the memory cells arrays.

The degradation mechanism for Fe has been identified as thinning of gate oxide by $FeSi_2$ precipitates [6], Fig. 9.30a and b demonstrate that similar mechanisms are also operative for copper. In Fig. 9.30a an SEM picture of a gate-oxide surface shows a mushroom like protrusion, where something has obviously lifted the gate oxide from underneath. A TEM study with energy dispersive X-ray analysis on the same type of wafer demonstrated that the "mushrooming" of the gate oxide was caused by a Cu precipitate under the gate oxide, which led a local thinning of the gate oxide, as seen in the cross sectional TEM picture in Fig. 9.30b.

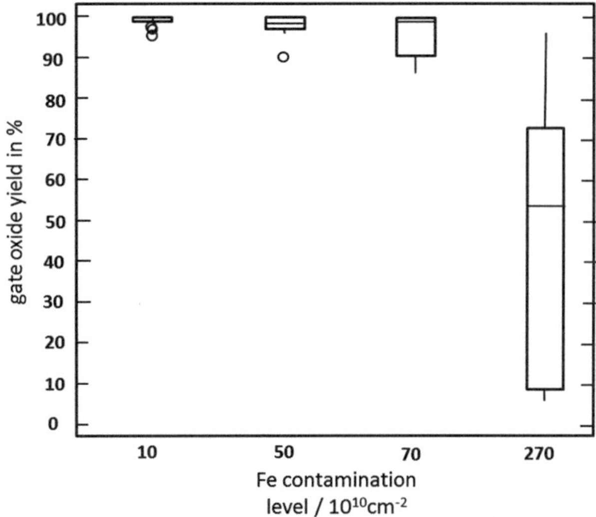

Fig. 9.29 Correlation between the surface Fe contamination of gate oxide test wafers and the yield of a standard gate oxide quality test on special test wafers

The detrimental effect of local thinning of the gate oxide on the gate oxide breakdown can be explained in the following straightforward manner: The normal operating field strength across the gate oxide is usually about 5MV/cm, which can be calculated from the applied voltage, e.g. 5 V and the average gate oxide thickness, e.g. 10 nm. If above the Cu silicide, the thickness is only 7.5 nm, the field strength is 7.5 MV/cm, which is close to the Fowler Nordheim tunneling threshold at 8–9 V [34]. Due to the excessive injection of charge carriers into the gate oxide at the thinned location, the normal wear-out of the gate oxide is dramatically accelerated.

In order to investigate what level of contamination is tolerable for a given contamination element, data have to be gathered for a correlation of the metal impurity concentration and a gate oxide integrity test for a particular device process. The critical threshold can indeed depend on the particular device process and device type, and the wafer type. Fortunately, as a rule of thumb, degradation thresholds for contamination levels are fairly consistently around concentrations of 10^{11} cm^{-3}.

9.6.2 Extended Defect Formation by Thermal or Layer Stress or Silicon Interstitial Super Saturation

The phenomenon of OISF formation by oxidation has already been mentioned above. Stacking faults can also form via a similar mechanism after ion implantation, since ion implantation creates a large number of intrinsic defects, so in both cases the super saturation of mainly self-interstitials results in the formation of extrinsic

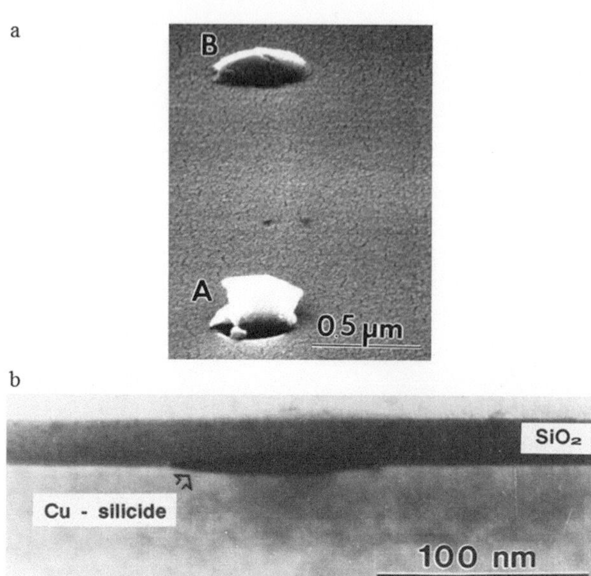

Fig. 9.30 (**a**) SEM picture of a gate oxide surface on a silicon wafer which has been intentionally contaminated. At both sites *A* and *B* the gate-oxide is lifted by a Cu precipitate underneath the oxide. These are particularly drastic examples. The precipitate in cross section Fig. 9.30b hardly lifts the gate-oxide (but thins it!). (**b**) TEM cross section of a Cu-silicide precipitate (as indicated by the *arrow*). It is clearly visible that the lens-shape silicide precipitate locally thins the gate oxide, which implies an increased field strength, which dramatically accelerates the wear out of the gate oxide by local Fowler-Nordheim tunneling currents which ultimately lead to a local gate oxide breakdown

stacking faults. Similar arguments apply to reactive ion etching, which among other effects results in the formation of crystal defects.

The formation of thermally induced dislocations is visible in Figs. 9.14 and 9.16 where the rapid thermal anneal step of the haze test introduces thermal stress which forms dislocations visible as thin lines in the <100> directions, which are glide lines caused by slip through the glide of dislocation [6]. This phenomenon can occur both in contaminated and in uncontaminated areas and is caused by pure thermal stress, not by the contamination alone (although contamination can facilitate the process). For the geometry of the dislocations, also consider Fig. 9.7.

Figure 9.31 provides more evidence that dislocation formation by thermal stress is not only possible in the untypical high stress situation of a rough rapid thermal anneal process, but also occurs in a regular device process, if the silicon material is and/or the process are not optimized. Dislocation arrays on {111} glide planes are visible In the critical <100> directions, near the wafer flat further dislocation arrays have formed with the additional "assistance" of the damage caused by the laser ID marking of the wafer and/or slight mechanical damage of the wafer edge at the wafer flat (this part of the wafer edge is handled particularly often during the alignment process of the wafer in equipment, and therefore some light

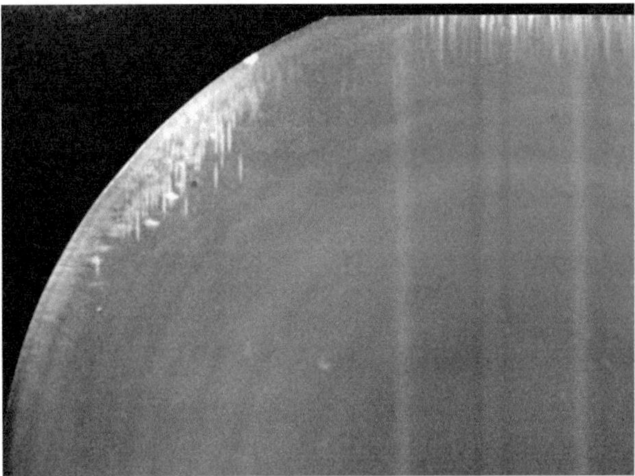

Fig. 9.31 X-ray topograph of part of a 150 mm silicon wafer after a partial regular 4 Mbit device process. The white crisscross lines are due to arrays of dislocations which have formed by excessive thermal stress in the critical <100> directions (at 45° angle with respect to the wafer flat), caused by too high temperature gradients in furnace during move-in or during ramp-up. Additional dislocation arrays on glide planes have formed near the wafer flat, due to the damage from laser marking the wafer ID number and some mechanical damage from handling the wafer in the device process

mechanical damage at the rounded edge is created which lowers the nucleation barrier for dislocation formation). In the 1970s, wafers for microelectronics did not have edge rounding yet, so the edge was very vulnerable to severe mechanical damage, with strong subsequent dislocation formation [4]. The introduction of edge rounding dramatically reduced dislocation formation in integrated circuit processing technology.

An example for dislocations generated by layer stress (rather than thermal stress) is shown in Fig. 9.32, where stress between silicon and silicon nitride (= the mask layer of local oxidation of silicon) has led to the formation of many dislocations at the oxide edges, where the layer stress had its maximum [4]. In the X-ray topograph of a 150 mm production wafer, the dislocations give rise to the white contrast which coincides with some of the structures of the 4 Mbit DRAMs devices, formed during field oxidation. Compared to the X-ray topograph of Fig. 9.31, there are no indications of the dislocations formed by excessive thermal gradients in furnaces.

As repeatedly mentioned, the task of defect engineering is to identify the factors which can prevent the defect formation. Examples for such process measures are:

- Avoiding nucleation centers (like damage at the wafer edge),
- reducing the ramp rates and the move-in temperature
- the addition of an SiO_2 layer (pad oxide) between silicon and silicon nitride as "lubrication" between the two layers with a large difference in the thermal expansion coefficients [4]

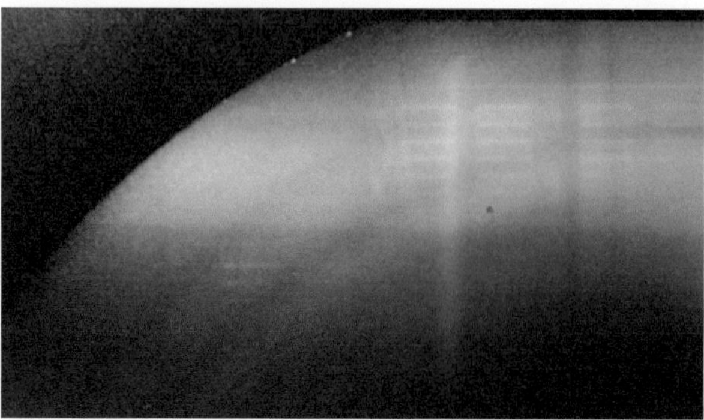

Fig. 9.32 Example for dislocations formed by layer stress, as visible from the fact that the sense amplifier structures from the 4 Mbit DRAM process wafers show up in the X-ray topograph, which implies that in those particular regions of the integrated the layer stress during one of the high temperature steps exceeded the critical yields stress

In more modern devices with design rules below quarter micron, instead of local oxidation of silicon (by the LOCOS process) the lateral insulation of devices has been implemented by etching shallow trenches into the silicon substrate and filling them with deposited oxide. Predictably, these processes also have led to problem of dislocation formation initially, but the problem could be solved e.g. by the appropriate deposition conditions and a suitable temperature time profiles to induce stress relaxation in the deposited material and to keep thermal stresses below the threshold for dislocation formation [4, 35, 36].

9.6.3 *Interaction of Defects After Initial Formation During Subsequent Processes*

It is obvious that a combination of metal contamination and stresses, as indicated in Fig. 9.11, can lead to complex phenomena via different defect mechanism paths over the course of the many consecutive process steps of microelectronics manufacturing, as already illustrated in Fig. 9.12.

An example from a real device process of the path in Fig. 9.12 where some stress, e.g. during oxidation, have led to the formation of a stacking fault, and the subsequent decoration of the stacking fault by contamination not during the actual device process but during the burn-in operation of the DRAM is represented in Fig. 9.33: The retention time fail of a particular DRAM cell in a 64 kB DRAM (after results from [4]) and the subsequent physical failure analysis showed that the failed cell had a stacking fault attached to it, as revealed by Secco defect etching. The remarkable fact is that during the initial test on a wafer at the end of the device process, that particular memory cell had shown no problems, in other words the stacking fault had formed but was not decorated by metal contamination [4]. After

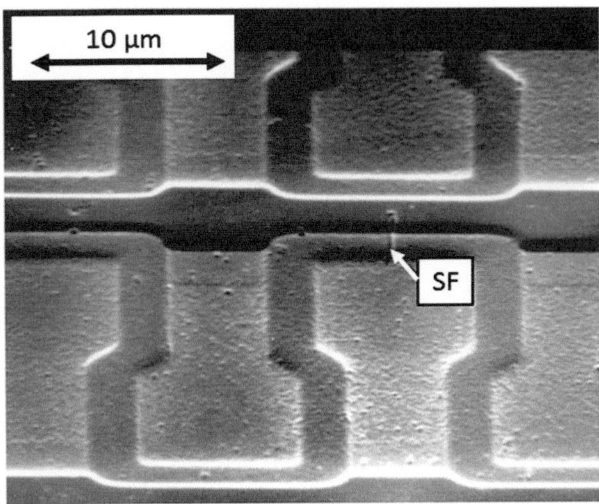

Fig. 9.33 Stacking fault in the memory cell array of a 64 kbit DRAM. The cell had been functional immediately after processing, but developed a retention time fail due to excessive leakage currents after the burn-in (i.e. operation at higher than normal operating temperature [up to 150 °C] and at an overvoltage compared to the normal operating voltage, to accelerate aging by orders of magnitude. To subject the DRAMS to burn-in is a standard procedure before they are sent to customers)

several hours of burn-in testing, however (i.e. the equivalent of months to years of operation), the cell started to degrade until the leakage current caused by the stacking fault got so large that the charge stored in the memory cell capacitor leaked before the charge was replaced in the memory refresh cycle for a DRAM. The only reasonable explanation of this finding is that during operation, impurity atoms like Fe, Ni, Co, Cu who all can diffuse even at room temperature gradually decorated the stacking fault more and more, leading to a higher and higher leakage current for the affected cell. The ability for diffusion at room temperature can be concluded from an extrapolation of the diffusion coefficient to room temperature, it is also obvious from the fact that the SPV method for the determination of Fe concentrations in Si works at all [32]: The dissociation of Fe B pairs around 150 °C implies that the interstitial Fe atoms have to move away from the Boron atoms they had formed pairs with. In fact, the time constant for re-pairing is consistent with the extrapolation of the diffusion coefficient to room temperature [37]. This example also shows directly that the complex interaction of extended defects and metal impurities is not only a yield detractor, but also a reliability and durability problem, which is the worst and often most difficult problem in defect engineering.

Figure 9.34 provides further direct evidence of the detrimental electrical effects of stacking faults by causing leakage current via the joint electrical effect of metal contamination and stacking faults is illustrated:

By a spatial correlation of stacking fault densities and the measured leakage current in the power MOS devices in one whole wafer it could be established that the

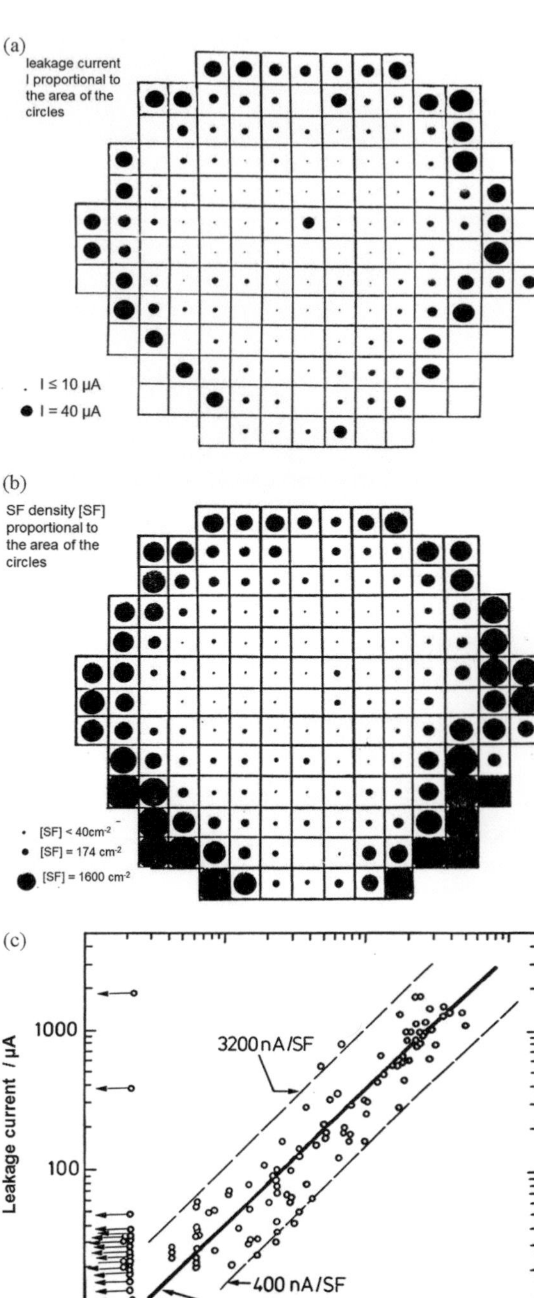

Fig. 9.34 Comparison of the spatial distribution of leakage currents of MOS power devices (**a**), stacking fault densities of (**b**), and a correlation between them number of stacking faults in one devices and its leakage current (**c**)

leakage current caused by one stacking fault was approximately 1 μA. This is orders of magnitude more than for uncontaminated stacking faults: General experience shows that clean stacking faults will cause leakage currents of the order of picoamps or a few nanoamps. The observed comparatively high values and first and foremost, the scatter are both a strong indicator for decoration of the stacking faults by metal contamination.

Another detrimental interaction of metal contamination and extended defects has already been shown in Fig. 9.18, where the formation of stacking faults during oxidation is dramatically enhanced if before the oxidation step, the metal precipitates are formed in a preceding high temperature step. The precipitates subsequently lower the nucleation barrier for stacking fault nucleation by a significant amount for all the following process steps.

It is thus clear that metal contamination has a two-fold risk potential for microelectronics production:

- the enhancement of nucleation of extended lattice defects (stacking faults, dislocations) by lowering the nucleation barrier
- The subsequent decoration by metal impurities to render these defects highly electrically active

In other words, this particular yield detraction path proceeds in three steps:

Metal contamination and precipitation → formation of extended defects → decoration of those defects by metals

To further illustrate this important mechanism, Fig. 9.35 demonstrates that the leakage current from stacking faults increases with the Fe impurity concentration.

The interactions during a complete device process which can consist of up to 1000 individual process steps can be quite complex. Another fairly common and typical example is the interaction between a super saturation of self-interstitials from oxidation with residual implantation damage, as shown in Fig. 9.36. The implantation of boron enhances the formation of stacking-fault-like defects in a subsequent well oxidation step, in this case the residual damage from boron implantation lowers the nucleation barrier for stacking fault formation.

9.6.4 Monitoring and Process Control During Mass Production on the Product Level

As mentioned in the previous section, metal contamination is a considerable and ubiquitous risk for yield and reliability issues in microelectronics production. Therefore, additional monitoring not only on the equipment level, but on the product level is appears certainly advisable, since this monitors all process steps and all potential defect scenarios. It will be demonstrated in this section, which introduce two additional aspects not previously mentioned, that this is not only advisable, but clearly necessary:

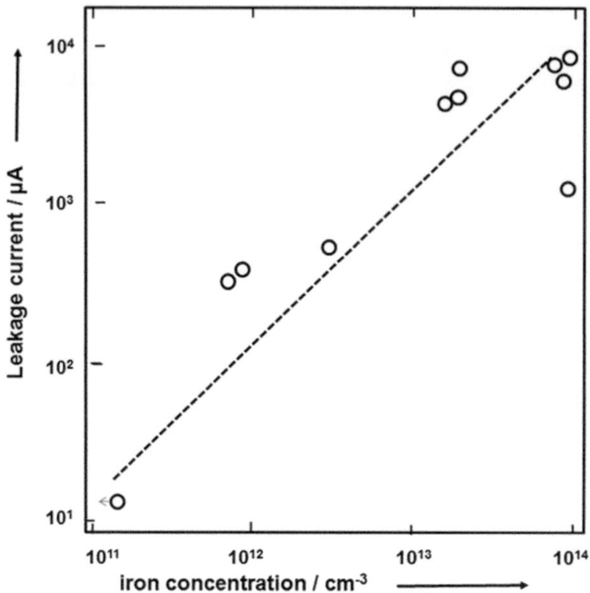

Fig. 9.35 Leakage current caused by one stacking fault as a function of the average Fe concentration in the silicon material

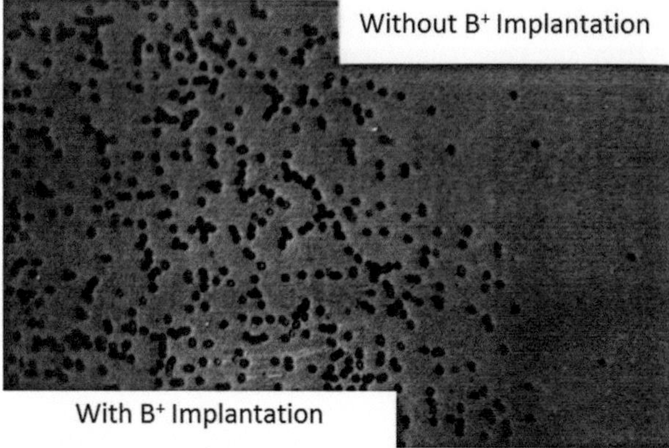

Fig. 9.36 Stacking-fault-like defects formed after boron implantation and subsequent oxidation. The area which has not been exposed to any implantation is free of defects. The residual damage from the boron implantation has caused the formation of defects during oxidation, by coalescence of excess self-interstitials

- The emergence of three-dimensional rather than planar device structures introduce additional failure paths not captured in the failure path map of Fig. 9.11, so a screening on the product levels is not only advisable but indispensable, since this defect mechanism cannot be detected by exclusive equipment monitoring.

- Control of defect formation and metal contamination on the product level can be done in a very efficient and cost effective manner via the electrical test results at the end of the production process, if it is known which of the tested performance parameters are especially sensitive to metal contamination and defects. This has the advantage that electrical tests which are performed on a routine basis for 100 % of the product can be used to detect any significant metal contamination and/or crystal defect formation that has escaped the equipment screening process. The reason that this is a "fail-safe" way to check for metal contamination and/or crystal defect formation is that such electrical testing is done for 100 % of the integrated circuits produced, so it is impossible that such contamination or crystal defect formation events can remain undetected. Also, since the electrical tests are performed after the complete production process, the test results it will represent effects of the accumulated contamination and crystal defects, including their interactions.

To elucidate these aspects, we consider the example of trench cells for 4 Mbit DRAMS: With the introduction of trench capacitors (which look more like a bore hole than an elongated trench, see e.g. Fig. 9.37a), it was noticed that a small number of cells were affected by dislocations which had developed in the immediate vicinity of the electrically defective memory cell, due to the stress concentrations in the corners of the 3D-structure [35, 36]. Figure 9.37a shows the etch figure of such a trench-induced dislocation. To unambiguously prove that the observed electrical retention time fails were due to such trench-induced dislocations, the number of dislocations revealed by defect etching in a complete 256 Kbit memory cell block were counted and compared to the number of retention time fails in that block. Figure 9.37b shows that there is indeed a very good correlation.

The production lots from the initial phases of process development had a high enough number of such defects to generate such statistical evidence that the trench-induced dislocations were indeed the cause for the retention time fail cells. In routine full-scale production, the situation is very different: There will be very few such retention time fail cells in one 4 Mbit DRAM memory, leave alone in one of the sixteen 256 k cell blocks. In routine monitoring, the occurrence of such very few single cells with retention time fails will be analyzed statistically per production lot (which comprises typically 10,000–20,000 DRAMs), and if any increase beyond the normal statistical variation of the number of fail cells per production lot is observed, physical failure analysis on that particular cell will be employed to find out whether trench-induced dislocations or other crystal defects are responsible for the retention time fail through increased leakage currents.

Another common procedure to monitor whether any abnormal contamination and/or crystal defect levels have occurred by electrical test results is to correlate the electrical test results to the "usual suspects" in terms of process steps or equipment, i.e. to see whether there e.g. a significant difference in the level of electrical failure classes between he different essentially identical pieces of equipment used for the same process. The idea behind this analysis strategy is, that if a particular process step is known to potentially cause that failure mode, then there is a high probability

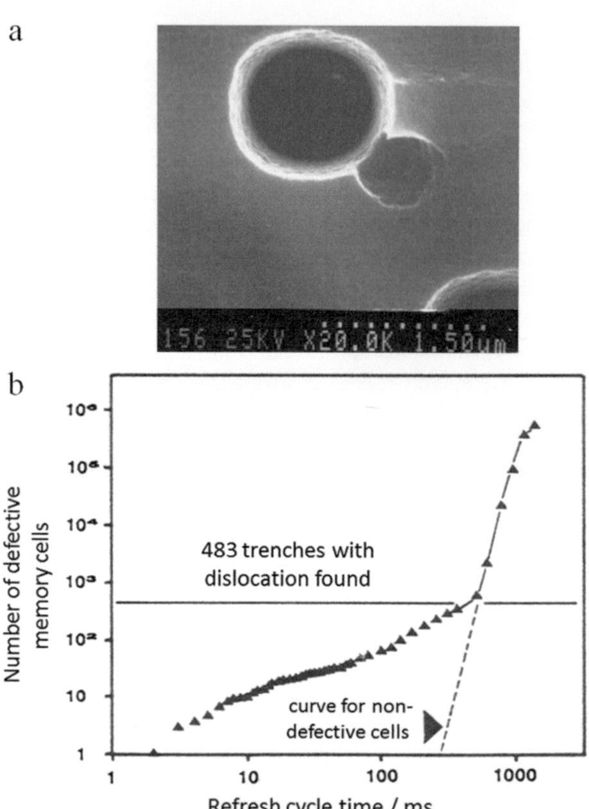

Fig. 9.37 (**a**) SEM micrograph of the etch pit of a trench-induced dislocation, as delineated by Secco defect etching. The round holes are the dry etched holes for trench capacitors, the "irregular" etch pits near the trench delineate the dislocation. (**b**) Plot of the cumulative statistics of the refresh time of the 256 K cells of a 256 K cell block. The portion of the *curve* with the *shallow slope* is representative of defective cells with too short a refresh time, the steep part of the *curve* is representative of the natural distribution of refresh times for cells without any defects, i.e. with design related leakage paths rather than leakage caused by a decorated dislocation. The number of cells in the defect branch tallies very well with the count of trench induced dislocations [35, 36]

that the extent of the failure will differ from machine to machine. The differences can be expected due to the complex nature of defect formation. In addition, there can be different equipment states that influence the defect/contamination level, which can be due to either the state of maintenance or the usage for different processes other than the suspect process. Figure 9.38 shows an example.

Similar principles to trace electrical performance problem due to either silicon wafer issues or to processing problems/equipment issues during the photovoltaic cell manufacturing process are being used on PV mass production. A particularly instructive example of how SPC can help to stabilize a process and improve the manufacturing process has been reported in the PhD Thesis of Dinkel [16]. He could

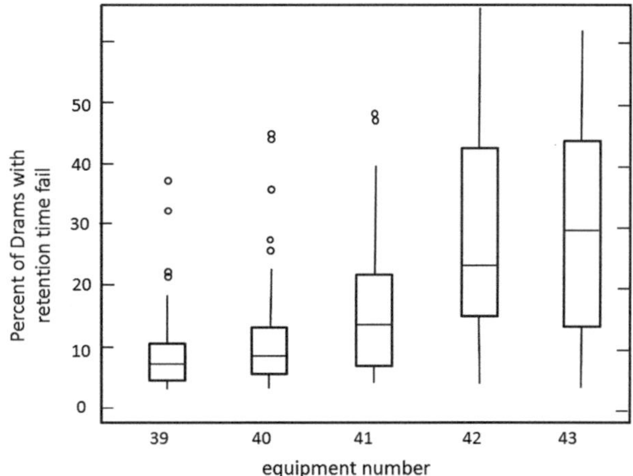

Fig. 9.38 Box plot of the probability of finding at least one retention time-affected cell in a 4 Mbit DRAM, as sorted by the oxide etching equipment (five identical machines were used for the particular process step). It is obvious that there are significant differences, and that the equipment 39 is the best, and equipments 42/43 are about a factor of four worse. It was possible, by contamination studies, to identify the root cause: The higher incidence of retention time fails was due to a Molybdenum contamination, after removal of the source for that contamination, all machines were on the level of 39, i.e. the residual occurrence of retention time fails was now mainly due to other reasons

show that by the application of SPC, the variability in the most important process parameters in cell production was reduced by a factor of 2, and that this led to a much faster improvement in the average efficiency of the cells compared to a reference production line with business as usual, compare Fig. 9.39a and b.

Although this study in the photovoltaics industry has been focused not primarily on metal contamination and defect formation, it is representative for any efforts to improve processes. A more stable process will in itself, as a side effect, also improve that situation regarding crystal defects and contamination. It will, in particular allow in a much better way to identify significant contamination or defect sources, by the "equipment benchmarking" method demonstrated in Fig. 9.38.

The recently discovered potential-induced degradation (PID) phenomenon [38] has not been understood yet, although many different hypotheses have been put forward [39]. What emerges at this stage, that it appears highly likely, that mechanisms and principles of defect formation that have been identified over decades of research in the microelectronics industry also apply here [17, 38]. It has been shown that there is a strong similarity to the mechanism describe in Sect. 9.6.2, in which dislocations are generated by the layer stress caused by the different thermal expansion coefficients of silicon and silicon nitride. Exactly that situation is found in most photovoltaic cells: The silicon nitride antireflective coating is directly on top of the silicon wafers. As mentioned in Sect. 9.6.2, the solution to this problem was the introduction of a silicon dioxide buffer layer between the silicon nitride

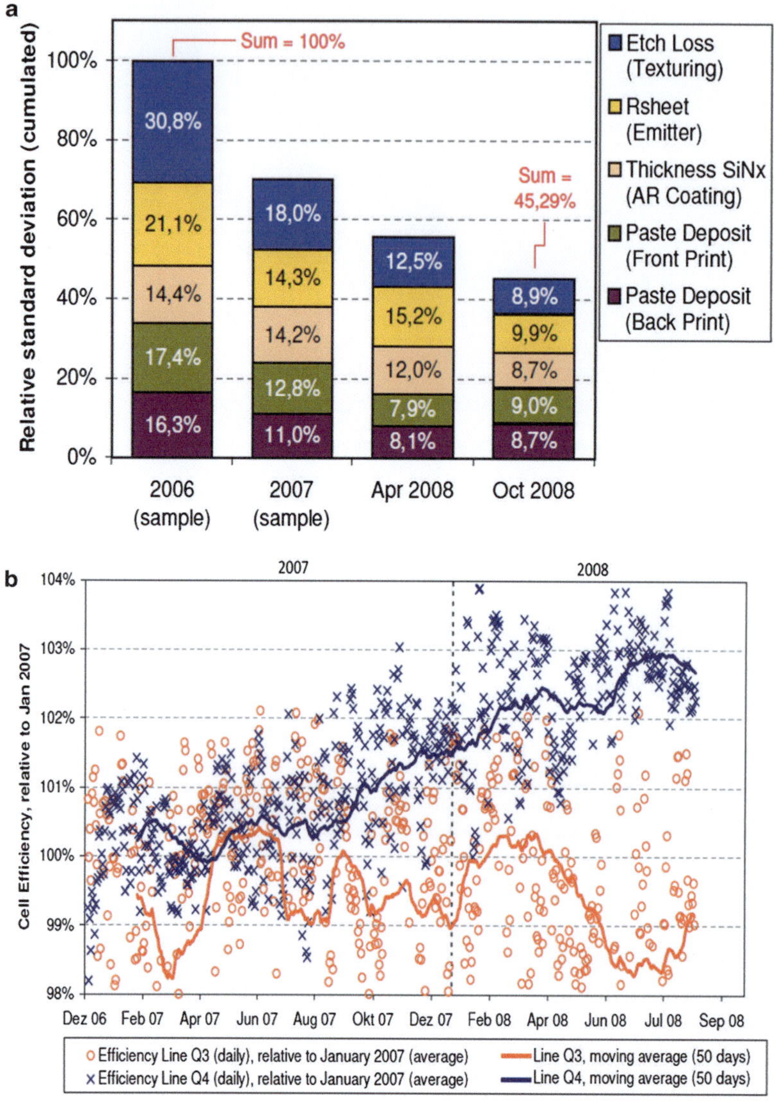

Fig. 9.39 (**a**) In photovoltaic cell production the five most important parameters which cause variability in the energy conversion efficiency of the PV cells are the etch loss during texturing the as-sawn raw wafer, the sheet resistance of the emitter, the thickness of the antireflective silicon nitride coating and the silver paste deposit steps for both the front side and the back side metallization. By the introduction of SPC, the variability in these key control characteristics could be reduced by more than a factor of two. It should be noted that another key control characteristic is the minority carrier lifetime of the silicon wafer, this was not the scope of the work [16]. (**b**) Measured cell efficiency (daily average of the complete production volume in two different production lines Q3 and Q4, which contained practically identical equipment). Line Q3: without SPC, Line Q4: with SPC. It is clear that a systematic and continuous improvement of the efficiency was only possible for reduced variability of the key control characteristics, as listed in Fig. 9.39a. It is not directly visible that among other things, the defect level was positively affected by the process improvements, hence an increase in the efficiency was achieved [16]

and the silicon, which served as a kind of lubricant to mitigate the layer stress. If for photovoltaic cells which are prone to the PID effect, such a buffer layer is introduced, the PID effect is absent. It appears therefore likely that one important root cause of the potential induced degradation is the formation of dislocations and stacking faults by the layer stress between silicon and silicon nitride.

Returning to the microelectronics industry, there is the severe problem that Fe, Co, Ni and Cu can even diffuse at room temperature, so that a gradual decoration of the extended lattice defects during device operation is possible, as demonstrated by the examples described earlier (see also [4]). The general conclusion is that a high additional risk for early failure after 10–1000 h of operation, i.e. reliability and durability risks, can be caused by these failure mechanisms. In other words, the "synergetic action" of metal contaminants and crystal defect will lead to the worst problem for any manufacturer, namely reliability issues, and therefore preventive countermeasures are mandatory. Incidentally, there is strong evidence [39] that the PID effect in PV, which is a reliability issue, is also associated with the decoration of the partial dislocations bounding stacking faults.

Since it is next to impossible under mass production environments to completely avoid metal contamination, in addition to monitoring an minimizing metal contamination and the driving forces for crystal defect formation, an additional strategy is normally implemented, which is similar to the ideas behind taking out an insurance policy: Extra process measures are taken, which will mitigate or suppress the detrimental effects of any residual metal contamination. Such process measures are called gettering, the different gettering types that have been invented and implemented are described in the following Sect. 9.6.5.

9.6.5 Gettering of Metals by Intentionally Introduced Defects

As mentioned in the previous section, metal contamination is a considerable risk for yield and the reliability of products in microelectronics production. In real life, such small but unavoidable risks with a large damage potential are mitigated by taking out an insurance policy. Exactly the same principle applies in microelectronics: Several gettering schemes have been developed over the years, an overview inspired by a figure in Chap. 13 [7] is shown in Fig. 9.40.

The two most frequently used gettering schemes used in the microelectronics industry involve the intentional formation of extended defects, namely

1. backside damage gettering (extended defects on the wafer back surface are provoked by slight controlled mechanical damage by proprietary, sometimes accompanied by a low temperature oxide LTO) and
2. intrinsic gettering, which involves the intentional and controlled formation of oxygen precipitates and bulk microdefects (stacking faults) in the bulk of the wafer, but not at the surface. This is illustrated in Fig. 9.41.

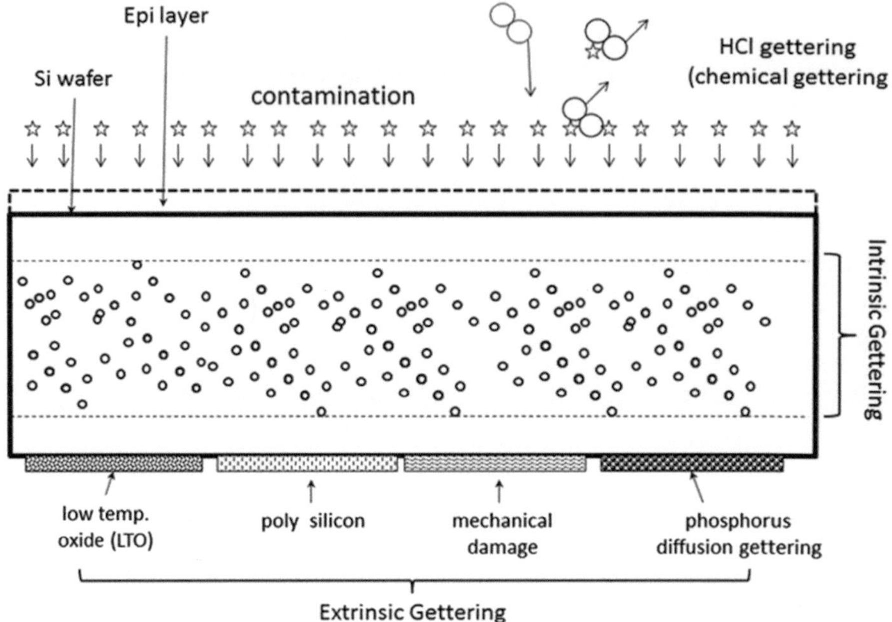

Fig. 9.40 Overview of different gettering schemes (after Shimura [7], Fig. 14, page 595) to reduce metal contamination and reduce the associated risks. Chemical gettering by chlorine containing gases in furnaces tries to remove the metal before they can even enter the silicon wafer, since chlorides of the important metals are volatile. Once the metals are in the silicon, they have to be kept away from residing in the top surface layer where the active device areas are. This is either accomplished through various intentional damage creation schemes on the wafer back surface or it is using the natural tendency in CZ-silicon for oxygen precipitates to form in the bulk, but not near the surface. In all cases, the tendency of metal contamination to aggregate at defects is used to keep them away from active device areas. Naturally this technique cannot be used in power devices if the whole wafer is active in the device operation. Not directly shown in this graph: Gettering by solubility enhancement via high doping levels

Like with any process, some control has to be established that the intended "insurance policy" is really effective. This can be either, through only the monitoring of the intentional defect density, or better by correlation of the gettering type and intensity with the electrical results on devices.

Figure 9.42 is an example of the latter approach: The density bulk micro defects (BMDs) which are generated as a result of oxygen precipitation (oxygen precipitates, stacking faults and dislocation all three defect types are visible via their respective etch artefacts in Fig. 9.41). Below a density of 10^3 cm^{-2}, there is no appreciable positive effect on the yield of gate oxide test structures (the BMD density is measured as an area density, as visible on the etched surface, from the thickness of the etched layers the volume density can be calculated, but it is not needed for such a correlation). Between 10^3 and 10^4 cm^{-2} there is a transition

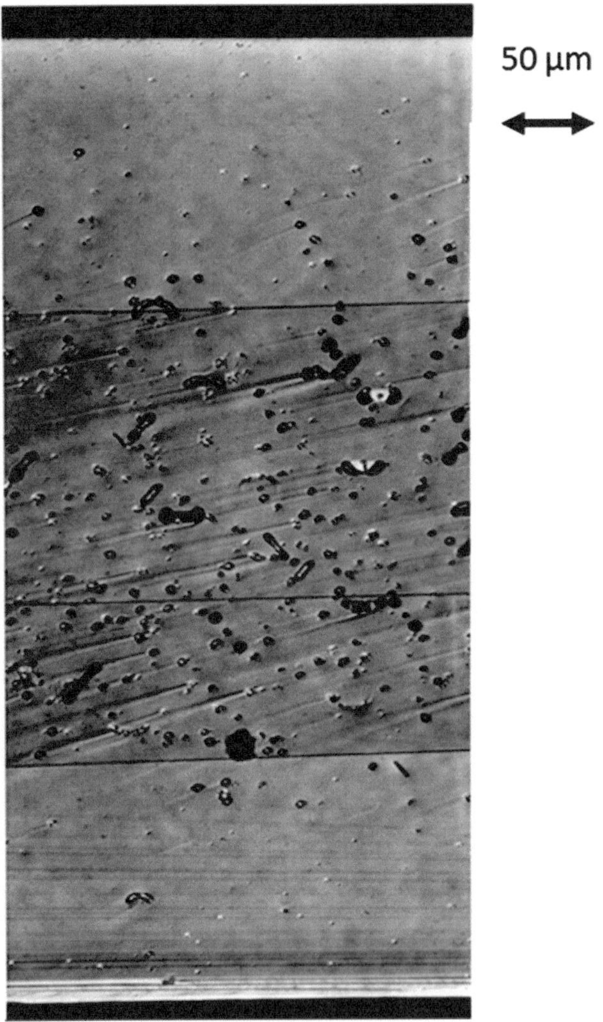

Fig. 9.41 Cross section of a wafer from the front to the back surface, which shows that the denuded zone which also forms also on the back surface. It is also visible that there can be small residual defects in the denuded zone

zone, and for higher densities, the gate-oxide-yield approaches 100 %. The residual variation is due to the variability of the impurity concentration.

A slightly different approach to testing the effectiveness of intrinsic gettering using spatial correlations has been used in an experiments by Falster and Bergholz [33], in which the most common impurities Fe, Ni, Cu, Co and in addition for Pd have shown the intrinsic gettering action of oxygen precipitates directly. To this end, wafers were prepared which were oxidized once or had received a partial or a full CMOS heat process (without the actual device process, with respect to the oxygen

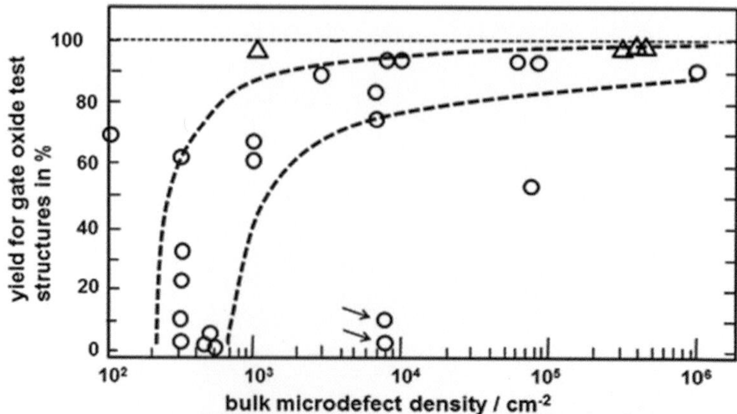

Fig. 9.42 Positive effect of intrinsic gettering. The bulk microdefects (BMDs) can prevent the detrimental effect of a significant metal contamination on the gate oxide integrity test structures, provided the density is higher than about 10^4 cm^{-2}. The *arrowed* points are from lots with more heavy metal contamination

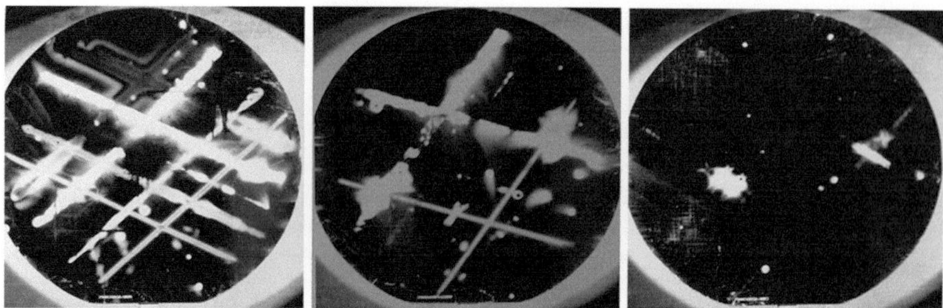

Fig. 9.43 Haze test on wafers contaminated in a similar manner on the wafer back surface as shown in Fig. 9.24. The leftmost wafer has received an initial oxidation after crystal growth and wafer manufacturing (i.e. negligible oxygen precipitation), the center wafer has received a partial CMOS heat process simulation (i.e. moderate oxygen precipitation), the wafer on the right has received a full CMOS process simulation so that gettering is strong. Also visible: significant dislocation formation in the latter wafer which indicates a softening of the material, too soft for the harsh haze test. All wafers had been intentionally contaminated by Fe, Ni, Cu, Co and Pd as indicated in Fig. 9.24

precipitation, the partial or full CMOS thermal simulation is equivalent to the actual device process). Subsequently the three types of wafers with four levels of "built-in" intrinsic gettering were contaminated in the same manner as shown in Fig. 9.24. From Fig. 9.43 it can be seen that for the wafer that had only received one oxidation, there is little gettering visible in this haze test, whereas wafers with the partial and full CMOS heat process simulation show strong gettering via the reduction of the incidence of haze on the front surface: The impurities have been gettered by the bulk microdefects before they could precipitate near the surface.

In addition to defect-induced gettering via oxygen precipitates in the bulk or intentional damage at the wafer back surface, gettering by heavily doped layers is also commonly used, as already mentioned before, this gettering type utilizes the solubility enhancement by pairing of charged impurities with dopant atoms, by the introduction of additional species for the impurity in question and last but not least by dynamic effects, such as the Si self-interstitial injection during e.g. phosphorus diffusion, which enhances the formation of SiP precipitates [40]. The increase in the solubility can be several orders of magnitude in total [22], which is certainly one reason that epitaxial wafers on a heavily doped substrate enhance significantly the robustness of CMOS processes against contamination.

In photovoltaics, gettering is also very important, although it was presumably never introduced intentionally, but came as a beneficial side effect of the cell and process design: Phosphorus diffusion gettering is highly effective during the emitter diffusion in cell which use p-type silicon as the starting wafer. Equally important, the gettering of a heavily aluminium-doped back surface layer (from the aluminium paste covered back surface, for reliable electrical contacts) is beyond any doubt contributing significantly to very effective gettering and thus results in a high resilience of the cells to metal contamination. It is well known and a consequence of the need for low cost production, that there are much higher metal contamination levels in PV production than in microelectronics. As already briefly mentioned before, the more modern cell designs (lower doping of the P-doped emitter, selective emitter doping both to improve the conversion efficiency for blue light) and the introduction of dielectric layers on the back surface can be expected to reduce the gettering efficiency. Thus predictably, contamination-related new defect types will come up in mass production of modern cell designs, and will have to be addressed along the principles explained in this chapter. Thus, this additional risk factor has to be taken into account when converting mass production from the simple cells to the more sophisticated cell concepts. It is clear that this will need additional efforts in terms of defect and contamination control.

Also, going by the general characteristics of the potential induced degradation in photovoltaics (PID effect, [17]), and by direct evidence [39] it appears certain that fast diffusing metal contamination is involved in the gradual deterioration of the pn-junction over months, which is typical for the PID effect. In an intentional contamination experiment, Raykov et al. [38] could show however, that Na, Li, K and Ca do not lead to the PID effect, if they are deposited on top of Si cell at an intermediate process stage. The role of ubiquitous metals such as Fe, Ni, Cu will have to be studied in order to clarify their potential role in the PID effect.

Many researchers have demonstrated the effectiveness of gettering directly in production. One particularly instructive example has been published by Jastrzebski et al. [41]. He could show that the yield of bipolar devices drops in a period of high contamination both with and without intrinsic gettering. However, the drop is much less with intrinsic gettering, so although the effect of metal contamination is not completely suppressed, it is mitigated to such a degree that it is close to the "natural" process yield variations.

Another excellent example to demonstrate the effectiveness of intrinsic gettering has been published by Hourai et al. [27]. The dramatic enhancement of OISF formation by metal contamination has already been shown in Fig. 9.18. In the same series of experiments the authors could show how intrinsic gettering can counteract the enhancement of OISF formation by metal contamination.

Traditionally, the oxygen content of silicon wafers has been regarded as the main parameter to be adjusted to a particular device manufacturing process, depending on the thermal profile of that process. From the work of Hourai [27] it is obvious that a minimum amount of oxygen has to precipitate, in "traditional" silicon wafers, a minimum oxygen content is needed to ensure this. So too low an oxygen content is not suitable to ensure the effectiveness of intrinsic gettering, but too high an oxygen content can be detrimental: Excessive oxygen precipitation will lead to warp the wafer after the production process by plastic deformation, i.e. the high large oxygen precipitates reduce the critical shear significantly.

A method pioneered by Falster et al. [42] that consists of a suitable thermal treatment with specific gas atmospheres, enables the manipulation of the oxygen precipitate nuclei in such a manner that the oxygen precipitation can be made almost independent of the oxygen content and the thermal history during crystal growth. This is achieved through erasing all existing oxygen precipitate nuclei (which are at an embryonic stage) by a rapid thermal anneal step and then using a second high proprietary temperature treatment in a suitable ambient to form nuclei in a preset density with a well-controlled denuded zone. To achieve this, a detailed understanding of defect dynamics was needed, the description of this process is beyond the scope of this chapter.

While this subsection has explained the fundamentals of intrinsic gettering without too much regard for the latest state-of-the-art microelectronic processes, in Chap. 6 Kissinger describes in detail oxygen precipitation, the delineation of oxygen-related defects by defect etches, and the impact of modern CMOS process with drastically reduced thermal budgets.

9.7 Summary and Conclusions

The dual task of defect engineering is to prevent the formation of crystal defects and to intentionally create defects for gettering in a controlled fashion.

The reduction of the minimum feature size in microelectronics and the introduction of new device designs and processes in photovoltaics have to be accompanied by vigilance regarding new contamination sources and/or new mechanisms for the formation of extended defects, such as the onset of dislocation formation at corners in 3D-device structures.

Over many past device and wafer generations, defect formation has been brought under control by weakening the "driving forces" (i.e. reduction of thermal/thermal stress and avoiding too high supersaturation of point defects) and strengthening the "impeding forces" (i.e. elimination of nucleation centers and reduction of processing temperatures), in spite of the emergence of new defect mechanisms emerging as the technologies progressed. These principles have held over 40 years of microelectronic manufacturing. Control of defects is achieved on the one hand by a thorough scientific understanding of the very often complex defect formation and device degradation mechanisms; in addition it is necessary to employ quality engineering principles and quality management tools (such as SPC) to reliably control the defect scenarios.

The successful applicability of the general principles of defect engineering to photovoltaics and novel types of silicon wafers (perfect silicon, silicon on insulators and other innovative wafer types) can be confidently predicted. As mentioned in the last sections, there is strong evidence that the recently discovered PID effect is due to defect mechanisms in conjunction with metal impurities in the bulk of, or near the pn-junction of silicon-based solar cells, and that the formation mechanism is very similar to the formation mechanism of dislocations via layer stress in microelectronics. The analogy even includes that the solution to the problem, namely the introduction of a silicon dioxide buffer layer between silicon and silicon nitride. Moreover it is clear that with the advent of more sophisticated cell concepts (such as the PERC cell) the susceptibility to contamination and/or defect formation will increase.

It has been the overall objective of this chapter to "build a bridge" between the science behind defects in silicon and how to apply this knowledge effectively in R&D in the microelectronics, micromechanical and photovoltaic industries. In this endeavor, it was necessary to introduce the essentials of process and quality management and reliability engineering. We hope that electrical engineering and information technology students will be capable to apply materials and defect sciences effectively in their future assignments.

References

1. Falster, R., Voronkov, V.V., Quast, F.: On the properties of intrinsic defects in silicon: a perspective from crystal growth and wafer processing. Phys. Stat. Sol (B) **222**, 219 (2000)
2. SEMI M1 Standard for silicon wafers. http://ams.semi.org/ebusiness/standards/SEMIStandardDetail.aspx?ProductID=1948&DownloadID=3469
3. International Roadmap for Semiconductors ITRS. http://www.itrs.net/Links/2013ITRS/2013Chapters/2013Overview.pdf and International Technical Roadmap for Photovoltaics, http://www.itrpv.net/Reports/Downloads/
4. Kolbesen, B.O., Strunk, H.: VLSI electronics: microstructure science. In: Einspruch, N.G., Huff, H.R. (eds.) Silicon Materials, vol. 12, p. 143. Academic Press, New York (1985)
5. http://www.semi.org/Standards
6. Bergholz, W., Zoth, G., Wendt, H., Sauter, S., Asam, G.: Metal contamination control in silicon VLSI technology: fundamentals. Siemens Forsch- und Entwickl Ber **16**, 241 (1987)
7. Shimura, F.: Oxygen in Silicon. Semiconductor and Semimetal Series, vol. 42. Academic Press, New York (1994)
8. Bergholz, W.: Grown-in and process induced defects. In: Shimura, F. (ed.) Oxygen in Silicon. Academic Press, Boston (1994)

9. http://www.itrpv.net/Reports/Downloads/2015/
10. http://www.oecd-ilibrary.org/energy/world-energy-outlook_20725302;jsessionid=3y9ykh6ljjtq.x-oecd-live-03
11. Bergholz, W., Gilles, W.: Impact of research on defects in silicon on the microelectronics industry. Phys. Stat. Sol (B) **222**, 5 (2000)
12. Obry, M., Bergholz, W., Cerva, H., Kürner, W., Schrems, M., Sachse, J.U., Winkler, R.: The role of metal contamination and crystal defects in quarter micron technology. In: Abe, T., Bullis, W.M., Kobaysashi, S., Lin, W., Wagner, P. (eds.) Proceedings of the 3rd International Symposium Defects in Silicon, p. 133. The Electrochemical Society, Pennington (1999)
13. Kolbesen, B.O., Cerva, H., Gelsdorf, F., Zoth, G., Bergholz, W.: Process-induced defects in silicon VLSI technology. In: Proceedings of the Semicon Europe, Zürich, Mar 1992
14. Bergholz, W., Landsmann, D., Schauberger, P., Schöpperl, B.: Contamination monitoring and control in device fabrication. In: Kolbesen, B.O., Claeys, C., Stallhofer, P., Tardif, F. (eds.) Crystalline Defects and Contamination: Their Impact and Control in Device Manufacturing, p. 69. The Electrochemical Society, Pennington (1993)
15. Bergholz, W., Zoth, G., Gelsdorf, F., Kolbesen, B.: Metal contamination in ULSI technology. In: Bullis, W.M., Gösele, U., Shimura, F. (eds.) Defects in Silicon II, p. 21. The Electrochemical Society, Pennington (1991)
16. Dinkel, T.: Integrated efficiency engineering in solar cell mass production. PhD Thesis, Jacobs University Bremen (2010)
17. Raykov, A.: Potential-induced degradation – a multi-level problem. PhD Thesis, Jacobs University Bremen (2015)
18. Graff, K.: Metal Impurities in Silicon-Device Fabrication. Springer Series in Materials Science (Book 24). Springer, Berlin (1990)
19. Pyztek, T., Keller, P.: The Handbook for Quality Management. Mc GrawHill, New York (2013)
20. Graff, K.: Transition metals in silicon and their gettering behaviour. Mater. Sci. Eng. B **4**, 63–69 (1989)
21. Tuck, B.: Introduction to Diffusion in Semiconductors. IEE Monograph Series, vol. 16. Peter Peregrinus Ltd., on behalf of the Institution of Electrical Engineers, Stevenage (1974)
22. Gilles, D., Schröter, W., Bergholz, W.: Impact of the electronic structure on the solubility and diffusion of 3d transition elements in silicon. Phys. Rev. **B41**, 5770 (1990)
23. Winkler, R., Behnke, G.: Gate oxide quality related to bulk properties and its influence on DRAM device performance. Semicond. Silicon. **94**, 673 (1994)
24. Hannay, N.B.: Bell Lab reports (1958)
25. Pell, E.M.: Ion drift in an n-p junction. J. Appl. Phys. **31**, 291 (1960)
26. Bergholz, W.: Analysis of extended defects. In: Schulz, M. (ed.) Landolt Börnstein Handbook of Physics, Neue Serie 22B, p. 126. Springer, Berlin (1988)
27. Hourai, M., et al.: Behavior of defects induced by metallic impurities on Si (100) surfaces. Jap. J. Appl. Phys. **28**, 2413 (1989)
28. Nikkei Microdevices May 1990
29. Lehmann, V., Föll, H.: Minority corner diffusion length measurements in silicon wafers using a Si-electrolyte contact. J. Electrochem. Soc. **135**, 2831 (1988)
30. Hellmann, D., Rother, M., Hill, M., Bergholz, W., Riedlbauer, M.: Influence of quartzglass on silicon wafers application studies and examples from device production. In: Schidt, D.N. (ed.) Contamination Control and Defect Reduction in Semiconductor Manufacturing III, p. 285D. The Electrochemical Society, Pennington (1994)
31. Bergholz, W., Landsmann, D., Schauberger, P., Wittman, J., Hoffmann, H.: Relevance and effects of metal contamination on device processes and parameters. In: Proceedings of the Technical Conference SEMICON Europa, Geneva, 5 Apr 1995
32. Zoth, G., Bergholz, W.: A fast, preparation-free method to detect iron in silicon. J. Appl. Phys. **67**, 6764 (1990)
33. Falster, R., Bergholz, W.: The gettering of transition metals by oxygen-related defects in silicon. J. Electrochem. Soc. **137**, 1548 (1990)

34. Bergholz, W., Mohr, W., Drewes, W., Wendt, H.: Defect-related gate oxide breakdown. Mater. Sci. Eng. **B4**, 359 (1989)
35. Dellith, M., Gelsdorf, F., Bergholz, W., Booker, G.R., Kolbesen, B.O.: TEM, etching studies of fabrication-induced defects in 4 M DRAMs. Inst. Phys. Conf. Ser. **117**, 169 (1991)
36. Dellith, M., Booker, G.R., Kolbesen, B.O., Bergholz, W., Gelsdorf, F.: On the formation of trench-induced dislocations in dynamic random access memories (DRAMs). Inst. Phys. Conf. Ser. **134**, 235 (1993)
37. Zoth, G., Bergholz, W.: Metal contamination control by diffusion length measurements – principles and practice. In: Proceedings of the Technical Conference Productronica, Munich, Nov 1997
38. Raykov, A., Hahn, H., Stegemann, K.-H., Kutzer, M., Storbeck, O., Neuhaus, H., Bergholz, W.: Towards a Root Cause Model for the Potential-Induced Degradation in Crystalline Silicon Photovoltaic Cells and Modules. PV EUSEC, Paris, p. 2998 (2013); Raykov, A., Stegemann, K.-H., Hahn, H., Bitnar, B., Kutzer, M., Neuhaus, H., Bergholz, W.: On the PID inhomogeneities. Presented at the EU PVSEC, Amsterdam (2014)
39. Naumann, V., Lausch, D., Hähnel, A., Bauer, J., Breitenstein, O., Graff, A., Werner, M., Swatek, S., Großer, S., Bagdahn, J., Hagendorf, C.: Explanation of potential-induced degradation of the shunting type by Na decoration of stacking faults in Si solar cells. Sol. Energy Mater. Sol. Cells **120**(Part A), 383–389 (2014)
40. Schröter, W., Kühnapfel, R.: Model describing phosphorus diffusion gettering of transition elements in silicon. Appl. Phys. Lett. **56**, 2207 (1990)
41. Jastrzebski, L., Soydan, R., McGinn, J., Kleppinger, R., Blumenfeld, M., Gillespie, G., Armour, N., Goldsmith, B., Henry, W., Vecrumba, S.: A comparison of internal gettering during bipolar, CMOS, and CCD (high, medium, low temperature) processes. J. Electrochem. Soc. **134**, 1018–1025 (1987)
42. Falster, R., Voronkov, V.V.: The engineering of intrinsic point defects in silicon wafers and crystals. Mater. Sci. Eng. B **73**, 87 (2000)

MIX
Papier aus verantwortungsvollen Quellen
Paper from responsible sources
FSC® C105338

If you have any concerns about our products,
you can contact us on
ProductSafety@springernature.com

In case Publisher is established outside the EU,
the EU authorized representative is:
Springer Nature Customer Service Center GmbH
Europaplatz 3, 69115 Heidelberg, Germany

Printed by Libri Plureos GmbH
in Hamburg, Germany